GEOMETRIC STURMIAN THEORY of NONLINEAR PARABOLIC EQUATIONS and APPLICATIONS

CHAPMAN & HALL/CRC APPLIED MATHEMATICS AND NONLINEAR SCIENCE SERIES

Series Editors *Goong Chen and Thomas J. Bridges*

Published Titles

Geometric Sturmian Theory of Nonlinear Parabolic Equations and Applications, Victor A. Galaktionov

Mathematical Methods in Physics and Engineering with Mathematica, Ferdinand F. Cap

Optimal Estimation of Dynamic Systems, John L. Crassidis and John L. Junkins

Forthcoming Titles

Introduction to Partial Differential Equations with MATLAB®, Matthew P. Coleman

Mathematical Theory of Quantum Computation, Goong Chen and Zijian Diao

Stochastic Partial Differential Equations, Pao-Liu Chow

CHAPMAN & HALL/CRC APPLIED MATHEMATICS
AND NONLINEAR SCIENCE SERIES

GEOMETRIC STURMIAN THEORY of NONLINEAR PARABOLIC EQUATIONS and APPLICATIONS

Victor A. Galaktionov

University of Bath, England

and

Keldysh Institute of Applied Mathematics, Moscow

CHAPMAN & HALL/CRC

A CRC Press Company

Boca Raton London New York Washington, D.C.

Library of Congress Cataloging-in-Publication Data

Galaktionov, Victor A.
Geometric sturmian theory of nonlinear parabolic equations and applications / Victor A. Galaktionov
p. cm. — (Chapman & Hall/CRC applied mathematics and nonlinear science series ; 3)
Includes bibliographical references and index.
ISBN 1-58488-462-2 (alk. paper)
1. Differential equations, Parabolic. I. Title. II. Series.

QA377.G222 2004
515'.3534—dc22 2004042809

Direct all inquiries to CRC Press LLC, 2000 N.W. Corporate Blvd., Boca Raton, Florida 33431.

International Standard Book Number 1-58488-462-2
Library of Congress Card Number 2004042809
Printed in the United States of America 1 2 3 4 5 6 7 8 9 0
Printed on acid-free paper

To my parents, Lidija Ivanovna

and Alexander Alekseevich Galaktionov

Contents

Introduction: Sturm Theorems and Nonlinear Singular Parabolic Equations

This book is devoted to nonlinear second-order parabolic equations including well-known reaction-diffusion-absorption models from combustion, heat conduction and nonstationary filtration theory. We consider typical questions such as existence, nonexistence, uniqueness and regularity properties of solutions to nonlinear equations admitting blow-up, extinction, or other types of evolution singularities with finite propagation and free boundaries. The ideas and techniques used in the analysis are purely geometric and their cornerstone is the Sturm theory of zeros of solutions of one-dimensional linear parabolic equations.

In 1836 **C. Sturm*** published two celebrated papers in the first volume of J. Liouville's *Journal de Mathématique Pures et Appliquées*. The first paper [323] on zeros of solutions $u(x)$ of second-order ordinary differential equations such as

$$u'' + q(x)u = 0, \quad x \in I\!R, \tag{0.1}$$

very quickly exerted a great influence on the general theory of ODEs. Then and nowadays Sturm oscillation, comparison and separation theorems can be found in most textbooks on ODEs with various generalizations to other equations and systems of equations. Such theorems classify and compare zeros and zero sets $\{x \in I\!R : u(x) = 0\}$ of different solutions $u_1(x)$ and $u_2(x)$ of (0.1), or solutions of equations with different continuous ordered potentials $q_1(x) \geq q_2(x)$.

The second paper [324] was devoted to the *evolution* analysis of zeros and zero sets $\{x : u(x,t) = 0\}$ for solutions $u(x,t)$ of partial differential equations of parabolic type, for instance, of the heat equation with a linear term as in (0.1)

$$u_t = u_{xx} + q(x)u, \quad x \in [0, 2\pi], \; t > 0, \tag{0.2}$$

with the Dirichlet boundary condition $u = 0$ at $x = 0$ and $x = 2\pi$ and given smooth initial data at $t = 0$. Sturm results on PDEs such as (0.2) can be stated as follows:

First Sturm Theorem: *nonincrease with time of the number of zeros (or sign changes) of solutions,*

Second Sturm Theorem: *a classification of blow-up self-focusing formations and collapses of multiple zeros.*

We will refer to both Sturm Theorems together as the *Sturmian argument* on zero sets.

Most of Sturm's PDE paper [324] was devoted to the second Theorem on strik-

* **Jacques Charles François Sturm, 1803–1855.**

ing evolution "dissipativity" properties of zeros of solutions of linear parabolic equations, where a detailed backward-forward continuation analysis of the collapse of multiple zeros of solutions was performed. The first Theorem was formulated as a consequence of the second one (it is a form of the strong Maximum Principle for parabolic equations). As a by-product of the first Theorem, Sturm presented an evolution proof of bounds of the number of zeros of eigenfunction expansions. For finite Fourier series

$$f(x) = \sum_{L \le k \le M} (a_k \cos kx + b_k \sin kx), \quad x \in [0, 2\pi], \tag{0.3}$$

by using the PDE (0.2), where $q \equiv 0$ (with periodic boundary conditions), he showed that $f(x)$ has at least $2L$ and at most $2M$ zeros.† Sometimes the lower bound on zeros is referred to as the Hurwitz Theorem, which, possibly, was better known than the first Sturm PDE Theorem. This *Sturm-Hurwitz Theorem* is the origin of many striking results, ideas and conjectures in topology of curves and symplectic geometry.

Unlike the classical Sturm theorems on zeros of solutions of second-order ODEs, Sturm's evolution zero set analysis for parabolic PDEs did not attract much attention in the nineteenth century and, in fact, was forgotten for almost a century. It seems that G. Pólya (1933) [296] was the first mathematician in the twentieth century to revive interest in the first Sturm Theorem for the heat equation. The earlier extension by A. Hurwitz (1903) [200] of Sturm's result on zeros of (0.3) to infinite Fourier series with $M = \infty$ did not use parabolic PDEs. Since the 1930s the Sturmian argument has been rediscovered in part several times. For instance, a key idea of the Lyapunov monotonicity analysis in the famous KPP-problem by A.N. Kolmogorov, I.G. Petrovskii and N.S. Piskunov (1937) [226] on the stability of travelling waves (TWs) in reaction-diffusion equations, was based on the first Sturm Theorem in a simple geometric configuration with a single intersection between solutions. This was separately proved there by the Maximum Principle.

From the 1980s the Sturmian argument began to penetrate more and more into the theory of linear and nonlinear parabolic equations and was found to have several fundamental applications. These include asymptotic stability theory for various nonlinear parabolic equations, orbital connections and transversality of stable-unstable manifolds for semilinear parabolic equations as Morse-Smale systems, unique continuation theory, Floquet bundles and a Poincaré-Bendixson Theorem for parabolic equations, problems of symplectic geometry and curve shortening flows. A survey on Sturm's ideas in parabolic PDEs is presented in Chapter 1, where we include the proofs of both Sturm Theorems and describe further related results and generalizations achieved in the twentieth century.

Among many reasons stimulating the fundamental importance of the Sturmian argument to be characterized later on, we emphasize the geometric and the asymptotic stability aspects related to the context of the book. In this period, a number of essentially nonlinear reaction-diffusion equations from different areas of applications in mechanics and physics attracted the attention of mathematicians. A key feature of many such models is their nonlinear character and highly nonstationary

† Sturm also presented an ODE proof.

behaviour of solutions, leading to the formation of free boundaries and finite-time singularities like blow-up, extinction, quenching, self-focusing, etc. On the other hand, as models from mechanics and physics based on fundamental conservation laws, these equations often inherit scaling invariance and admit groups of transformation, and hence different particular exact invariant solutions describing singularity formation phenomena. Then it is necessary to prove that such exact solutions are stable near singularities and are attractors of a wide class of more arbitrary solutions. The stability concept is extremely fruitful in evolution PDEs, but very often rigorous proofs are extremely hard even for simple nonlinear models, especially if stability of a singular blow-up process is studied.

In view of their clear geometric nature, Sturm zero-set ideas became a powerful tool of the asymptotic analysis of parabolic PDEs. It turned out that the structure and time-evolution of intersections of pairs of different solutions (that are zeros of the differences satisfying linear parabolic equations) can reveal the actual asymptotic behaviour of general solutions. In other words, some important properties of general solutions can be described by using the *intersection comparison* with families of particular exact solutions. Ideas of comparison were always important in the theory of nonlinear parabolic equations. Particular solutions or super and subsolutions satisfying the corresponding partial differential inequalities are used in the *barrier* analysis, yielding *a priori* bounds on classes of solutions by the Maximum Principle. Effective barrier approaches form the basis of the classical theory of nonlinear parabolic equations. But in the case of singular finite-time blow up behaviour, the usual comparison is not sufficient and, actually, is good for nothing: no pairs of solutions to be compared in the usual sense exist since these must always intersect each other at any moment in time. Instead, the ideas of the intersection comparison with the control of the number of intersections and also their characters then begin to play a key role. This part of the geometric analysis is covered by Sturm Theorems.

In this book, we concentrate on geometric aspects of the intersection comparison for nonlinear models creating *finite-time singularities*. A general principle of the analysis is as follows: given a sufficiently wide (we call it complete, in a natural geometric sense) subset of particular solutions, we perform intersection comparison analysis based on the Sturmian argument involving an infinite number of such "characteristic" solutions. We call such analysis the *geometric intersection theory* or the *G-theory*, for short, of nonlinear singular parabolic equations. Our goal is to show that, for a wide class of such equations, the existence, nonexistence and a substantial part of the regularity theory can be reconstructed by means of known proper functional subsets of characteristic solutions. This is done by using simple geometric principles of comparison, transversality and convexity. These principles of *ordered geometric flows* (explained in Chapter 1 after two Sturm's Theorems) are the only machinery of intersection comparison we are going to use here. We do not apply other techniques that can be efficient for more specific classes of equations.

Let us introduce the basic evolution parabolic equations to be studied. We explain next in greater detail the main methodology of the G-theory.

1. Singular nonlinear parabolic models. In the second half of the twentieth century, quasilinear evolution equations of parabolic type became one of the major directions of the general theory of nonlinear PDEs of Mathematical Physics. Such equations occurred in the 1930s–1950s as models describing important applications in mechanics, plasma physics, filtration, combustion, explosion and nuclear reactors theory intensively developed in that period. More recently, in the 1960s and 1970s, other kinds of nonlinear evolution models were formulated in related branches of synergetics, theory of dissipative structures and self-organization in nonlinear media.

The porous medium equation and others. The most well-known and nowadays the classical example is the *porous medium equation* (PME) with two independent variables formulated here for nonnegative solutions $u = u(x,t) \geq 0$,[‡]

$$\boxed{u_t = (u^m)_{xx} \quad \text{in} \;\; S = I\!R \times I\!R_+,} \tag{0.4}$$

where $m > 1$ is a fixed exponent. In the filtration theory the quadratic equation $u_t = (u^2)_{xx}$, i.e., $m = 2$, plays a special role. The exponent $m = 1$ gives the classical *heat equation* (HE)

$$u_t = u_{xx}. \tag{0.5}$$

This is an example of a *uniformly parabolic equation*, where the coefficient of the higher-order derivative u_{xx} is 1 being uniformly positive and bounded. The classical, C^∞ (infinitely many times differentiable) and analytic solutions of the HE have been studied in detail and have been well understood since the nineteenth century.

In spite of its simple form, the PME (0.4) represents an example of a *singular* quasilinear equation, where the classical parabolic theory fails. This is easily seen if we differentiate in the right-hand side:

$$u_t = mu^{m-1}u_{xx} + m(m-1)u^{m-2}(u_x)^2, \tag{0.6}$$

to conclude that the coefficient mu^{m-1} of the higher-order derivative u_{xx} vanishes at the zero set of the solution $\{(x,t) \in S : u(x,t) = 0\}$ denoted later on by $\{u = 0\}$. The solutions are not expected to behave here as smoothly as those of uniformly parabolic equations like the HE (0.5). Such a nonlinear *degeneracy* of the equation at $u = 0$ means that the solutions are of finite smoothness and do not even have the derivatives prescribed by the differential expressions in (0.6). In fact, equation (0.6) at the *singularity level* $\{u = 0\}$ ceases to exist. The equation admitting weak solutions is then understood in the sense of distributions using the divergence form of the right-hand side of (0.4) and integration by parts. In particular, (0.4) describes the finite propagation on the singularity level, and actually a *free-boundary problem* (FBP) with unknown *a priori interface equations* occurs. First results for the PME on existence, uniqueness, regularity, comparison and finite propagation for weak solutions were obtained in the 1950s. Interface equations, C^∞ and analyticity of solutions after waiting time were proved in the middle of the 1980s. A complete mathematical theory of the PME took about

‡ We put boxes around the main linear and nonlinear equations that are to be studied in greater detail.

thirty years of intensive work, striking innovations and discoveries of new mathematical ideas and methods. It is not an exaggeration to say that, in this period, the PME became, and still remains, a crucial test example, which laid the foundation of the theory of nonlinear degenerate parabolic equations.

Further singular phenomena are described by the *filtration equation*

$$\boxed{u_t = (\varphi(u))_{xx},} \tag{0.7}$$

with a sufficiently arbitrary function φ satisfying $\varphi'(0) > 0$ for $u > 0$ and $\varphi'(0) = 0$. The N-dimensional counterpart of the PME and (0.7) are

$$\boxed{u_t = \Delta u^m \quad \text{and} \quad u_t = \Delta\varphi(u) \quad \text{in } I\!\!R^N \times I\!\!R_+,} \tag{0.8}$$

where Δ is the Laplace operator. The singular subset then consists of moving surfaces in $I\!\!R^N$ and the regularity singular propagation phenomena become essentially more complicated.

Another well-known example of a singular quasilinear parabolic equation is the *p-Laplacian equation* with gradient dependent diffusivity

$$\boxed{v_t = (|v_x|^{p-2}v_x)_x \equiv (p-1)|v_x|^{p-2}v_{xx},} \tag{0.9}$$

which for $p > 2$ degenerates on the set $\{v_x = 0\}$, where the gradient vanishes and the derivatives in (0.9) are not well defined. This comes from the theory of nonstationary filtration of non-Newtonian (or dilatant) fluids, occurs in combustion of solid fuels and has various other applications. Differentiating formally in x and setting $v_x = u$ reduces it to the *sign PME* for *solutions of changing sign*

$$\boxed{u_t = (|u|^{m-1}u)_{xx} \quad (m = p-1).} \tag{0.10}$$

The p-Laplacian equation in the N-dimensional geometry takes the form

$$\boxed{u_t = \nabla \cdot (|\nabla u|^{p-2}\nabla u) \quad \text{in } I\!\!R^N \times I\!\!R_+ \quad (\nabla = \text{grad}_x).} \tag{0.11}$$

Singular diffusion equations with absorption terms. Let us include into the PME a zero-order operator describing the additional absorption process. We then arrive at the *PME with absorption*

$$\boxed{u_t = (u^m)_{xx} - u^p \quad (u \geq 0),} \tag{0.12}$$

where the exponent p satisfies the inequality $p > -m$ (actually, this is the criterion of the existence of nontrivial solution with finite interfaces). Such equations occur in filtration theory where absorption is due to a permeability phenomenon. In combustion processes such volumetric absorption describes energy radiation effects. We then observe a nonlinear interaction between two different singular operators, the diffusion $(u^m)_{xx}$ and the absorption $-u^p$, which, though of lower differential order, can be *very singular* at $u = 0$ if $p < 0$. Such an interaction is promising even for a simpler case of the *heat equation with absorption*

$$\boxed{u_t = u_{xx} - u^p, \quad p < 0 \quad (u \geq 0).} \tag{0.13}$$

Existence and regularity estimates near interfaces and the second-order interface equation for $p \in (-1, 0)$ and nonexistence for $p \le -1$ were unknown for a long period and will be explained as a part of a general geometric theory of such equations. The N-dimensional PME with absorption for solutions of changing sign takes the form

$$\boxed{u_t = \Delta |u|^{m-1}u - |u|^{p-1}u.} \tag{0.14}$$

The p-Laplacian equations with absorption in $I\!R$ and $I\!R^N$ are

$$\boxed{u_t = (|u_x|^\sigma u_x)_x - |u|^{p-1}u, \quad u_t = \nabla \cdot (|\nabla u|^\sigma \nabla u) - |u|^{p-1}u,} \tag{0.15}$$

where $p > -1$. For convenience, we use parameter $\sigma > 0$ in the gradient diffusivity operator.

As a natural extension of such examples of a finite singular interface propagation, one can present a multi-parametric equation with doubly nonlinear diffusion and an absorption-like term

$$u_t = [|(u^m)_x|^\sigma (u^m)_x]_x - u^p |u_x|^q \tag{0.16}$$

with four real parameters and two different singular operators. The level $\{u = 0\}$ can be singular for both diffusion and absorption-like operators, especially when $p \in (0, 1)$ (strong absorption) or $p < 0$ (singular absorption). The diffusion operator in (0.16) includes the PME one of $\sigma = 0$ and the p-Laplacian if $m = 1$. Such models lead to a more general 1D quasilinear singular equation with the divergent second-order operator

$$u_t = [\varphi(u, u_x)]_x + \psi(u, u_x), \quad \text{where} \quad \varphi_q(p, q) > 0 \quad \text{for} \quad p > 0, \tag{0.17}$$

describing singular free-boundary propagation on $\{u = 0\}$.

Singular diffusion equation with reaction terms: blow-up. Another important example of very singular propagation is presented by *blow-up* solutions of nonlinear parabolic equations from combustion theory. The classical example is the nonstationary *Frank-Kamenetskii equation* (1938) [120]

$$u_t = \Delta u + e^u. \tag{0.18}$$

This equation is also known under the name of *solid-fuel model* [346] and plays a special role in the generalization of *Semenov's Chain Reaction Theory* and *thermal explosion* developed in the 1930s. Bearing in mind nonlinear diffusion operators, we introduce quasilinear reaction-diffusion equations

$$\boxed{u_t = (|u_x|^\sigma u_x)_x + e^u \quad \text{or} \quad u_t = (|u|^{m-1}u)_{xx} + |u|^{p-1}u.}$$

In $I\!R^N$, the sign PME (or the fast diffusion equation if $m \in (0, 1)$) with source is

$$\boxed{u_t = \Delta |u|^{m-1}u + |u|^{p-1}u, \quad p > 1.} \tag{0.19}$$

In the above equations, the reaction terms e^u or u^p are positive for $u > 0$ and are superlinear for $u \gg 1$. The *Osgood criterion* (1898) [281] for the ODE solutions $u = u(t)$ independent of x satisfying $u' = e^u$ or $u' = u^p$, and the comparison for parabolic PDEs guarantee that sufficiently large solutions can blow up, i.e., reach the singularity level $\{u = \infty\}$ at a finite time $t = T$. If blow-up occurs, in some cases for $t > T$, we can obtain a FBP with singular *blow-up interfaces* that bound the regions with infinite values of the solution ("burnt zones").

Using the transformation $u \mapsto \frac{1}{u}$, one can reduce the blow-up propagation for the general equation (0.17) to the propagation on the zero level $\{u = 0\}$ for the quasilinear equation with absorption of a non-divergent form

$$u_t = -u^2[\varphi(\tfrac{1}{u}, -\tfrac{u_x}{u^2})]_x - u^2\psi(\tfrac{1}{u}, -\tfrac{u_x}{u^2}) \quad (u \geq 0). \tag{0.20}$$

This creates a class of other equations with various nonlinear terms and operators.

Equations of mean curvature and curve shortening flows. This area is the origin of many quasilinear and fully nonlinear parabolic equations. For instance, after suitable parameterization, the equation

$$u_t = \frac{u_{xx}}{1 + (u_x)^2} - \frac{N-2}{u} \tag{0.21}$$

describes the evolution of symmetric hypersurface in $I\!R^N$ for $N \geq 3$ driven by mean curvature. Then a typical problem of singular interface propagation on $\{u = 0\}$ appears. In equations of curve shortening flows the normal velocity of propagation can be a general function of the mean curvature and lower-order operators, leading to a wide variety of nonlinear parabolic equations.

Fully nonlinear singular models. Similar singular propagation phenomena at $\{u = 0\}$ or at $\{u = \infty\}$ occur in *fully nonlinear equations* such as

$$u_t = g(u_{xx}) \pm \psi(u), \quad \text{where } g'(r) > 0 \text{ for } r > 0. \tag{0.22}$$

The power function $g(u_{xx}) = |u_{xx}|^{m-1}u_{xx}$ with $m > 1$ corresponds to the *dual porous medium operator* of the *dual PME* $u_t = |u_{xx}|^{m-1}u_{xx}$, while $m \in (0,1)$ is associated with fast diffusion. Equations with absorption

$$u_t = g(uu_{xx}) - (u_x)^2 - \psi(u) \tag{0.23}$$

are known in the *detonation theory*, where typical nonlinearities

$$g(s) = \log(\tfrac{e^s - 1}{s}) \quad \text{and} \quad \psi(u) = -\log u \quad \text{for } u \in (0,1)$$

describe the instability of Zel'dovich-von Neumann-Doering square wave in detonation in a duct (*equation of ZND detonation*).

A natural fully nonlinear generalization of equations (0.14) and (0.19) is

$$\boxed{|u_t|^{\gamma-1}u_t = \Delta|u|^{m-1}u \pm |u|^{p-1}u, \quad \gamma > 0.} \tag{0.24}$$

A general fully nonlinear singular parabolic equations. Thus, for a large variety of quasilinear and fully nonlinear parabolic equations with singularities localized at the level $\{u = 0\}$, we arrive at questions of existence or nonexistence of nontrivial solutions, uniqueness, optimal regularity and deriving dynamic equations of singular free boundaries. Therefore, it is convenient to study solutions $u(x,t) \geq 0$ of a fully nonlinear parabolic PDE of the most general form

$$u_t = \mathbf{F}(u) \equiv F(u, u_x, u_{xx}) \quad \text{in } S = I\!R \times I\!R_+, \tag{0.25}$$

where $F(p,q,r)$ is a given function that is sufficiently smooth for $p > 0$ and satisfies the *parabolicity condition* away from the singularity level:

$$F_r(p,q,r) > 0 \quad \text{for } p > 0 \text{ and } q, r \in I\!R. \tag{0.26}$$

The singularity analysis is essentially local and, without loss of generality, we pose for (0.25) the Cauchy problem with bounded continuous initial data $u(x,0) = u_0(x) \geq 0$ in $I\!R$. We assume that, in any non-singular domain $\{p \geq \delta > 0\}$, the function $F(p,q,r)$ satisfies all necessary assumptions, which guarantee the existence of classical, bounded, positive, smooth solutions. A typical general multi-dimensional equation is then

$$u_t = \mathbf{F}(u) \equiv F(u, |\nabla u|, \Delta u) \quad \text{in } S = I\!R^N \times I\!R_+. \tag{0.27}$$

As a common feature, equations presented above are singular and admit *weak* (for equations of divergent form) or, in general, *proper* solutions describing propagation on the singular level $\{u = 0\}$, where operators $\mathbf{F}$ are not well defined, or, at least, are not uniformly parabolic. We first consider proper *maximal* solutions constructed by monotone regular approximations of the singular problems. In this case equations of singular interfaces are not given *a priori*. Other types of non-maximal solutions correspond to generalized FBPs where the interface equation is to be prescribed in the problem statement. To this end, we will describe a class of well-posed FBPs including Stefan, Florin or other generalized ones.

The main problem of concern is how to prove, explain and describe the results for general singular PDEs such as (0.25) without specifying particular features of the fully nonlinear operator $\mathbf{F}$. This is done by reducing the corresponding parabolic theory to the ordinary differential one what we call a *PDE-ODE duality* for (0.25).

2. Duality of a second-order PDE and a family of ODEs by Sturm's Theorem. In the rigorous sense, PDE–ODEs dualities are well known for quasilinear first-order equations or for some linear and semilinear second-order ones, where the general solution can be expressed via a system of ODEs for characteristic curves. Let us explain what is meant by duality via the G-theory for nonlinear second-order parabolic equations such as (0.25).

Main questions to be studied. The main difficulty of the above nonlinear FBPs (admitting maximal or non-maximal solutions) is associated with the delicate differential properties of solutions near unknown interfaces creating evolution singularities. On subsets uniformly bounded away from singular interfaces, where the classical interior regularity theory for smooth uniformly parabolic equations is assumed to apply, the solutions stay sufficiently regular unless the singularity destroys those completely (not a rare opportunity meaning *nonexistence*). In other words, extensions of the solutions beyond singularities are straightforward provided that we can reconstruct those precisely in small neighbourhoods of singularities. This sounds similar to the phase-plane analysis of nonlinear first-order ODEs, where the local structure of the flow in small neighbourhoods of singular points plays a crucial role in describing of global properties of the flow portrait. As we show, this similarity with ODEs is not completely formal.

Various estimates in special functional classes play a fundamental role in the theory of nonlinear evolution equations and this is the subject of the well-known mathematical literature. There are many examples where these crucial estimates are refined and involved even for some particular semilinear or quasilinear equations such as the PME or the p-Laplacian one with lower-order operators having power-type nonlinearities. In these cases we typically expect to have weak, Hölder continuous solutions (though this is not always the case). What are the optimal Hölder continuity exponents? Which estimates near singularity subsets do exist for equations with other more general nonlinearities? Is it possible to describe optimal regularity estimates of solutions of equations stated in the most general fully nonlinear form such as (0.25)? These are the questions of the geometric theory devoted to treating those in a maximal generality.

Thus we will study the following standard questions (here we mean the maximal solutions with unknown *a priori* interface equations):

(Q1) *Existence or nonexistence and uniqueness of a nontrivial solution,*
(Q2) *Regularity: optimal Bernstein or higher-order differential estimates near singularities,*
(Q3) *Interface equations or systems and the interface regularity.*

Our main strategy is to show that, for a wide class of nonlinear one-dimensional parabolic equations (0.25), these and some other related questions are essentially the questions of ODE theory in the sense that the answers are based on the properties of a "proper" set B of particular characteristic solutions generated by a family of the second-order ODEs associated with the nonlinear parabolic PDE. The G-theory translating properties of the family of the ODEs to the PDE uses intersection comparison techniques based on the Sturmian argument. Namely, we compare a sufficiently arbitrary solution $u(x,t)$ with particular ones $V(x,t) \in B$ and take into account the first Sturm Theorem implying that the finite number of their intersections $\mathrm{Int}(t,V)$ satisfies

$$\mathrm{Int}(t,V) \quad \text{is nonincreasing in} t \;\; \text{for any} \;\; V \in B. \tag{0.28}$$

Actually, this is the main and the only principle of the geometric intersection theory and actually, the only "trace" of evolution parabolic equations under consideration. Of course, then the results and, what is most important, the sharpness of final estimates crucially depend on the choice of the characteristic set B. In general, any suitable complete set B of particular solutions of the nonlinear PDE generates the existence-uniqueness-regularity theory in suitable functional B-classes.

For the autonomous PDEs (0.25), the set B can be composed of the simplest travelling wave (TW) solutions

$$V(x,t) = f_\lambda(x - \lambda t)$$

propagating with a fixed speed $\lambda \in I\!R$ for which the straight lines $x - \lambda t = \text{const}$ are characteristics. Then f_λ solves the second-order ODE

$$F(f, f', f'') + \lambda f' = 0 \quad \text{in } I\!R, \;\; \text{with a parameter } \lambda \in I\!R. \tag{0.29}$$

TW solutions often play a key role in the proof of remarkable results for nonlinear parabolic PDEs. For a long period, such simplest particular solutions of autonomous parabolic PDEs and systems have been an important tool for understanding different local, global and asymptotic properties of general solutions. We mention again the KPP-paper [226], where the convergence to a "minimal" TW profile was proved by intersection comparison with a continuous family of TWs. In the paper [277] (1958) by O.A. Oleinik, A.S. Kalashnikov and Chzhou Yui-Lin' (the foundation of the PME theory) the criterion of finite propagation for a general filtration equation (0.7) was proved by comparison with finite interface TWs and later on this approach was extended to many other nonlinear equations.

It turns out that a complete set B of TWs contains sufficient information to answer the questions **(Q1)–(Q3)**. More precisely, the duality between the singular parabolic PDE and the corresponding one-parameter family of ODEs via the geometric theory, which we write in the form

$$\{F(f, f', f'') + \lambda f' = 0, \ \lambda \in I\!R\} \iff u_t = F(u, u_x, u_{xx}), \tag{0.30}$$

can be expressed as follows: the answers to questions **(Q1)–(Q3)** are essentially the same for the PDE (0.25) and for the one-parameter family of ODEs (0.29) generating the set B, i.e., existence, nonexistence, typical regularity and interface equations for the PDE and for the ODEs *coincide*. The G-theory establishes that, for a general nonlinear parabolic PDE (0.25), a complete set B of particular solutions viewed as heteroclinic connections of the singular level gives basic existence-regularity results (uniqueness can also be associated with the subset B). A partial weaker form of such a PDE–ODEs duality concerning existence and nonexistence is observed for singular equations in $I\!R^N$.

The geometric approach based on TWs can be understood as a certain version of an asymptotic "characteristic method" for the nonlinear second-order singular parabolic PDEs. The set B generates the corresponding Rankine–Hugoniot condition of the TW-propagation of singularities, which is shown to be valid for general proper solutions. Note that manipulations with TW-like structures rigorously solve the first-order PDEs such as

$$u_t = F(u, u_x) \tag{0.31}$$

by the method of characteristics, where the general solution is expressed as $u = f(x - \lambda t)$ with algebraic relations between λ, f and u. Of course, characteristic methods cannot solve second-order PDEs such as (0.25) in any explicit way. Nevertheless, the idea of the straight line characteristic propagation is fruitful for such parabolic PDEs, where the TW characteristics are proved to be dominant on the singularity level. Complete sets of TW solutions with the straight line characteristics describe the singular propagation of arbitrary solutions and establish their first regularity properties near singularities.

It is important that purely geometric techniques exhibiting the duality (0.30) are able to treat at once important questions for general singular PDEs. The actual structure of nonlinear operators in PDEs plays no role and one needs only to know properties of the corresponding family of ODEs. Then the existence-uniqueness-regularity questions for singular FBPs for such PDEs can be answered by using the "lower level" ODE-language. Dealing with general fully nonlinear equations with arbitrary nonlinear operators, we will try to prove as much as possible in the existence-uniqueness-regularity theory on the basis of intersection comparison with sets B of particular solutions only. The G-theory deals with a *geometrically ordered evolution* on given "characteristic" curves (particular solutions) and neither nonlinear PDEs nor different concepts of solutions are essentially involved in or play a role for applications. Our analysis is based on the idea that in 1D complete proper solution subsets B can generate their own basic existence (or nonexistence), uniqueness and regularity results.

3. Plan of chapters. Chapter 1 is devoted to the original Sturm zero set results for linear parabolic equations and some generalizations. In Chapters 2–6, using particular singular parabolic models, we introduce basic concepts and notions of the geometric analysis. For special equations having extra invariant properties and exact solutions, we often prove better estimates.

The main concepts and regularity results of the geometric theory are presented in Chapters 7–9, where we consider general singular equations (0.25) and (0.27), and introduce the geometric notions related to the regularity and the interface propagation of solutions. Some of them are new (second-order Rankine–Hugoniot interface conditions, Bernstein gradient estimates, continuity moduli, etc.), others summarize the ideas and techniques of intersection comparison developed in the previous chapters for particular quasilinear models. In the general setting, we describe the main aspects of the PDE–ODEs duality (0.30). Here we use a full machinery of intersection comparison with complete B-bundles of TW solutions, proving existence and nonexistence theorems, establishing uniqueness and optimal Bernstein-type estimates (and hence optimal moduli of Hölder continuity in x and t) and deriving interface equations including the case of higher-order equations, when the evolution of singular interfaces is governed by systems of equations (unlike the single *Darcy law* for the PME).

In Chapter 7 we perform a detailed geometric analysis of the nonnegative maximal solutions. In Chapter 8 the G-theory applies to classes of non-maximal solutions satisfying generalized FBPs. This implies no novelties in the regularity estimates by geometric techniques. Given a singular 1D equation, we show that its complete set B of TWs makes it possible to classify various one-phase or two-phase FBPs that can be well-posed for the equation. We study FBPs with second or higher-order interface equations. In Chapter 9 geometric techniques apply to solutions of changing sign. In the last Chapters 10 and 11 we consider some special aspects of discontinuous and continuous limit semigroups generated by singular parabolic equations. In particular, we study the *singular Zhang equation* in the theory of complex directed polymers

$$u_t = u_{xx} + \log|u_x| \quad \text{in } (-1,1) \times I\!R_+, \quad u(\pm 1, t) = 0 \quad \text{for } t > 0.$$

This equation represents a special case of non-localized singularity subsets $\{u_x = 0\}$ consisting of extremum or inflection points, where $u_x = 0$. We describe a discontinuity phenomenon of the *instantaneous shape simplification* in $I\!R$ and $I\!R^N$ occurring at the initial moment $t = 0$. This is a special kind of discontinuity of the proper limit semigroup. Instead of TWs, we use a set of self-similar solutions describing local singularity formation for this equation. We also study discontinuous semigroups for equations with extra spatial singularities including the fully nonlinear equation (the dual PME if $m > 1$) with source

$$u_t = |\Delta u|^{m-1}\Delta u + \frac{u^p}{|x|^{2m}},$$

where $m > 0$ and $p > 1$. Then, in addition to blow-up due to the superlinear source u^p for $u \gg 1$, a singularity is generated at the origin $x = 0$.

Prerequisites. This book assumes a graduate level in theory of ODEs and second-order parabolic PDEs. The cornerstone of the parabolic theory is the Maximum Principle, which is perfectly explained in a number of books. We recommend [121], [233], [299] or [317]. In what follows we deal with a number of various degenerate nonlinear parabolic equations originated from the PME. Typical local and global existence, uniqueness and differential properties of weak solutions, as well as key mathematical techniques developed in the last fifty years can be found in the books [99] and [122] and in a detailed survey [213]. We also refer to introductory chapters in [170], [306], where basics of the asymptotic theory for parabolic equations applied to blow-up singularity formation phenomena can be found.

Acknowledgements. The book reflects the author's interests in the intersection comparison area from the beginning of the 1980s. During this period he had a privilege of inspiring discussions of many related subjects with several experts in PDEs, nonlinear mechanics and plasma physics.

The author would like to thank my senior colleagues from the Keldysh Institute of Applied Mathematics, Moscow, S.P. Kurdyumov, A.P. Mikhailov and A.A. Samarskii for collaboration and systematic discussions in the 1970s and 80s of singularity blow-up phenomena in nonlinear media and reaction-diffusion equations where intersection ideas turned out to be of crucial importance. The author thanks S.I. Pohozaev from the Steklov Institute of Mathematical, Moscow, and permanent participants of his seminar on nonlinear equations at the Department of Mathematics of the Moscow Energy Institute, where the author reported his intersection comparison results of the 1980s.

The author thanks colleagues and co-authors of later papers, where intersection ideas were used or at least discussed, J. Bebernes, A. Bressan, C.J. Budd, P.J. Harwin, R. Kersner, J.R. King, A.A. Lacey, L.A. Peletier, S.A. Posashkov and S.I. Shmarev. Especially, the author would like to thank J.L. Vazquez for discussions and a long-term collaboration in the 1980s-90s, which clarified new trends in the intersection comparison. The author thanks S. Angenent and H. Matano for discussions and comments on applications of Sturm's analysis in dynamical systems, mean curvature flows and in the geometry of curves. The author acknowledges A.S. Shvedov, S.R. Svirshchevskii, D.G. Vassiliev and N.N. Vorobjov for consultations at the final stage of writing this book. It was mainly completed at the Isaac Newton Institute for Mathematical Sciences, Cambridge, to which the author is thankful for the hospitality and thanks J.F. Toland for the kind invitation to participate in the program on nonlinear PDEs in winter-spring 2001.

Victor A. Galaktionov
Bath, January 2004

CHAPTER 1

Sturm Theorems for Linear Parabolic Equations and Intersection Comparison. B-equations

In this chapter we state and prove two fundamental Sturm Theorems on zero sets for linear parabolic equations trying, whenever possible, to keep Sturm's original notations and calculations. We next describe necessary extensions of Sturm's Theorems to wider classes of equations and boundary-value problems. Furthermore, we present a geometric interpretation of flows satisfying Sturmian intersection principles and introduce a class of the so-called B-equations preserving Sturmian intersection properties with respect to a given complete functional subset B playing the role of exact characteristic solutions.

The last section contains a survey on applications of Sturm Theorems and ideas in various areas of the ODEs and PDEs, as well as in geometric and other problems.

1.1 First Sturm Theorem: Nonincrease of the number of sign changes

Let D and J be open bounded intervals in $I\!R$. Consider in $S = D \times J$ a linear parabolic equation

$$u_t = a(x,t)u_{xx} + b(x,t)u_x + c(x,t)u. \tag{1.1}$$

Given a constant $\tau \in J$, let ∂S_τ be the *parabolic boundary* of the domain $S_\tau = S \cap \{t < \tau\}$, i.e., the *lateral boundary* and the bottom of the boundary of S_τ. Given a solution $u = u(x,t)$ defined on S_τ, the positive and negative sets of u are defined as follows:

$$U^+ = \{(x,t) \in S_\tau : u(x,t) > 0\}, \;\; U^- = \{(x,t) \in S_\tau : u(x,t) < 0\}. \tag{1.2}$$

A *component* of U^+ (or U^-) is a maximal open connected subset of U^+ (or U^-).

Given a $t \in \overline{J}$, the number (finite or infinite) of components of $\{x \in D : u(x,t) \neq 0\}$ minus one is called the *number of sign changes* of $u(x,t)$ and is denoted by $Z(t,u)$. Alternatively,

$$Z(t,u) = \sup\{k\},$$

where the subset $\{k\}$ consists of integer $k \geq 0$ such that there exist $k+1$ points from D, $x_0 < x_1 < ... < x_k$, satisfying

$$u(x_j,t) \cdot u(x_{j+1},t) < 0 \quad \text{for all } \; j = 0,1,...,k-1.$$

Theorem 1.1 (First Sturm Theorem: sign changes) *Let a, b, c be continuous,* *bounded and* $a \geq \mu > 0$ *in* S. *Let* $u(x,t)$ *be a solution of* (1.1) *in* S *that is continuous on* $\overline{S}$.

(i) Suppose that on ∂S_τ *there are precisely* n *(respectively* m*) disjoint intervals where* u *is positive (resp. negative). Then* U^+ *(resp.* U^-*) has at most* n *(resp.* m*) components in* S_τ *and the closure of each component must intersect* ∂S_τ *in at least one interval.*

(ii) The number of sign changes $Z(\tau, u)$ *of* $u(x,\tau)$ *on* D *is not greater than the number of sign changes of* u *on* ∂S_τ.

Proof. The proof is based on the Maximum Principle (the MP for short).

(i) Let $I \subset \partial S_\tau$ be the maximal interval where $u > 0$. Suppose that two open connected subsets F_1, $F_2 \subset U^+$ intersect ∂S_τ in disjoint open intervals I_1, $I_2 \subset I$. Since u is continuous in $\overline{S}_\tau$, there exists an open neighbourhood $G \subset U^+$ whose closure in $\overline{S}_\tau$ contains I. Then G must contain points of both F_1 and F_2, so that these must belong to the same open component of U^+. Thus at most one component of U^+ intersects each of the n open intervals on ∂S_τ where $u > 0$. The same result holds for the components of U^-.

Therefore, it suffices to show that every component of U^+ (or U^-) intersects ∂S_τ in an interval. We can assume that $c \leq 0$ in S_τ. Otherwise, we set

$$u = e^{\lambda t} v$$

(then $U^\pm$ stay the same for v), where v then solves equation (1.1) with the last coefficient c on the right-hand side replaced by $c - \lambda$ and we can choose the constant $\lambda \geq \sup c$.

Let $F \subset U^+$ be a component in S_τ. Since u is continuous, it must attain a positive maximum on $\overline{F}$. Then $c \leq 0$ implies

$$u_t \leq a u_{xx} + b u_x \quad \text{in } F,$$

and, by continuity, $u = 0$ at any boundary point of F that is interior to S_τ. By the MP, u cannot attain its maximum at an interior point of F or on the line $\{t = \tau\}$. Hence, F must have a boundary point $Q \in \partial S_\tau$ such that $u(Q) > 0$ and by continuity u is positive in an interval of ∂S_τ about Q.

(ii) is a straightforward consequence of (i). □

The Sturm Theorem is true for wider classes of linear parabolic equations, which are regular enough to have the strong MP applied. In particular, an important example is the radial parabolic equation in $I\!R^N$ with continuous bounded coefficients and $a \geq \mu > 0$,

$$u_t = a(r,t)\Delta u + b(r,t)u_r + c(r,t)u, \tag{1.3}$$

where $r = |x| \geq 0$ denotes the single spatial variable. The radial Laplace operator

$$\Delta u = u_{rr} + \tfrac{N-1}{r}\, u_r$$

is formally singular at $r = 0$. Bearing in mind that we consider smooth bounded solutions satisfying the symmetry condition at the origin,

$$u_r(0,t) = 0 \quad \text{for } t \in J,$$

the MP applies to equation (1.3) in $S = D \times J$, where $D = \{r < R\}$ is a ball in $I\!R^N$, and the first Sturm Theorem holds.

1.2 Second Sturm Theorem: Evolution formation and collapse of multiple zeros

Results in the class of C^∞ functions

Before stating the result, we note that parabolic equations with analytic coefficients admit analytic solutions and then any zero of $u(x,t) \not\equiv 0$ has finite multiplicity.

Theorem 1.2 (Second Sturm Theorem: multiple zeros) *Let $O = (0,0) \in S$ and $u \in C^\infty(S) \cap C(\overline{S})$ be a solution of equation (1.1), with C^∞-coefficients a, b, c, where $a \geq \mu > 0$ in S. Assume that $u(x,t)$ does not have sign changes on the lateral boundary of S, and let $u(x,0)$ have a zero of multiplicity $m \geq 2$ at the origin $x = 0$, i.e.,*

$$D_x^k u(0,0) = 0 \quad \textit{for } k = 0, 1, ..., m-1 \quad \textit{and } D_x^m u(0,0) = Am! \neq 0. \tag{1.4}$$

Then $Z(t,u)$ drops at $t = 0$ and, for any $t_1 < 0 < t_2$ near $t = 0$, there holds

$$Z(t_1, u) - Z(t_2, u) \geq \begin{cases} m, & \textit{if } m \textit{ is even,} \\ m-1, & \textit{if } m \textit{ is odd.} \end{cases} \tag{1.5}$$

Proof. We follow the original Sturm calculations in [324, p. 417-427]. By Taylor's formula in x near the origin, we have

$$u(x,0) = Ax^m + O(x^{m+1}). \tag{1.6}$$

By Taylor's expansion in t,

$$u(x,t) = u(x,0) + u_t(x,0)t + \tfrac{1}{2!}\, u_{tt}(x,0)t^2$$
$$+ ... + \tfrac{1}{n!}\, D_t^n u(x,0)t^n + O(t^{n+1}), \tag{1.7}$$

where $n = \frac{m}{2}$, if m is even, and $n = \frac{m-1}{2}$, if m is odd. Let us estimate the coefficients in (1.7). Denote

$$d_j = \frac{m!}{(m-2j)!}, \quad j = 0, 1, ..., n.$$

It follows from the parabolic equation (1.1) and (1.6) that $u_t(x,0)$ has the following Taylor expansion:

$$u_t(x,0) = a(x,0)u_{xx}(x,0) + b(x,0)u_x(x,0) + c(x,0)u(x,0)$$
$$= a_0 A d_1 x^{m-2} + O(x^{m-1}),$$

where $a_0 = a(0,0)$ and $a(x,0) = a_0 + O(x)$. Differentiating the equation and using expansion (1.6) again we obtain, keeping the leading terms only,

$$u_{tt}(x,0) = a u_{txx} + ... = a_0^2 A d_2 x^{m-4} + O(x^{m-3}),$$

and finally

$$D_t^n u(x,0) = D_t^{n-1} a_0 u_{xx}(0,0) + ... = a_0^n A d_n x^{m-2n} + O(x^{m-2n+1}).$$

Taylor's expansion in both independent variables x and t is now

$$u(x,t) = A(x^m + a_0 d_1 x^{m-2} t + \tfrac{1}{2!} a_0^2 d_2 x^{m-4} t^2 + ... + \tfrac{1}{n!} a_0^n d_n x^{m-2n} t^n) + O(\cdot) \tag{1.8}$$

with remainder $O(\cdot) = O(|x|^{m+1} + |x|^{m-1}|t| + ... + |x|^{m-2n+1}|t|^n + |t|^{n+1})$.

(i) Backward continuation. Consider the behaviour for $t \approx 0^-$. The dimensional structure of the right-hand side of (1.8) suggests rewriting this expansion in terms of the rescaled *Sturm backward continuation variable*

$$z = x/\sqrt{a_0(-t)} \quad \text{for } t < 0. \tag{1.9}$$

Substituting $x = z\sqrt{a_0(-t)}$ yields

$$A^{-1} a_0^{-m/2} (-t)^{-m/2} u(x,t) = P_m(z) + O(\sqrt{-t}(1 + |z|^{m+1})), \tag{1.10}$$

$$\text{where } P_m(z) = \textstyle\sum_{j=0}^n (-1)^j \dfrac{d_j}{j!} z^{m-2j}.$$

The function $P_m(z)$ is indeed the mth order *Hermite polynomial* $H_m(z)$ (up to a constant multiplier, which we omit in what follows). Each orthogonal polynomial $H_m(z)$ has exactly m simple zeros $\{z_i,\ i = 1, ..., m\}$, with $H_m'(z_i) \neq 0$, i.e., the intersections with zero are always *transversal*. Sturm proved this separately on p. 426. This is nowadays a well-known fact in the classical theory of orthogonal polynomials; see G. Szegö's book [328, Chapter 6].

A similar expansion for the derivative $u_x(x,t)$ shows that expansion (1.10) can be differentiated in x, yielding the derivative $P_m'(z)$ on the right-hand side. It follows from the expansions of $u(x,t)$ and $u_x(x,t)$ near the multiple zero that, for any $t \approx 0^-$, the solution $u(x,t)$ has m simple zeros $\{x_i(t),\ i = 1, ..., m\}$ such that $u_x(x_i(t), t) \neq 0$, with the following asymptotic behaviour:

$$x_i(t) = z_i\sqrt{-t} + O(-t) \to 0 \quad \text{as } t \to 0^-,$$

so exactly m smooth zero curves intersect each other at the origin $(0,0)$. For instance, on Figure 1.1, $u(x,t)$ for $x \approx 0$, $t \approx 0^-$ has the structure of the fifth Hermite polynomial as in (1.10).

(ii) Forward continuation. Following Sturm's analysis, we now consider the behaviour of the solution $u(x,t)$ as $t \to 0^+$. Introducing the heat kernel rescaled variable of the *forward continuation*

$$z = x/\sqrt{a_0 t} \quad \text{for } t > 0, \tag{1.11}$$

instead of (1.10) we obtain another polynomial on the right-hand side

$$A^{-1} a_0^{-m/2} t^{-m/2} u(x,t) = Q_m(z) + O(\sqrt{t}\,(1 + |z|^{m+1})), \tag{1.12}$$

$$\text{where } Q_m(z) = \textstyle\sum_{j=0}^n \dfrac{d_j}{j!} z^{m-2j}.$$

The mth order polynomial $Q_m(z)$ has positive coefficients. If m and n are odd, then $Q_m(z)$ is strictly increasing and $Q_m(0) = 0$. If m and n are even, this polynomial has a single positive minimum at $z = 0$. Therefore, (1.12) implies

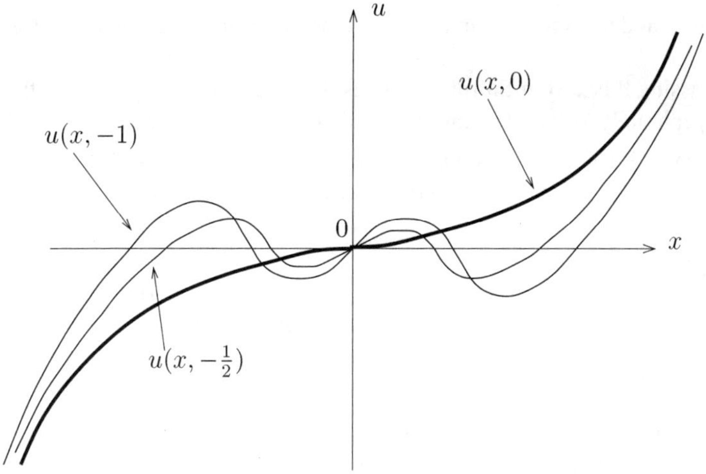

Figure 1.1 *Formation as* $t \to 0^-$ *of a multiple zero (a higher-order inflection) of* $u(x,0)$ *from five transversal zeros of* $u(x,t)$ *for* $t < 0$.

that, for small $t > 0$ on compact subsets $\{|x| \le c\sqrt{t}\}$ with any $c > 0$, the solution $u(x,t)$ has a unique simple zero $\tilde{x}_1(t) = O(t)$ if m is odd, and no zeros if m is even. This is Sturm's analysis on p. 423.

In order to complete the proof, it suffices to observe that, if m is even and, say, $A > 0$, there exists a small interval $(-\varepsilon, \varepsilon)$ such that $u(x,t)$ becomes strictly positive on $(-\varepsilon, \varepsilon)$ for all small $t > 0$. This follows from the strong MP. Hence, at least m zero curves disappear at $(0,0)$. If m is odd and $A > 0$, then applying Theorem 1.1 to the domain $S = (-\varepsilon, \varepsilon) \times (0, \varepsilon)$ we have that on $(-\varepsilon, \varepsilon)$ for $t > 0$ there exists a unique continuous curve of simple zeros $\tilde{x}_1(t)$ starting from $(0,0)$. In this case at least $m - 1$ zero curves disappear at the origin as $t \to 0^-$. Since, by the assumption, new zeros of $u(x,t)$ do not appear on the lateral boundary of S and cannot occur in S by Theorem 1.1, this completes the proof of (1.5). □

Such a complete analysis of the evolution of multiple zeros in one dimension applies to more general parabolic equations. In particular, in the N-dimensional geometry similar results are true for radial solutions $u = u(r,t)$ of parabolic equations (1.3) with C^∞ (or analytic) coefficients, where the Laguerre polynomials occur in asymptotic expansions instead of the Hermite ones; see Remarks.

Comments on Sturm's evolution analysis of zeros

Sturm's proof consisting of parts (i) and (ii) exhibits typical features of a detailed asymptotic evolution analysis for general linear uniformly parabolic equations, which can be classified as follows:

(i) A finite-time creation of a multiple zero as $t \to 0^-$ as a *singularity formation* (single point blow-up self-focusing of zero curves), and

(ii) Disappearance of multiple zeros at $t = 0^+$, i.e., instantaneous *collapse of singularity* and a unique continuation (extension) of the solution beyond singularity.

Discussing this part of Sturm's analysis, for convenience, we present the above results separately as follows, stating the precise asymptotic behaviour of solutions.

Corollary 1.3 *Under the assumptions of Theorem* 1.2, *there holds:*

(i) As $t \to 0^-$, *the rescaled solution converges to the* m*th order Hermite polynomial,*

$$A^{-1}a_0^{-m/2}(-t)^{-m/2}u(x,t) \to H_m(z) \tag{1.13}$$

uniformly on compact subsets in the variable z *given by* (1.9).

(ii) As $t \to 0^+$, *the rescaled solution converges to the non-oscillating* m*th order polynomial,*

$$A^{-1}a_0^{-m/2}t^{-m/2}u(x,t) \to Q_m(z) \tag{1.14}$$

uniformly on compact subsets in z *given by* (1.11).

Phenomena of singularity blow-up formation, collapse and solution extensions beyond singularities are important subjects of the general PDE theory. In applications to semilinear and quasilinear parabolic equations of reaction-diffusion type, the perturbation techniques for infinite-dimensional dynamical systems often play a key role. We now briefly comment on Sturm's analysis using the standard perturbation theory of linear operators revealing its other aspects that are important for further extensions.

(i) Formation of multiple zeros: backward continuation. Using Sturm's backward rescaled variable (1.9), we introduce the rescaled solution

$$u(x,t) = \theta(z,\tau), \quad z = z/\sqrt{a_0(-t)}, \tag{1.15}$$

where we define the new time variable as follows:

$$\tau = -\log(-t), \quad \text{so } \tau \to +\infty \quad \text{as } t \to 0^-.$$

Hence, in this variable, Sturm's multiple zeros formation analysis as $t \to 0^-$ reduces to the infinite time, as $\tau \to \infty$, asymptotic behaviour problem for the rescaled solution $\theta(z,\tau)$. Substituting (1.15) into equation (1.1) yields the following rescaled equation:

$$\theta_\tau = \mathbf{B}\,\theta + \mathbf{C}(\tau)\theta, \tag{1.16}$$

where $\mathbf{B}$ is the linear symmetric operator

$$\mathbf{B} = \frac{\mathrm{d}^2}{\mathrm{d}z^2} - \tfrac{1}{2}\,z\frac{\mathrm{d}}{\mathrm{d}z} \equiv \frac{1}{\rho}\frac{\mathrm{d}}{\mathrm{d}z}\left(\rho\frac{\mathrm{d}}{\mathrm{d}z}\right) \quad \text{with weight } \rho(z) = e^{-z^2/4}. \tag{1.17}$$

The non-autonomous perturbation in (1.16) has the form

$$\mathbf{C}(\tau)\theta = \left(\tfrac{a-a_0}{a_0}\right)\theta_{zz} + e^{-\tau/2}\tfrac{b}{\sqrt{a_0}}\theta_z + e^{-\tau}c\,\theta,$$

where, for the regular coefficient $a(x,t)$ on compact subsets in z,

$$\tfrac{1}{a_0}\left(a(x,t)-a_0\right) \equiv \tfrac{1}{a_0}\left(a(z[a_0(-t)]^{1/2},t)-a_0\right) = O(e^{-\tau/2}).$$

This implies that for smooth solutions, the perturbation

$$\mathbf{C}(\tau)\theta = e^{-\tau/2}\left[\theta_{zz}O(1) + \tfrac{b}{\sqrt{a_0}}\,\theta_z + e^{-\tau/2}c\,\theta\right]$$

is exponentially small as $\tau \to \infty$. Hence, the equation (1.16) is an exponential small perturbation of the autonomous equation

$$\theta_\tau = \mathbf{B}\,\theta. \tag{1.18}$$

The symmetric operator $\mathbf{B}$ is the classical singular Sturm-Liouville operator, [328], [57]. It is known to be self-adjoint in the weighted space $L^2_\rho(I\!R)$ with the inner product

$$(v,w)_\rho = \int_{-\infty}^{\infty} \rho(z)v(z)w(z)\,\mathrm{d}z.$$

Its domain $\mathcal{D}(\mathbf{B}) = H^2_\rho(I\!R)$ is a Hilbert space of functions v satisfying $v, v', v'' \in L^2_{\mathrm{loc}}(I\!R)$ with the inner product $\langle v, w\rangle_\rho = (v,w)_\rho + (v',w')_\rho + (v'',w'')_\rho$ and the induced norm $\|v\|^2_\rho = \langle v, v\rangle_\rho$. $\mathbf{B}$ has compact resolvent and its spectrum consists of eigenvalues

$$\sigma(\mathbf{B}) = \{\lambda_k = -\tfrac{k}{2},\;\; k = 0, 1, ...\}.$$

The eigenfunctions are orthonormal Hermite polynomials $\tilde{H}_k(z) = c_k H_k(z)$, where c_k are normalization constants, and the eigenfunction subset $\{H_k\}$ is complete and closed in $L^2_\rho(I\!R)$. These are classical results of the theory of linear self-adjoint operators in Hilbert spaces; see first chapters of M.S. Birman and M.Z. Solomjak's book [57]. Using eigenfunction expansions and semigroup estimates (see Remarks) yields that the exponentially perturbed dynamical system (1.16) on $L^2_\rho(I\!R^N)$ admits a discrete subset of different asymptotic patterns. They coincide with those for the unperturbed equation (1.18) and exhibit the asymptotic behaviour on tangent stable ($\lambda_m < 0$) eigenspaces of $\mathbf{B}$. This implies (1.13) and as $\tau \to \infty$, uniformly on compact subsets in z, there holds

$$\theta(z,\tau) = Ce^{\lambda_m \tau} H_m(z) + o(e^{\lambda_m \tau}) \quad \text{with a constant } C \neq 0. \tag{1.19}$$

(ii) Collapse of multiple zeros on spatial structure of adjoint polynomials: forward continuation. For $t > 0$, the forward rescaled variable (1.11) is used. Similarly, the rescaled function $u(x,t) = g(z,s)$, where the time variable is now $s = \log t \to -\infty$ as $t \to 0^+$, solves the exponentially perturbed equation

$$g_s = (\mathbf{B}^* - \tfrac{1}{2}I)g + \mathbf{C}(s)g \quad \text{as } s \to -\infty, \tag{1.20}$$

where $\mathbf{B}^*$ is the adjoint differential operator (I denotes the identity)

$$\mathbf{B}^* = \frac{\mathrm{d}^2}{\mathrm{d}z^2} + \tfrac{1}{2}\,z\frac{\mathrm{d}}{\mathrm{d}z} + \tfrac{1}{2}\,I \equiv \frac{1}{\nu}\frac{\mathrm{d}}{\mathrm{d}z}(\nu\frac{\mathrm{d}}{\mathrm{d}z}) + \tfrac{1}{2}\,I$$

with the weight function $\nu(z) = e^{z^2/4} = \frac{1}{\rho(z)}$. As in the backward analysis, the perturbation term

$$\mathbf{C}(s)g = O(e^{s/2}) \to 0 \quad \text{as } s \to -\infty$$

is exponentially small on compact subsets for smooth solutions. $\mathbf{B}^*$ is also a classical singular Sturm-Liouville operator and is self-adjoint in $L^2_\nu(I\!R)$, $\mathcal{D}(\mathbf{B}^*) = H^2_\nu(I\!R)$, with the same point spectrum $\sigma(\mathbf{B}^*) = \sigma(\mathbf{B})$ and a complete and closed subset of orthonormal eigenfunctions, and so on.

But as often happens in the evolution singularity theory, unlike the phenomenon of the *evolution* blow-up formation of multiple zeros, in the complementary asymptotic analysis beyond blow-up, as $t \to 0^+$ ($s \to -\infty$), spectral properties and eigenfunctions of $\mathbf{B}^*$ play no role. The limit $t \to 0^+$ corresponds to the collapse of the *initial* "singularity" created by the preceding singularity formation as $t \to 0^-$.

Obviously, for such parabolic equations, the behaviour of $u(x,t)$ as $t \to 0^+$ is uniquely determined by initial data $u(x,0)$. Consider (1.19) for $|z| \gg 1$. Since $P_m(z) \equiv H_m(z) \sim z^m + ...$ as $z \to \infty$, it can be shown (a standard scaling compactness argument is necessary at this step of extending the behaviour from compact subsets $\{|z| \le c\}$ to $\{0 < |x| \ll 1\}$; see references in Remarks) that, as $t \to 0^-$,

$$u(x,t) = C(-t)^{-\lambda_m} \frac{x^m}{a_0^{m/2}(-t)^{m/2}} + ... \to C a_0^{-m/2} x^m (1 + o(1)), \tag{1.21}$$

where the right-hand side will be denoted by $u(x,0)$. The solution $g(z,s)$ of the rescaled equation (1.20) with initial data calculated in (1.21) has the expansion

$$g(z,s) = \tilde{C} e^{-\lambda_m s} Q_m(z) + ..., \quad \tilde{C} \ne 0, \tag{1.22}$$

where Q_m is the polynomial solution of the linear equation

$$(\mathbf{B}^* - \tfrac{1}{2} I) Q_m = \tfrac{m}{2}\, Q_m.$$

This gives the linear problem for the "adjoint" polynomials $\{Q_m\}$. Notice that these have nothing to do with the orthogonal subset of eigenfunctions $\{H_m(z)e^{-z^2/4}\}$ of the adjoint operator $\mathbf{B}^*$. Moreover $Q_m \notin L^2_\nu(I\!R)$. In order to match (1.22) and the initial condition (1.21), by a similar local extension to $\{0 < |x| \ll 1\}$, we have that

$$g(z,s) = \tilde{C} t^{-\lambda_m} \frac{x^m}{a_0^{m/2} t^{m/2}} + ... \to \tilde{C} a_0^{-m/2} x^m + ... \quad \text{as } t \to 0^+.$$

By matching with (1.21), this uniquely determines the constant $\tilde{C} = C$ in (1.22) and completes the asymptotic analysis of both the backward and forward evolution of multiple zeros.

Results in classes of finite regularity

Fix a finite $T > 0$ and set $J = (0,T)$. As we mentioned, if $u(x,t) \not\equiv 0$ is a solution analytic in x of the linear parabolic equation (1.1) that has analytic coefficients a, b, c, then, for any $t \in (0,T)$, all zeros of $u(x,t)$ are isolated and hence the number of sign changes $Z(t,u)$ is finite even if $Z(0,u) = \infty$. It turns out that a similar conclusion remains valid in classes of equations and solutions of finite regularity. We present here the two most general results by S. Angenent [11]; more references are given in Remarks. We begin with initial-boundary value problems.

Theorem 1.4 *Let* u *be a bounded solution of* (1.1) *in* $S = D \times (0,T)$, *which does not change sign on the lateral boundary of* S. *Assume that the coefficients* a, b *and* c *of the equation are such that*

$$a,\ a^{-1},\ a_x,\ a_{xx},\ b,\ b_t,\ b_x,\ c \in L^\infty(S). \tag{1.23}$$

Then the number of sign changes of $u(\cdot, t)$ *satisfies:*

(i) $Z(t, u)$ *is finite and nonincreasing on* $(0, T)$.

(ii) If $x = x_0 \in D$ *is a multiple zero of* $u(x, t_0)$ *for some* $t_0 \in (0, T)$, *then, for all* $0 < t_1 < t_0 < t_2 < T$, *the strict inequality* $Z(t_1, u) > Z(t_2, u)$ *holds, so* $Z(t, u)$ *is strictly decreasing at* $t = t_0$.

As a consequence, any global solution $u(x, t)$ defined in $S = D \times \mathbb{R}_+$ has only simple zeros for all $t \gg 1$. A similar result is valid for parabolic equations in unbounded domains if we restrict the analysis to classes of functions with a fixed growth at infinity, similar to Tikhonov's classes of uniqueness. Let $D = \mathbb{R}$, and consider the following linear parabolic equation:

$$u_t = u_{xx} + q(x, t)u \quad \text{in } S = \mathbb{R} \times (0, T). \tag{1.24}$$

Theorem 1.5 *Let*

$$q \in L^\infty(S), \tag{1.25}$$

and let $u(x, t)$ *be a solution of* (1.24) *in the class* $\{|u(x, t)| \le Ae^{Bx^2} \text{ in } S\}$ *for some positive constants A and B. Then:*

(i) For each $t \in (0, T)$, *the zero set of the solution* $\{x \in \mathbb{R} : \ u(x, t) = 0\}$ *is a discrete subset of* $\mathbb{R}$.

The following conclusion is a consequence.

(ii) If $x = \pm\infty$ *are not accumulation points of zeros of* $u(x, 0)$, *then statements (i) and (ii) of Theorem* 1.4 *hold.*

This theorem is true for more general equations such as (1.1) in unbounded domains in suitable classes of uniqueness. Indeed, equation (1.1) can be reduced to (1.24) by the Liouville transformation. Using the new spatial coordinate

$$y = \int_0^x \frac{\mathrm{d}s}{\sqrt{a(s,t)}},$$

we have that $u = u(y, t)$ satisfies the equation

$$u_t = u_{yy} + \tilde{b}(y, t)u_y + \tilde{c}(y, t)u.$$

Substituting

$$v(y, t) = \exp\left\{\tfrac{1}{2}\int_0^y \tilde{b}(s, t)\,\mathrm{d}s\right\} u(y, t),$$

yields equation (1.24) for $v(y, t)$ with a potential $\tilde{q}(y, t)$. Checking necessary properties of $\tilde{q}(y, t)$, we deduce that Sturm's results are valid in the corresponding uniqueness classes.

1.3 First aspects of intersection comparison of solutions of nonlinear parabolic equations

Main results of the G-theory are proved by intersection comparison of solutions $v = v(x, t)$ of nonlinear parabolic equations of the general form

$$v_t = \mathbf{F}(v) \equiv F(x, t, v, v_x, v_{xx}) \quad \text{in } S, \tag{1.26}$$

where $F(x,t,p,q,r)$ is nondecreasing relative to the last argument $r \in I\!R$ (the parabolicity condition), with a subset $B = \{V(x,t)\}$ of some particular solutions of the same equations. In general, these are not classical solutions but some weak or proper, minimal or maximal ones constructed by smooth and, if possible, monotone approximations (this is explained in detail in Chapter 6). We always deal with solutions $v(x,t)$ and $V(x,t)$ that are continuous in x, so that we can define the number of intersections as the number of sign changes (Section 1.1) of the difference $w(x,t) = v(x,t) - V(x,t)$:

$$\mathrm{Int}(t,V) = Z(t,w) \quad \text{for } t \in [0,T).$$

If F in (1.26) is sufficiently smooth, the difference w satisfies a linear parabolic equation

$$w_t = aw_{xx} + bw_x + cw, \tag{1.27}$$

where the coefficients are given by Hadamard's formulae

$$\begin{aligned} a &= \textstyle\int_0^1 F_r(x,t,v,v_x,\theta v_{xx} + (1-\theta)V_{xx})\,\mathrm{d}\theta \geq 0, \\ b &= \textstyle\int_0^1 F_q(x,t,v,\theta v_x + (1-\theta)V_x, V_{xx})\,\mathrm{d}\theta, \\ c &= \textstyle\int_0^1 F_p(x,t,\theta v + (1-\theta)V, V_x, V_{xx})\,\mathrm{d}\theta. \end{aligned}$$

If these coefficients satisfy conditions of Sturm Theorems for linear equations, then the number of intersections $\mathrm{Int}(t,V)$ of two solutions v and V of the parabolic equation (1.26) obeys the same properties as the number of sign changes of the difference w satisfying (1.27).

At this stage the intersection comparison approach uses the fact that the non-increase with time or collapse of multiple intersections hold with respect to *any* fixed solution $V(x,t) \in B$. In a simple case, we study the evolution of *tangency points* or *inflection points* of solution profiles. In other words, the intersection comparison with the set B means that we apply Sturm Theorems to an infinite number of different linearized parabolic equations. The main ingredient of the G-theory is to organize such intersection comparison in an effective way. First of all, the subset B of particular solutions will be assumed to be *complete* in a natural geometric sense in order to "exhaust" necessary spatial shapes of the general solution $v(x,t)$ under consideration. We also need some continuity, monotonicity and compactness properties of the subset B to be defined and checked for a number of problems in the next chapter.

If equation (1.26) is essentially singular and the only known regularity of $v(x,t)$ is expected to be just continuity in x for $t > 0$ (this happens for a number of weak solutions to degenerate quasilinear equations), then typically we consider maximal or minimal solutions v constructed as the limit

$$v = \lim_{n \to \infty} v_n$$

of a monotone sequence $\{v_n\}$ of classical solutions of regularized uniformly parabolic equations

$$v_t = F_n(x,t,v,v_x,v_{xx}), \quad \text{where } (F_n)_r \geq \mu_n > 0.$$

The sequence $\{F_n\}$ is assumed to converge to F monotonically and uniformly on compact subsets bounded away from the singularity. The initial and boundary conditions are assumed to be regularized in a suitable monotone manner. Then such a *proper* solution is unique and does not depend on monotone approximations (Chapter 6).

Performing a similar construction for any characteristic solution $V \in B$,

$$V = \lim_{n \to \infty} V_n,$$

and organizing a suitable comparison of each pair v_n and V_n on the parabolic boundary, we obtain Sturm's results on sign changes $Z(t, w_n)$ of the difference $w_n = v_n - V_n$ satisfying a linear parabolic equation with sufficiently smooth coefficients. Passing to the limit $n \to \infty$ yields estimates on the number of intersections $\mathrm{Int}(t, V) = Z(t, w)$ of proper continuous solutions v and any $V \in B$.

Therefore, the intersection comparison principles can be applied to classes of "weak" (generalized) solutions of parabolic PDEs with, possibly, unknown precise regularity properties.

1.4 Geometrically ordered flows: Transversality and concavity techniques

We now focus on crucial ideas of the intersection regularity analysis. Techniques of the G-theory are based on general concepts of geometrically ordered evolution of curves and are not necessarily connected with nonlinear parabolic PDEs. The functions and coefficients involved manipulations below are assumed to be sufficiently smooth.

Regularity in ordered smooth geometric evolution. Let $B = \{V\}$ be a finite-parameter set of given sufficiently smooth "characteristic" functions $V = V(x,t)$ defined in $S = I\!\!R \times I\!\!R_+$, where $x \in I\!\!R$ denotes a single space variable and $t \geq 0$ is the evolution time variable. We suppose that the functions from B satisfy a hypothesis of strong separability and transversality: given any pair of different functions V, $\tilde{V} \in B$ and any fixed moment $t \geq 0$, there holds:

(i) $V(x,t) \not\equiv \tilde{V}(x,t)$, and

(ii) each intersection of $V(x,t)$ and $\tilde{V}(x,t)$ as functions of x is transversal, i.e., $V_x \neq \tilde{V}_x$ at any intersection point, where $V = \tilde{V}$.

The main assumption imposed on functions from B is the following intersection principle of the geometric order borrowed from the first Sturm Theorem. Given arbitrary pairs $V(x,t)$ and $\tilde{V}(x,t)$ from B, the number of intersections in x denoted by $\mathrm{Int}(t, V, \tilde{V})$ satisfies

$$\mathrm{Int}(t, V, \tilde{V}) \ \text{ is finite and nonincreasing in } \ t \geq 0. \tag{1.28}$$

Then the geometric structure of B can be characterized by its *intersection index*

$$i(B) = \sup_{V, \tilde{V} \in B;\, t \geq 0} \mathrm{Int}(t, V, \tilde{V}) \equiv \sup_{V, \tilde{V} \in B} \mathrm{Int}(0, V, \tilde{V}).$$

Let us now define a smooth geometrically ordered evolution by means of the characteristic functions B. Namely, let a smooth function $v = v(x,t) \notin B$ satisfying the strong separability property, $v(x,t) \not\equiv V(x,t)$ for any $t \geq 0$, evolve on B. We assume that the Sturmian property (1.28) holds for the number of intersections $\mathrm{Int}(t, V, v)$ between $v(x,t)$ and any function $V(x,t) \in B$. In fact, this is the only "evolution equation" describing the corresponding geometrically ordered flow.

We then arrive at the following question concerning this geometric evolution: how many and which evolution properties of the set B composed of *particular* characteristic functions V can be translated to *arbitrary* functions v?

First of all, one can see that (1.28) assumes the comparison principle.

(i) **Comparison: a standard order-preserving property.** The evolution on B satisfies the comparison theorem: for any function $V \in B$ such that $v(x,0) \not\equiv V(x,0)$, there holds

$$v(x,0) \le V(x,0) \quad \text{in } I\!\!R \implies v(x,t) \le V(x,t) \quad \text{in } S. \tag{1.29}$$

Indeed, Sturmian property (1.28) gives $\mathrm{Int}(0, v, V) = 0$, hence $\mathrm{Int}(t, v, V) = 0$ for all $t > 0$ and comparison follows by separability.* A similar comparison from below is also true. As usual in the parabolic theory, comparison gives L^∞-estimates of the solutions.

We can obtain more if we know more about the internal geometric structure of the set B. Further regularity of the geometric evolution uses two geometric principles of *transversality* and *tangency*.

We begin with an easy transversality analysis. For simplicity we assume that each $V(x,t) \in B$ is monotone increasing with x and is unbounded as $|x| \to \infty$ (i.e., $V(x,t) \to \pm\infty$ as $x \to \pm\infty$), while $v(x,t)$ is a uniformly bounded function with smooth initial data $v(x,0)$ satisfying

$$|v(x,0)| \le 1, \;\; |v_x(x,0)| \le 1 \text{ and } |v_{xx}(x,0)| \le 1 \quad \text{in } I\!\!R. \tag{1.30}$$

Then new intersections between $v(x,t)$ and $V(x,t)$ cannot appear for $|x| \gg 1$.

(ii) **Transversality: a first-order estimate.** Given initial data $v(x,0)$, we choose a subset $B_1^+ \subset B$ of "steep" enough increasing with x functions intersecting $v(x,t)$ at $t = 0$ exactly once and transversally. For instance, in view of the second bound in (1.30), $V \in B_1^+$ if $V_x(x,0) > 1$ for all $x \in I\!\!R$. Then (1.28) implies that, for any $V \in B_1^+$,

$$\mathrm{Int}(0, v, V) \le 1 \implies \mathrm{Int}(t, v, V) \le 1 \;\text{ for }\; t > 0. \tag{1.31}$$

One can see that the transversality of all intersections gives an estimate of the spatial derivative v_x in terms of v,

$$v_x(x,t) \le V_x(x,t) \;\text{ at any intersection point of } v \text{ and } V \in B_1^+, \tag{1.32}$$

as shown on Figure 1.2. We argue by contradiction. If for some $t > 0$, $v_x(x,t) > V_x(x,t)$ at the intersection point with a $V \in B_1^+$, then, since $V(x,t) \gg v(x,t)$ for $x \gg 1$ and $V(x,t) \ll v(x,t)$ for $x \ll -1$, there must exist, at least, two more intersections between solutions. Hence, $\mathrm{Int}(t, v, V) \ge 3$ contradicting (1.31). Choosing a subset B_1^- of steep decreasing functions yields a similar bound from

* No rigorous or any detailed proofs are necessary at this moment. We will shortly pass to parabolic equations where the conclusions are guaranteed by the Maximum Principle.

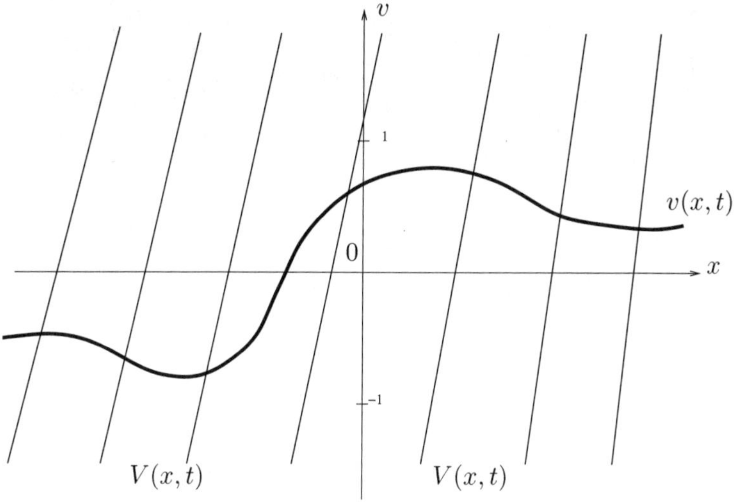

Figure 1.2 *The transversality principle: at any intersection point $v_x \leq V_x$.*

below, so the transversality principle gives a *gradient estimate* of $v(x, t)$ in terms of the functional subset B.

(iii) **Convexity: a second-order estimate.** Assume now that we can choose another sufficiently large subset $B_2^+ \subseteq B$ of functions exhibiting strict *convexity*. Namely, we assume that

$$V_{xx} > 1 \quad \text{in } S \text{ for any } V \in B_2^+. \tag{1.33}$$

Then $V(x, t) \to +\infty$ as $x \to \pm\infty$ and again new intersections cannot occur at infinity. Since, by the third bound in (1.30), the difference $V(x, 0) - v(x, 0)$ is strictly convex, it has at most two zeros, so, for all $V \in B_2^+$,

$$\text{Int}(0, v, V) \leq 2 \implies \text{Int}(t, v, V) \leq 2 \text{ for } t > 0. \tag{1.34}$$

This version of intersection comparison gives an estimate of the second-order derivative v_{xx} in terms of the lower-order ones v_x and v:

$$v_{xx} \leq V_{xx} \text{ at any tangency point with } V \in B_2^+, \tag{1.35}$$

where $V = v$ and $V_x = v_x$, as Figure 1.3 shows. Indeed, assuming that $v_{xx} > V_{xx}$ at a tangency point, we then find a function $\tilde{V} \in B_2^+$, $\tilde{V} \approx V$ such that $\text{Int}(t, v, \tilde{V}) \geq 4$ contradicting (1.34). The existence of such $\tilde{V}$ (being a "small perturbation" of V) is to be guaranteed by a certain *completeness* of set B_2^+, which is assumed to be sufficiently wide and "dense" in a geometric sense to be precisely stated in Chapter 2.

A similar analysis is performed by intersection comparison with complete subsets $B_2^- \subset B$ of strictly concave functions. This convexity/concavity study leads to two-sided estimates on the second derivative v_{xx}.

Let us comment on some extensions of the geometric techniques. The transversality and convexity estimates crucially depend on the chosen characteristic set

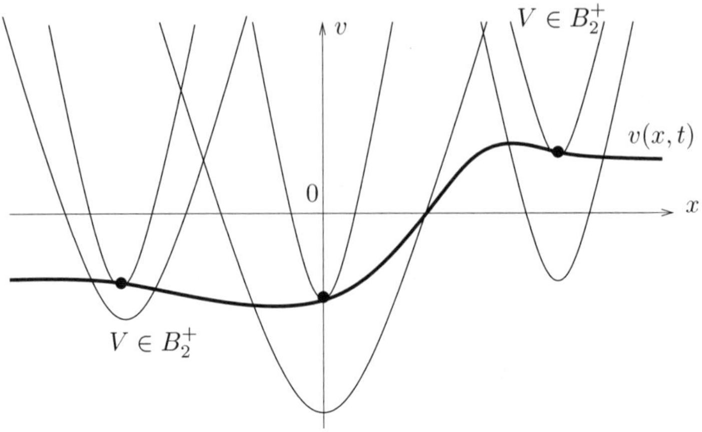

Figure 1.3 *The convexity principle: at each tangency point $v_{xx} \le V_{xx}$.*

B. A proper set B defines the concepts of transversality and *B-convexity* (*B-concavity*) in the spatial variable x, i.e., convexity (concavity) relative to characteristic functions $V(x,t) \in B$. We will show that these properties are preserved with time (are invariants of the geometric evolution) or can occur *eventually* as t increases for a class of more general functions. Wider sets B of reach geometric structures with, at least, $i(B) = 2$, can provide us with more general bounds on the second derivative v_{xx} in terms of v and v_x. Such *a priori* bounds on the spatial derivatives establish a natural next step towards evolution PDEs that are not necessarily of the parabolic type, as examples in the next section show. The geometric analysis becomes more delicate for the case where the characteristic functions $\{V\}$ determine a singular evolution, i.e, each pair of them, V and $\tilde{V}$, can create a singular intersection understood as a finite-time focusing of regular intersection curves. Then, at each time, a one-parameter family of functions from B having the same singularity point forms a *B-bundle*, which evolves with time and describes the singularity evolution of general curves. These are the main subjects of the geometric theory to be studied in detail.

1.5 Evolution B-equations preserving Sturmian properties

In the next chapters we will use the intersection comparison of a general solution $v(x,t)$ with specially chosen families $B = \{V(x,t)\}$ of particular characteristic solutions. We are going to use special functions V that are explicit solutions invariant under scaling or other groups of transformations, or at least these are exact solutions described by finite-dimensional dynamical systems. In any case, B is a low dimensional functional family depending on two or, at most, on three parameters. Therefore, we need Sturmian intersection properties to be valid not relative to any other solution $v(x,t)$ of the equation under consideration, but relative to 2D or 3D subsets B of particular solutions only.

In order to describe a wider class of evolution equations obeying the Sturmian property relative to a fixed subset B, we temporarily forget about second-order parabolic equations and, as in the previous section, consider a geometrically ordered evolution of a smooth function $v(x,t)$ on suitable subset $B = \{V(x,t)\}$ of given characteristic functions.

What kind of evolution equations is prescribed by such a complete subset B provided that the first Sturm Theorem is valid relative to any solution $V \in B$?

Intersection comparison in the tangential space $I\!R^3$

We first consider the case when the set B is *complete* in the *tangential* space $I\!R^3$ in the sense that, for any point $P_0 = (x_0, t_0) \in S$ and any $(\mu, \nu, \rho) \in I\!R^3$, there exists a unique *tangent* function $V(x,t) \in B$ satisfying the three tangency conditions relative to the spatial variable x

$$V = \mu, \quad V_x = \nu \quad \text{and} \quad V_{xx} = \rho \quad \text{at } P_0. \tag{1.36}$$

This is the second-order tangency taking into account curvature of curves.

Proposition 1.6 *(i) If B is complete in the tangential space $I\!R^3$, there exists a function $F : S \times I\!R^3 \to I\!R$ such that B is determined by the second-order evolution equation*

$$V_t = F(x, t, V, V_x, V_{xx}) \quad in \ S. \tag{1.37}$$

(ii) Fix a pair V, $\tilde{V}$ of two different functions from B having tangency point at $P_0 = (x_0, t_0)$: $V = \tilde{V}$, $V_x = \tilde{V}_x$ and $V_{xx} \neq \tilde{V}_{xx}$ by completeness. Assume that, for any such pair and any point P_0, new intersections cannot occur for $t \approx t_0^+$ at $x \approx x_0$. Then equation (1.37) is parabolic, i.e.,

$$F_r(x, t, p, q, r) \geq 0 \quad in \ S \times I\!R^3. \tag{1.38}$$

Proof. (i) Since (1.36) determines a unique function V in terms of five values $(x_0, t_0, V, V_x, V_{xx}) \in S \times I\!R^3$, this gives the time derivative V_t as a function F of these five arguments. (ii) If new intersections cannot occur at any tangency point $P_0 \in S$, we must have, at least, that $(V - \tilde{V})_t$ has the same sign as $(V - \tilde{V})_{xx}$ at P_0, and the parabolicity condition (1.38) follows. □

Under this completeness assumption, the subset B uniquely determines the 1D parabolic equation of its evolution. For such a tangency construction in $I\!R^3$, extensions to other classes of non-parabolic equations are not possible.

Intersection comparison in the hodograph plane $I\!R^2$

We now restrict our attention to separable, transversal subsets B of lower dimensions. Let us show that 2D subsets B yield a wider class of evolution equations than purely parabolic ones.

We say that a set B is *complete* in the *hodograph plane* $I\!R^2$ if, for any $P_0 = (x_0, t_0) \in S$ and any point $(\mu, \nu) \in I\!R^2$, there exists a unique *tangent* function $V(x,t) \in B$ satisfying the tangency conditions

$$V = \mu \quad \text{and} \quad V_x = \nu \quad \text{at } P_0. \tag{1.39}$$

One can see that such tangency conditions do not prescribe a parabolic evolution of B, since, at any tangency point, the time-derivative V_t is now uniquely determined by the four arguments (x, t, V, V_x), and we arrive at the following first-order evolution.

Proposition 1.7 *If B is complete in the hodograph plane $I\!R^2$, there exists a function $G : S \times I\!R^2 \to I\!R$ such that B is governed by the Hamilton-Jacobi equation*

$$V_t = G(x, t, V, V_x) \quad in \quad S. \tag{1.40}$$

Therefore, we can prescribe a class of evolution equations satisfying the intersection property relative to any subset of its particular solutions $V \in B$. We call them *Sturmian B-equations* emphasizing their dependence on the given complete subset B. In the further study below, we deal with standard local applications of the strong MP only and do not take into account in which functional classes the Cauchy problems are well posed (this can be difficult).

Proposition 1.8 *Let the evolution equation*

$$v_t = \mathbf{F}(v) = F(x, t, v, v_x, v_{xx}, ...) \quad in \quad S, \tag{1.41}$$

where F can depend on higher-order spatial derivatives of v and other operators, admit a suitable subset $B = \{V\}$ of solutions that is complete in the hodograph plane $I\!R^2$. Then the number of intersection of its sufficiently regular solution v with any $V \in B$ satisfies the first Sturm Theorem if the function

$$\Phi(x, t, w, w_x, w_{xx}, ...) \equiv F(x, t, V + w, V_x + w_x, V_{xx} + w_{xx}, ...)$$
$$- F(x, t, V, V_x, V_{xx}, ...)$$

satisfies

$$a(\cdot) \equiv \tfrac{1}{w_{xx}} \Phi(\cdot)|_{w=w_x=0} \geq a_0 = \text{const} > 0 \quad for\ any \quad w_{xx} \neq 0. \tag{1.42}$$

Proof. Let us show that given v and $V \in B$, the difference $w = v - V$ satisfies a linear smooth uniformly parabolic equation. Indeed, w solves

$$w_t = \Phi(x, t, w, w_x, w_{xx}, ...) \quad \text{in} \quad S, \tag{1.43}$$

where by (1.42), $\Phi(x, t, 0, 0, w_{xx}, ...) = a(\cdot)w_{xx}$. Then, by Taylor's formula with Lagrange's form of the remainder,

$$\Phi(x, t, w, w_x, w_{xx}, ...) = a(\cdot)w_{xx} + b(\cdot)w_x + c(\cdot)w,$$

and hence for smooth solutions, (1.43) can be written as a linear parabolic equation with $a \geq a_0 > 0$. □

Let us present a simple example of a higher-order equation satisfying Sturmian intersection properties.

Example 1.1 Setting $u = e^v$ in the heat equation $u_t = u_{xx}$ yields the semilinear equation

$$v_t = v_{xx} + (v_x)^2.$$

It admits a subset of linear travelling wave solutions $B = \{V(x, t) = \lambda^2 t + \lambda x + a,\ \lambda, a \in I\!R\}$ satisfying the quadratic Hamilton-Jacobi equation $V_t = (V_x)^2$. B

is complete in the hodograph plane $I\!R^2$, the tangential system (1.39) has a unique solution $\lambda = \nu$, $a = \mu - \nu^2 t_0 - \nu x_0$ and hence the tangent solution at any P_0 always exists and is unique. Then Proposition 1.8 gives a wide class of evolution B-equations generated by the given solutions subset B. For instance, consider the equation

$$v_t = v_{xx} + (v_x)^2 + H(\cdot)v_{xx}, \tag{1.44}$$

with an arbitrary smooth coefficient $H = H(x, t, v, v_x, v_{xx}, ..., D_x^k v, ...) \geq 0$ possibly depending on several higher-order derivatives $D_x^k v$ with $k > 2$. Since functions $V(x,t)$ are linear in x, the extra higher-order operator Hv_{xx} vanishes on B, and so any V is also a solution of (1.44). Consider another smooth solution $v(x,t)$ of (1.44) (if any). The difference $w = v - V$ solves a linear uniformly parabolic equation with smooth coefficients

$$w_t = aw_{xx} + bw_x, \quad \text{where } a = 1 + H(\cdot) \geq 1 \text{ and } b = 2V_x + w_x. \tag{1.45}$$

Therefore, the first Sturm Theorem is valid under the assumption that new zeros cannot occur at $x = \infty$.

Given a complete B, the corresponding class of B-equations often looks rather artificial and, possibly, does not include equations having reasonable applications. Moreover, for such higher-order B-equations the existence, uniqueness and higher-order regularity theory are not straightforward though the equation obeys the strong MP, which takes into account the first derivatives v_t, v_x and v_{xx} only. Let us mention again that, for a number of quasilinear and fully nonlinear singular parabolic equations, the intersection comparison with different complete subsets B of particular solutions implies special *a priori* regularity estimates on the general solution. Such Bernstein-type analysis in functional B-classes driven by the geometric properties of the subset B follows from the transversality principle. We state a simple version of such gradient estimates where conventionally we assume the same hypotheses as in Section 1.4 (though these are justified by the MP for the linear equation (1.45)).

Proposition 1.9 *Let $v(x,t)$ be a classical bounded solution of the Cauchy problem for* (1.44) *with smooth initial data $v(x,0) = v_0(x)$ satisfying*

$$|v_0(x)| \leq 1 \quad \textit{and} \quad |v_0'(x)| \leq 1 \quad \textit{in } I\!R. \tag{1.46}$$

Then

$$|v| \leq 1 \quad \textit{and} \quad |v_x| \leq 1 \quad \textit{in } S. \tag{1.47}$$

Proof. Taking $V \equiv \pm 1$ from B and using the intersection property $\text{Int}(0, \pm 1) = 0$ implies $\text{Int}(t, \pm 1) \equiv 0$ for $t > 0$ and we obtain that $-1 \leq v \leq 1$ in S (indeed, this is the standard comparison for the parabolic equation (1.45)).

Now assume that there exists a point $P_0 = (x_0, t_0) \in S$ such that, say, $c = v_x(P_0) > 1$. By completeness of B, we choose a solution $V \in B$ such that $V = v$ and $V_x = \lambda \in (1, c)$ at P_0. Since $v(x, t_0)$ is uniformly bounded and $V(x, t_0)$ is unbounded as $x \to \pm\infty$, as on Figure 1.2, one concludes that $\text{Int}(t_0, V) \geq 3$. This contradicts the first Sturm Theorem since the assumption $|v_0'| \leq 1$ implies that $\text{Int}(0, V) = 1$ for any $V \in B$ with $|\lambda| > 1$. □

A priori estimates of the second derivative v_{xx} are more delicate and can be obtained by another extended subset B of particular solutions being complete in the tangential space $I\!R^3$. We will consider examples of such a regularity analysis in Chapter 5. Thus, under suitable hypotheses, main conclusions of the geometric analysis remain valid for a class of Sturmian B-equations.

Remarks and comments on the literature

ODEs: Sturm theorems and generalizations. Two papers of Sturm on zeros for second-order ODEs and PDEs were published in the first volume of *Journal de Mathématique Pures et Appliquées* in 1836, [323] (ODEs) and [324] (PDEs). Sturm Theorems on zeros for second-order ODEs (a whole list includes the first, second, fundamental, comparison, separation, alternation and oscillation theorems) and the corresponding methods can be found in many text books on the theory of ODEs and are widely used in mathematical monographs. See E.L. Ince [203, Chapter 10], P. Hartman [185, p. 333], L. Cesari [81, Chapter 2], G. Birkhoff and G.-C. Rota [56, Chapters 2 and 10], G. Sansone and R. Conti [308], M.H. Protter and H.F. Weinberger [299, Chapter 1], B.M. Levitan and I.S. Sargsian [246, Chapter 1], G. Szegö [328, Chapters 1, 6]. The book by W.T. Reid [301] is entirely devoted to generalizations and applications of Sturm ideas and theorems in the ODEs theory. The books [301] and [325] contain a detailed description of the results, historical comments and extensive lists of earlier references.

Classical Sturm results on zeros for a single second-order ODE such as

$$y'' + q(t)y = 0, \quad t \in (0, 2\pi), \tag{1.48}$$

can be stated in the form of a topological nature describing rotations in the phase space of equations (this form is convenient for extensions to higher-order equations, see below). Let

$$Y(t) = \begin{pmatrix} y_1(t) & y_2(t) \\ y_1'(t) & y_2'(t) \end{pmatrix} \quad \text{satisfying } Y(0) = E_2 = \begin{pmatrix} 1 & 0 \\ 0 & 1 \end{pmatrix}$$

be a matrix solution of (1.48), where $y_1(t)$ and $y_2(t)$ are linearly independent. Then the vector $z(t) = y_1(t) + iy_2(t)$ moves counterclockwise on the complex plane. Indeed, since, by construction, the Wronskian $W(y_1, y_2)(t) = \det Y(t) \equiv 1$, we have that $\arg z(t) = \tan^{-1}(y_2(t)/y_1(t))$ satisfies

$$\tfrac{\mathrm{d}}{\mathrm{d}t} \arg z(t) = \tfrac{W(y_1,y_2)(t)}{y_1^2(t)+y_2^2(t)} = \tfrac{1}{y_1^2(t)+y_2^2(t)} > 0.$$

Sturm Theorems follow from this monotonicity property.

The first generalizations of Sturm Theorems to the case of vector-valued operators and to systems (1.48) with symmetric matrices $q(t)$ are due to M. Morse (1930) [269], [270], where variational methods are applied.

Oscillatory theorems for general canonical systems of $2k$th order were first established by V.B. Lidskii (1955) [247] who studied the systems

$$y' = IH(t)y, \quad \text{with } I = \begin{pmatrix} 0 & E_k \\ -E_k & 0 \end{pmatrix},$$

where E_k is the $k \times k$ identity matrix and $H(t)$ is a $2k \times 2k$ real continuous symmetric one (the Hamiltonian). Then, similarly, let $Y(t)$ with $Y(0) = E_{2k}$ be a matrix solution. Hence $Y(t)$ is symplectic, i.e., $Y^*IY \equiv I$. Denote

$$H(t) = \begin{pmatrix} h_{11}(t) & h_{12}(t) \\ h_{21}(t) & h_{22}(t) \end{pmatrix} \quad \text{and} \quad Y(t) = \begin{pmatrix} y_{11}(t) & y_{12}(t) \\ y_{21}(t) & y_{22}(t) \end{pmatrix},$$

where $h_{ij}(t)$ and $y_{ij}(t)$ are $k \times k$ blocks. Consider the non-singular matrix $z(t) = y_{11}(t) + iy_{12}(t)$ (cf. the case $k = 1$ above), and set

$$u(t) = (\overline{z}(t))^{-1} z(t).$$

Then $u(t)$ is unitary and symmetric. The alternation theorem by Lidskii is as follows. Let

$$h_{22}(t) > 0$$

($h_{22} \equiv 1$ for (1.48) with $k = 1$). Then the eigenvalues $\rho_1(t)$,..., $\rho_k(t)$ of $u(t)$ move counterclockwise around the unit circle, i.e.,

$$\tfrac{\mathrm{d}}{\mathrm{d}t} \arg \rho_s(t) > 0 \quad \text{for} \quad s = 1, ..., k.$$

For $\rho_s(t) = -1$ (resp., $\rho_s(t) = +1$), the matrix $u(t)$ has the same zero subspace as $y_{11}(t)$ (resp., $y_{12}(t)$), i.e., the "zeros" of the matrices $y_{11}(t)$ and $y_{12}(t)$ alternate. Lidskii also proved the following analogue of the Sturm comparison theorem. Consider two canonical systems

$$Y_1' = IH_1(t)Y_1 \quad \text{and} \quad Y_2' = IH_2(t)Y_2, \qquad \text{where } H_1(t) > H_2(t).$$

Then specially enumerated eigenvalues $\rho_s^{(1)}(t)$ and $\rho_s^{(2)}(t)$ of the unitary matrices $u^{(1)}(t)$ and $u^{(2)}(t)$ satisfy

$$\arg \rho_s^{(1)}(t) > \arg \rho_s^{(2)}(t) \quad \text{for} \quad s = 1, ..., k,$$

i.e., $\rho_s^{(1)}(t)$ moves "ahead" of $\rho_s^{(2)}(t)$.

Variational approaches to Sturm theorems for self-adjoint linear $2k$th order systems were also developed by R. Bott (1959) [61] and by H.H. Edwards (1964) [101]. See detailed presentation in the books [301] and [33]. These results were also related to the *Maslov index* [259]. In 1985 V.I. Arnold characterized this as follows [21]: "...numerous authors writing about the Maslov index, symplectic geometry, geometric quantization, Lagrangian analysis, etc., starting with [20], have not noticed the earlier works by Lidskii [247], as well as the earlier works of Bott [61] and Edwards [101], in which a Hermitian version of the theory of the Maslov index and Sturm intersections were constructed."

A survey of earlier results concerning distribution and alternation of zeros for the nth order linear ODEs can also be found in [242], where, as well as in the books mentioned above, various links to other related subjects are described in detail. These include S.A. Chaplygin's comparison theorem (1932) [82] closely connected with the theory of positive operators, W.A. Markov's theorem (1916) [258] on the conservation of the alternation of zeros of polynomials under differentiation, C. de la Vallée-Poussin's theorem (1929) [96] and G. Pólya's criterion (1924) [295] on non-oscillation (the first non-oscillation test of the best-possible character is found in N.E. Zhukovskii (1892) [350]), F.R. Gantmakher's (1936) [171] and M.G. Krein's (1939) [229] theory of oscillating kernels [172] (a direction originated with O.D. Kellog's work (1922) [221] on symmetric kernels), S.N. Bernstein's results (1938) [51] on connections between Chebyshev and Cartesian systems, etc.

Sturmian methods for ODEs can be adjusted to explorations in the complex plane; see [191, Chapter 8]. The classical Sturm comparison theorem for ODEs admits special extensions to linear and quasilinear elliptic and parabolic PDEs; see first results in [289], the book [327] and [2]–[4], as well as to ODEs in Hilbert spaces [206].

Sturm's Theorem on the number of the distinct real roots of polynomials by computing the number of sign changes in Sturm sequences (1835) [322] is well known in Algebra; see e.g. [240] and [60]. Composing Sturm sequences, the first step is differentiation establishing a link to ODEs (Sturm's comparison or oscillation theorems).

As a relation to ODEs, Sturm's ideas have applications in the classical problem on zeros of complete Abelian integrals defined by means of a planar Hamiltonian flow, which is closely related to the Hilbert 16th problem (the so-called weakened, infinitesimal or tangential Hilbert problem). Abelian integrals were known to satisfy a system of Picard-Fuchs ODEs [178]; see also [202] for further references. This is a part of a general problem on zeros of Pfaffian functions and the fewnomials theory, [223], [224], where the eventual reduction to polynomial structures is used. In particular, algorithmic consistency problems of systems of Pfaffian equations and inequalities occur (with applications to computer sciences); see [124] and references therein.

Let us return to the Sturm-Hurwitz Theorem establishing that the finite Fourier series

$$f(x) = \sum_{L \le k \le M} (a_k \cos kx + b_k \sin kx), \quad x \in [0, 2\pi],$$

has at least $2L$ and at most $2M$ zeros. On pp. 436-444 of the PDE paper [324] Sturm presented an ODE proof of the result. Sturm's ODE proof, as well as Liouville's one in [250] published in the same volume, exhibit certain features of a discrete evolution analysis (to be compared with Sturm's PDE proof via evolution parabolic equation with continuous time variable). A. Hurwitz (1903) [200] extended this result to Fourier series with $M = \infty$.

Further extension is due to O.D. Kellog (1916) [220] who proved oscillation theorems for linear combinations of real continuous orthonormal functions $\phi_0(x)$, $\phi_1(x)$, ... , $\phi_n(x)$ on the interval $(0, 1)$. Note that these are not eigenfunctions of a Sturm-Liouville problem. The main assumption is as follows (we keep the original notation). For any $n \ge 1$, let the determinants

$$D(x_0, x_1, ..., x_n) = \begin{vmatrix} \phi_0(x_0) & \phi_1(x_0) & ... & \phi_n(x_0) \\ \phi_0(x_1) & \phi_1(x_1) & ... & \phi_n(x_1) \\ ... & ... & ... & ... \\ \phi_0(x_n) & \phi_1(x_n) & ... & \phi_n(x_n) \end{vmatrix}$$

be positive for any $0 < x_0 < x_1 < ... < x_n < 1$ ($D_0(x_0)$ being understood as $\phi_0(x_0)$). Let

$$\Phi_{m,n}(x) = c_m \phi_m(x) + ... + c_n \phi_n(x).$$

Then, among other results, it is established that:

(i) $\Phi_{0,n}(x)$ cannot vanish at $n+1$ distinct points in $(0,1)$ without vanishing identically;

(ii) $\phi_n(x)$ vanishes exactly n times and changes sign at each zero;

(iii) every continuous function $\psi(x)$ orthogonal to $\phi_0(x)$, ... , $\phi_n(x)$ changes sign at least $n+1$ times;

(iv) $\Phi_{m,n}(x)$ changes sign at least m times and at most n times.

The infinitesimal version of the discriminants with $x_{k+1} - x_k \to 0$ for $k = 0, 1, ..., n-1$, deals with the Wronskian of the given functions. Hence, some of the assumptions are valid for eigenfunctions of regular Sturm-Liouville problems. On the other hand, Kellog's results do not cover some of Sturm's; see [220, p. 5].

The Sturm-Hurwitz Theorem plays a fundamental role in topological problems in wave propagation theory (topology of caustics and wave fronts), the geometry of plane and spherical curves and in general symplectic geometry and topology, see [21], [23], [24], [26] and references therein. Alternating, oscillating and non-oscillating Sturm theorems have multi-dimensional symplectic analogies and describe the rotation of a Lagrangian subspace of the phase space [21]. For instance, the Sturm-Hurwitz Theorem proves a generalization [329] of the classical four vertex theorem by S. Mukhopadyaya (1909) [271] and A. Kneser (1912) [225] asserting that a plane closed non-self-intersecting curve has at least four vertices (critical points of the curvature). It is pointed out in [24] that the same minimal number occurs in: (i) theorems on four casps of general caustics on every convex surface of positive curvature (the related conjecture goes back to C.G.J. Jacobi (1884) [205]), (ii) four casps of the envelope of the family of perturbed Larmor orbits of given energy, (iii) the "tennis-ball theorem" (a closed curve on the sphere without self-intersections, a smooth embedding $S^1 \to S^2$, dividing the sphere into two parts of equal area, has at least four points of spherical inflection with zero curvature), (iv) the four equilibrium points theorem, (v) the four flattening points theorem for perturbed convex curves of positive curvature on a plane lying in three-dimensional space, etc. Infinitesimal versions of such topological theorems (for infinitely small perturbations of curves) follow from the Sturm-Hurwitz Theorem. For finite perturbations, some of these results can be proved by means of the evolution Sturm Theorem on zeros for parabolic PDEs to be mentioned later on.

Half of Arnold's third lecture in the Fields Institute (1997) [25] was devoted to the Sturm theory about Fourier series, which "provides one of the manifestations of the general principle of economy in algebraic geometry" (related to Arnold's conjecture (1965) and the symplectification of topology). In particular, the Morse inequality (in the simplest version it says that the number of critical points of functions on the circle is at least 2) is the Sturm-Hurwitz Theorem with $L = 1$. Among others, let us also mention applications of the Sturm-Hurwitz Theorem to the "Hessian topology" of hyperbolic polynomials, see survey [27], and to topological characteristics of pseudoperiodic functions and manifolds [28].

The Sturm-Hurwitz Theorem was first proved by the PDE method [324, pp. 431-436] in the general form including any (finite) series composed of eigenfunctions of a Sturm-Liouville problem. These evolution Sturm ideas have many other applications to be discussed below in the parabolic part of the survey.

Extensions of Sturm results on zeros (nodal sets) of linear combinations of eigenfunctions to standard self-adjoint elliptic operators (e.g., the Laplacian Δ) in bounded smooth domains $\Omega \subset I\!R^N$, $N \geq 2$, are unknown; see [24] and [25]. In particular, the so-called Herman theorem announced in [90, p. 454]: a linear combination of the first n eigenfunctions divides the domain, by means of nodes (piecewise smooth nodal surfaces), into not more than n subdomains, is wrong for the spherical Laplacian [25]. By Courant's Theorem on p. 452, the nodes of the nth eigenfunction divide the domain into no more than n subdomains.

In dimensions $N \geq 2$, given a linear combination $f(x)$ of eigenfunctions of Δ, the structure of the nodal set itself $\mathcal{N}(f) = \{x \in \Omega : \; f(x) = 0\}$ (it can be viewed as the intersection of $f(x)$ with the trivial solution $V \equiv 0$ of $\Delta V = 0$ in Ω) is not sufficient to define a kind of a Sturmian "index" of the surface $z = f(x)$, similar to the number of zeros in 1D, which can inherit a certain numerical property (say, a lower bound) from the lowest harmonic of the series. Such an index is supposed to depend on global properties of $f(x)$ at all points $x \in \Omega$ including those far away from $\mathcal{N}(f)$. It seems reasonable that for a proper definition of a Sturmian index, it is necessary to control intersections of $f(x)$ with a finite-dimensional subset $B = \{V_\nu(x)\}$ of functions associated with operator Δ (i.e., with some higher-order eigenfunctions). Roughly speaking, this would mean that such a "local" characteristic as the number of zeros of $f(x)$ on an interval from $I\!R$ cannot work in $I\!R^N$, where any possible nonincreasing property of, say, the number of maximal connected subdomains of the positivity subset $\{f(x) > 0\}$ is supposed to include some global properties of the function formulated in an unknown way. In any case, a proper definition of Sturmian index of surfaces governed by parabolic equations in $I\!R^N$ is not expected to admit a simple formulation or such easy and effective applications as it has in the 1D case.

Parabolic PDEs and Sturm Theorems. As for Sturm's second paper [324] on zero set analysis for one-dimensional parabolic PDEs, I found a unique monograph by F.V. Atkinson (1964) [33] with a reference to this paper on p. 513 (though Sturm's PDEs results were not used there).

The Sturmian argument for 1D parabolic equations turns out to be an extremely effective technique in the study of different aspects of the theory of nonlinear parabolic equations. In the twentieth century the argument was partially and independently rediscovered several times. Let us mention some of the papers published at least twenty years ago. There are many other interesting and important papers published more recently, which are not referred to here. Some of them related to singular blow-up solutions of parabolic PDEs will be referred to in the following chapters.

G. Pólya (1933) [296] paid special attention to Sturm's zero number properties of periodic solutions to the heat equation. He studied the number of "Nullstellen" of $u(x,t)$, i.e., the number of $x \in [0, 2\pi]$ such that $u(x,t) = 0$, on the basis of Sturm's approach with a reference to [324]. Radial and cylindrical solutions were considered and zero properties of convolution integrals were also studied.

The celebrated KPP-paper (1937) [226] was devoted to the stability analysis of the minimal TW for a semilinear heat equation

$$u_t = u_{xx} + f(u) \quad \text{in } I\!R \times I\!R_+,$$

with the typical nonlinearity $f(u) = u(1-u)$. The construction of a geometric Lyapunov function in Theorem 11 was based on the following intersection comparison argument: the initial 1-step function, $u_0(x) = \{1 \text{ for } x > 0 \text{ and } 0 \text{ for } x \le 0\}$, intersects any smooth travelling wave profile exactly at a single point and there exists a unique intersection curve for $t > 0$. In our notation, this implies that $\text{Int}(t, V) \equiv 1$ for any TW $V(x,t) = f(x - \lambda_0 t + a) \in B$ and any $t > 0$, where $\lambda_0 > 0$ is the minimal speed. In general, the number of intersections can be treated as a discrete nonincreasing Lyapunov function. On the other hand, the number of intersections gives a standard monotone Lyapunov function: on any fixed level $\{u(x,t) = c \in (0,1)\}$ the derivative $u_x(x,t) > 0$ is monotone decreasing with t and bounded from below. Then passage to the limit $t \to \infty$ establishes the convergence to the minimal TW profile in the hodograph plane $\{u, u_x\}$ or in the moving coordinate system on the $\{x, u\}$-plane.

K. Nickel's paper (1962) [274] (see also [275]) established nonincrease of the number of sign changes of solutions of parabolic equations (more precisely, of the number of relative maxima of a solution profile, i.e., the number of zeros of the derivative $u_x(x,t)$). Nickel's results are explained in detail relative to general fully nonlinear parabolic equations (1.26) in W. Walter's books [344] and [343, Section 27]. R.M. Redheffer and W. Walter (1974) [300] extended such results to more general classes of equations. For particular linear parabolic equations in $I\!R$, these results were proved by S. Karlin (1964) [217], whose analysis was based on ideas of total positivity of Green's functions and applied to Brownian motion processes. Related questions and techniques were discussed by I.K. Ivanov (1965) [204] (the number of changes of sign was considered), by E.K. Godunova and V.I. Levin (1966) [179] (a proof of the existence of a single maximum was based on the theory of probabilistic distributions; eventual single maximum distribution and eventual concavity of solutions were also established) and by E.M. Landis (1966) [239] (properties of evolution of level sets for (1.1) were investigated). D.H. Sattinger's results (1969) [310] on sign changes for linear parabolic equations are similar to those obtained by Nickel and Walter. Observe that, in the proof of Theorem 7 on exponential decay of total variation, Sattinger uses a reflection technique and studies zeros of the differences of $u(x,t)$ and the reflected solutions $u(2l - x, t)$; see [310, p. 88]. Such a combination of Sturm's Theorem and A.D. Aleksandrov's Reflection Principle and ideas (1960) [1] later became a power tool in the asymptotic theory for nonlinear parabolic equations. Papers by A.N. Stokes (1977) [320]† and [321] used the nonincrease of zero number with application to stability analysis of travelling waves. Here the basic idea of proving a Lyapunov monotonicity property in the hodograph plane is essentially the

† The title of this paper contains the term "intersections of solutions", which led me to introduce the notion of "intersection comparison" of solutions in the mid of the 1980s.

same as in the KPP-analysis [226]. A general stability analysis of TWs for analytic semilinear parabolic equations by zero set properties was performed in [17].

H. Matano (1978) [260] proved the first Sturm Theorem and applied it to establishing that the ω-limit set of any bounded solution to a semilinear parabolic equation

$$u_t = (a(x)u_x)_x + f(x,u) \quad \text{on } (0,L) \times I\!R_+, \;\; a \geq a_0 > 0,$$

with smooth coefficients and Robin boundary conditions contains at most one stationary point. At that time such a result was already known [348] for smooth uniformly parabolic equations $u_t = a(x,u,u_x)u_{xx} + b(x,u,u_x)$ with general nonlinear boundary conditions and was proved by constructing a standard (integral) Lyapunov functional by the method of characteristics, a fruitful idea which applies to 1D quasilinear parabolic equations. The geometric proof by Matano is more general and can be applied to fully nonlinear parabolic equations

$$u_t = F(x,u,u_x,u_{xx}). \tag{1.49}$$

More detailed results by Matano related to the first Sturm Theorem were published in [261]. A finite difference approach to some of these Sturmian properties was developed earlier by M. Tabata (1980) [330] .

My first application of intersection comparison (1982) [127] was related to blow-up solutions of quasilinear parabolic equations $u_t = (k(u)u_x)_x + Q(u)$, where using the estimate $\text{Int}(t,U) \leq 2$, the envelope of an unbounded family of stationary states $\{U\}$ was shown to give a lower bound for blow-up solutions (the method of stationary states; more details and references are available in [306, Chapter 7]). ‡

We continue our historical survey later on and now present comments to the first two sections of Chapter 1.

§ **1.1.** We notice again that the first Sturm Theorem is formulated in [324, p. 431] and is a by-product of the second one on evolution of multiple zeros. The present proof of Theorem 1.1 is taken from [310] (similar to that in [274]).

§ **1.2.** In the proof of Theorem 1.2 we follow the lines of the original Sturm's analysis in [324], which was done for the semilinear parabolic equation on a bounded interval

$$gu_t = (ku_x)_x - lu, \quad x \in (\mathrm{x}, \mathrm{X}), \; t > 0, \tag{1.50}$$

with smooth functions g, k and l depending on x and t. The main calculations were performed for g, k, l depending on x only. A comment on p. 431 extends the results to dependence on t. Third type (Robin) boundary conditions were incorporated,

$$ku_x - hu = 0 \quad \text{at } x = \mathrm{x}, \quad ku_x + Hu = 0 \quad \text{at } x = \mathrm{X}, \tag{1.51}$$

‡ I heard for the first time about Sturm's zero-set theory for PDEs in 1984 at K.I. Babenko's seminar at M.V. Keldysh Institute of Applied Mathematics, Moscow, where I reported some intersection comparison results. My colleague A.S. Shvedov mentioned that this Sturm's theory was explained in lectures by Professor V.M. Tikhomirov in the 1970s at the Mechanical-Mathematical Department of the Moscow State University.

where h, H are constants but can also depend on t; see p. 431. (Zero Dirichlet boundary conditions are also mentioned there.) Note that Sturm's analysis on pp. 428-430 includes the case of multiple zeros occurring at boundary points x or X.

Computations similar to Sturm's ones can be found in [17, Section 5]. For radial equations (1.3) with $N > 1$, such computations for $t < 0$ lead to Laguerre polynomials $L_m^\gamma(z)$ of order $\gamma = \frac{N}{2}$; see [12, Section 3]. Perturbation techniques for operator (1.17) were developed in [186], [11], [83]. Sturm's backward parabolic rescaling with $z = x/\sqrt{-t}$ plays an important role in continuation theorems and topology of nodal sets for linear parabolic equations in $I\!R^N$ [83]. A weak form of the continuation analysis [297] based on a monotonicity formula and weighted inequalities (this idea goes back to T. Carleman (1939) [80] with applications to elliptic equations), which are convolutions with the backward heat kernel, uses the same Sturm backward variable.

The evolution proof of the Sturm-Hurwitz Theorem on zeros of (finite) linear combinations of eigenfunctions $\{V_k(x),\ k = 1, 2, ...\}$, where each V_k has exactly $k - 1$ simple transversal zeros, of a Sturm-Liouville operator given by (1.50), (1.51),

$$Y(x) = C_i V_i(x) + C_{i+1} V_{i+1}(x) + ... + C_p V_p(x)$$

is given in [324, pp. 431-444] and is as follows (we keep the original notation). Consider the solution

$$u(x,t) = C_i V_i(x) e^{-\rho_i t} + C_{i+1} V_{i+1(x)} e^{-\rho_{i+1} t} + ... + C_p V_p(x) e^{-\rho_p t} \quad (1.52)$$

of the parabolic equation (1.50) with $u(x, 0) \equiv Y(x)$, where the sequence of eigenvalues $\{-\rho_k\}$ is strictly decreasing. Then, for $t \gg 1$, the first harmonic is dominant and hence $u(x,t)$ has exactly $i - 1$ zeros. Since the number of zeros of $u(x,t)$ does not increase, $u(x,t)$ has at least $i - 1$ zeros for all $t \in I\!R$, and hence at $t = 0$. On the other hand, for $t \ll -1$, the last harmonic in (1.52) is dominant, $u(x,t)$ has exactly $p - 1$ zeros, so that, by the Sturm Theorem, $u(x,t)$ has at most $p - 1$ zeros for all $t \in I\!R$.

On p. 436 Sturm compares his proof with that by J. Liouville [250] "...without using consideration of the auxiliary variable t..." (i.e., by means of an ODE argument). In Section XXVI Sturm presents his own ODE proof.

Corollary 1.3 is a paraphrase of Sturm's calculations. The proof of Theorems 1.4 and 1.5 are given in [11]. Finiteness of $Z(t,u)$ on $(0,1)$ for $t > 0$ was also established in [234] for coefficients $a \in H^1$, $b \in W^{1,\infty}$ and $c \in L^\infty$ depending on x only. The second Sturm Theorem on formation of multiple zeros remains valid for $W^{2,1}_{p,\mathrm{loc}}$ solutions ($p > 1$) from Tikhonov's uniqueness class for linear uniformly parabolic equations in $I\!R^N$ with bounded coefficients [83] (the proof uses Sturmian backward rescaling). The analytic case was treated in [17]. Eventual simplicity of zeros was first observed in [67].

An evolution approach to connections of equilibria for semilinear parabolic equations was developed by D. Henry [186], where a time-dependent Sturm-Liouville theory was rigorously established (note Theorem 4 on the completeness of asymptotic limits proved by Agmon's estimates). This theory was used in completing the proof that, under some hypotheses, a general semilinear parabolic equation

$$u_t = u_{xx} + f(x, u, u_x) \quad \text{in } (0,1) \times I\!R_+,$$

with Dirichlet or nonlinear boundary conditions represents a Morse-Smale system. It is established that given a heteroclinic connection $\bar{u}(x,t)$ of two hyperbolic linearly nondegenerate equilibria $\phi_\pm$, $\bar{u}(x,-\infty) = \phi_-(x)$ and $\bar{u}(x,+\infty) = \phi_+(x)$, the stable manifold $W^s(\phi_+)$ and the unstable one $W^u(\phi_-)$ meet transversally at $\bar{u}(\cdot,t)$ for each t. See also [10] for the case $f = f(x,u) \in C^2$. This transversality result was used in [186] to describe all connecting orbits between equilibria for the Chafee-Infante problem with $f = f(u)$, $f(0) = 0$. For earlier results on connections for parabolic equations see [183] and [66]. For more general $f \in C^2$, such connections were established in [68]. See also surveys in [112], [113] and [293].

A spectrum of Hermite polynomials occurred in the zero set analysis by D. Henry [186] and S.B. Angenent [11]. Zero set results played a role in the analyticity study of solutions of the PME [12]. A few years after papers [186], [10] and [11] on parabolic Morse-Smale systems, the same linearized operators, with eigenfunctions composed of separable Hermite polynomials, were obtained in the centre and stable manifold behaviour in the study of blow-up solutions of the semilinear parabolic equations from combustion theory

$$u_t = \Delta u + u^p, \ p > 1 \quad \text{and} \quad u_t = \Delta u + e^u$$

(the nonstationary Frank-Kamenetskii equation); see [341], [116], [188], [338] and [266]. See also references to Chapter 4, where blow-up problems are considered in greater detail and the survey on intersection comparison in blow-up will be continued.

The Sturm Theorems play a key role in the analysis of other aspects of behaviour in infinite-dimensional dynamical systems associated with nonlinear parabolic equations. These are convergence to periodic solutions and related questions for periodic equations [85], [69] (results apply to general 1D fully nonlinear equations), [112], [84] (transversality properties), [292], [190] and [86] (applied to N-dimensional semilinear parabolic equations by means of symmetrization and moving plane techniques) and [312] (almost periodicity). Zero set analysis is a leading ingredient of a Poincaré-Bendixson Theorem for semilinear heat equations, [17], [262], [114] (see survey [293]), and in the construction of G. Floquet bundles (see [117] and results by A.M. Lyapunov [255]) for linear parabolic equations in periodic and non-periodic cases (solutions $u_n(x,t)$ having exactly n zeros for all $t \in I\!R$) [88], [89] (a generalization of Sturm-Liouville theory to the time-dependent case, results include exponential dichotomies and other estimates). Such Floquet-type solutions $\{u_n(x,t), t > 0\}$ exist for the semilinear heat equation

$$\boxed{u_t = u_{xx} - |u|^{p-1}u \quad \text{in } I\!R \times I\!R_+}$$

with decay rate as $t \to \infty$ depending on n [268]. The nonincreasing number of zeros plays a key role in the problems of Morse decomposition [256] and connections of Morse sets [115] for the monotone feedback differential delay equation

$$\dot{u}(t) = f(u(t), u(t-1)), \quad u \in \mathbb{R}.$$

Nonincrease of the number of zeros per unit interval for such linear equations was first established by A.D. Myschkis (1955); see [272, Theorem 32]. It is also true for monotone cyclic feedback systems [257]

$$\dot{u}_i = f_i(u_i, u_{i-1}), \quad u_i \in \mathbb{R}, \quad i \bmod n.$$

Sturm's intersection ideas play a fundamental role in *curve shortening* or *flows by mean curvature* problems for curves on surfaces. For curves on a surface M with a Riemannian metric g, such a motion is described by the curve shortening equation

$$v^{\perp} = V(t, k), \tag{1.53}$$

where $v^{\perp}$ is the normal velocity of the curve, k is the curvature and V is a $C^{1,1}$ function satisfying $\frac{\partial V}{\partial k} > 0$. The reason that Sturm's results apply to such evolution problems (though some of the properties are intuitively obvious for intersections of curves) is that (1.53) reduces to a nonlinear parabolic equation for the curvature k or, after a suitable parameterization, for a function $u(x,t)$ satisfying a fully nonlinear parabolic equation (1.49), where F depends on V. See first results in [199], [311] and [125] (a parabolic curvature equation $k_\tau = k^2(k_{\theta\theta} + k)$ was derived for the flow $v^{\perp} = k$), and [106], [182]. A general approach to curve shortening flows by 1D parabolic equations was developed in [13], [14] (where Sturm's intersection theory is described); see also [309]. Note that the mean curvature flows can generate different types of singularities.

Parabolic properties of a curve shortening evolution can be used in a number of well-known problems concerning plane curves. As a first example, a Birkhoff curve shortening evolution was a basic idea in proving the theorem of the three geodesics (any Riemannian 2-sphere has at least three simple closed geodesics) by L.A. Lusternik and L.G. Schnirel'man (1929) [254]. A smooth evolution via curvature was applied in [182] on the basis of Uhlenbeck's suggestion of using the curvature flow.

Sturm's evolution PDE approach on zero sets can give a new insight to a number of topological problems of plane and spherical curves, caustics, and related topics of symplectic geometry briefly outlined above. For instance, three of Arnold's theorems [22] on the number of inflection points (at least four for any embedded curve in S^2 called the "tennis ball theorem", and at least three for any noncontractible embedded curve in $\mathbb{R}P^2$) and extatic points (at least six for any plane convex curve) can be proved by using a suitable parabolic mean curvature evolution (the affine one for extatic points); see [15] and comments in [25]. Namely, the asymptotic expansion of the solution $u(x,t)$ as $t \to \infty$ describing the convergence to limiting geodesics by using a 1D parabolic equation determines a minimally possible number of critical points. Then the result follows from Sturm's Theorem on nonincrease with time of the number of such points (e.g. inflections that are zeros of the curvature). While the Sturm-Hurwitz Theorem can deal with infinitesimal perturbations of curves (see above), the parabolic evolution Sturm analysis extends the results to any finite perturbation. It follows that the statements from [24, p. 14], " The tennis ball theorem asserts that the result remains true for finite perturbations, even very large ones," and "...the tennis-ball theorem may be considered as a generalization of Hurwitz' theorem to the case of multivalued functions" are covered by the first Sturm Theorem on zeros of single-valued functions (solutions of the PDE) once a suitable parabolic 1D evolution is available. The case of finite perturbations reduces via parabolic evolution to the infinitesimal one and then the Sturm Theorem establishes that the number of critical points (zeros, inflections, extatic points, etc.) cannot be less than the eventual, infinitesimal one for arbitrarily small perturbations where a standard linearization applies. If a suitable parabolic evolution exists, the Sturm-Hurwitz Theorem guarantees that the "infinitesimal geometric characteristic" of convergence (the number of critical points) is the optimal lower bound for any finite, arbitrarily large perturbations.

After a suitable surface parameterization, the quasilinear parabolic equation $u_t = \frac{u_{xx}}{1+(u_x)^2} - \frac{N-2}{u}$ describes the evolution of cylindrically symmetric hypersurfaces moving by mean curvature in $I\!\!R^N$ for $N \geq 3$, [110], [318], [5]. A similar singular lower-order term occurs in the Prandtl boundary layer equations, which by von Mises nonlocal transformation reduce to the PME with an extra term $u_t = (uu_x)_x + \frac{g(t)}{u}$ where g depends on the velocity of the potential flow (though in the original setting no singularities occur); see Section 30 in [344].

§ **1.3–1.5.** Monotone approximations by regularized uniformly parabolic equations will be used in the next chapters where further references are available. The MP, existence, uniqueness classes and comparison results for sufficiently smooth fully nonlinear equations can be found in [344, Chapter 4], see also [299, Chapter 3].

Let us now comment on the second Sturm Theorem explaining evolution of multiple zeros. A similar classification of multiple zeros holds for a system of parabolic inequalities. Rescaling by the Sturm backward variable shows that the asymptotic behaviour is true for $W^{2,1}_{p,\mathrm{loc}}$ solutions (from Tikhonov's class) of a system of parabolic inequalities

$$|u_t - u_{xx}| \leq M_1 |u_x| + M_0 |u|, \quad x \in I\!\!R, \ t \in J,$$

where M_0 and M_1 are positive constants. See [83], where such rescaling local analysis of nodal sets was performed for equations in $I\!\!R^N$. For instance, the heat equation

$$u_t = \Delta u \quad \text{in } I\!\!R^N \times (-\infty, 0)$$

in terms of the backward Sturm variable $z = x/\sqrt{-t}$ reduces to

$$u_\tau = \mathbf{B}u \quad \text{in } I\!\!R^N \times I\!\!R_+, \quad \text{where } \tau = -\log(-t), \tag{1.54}$$

with the symmetric second-order operator

$$\mathbf{B}u = \Delta u - \tfrac{1}{2}\, z \cdot \nabla u \equiv \tfrac{1}{\rho}\, \nabla \cdot (\rho \nabla u), \ \ \rho(z) = e^{-|z|^2/4}. \tag{1.55}$$

It is self-adjoint in $L^2_\rho(I\!\!R^N)$ with the domain $H^2_\rho(I\!\!R^N)$ and a point spectrum $\sigma(\mathbf{B}) = \{\lambda_\beta = -\frac{1}{2}|\beta|, |\beta| = 0, 1, ...\}$ ($\beta = (\beta_1, ..., \beta_N)$ is a multiindex, $|\beta| = \beta_1 + ... + \beta_N$). The eigenfunction subset $\Phi = \{H_\beta(z) = \frac{1}{\rho(z)}\, D^\beta \rho(z)\}$

consists of the separable orthogonal Hermite polynomials in $I\!R^N$ and is complete and closed in $L^2_\rho(I\!R^N)$, [57, p. 48]. Therefore, the asymptotic structures such as $Ce^{\lambda_\beta \tau}\psi_\beta(z)$ with any eigenvalue $\lambda_\beta < 0$ describe as $\tau \to \infty$ all possible types of multiple zeros of the heat equation in $I\!R^N$. This makes it possible to study general properties (e.g., Hausdorff dimension) of nodal sets of general solutions [83].

On spectra of multiple zeros in linear and quasilinear parabolic equations

Continuing a survey on the Sturm zero set analysis, we present some comments on multiple zeros for some other linear and quasilinear parabolic equations.

(i) Linear degenerate parabolic equations. The result on the finite number of zeros of solutions $u(x,t)$ for $t > 0$ remains valid for linear degenerate parabolic equations if the degeneracy is sufficiently weak. Consider first the following 1D equation with the Dirichlet boundary condition:

$$u_t = x^\gamma u_{xx} \quad \text{in } (0,1) \times I\!R_+, \quad u(0,t) = u(1,t) = 0 \quad \text{for } t > 0, \tag{1.56}$$

with a positive exponent $\gamma > 0$. The equation degenerates at $x = 0$ and therefore the Sturm result does not apply since zeros may concentrate in a small neighbourhood of the origin. We show that, for $\gamma < 2$, this is not the case. Setting $u(x,t) = y^\alpha v(y,t)$, $x = \alpha^{-\alpha} y^\alpha$, where $\alpha = \frac{2}{2-\gamma} > 0$ yields the radial heat equation

$$v_t = \Delta_N v \equiv v_{yy} + \tfrac{N-1}{y}\, v_y \tag{1.57}$$

with a positive, non-integer dimension $N = \frac{2(3-\gamma)}{2-\gamma} > 0$ if $\gamma \in (0,2)$. Hence, $y = 0$ is not singular for v, the symmetry condition $v_y = 0$ at $y = 0$ can be assumed, and $v(0,t)$ is bounded. We then pose the above Dirichlet boundary condition for u at the origin. If finiteness of zeros for any $t > 0$ is true for the regular Laplacian in (1.57) (this can be proved by analytic semigroup techniques [11] or by perturbation technique similar to [83]), then the same holds for equation (1.56) if $\gamma < 2$. By the interior regularity for (1.57), for nonnegative solutions $u \not\equiv 0$ of (1.56), the transversality result is also true for $\gamma < 2$, i.e., $u_x(0,t) = \alpha^\alpha v(0,t) > 0$ for $t > 0$. For elliptic equations this is known as O.A. Oleinik–E. Hopf Boundary Point Lemma [276], [195]. Similar to the case of the integer $N \geq 1$, a complete subset of Laguerre polynomials as eigenfunctions of the rescaled operator describe all possible types of evolution of multiple zeros for equation (1.57) and hence for (1.56).

This is a sharp result. In the critical case $\gamma = 2$ the equation

$$u_t = x^2 u_{xx}$$

admits separable solutions $u(x,t) = e^{-\lambda t}\phi(x)$, where ϕ solves the ODE $x^2\phi'' + \lambda\phi = 0$ on $(0,1)$, $\phi(0) = 0$. Fixing any $\lambda > \frac{1}{4}$, we obtain the solution $\phi(x) = \sqrt{x}\,\sin(b\log x)$, $b = \sqrt{\lambda - \frac{1}{4}}$, with infinite number of zeros as $x \to 0^+$ for any $t \in I\!R$. For the general equation

$$u_t = \rho(x) u_{xx} \quad \text{in } (0,1) \times I\!R_+,$$

with smooth $\rho(x) > 0$ for $x \in (0,1]$ and $\rho(0) = 0$, the transformation is $u =$

$\psi(y)v$, $x = \psi(y)$, where $\psi(y) > 0$, $\psi'(y) > 0$ for $y > 0$ is obtained from the ODE $\psi' = \sqrt{\rho(\psi)}$. We then assume that $\int_0 \frac{\mathrm{d}s}{\sqrt{\rho(s)}} < \infty$ and hence $\psi(0) = 0$. This gives $\gamma < 2$ in (1.56). The semilinear equation for v reads

$$v_t = \mathbf{A}v \equiv v_{yy} + \mu(y)v_y, \quad \text{where } \mu(y) = \tfrac{\psi'}{\psi}\left[2 - \tfrac{\psi\psi''}{(\psi')^2}\right](y).$$

Consider the symmetric representation of the operator $\mathbf{A} \equiv \frac{1}{\nu}\frac{\mathrm{d}}{\mathrm{d}y}\left(\nu\frac{\mathrm{d}}{\mathrm{d}y}\right)$ with the positive weight $\nu(y) = \frac{\psi^2}{\psi'}(y)$. Equation $\mathbf{A}v = \lambda v$ admits two types of asymptotics as $y \to 0$: the regular $v_1(y) \sim 1$ and the singular one $v_2: \;\; v' \sim \frac{1}{\nu(y)} \equiv \frac{\psi'}{\psi^2}(y)$, i.e., $v_2(y) \sim \frac{1}{\psi(y)} \to \infty$ as $y \to 0$. Then $v_1 \in L^2_\nu$ if $\int_0 \frac{\psi^2(s)}{\psi'(s)}\,\mathrm{d}s < \infty$, and $v_2(y) \sim \frac{1}{\psi(y)} \notin L^2_\nu$ if $\int_0 \frac{\mathrm{d}s}{\psi'(s)} = \infty$. Under these assumptions, in the singular Sturm-Liouville problem on $(0,1)$, the end-point $y = 0$ is in the limit-point case for the symmetric operator $\mathbf{A}$. Hence, this operator is self-adjoint in the weighted space $L^2_\nu((0,1))$ with a discrete spectrum [246]. The eigenfunctions forming a complete subset in $L^2_\nu((0,1))$ satisfy the symmetry condition at $y = 0$ and, by the Hilbert-Schmidt theory, we have that $v(y,t)$ satisfies $v_y(0,t) = 0$ and v is bounded at the origin. Since $u(x,t) = xv(y,t)$, the Dirichlet problem for u is well-posed and moreover $u_x(0,t) \equiv v(0,t)$ is bounded and satisfies $u_x(0,t) > 0$ for $t > 0$ if $u \geq 0$, $u \not\equiv 0$.

In this case the spatial structure of multiple zeros is asymptotically described by eigenfunctions of the rescaled linear second-order operator as in Sturm's case of uniformly parabolic equations. But the situation is entirely different for essentially nonlinear equation to be discussed next.

(ii) Degenerate quasilinear equations. Zeros of infinite order: limit-cycle similarity solutions. Infinite order zeros can be generated by nonlinear singular operators. Consider the sign PME

$$u_t = (|u|^\sigma u)_{xx} \quad \text{in } I\!R \times (-\infty, 0), \quad \sigma > 0, \tag{1.58}$$

which is a quasilinear equation degenerating at points where $u = 0$. The p-Laplacian equation $v_t = (|v_x|^\sigma v_x)_x$ degenerates on $\{v_x = 0\}$ and reduces to the sign PME by differentiating and setting $u = v_x$. Of course, since (1.58) exhibits finite speed of propagation, any compactly supported solution $u(x,t)$ has zeros of infinite order at any point of the subset $\{x : u(x,t) = 0\}$. Those are trivial zeros, and next we describe nontrivial infinite-order zeros, where $u \not\equiv 0$ in any arbitrarily small left or right-hand neighbourhood.

As in Sturm's analysis in Section 1.2, we begin with a self-similar blow-up formation of multiple zeros. Consider blow-up self-similar solutions of (1.58) of the form

$$u_*(x,t) = (-t)^n f(z), \quad z = x/(-t)^\beta, \quad \text{with } \beta = \tfrac{1}{2}(1 + n\sigma), \tag{1.59}$$

where n is a parameter. The backward rescaled variable z depends now on $\sigma > 0$ unlike the linear case $\sigma = 0$ with the Sturm variable $z = x/\sqrt{-t}$ (i.e., $\beta = \frac{1}{2}$ for any n). Here f solves the ODE

$$(|f|^\sigma f')' - \beta f' z + nf = 0, \quad z \in I\!R. \tag{1.60}$$

It is invariant under a group of scaling transformations, and setting $f(z) = z^{2/\sigma}\phi(\eta)$, $\eta = \log z$ and $P = \phi'$ reduces it to the first-order ODE

$$\frac{\mathrm{d}P}{\mathrm{d}\phi} = -\frac{1}{P|\phi|^\sigma}\left[\gamma_1|\phi|^\sigma\phi + \gamma_2|\phi|^\sigma P + \sigma|\phi|^{\sigma-2}\phi P^2 - \tfrac{1}{\sigma}\,\phi - \beta P\right], \quad (1.61)$$

with constants $\gamma_1 = \frac{2(2+\sigma)}{\sigma^2}$ and $\gamma_2 = \frac{3\sigma+4}{\sigma}$, which can be studied on the phase-plane. First examples of a detailed phase-plane analysis for global solutions with the forward variable $z = x/t^\beta$, $t > 0$, to the PME and more general equations from filtration theory were presented by G.I. Barenblatt (1952) [39]. See other techniques in [177] and [306, Chapter 3], where blow-up solutions were studied.

In order to describe a limit cycle solution with nontrivial infinite order zeros, we use other variables by setting $f = g'$ and then g satisfies the ODE

$$|g'|^\sigma g'' - \beta g' z + \alpha g = 0, \quad (1.62)$$

where $\alpha = n + \beta$. Setting as above $g = z^\gamma\varphi(\eta)$, $\eta = \log z$ with $\gamma = \frac{2+\sigma}{\sigma}$ and $\varphi' = P$ yields the following first-order ODE:

$$P\frac{\mathrm{d}P}{\mathrm{d}\varphi} = \frac{\beta P + a\varphi}{|\gamma\varphi + P|^\sigma} - b\varphi - cP, \quad (1.63)$$

with constants $a = \beta\gamma - \alpha$, $b = \frac{2\gamma}{\sigma}$, $c = \gamma + \frac{2}{\sigma}$. It is known [146], [72] that ODEs such as (1.63) can admit stable limit cycles around the origin $\varphi = P = 0$. The corresponding orbit $\{\varphi_*(\eta)\}$ has infinitely many sign changes as $\eta \to \pm\infty$, so that the same is true for the functions $g(z)$ and $f(z)$ as $z \to 0$. See Figure 1.4. This implies that the solution $u_*(x,t)$ has a nontrivial zero of infinite order at $x = 0$ for all $t < 0$. Limit cycle solutions can also be constructed without reduction to first-order ODEs (1.63) [335]. We say that such limit cycle solutions generate a discrete S-spectrum of zeros of a given nonlinear degenerate equation. In general, there exist other spectra denoted by P, Q and R.

On general structure of spectra of multiple zeros. The structure of multiple zeros for the 1D sign PME can be rather complicated. We refer to rigorous and qualitative results in [146], [147], [72] and [193]. Let us comment on distinguished and common peculiarities of the general classification of zeros for nonlinear degenerate equations with a typical scaling invariance. As we have pointed out, such a classification of multiple zeros inherits usual properties of formation of finite-time singularities in nonlinear parabolic PDEs, and occurs in the classical area of blow-up of solutions intensively developed last thirty years; see references in the books [306], [170] and [43]. Oscillation properties of nonlinear rescaled operators and corresponding linearized ones play a special role in the construction of countable spectra of blow-up patterns. This establishes a direct link to spectra of multiple zeros. We have seen that phenomena of finite-time collapse of multiple zeros exhibit typical features of finite-time singular limits in parabolic problems.

P-*spectrum: essentially nonlinear point spectrum.* This is a new spectral part, which is not available in linear equations. In the case when an infinite-order zero exists, i.e., the corresponding ODE such as (1.63) admits a limit cycle solution, there exists a countable spectrum of nonlinear similarity eigenfunctions

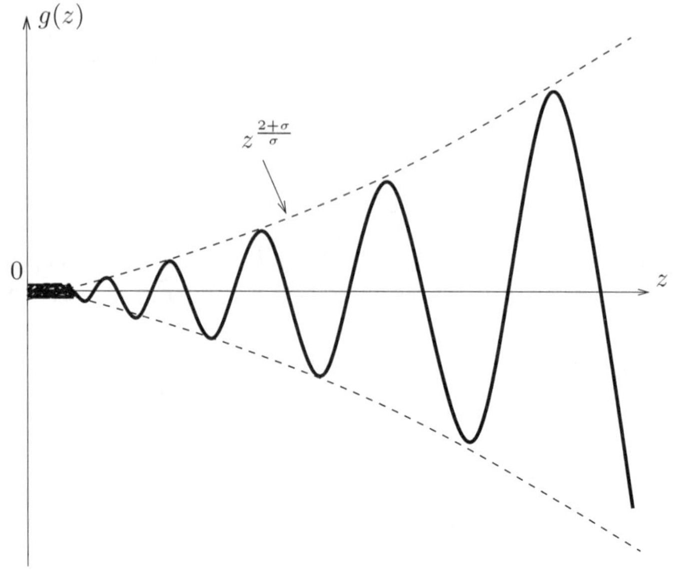

Figure 1.4 *Zero of infinite order given by the limit cycle solution* $g(z) = z^{\frac{2+\sigma}{\sigma}}\varphi(\log z)$ *of equation* (1.62).

$\mathrm{P}_N = \{f_k, k = 1, ...\}$ satisfying ODEs (1.60) with a countable subset of parameters $\{n_k\}$ (eigenvalues). Each function $f_k(z)$ has exactly k transversal zeros and describes the formation of kth order zero of the PDE as $t \to 0^-$. For the sign PME such as (1.58) such a nonlinear spectrum was constructed in [193]. Examples of other countable nonlinear spectra of blow-up patterns for different parabolic equations are given in [146] and [72]. A countable spectrum of global in time self-similar patterns (1.59), where $(-t)$ replaced by t (this changes signs of the last two terms in the ODE (1.60)) was proved to exist in [192].

P-*spectrum: linearized point part.* Assume that no limit-cycle similarity solutions exist and the nonlinear spectrum consists of a *finite* number of profiles $\mathrm{P}_N = \{f_k, k = 1, ..., M\}$. This can happen for weak degeneracy, i.e., for sufficiently small $\sigma > 0$ in ODEs such as (1.62) under some hypotheses on β and n [72]. Then, for a class of equations, other zero patterns P_L can be constructed by "linearization", which again leads to a nonlinear problem (note that the second-order operator in (1.62) does not admit linearization about a constant). A global construction of such non-self-similar nonlinear "linearized" patterns is performed by a matching procedure; see [72]. At the critical values $\sigma \in \{\sigma_k, k = 1, 2, ...\}$ describing the transition phenomenon between the nonlinear spectrum P_N and the linearized one P_L, there occur special patterns generated by logarithmically perturbed structures that correspond to the evolution near a "centre manifold" of the linearized (nonlinear) operator. We ascribe these to the linearized spectrum P_L.

There is a direct link to the Sturmian spectrum for linear operators where linearization leads to linear self-adjoint second-order operators. Then a stable manifold analysis shows that P_L is composed of the eigenfunctions, which are often

Hermite, Kummer or other polynomials $\{H_k, k = M+1, ...\}$. The linearized part of the spectrum of blow-up patterns (corresponding to the stable manifold behaviour) has been studied in detail [63], [116], [188], [266], [338], [341]. A description of spectra of blow-up patterns $\mathrm{P} = \mathrm{P}_N \cup \mathrm{P}_L$ is given in [72], where a survey of earlier results on nonlinear and linearized spectral parts is available.

Point Q-*spectrum.* These are zero structures $u(x,t)$ with a hole for all $t < 0$, where $u(x,t) \equiv 0$, which focuses at the origin $x = 0$ as $t \to 0^-$. The point Q-spectrum can be countable and is composed of self-similar solutions [146], [72], or can be constructed by matching and is not self-similar [193]. We expect that, in general, similar to the above P-spectrum, the Q-spectrum will be partially composed of nonlinear Q_N and linearized Q_L parts.

Thus the countable point spectrum Φ_p of multiple zeros for nonlinear equations is expected to be composed of three subsets, $\Phi_p = \mathrm{S} \cup \mathrm{P} \cup \mathrm{Q}$, where P and Q consist of essentially nonlinear and linearized nonlinear parts. In the classical uniformly parabolic Sturm case we have $\mathrm{P}_N = \mathrm{Q} = \mathrm{S} = \emptyset$. On the other hand, we referred to examples in [146], [72] where the linearized parts are empty and the point spectrum is completely nonlinear, i.e., $\Phi_p = \mathrm{S} \cup \mathrm{P}_N \cup \mathrm{Q}_N$. A classification of the point spectrum of zeros becomes extremely difficult for multi-dimensional equations such as the sign PME

$$u_t = \Delta(|u|^\sigma u) \quad \text{in } I\!R^N \times (-\infty, 0).$$

Nonlinear self-similar zero patterns (1.59) generate quasilinear elliptic equations

$$\Delta(|f|^\sigma f) - \beta\, z \cdot \nabla f + nf = 0 \quad \text{in } I\!R^N$$

with unknown spectra of suitable compactly supported nonlinear eigenfunctions from P_N. On the other hand, a countable subset of radially symmetric nonlinear eigenfunctions $\{f_k(|z|)\}$ satisfying ODEs is known to exist [192], [193].

R-*spectrum: continuous part.* In general, the point spectrum does not exhaust the classification of zero patterns. Examples of singularity blow-up formations considered in [146], [147], [193] show that under certain conditions, strongly degenerate nonlinear equations can admit continuous spectra of singularities. Namely, it was shown that ODEs such as (1.63) for blow-up similarity profiles can admit both discrete spectra P, Q and S, and a continuous one R with rather involved construction. For instance, relative to two parameters, R can be a countable subset of intervals [147]. Nevertheless, in spite of such sophisticated general continuous spectra, we expect that the point spectrum is more stable and contains the generic type of zero singularity formations in these singular nonlinear problems. A sophisticated local structure of multiple zeros can essentially affect the global structure of nodal sets. These problems remain open.

(iii) Higher-order parabolic equations. The main principles of Sturm's evolution analysis of multiple zeros remain valid for $2m$th order linear parabolic equations. Since the analysis is essentially local in shrinking zero neighbourhoods, without loss of generality we consider the canonical $2m$th order parabolic equation with constant coefficients

$$u_t = -(-\Delta)^m u \quad \text{in } I\!R^N \times (-\infty, 0), \quad m \geq 2.$$

The backward Sturm variable takes the form $z = x/(-t)^{1/2m}$ and we arrive at the equation (cf. (1.54))

$$u_\tau = \mathbf{B}u, \quad \text{where } \mathbf{B} = -(-\Delta)^m - \tfrac{1}{2m}\, z \cdot \nabla, \tag{1.64}$$

and as usual, $\tau = -\log(-t)$. For any $m > 1$, this operator is not self-adjoint unlike the second-order case $m = 1$. We introduce the space $L^2_\rho(I\!\!R^N)$ with the exponential weight $\rho(z) = e^{-a|z|^\alpha} > 0$ in $I\!\!R^N$, where $\alpha = \frac{2m}{2m-1} \in (1,2)$ and $a = a(m,N) > 0$ is a sufficiently small constant. For $m = 1$, we have $\alpha = 2$, $a = \frac{1}{4}$ and $\rho(z) = e^{-|z|^2/4}$ is the rescaled Gaussian kernel as in (1.55). In $L^2_\rho(I\!\!R^N)$ operator $\mathbf{B}$ with domain $H^{2m}_\rho(I\!\!R^N)$ being a weighted Sobolev space has the point spectrum $\sigma(\mathbf{B}) = \{\lambda_\beta = -\frac{1}{2m}\,|\beta| \le 0,\ |\beta| = 0, 1, ...\}$. The subset of eigenfunctions $\{\psi_\beta(z)\}$ (Kummer's polynomials in $I\!\!R^N$ of orders $|\beta|$) is complete in $L^2_\rho(I\!\!R^N)$ [102], [138]. For $m = 1$, these are the Hermite polynomials. In the space of eigenfunction expansions, where $\{\psi_\beta\}$ is closed, any solution of (1.64) from the existence class $\mathcal{U}$ given below has the eigenfunctions expansion $u(z,\tau) = \sum C_\beta\, e^{\lambda_\beta \tau} \psi_\beta(z)$. Hence, the complete subset of polynomials $\{\psi_\beta(z)\}$ describes in the rescaled form possible types of formation of multiple zeros occurring for this higher-order parabolic equation and describing local properties of nodal sets, [140]. Of course, the first Sturm Theorem in 1D (nonincrease of the number of zeros) is no longer available for $2m$th order equations, where new zeros can occur with evolution. On the other hand, for a class of higher-order, self-adjoint, positive, ordinary differential operators, eigenfunctions are known to obey typical Sturmian properties of zeros and extrema [103], and the same is true for countable subsets of solutions of some related semilinear $2m$th order ODEs with potential operators; see applications [303, 40] of the results of [103]. This is connected with Krasnosel'skii's genus version [228, p. 385] of the Lusternik–Schnirel'man category critical point theory [254].

The Sturmian evolution analysis of multiple zeros applies to more general linear $2m$th order parabolic equations, $m \ge 2$,

$$u_t = \textstyle\sum_{|\beta| \le 2m} a_\beta(x,t) D^\beta u,$$

where the real coefficients $\{a_\beta\}$ are bounded for $|\beta| < 2m$, continuous for $|\beta| = 2m$ and satisfy the parabolicity condition $(-1)^m \sum_{|\beta|=2m} a_\beta(x,t)\xi^\beta \le -\delta|\xi|^{2m}$ with a constant $\delta > 0$. The classification of multiple zeros by the eigenfunction subset $\{\psi_\beta\}$ applies [140] to classical, $C^{2m,1}_{x,t}$, solutions $u(x,t)$ in the standard existence-uniqueness class of locally measurable functions $\mathcal{U} = \{|u(x,t)| \le Ae^{a|x|^\alpha}\}$, A, $a > 0$, $\alpha = \frac{2m}{2m-1}$.

A little is known about blow-up multiple zero formations for nonlinear degenerate $2m$th order parabolic equations, where similarity analysis of the nonlinear spectrum even in 1D reduces to complicated higher-order ODEs and the subsets of such nonlinear eigenfunctions are unknown. Nevertheless, a classical nonlinear operator theory is expected to be useful to describe a "bifurcation" of nonlinear eigenfunctions from the known polynomial eigenfunctions of linear operators such as in (1.64).

CHAPTER 2

Transversality, Concavity and Sign-Invariants. Solutions on Linear Invariant Subspaces

In this chapter we describe first simpler aspects of the geometric analysis of nonlinear parabolic equations by using Sturm Theorems. In order to explain basic techniques of intersection comparison, it is convenient to begin with well-known equations such as the heat equation (HE), the porous medium equation (PME) or, at most, with the filtration equation having a single nonlinearity. These classical equations of Mathematical Physics often admit simple subsets of elementary particular similarity solutions that make applications of intersection comparison ideas easier. Note that such simple explicit solutions often play a significant role in the classical existence, uniqueness and regularity theory.

We begin with an intersection comparison study of concavity and convexity properties of solutions $u(x,t)$ with respect to the spatial variable x. We show that the concavity/convexity properties are preserved in time (i.e., are invariants of the evolution), if there exists a suitable subset B of particular piecewise solutions.

The main scheme for the geometric analysis is as follows. Given a nonlinear filtration-like equation, we perform the following steps.

(i) We fix a suitable subset B of particular (similarity) solutions.

(ii) Checking necessary straightforward properties of B such as completeness, continuity, etc., we analyze simpler transversality conditions, and next introduce a notion of the spatial B-concavity (or convexity) as concavity relative to functions from B.

(iii) By intersection comparison we prove that both properties are preserved with time for classes of general solutions or can occur eventually in time.

(iv) We show that transversality gives Bernstein-type first-order estimates, while the B-concavity is equivalent to a second-order differential inequality with a nonlinear operator called *sign-invariant* that preserves its signs on evolution parabolic orbits.

(v) Such geometric analysis can be repeated if sign-invariants can be found independently. Once such a sign-invariant is found, it is natural to determine the corresponding set B of particular solutions generating it via B-concavity (convexity) property. We then return to (i), and so on.

Thus we show that autonomous 1D parabolic equations preserve in time those concavity properties that are generated by suitable subsets B of particular solutions. Hence, each equation admits infinitely many invariant B-concavity properties in general.

2.1 Introduction: Filtration equation and concavity properties

Consider the general filtration equation

$$\boxed{u_t = (\varphi(u))_{xx} \quad \text{in } S = I\!R \times I\!R_+,} \tag{2.1}$$

with some regularity assumptions on an increasing function φ. Particular cases include the heat equation, $u_t = u_{xx}$, the PME

$$\boxed{u_t = (u^m)_{xx}} \tag{2.2}$$

with $m > 1$ and the fast diffusion equation with $0 < m < 1$. Furthermore, we will treat the p-Laplacian equation

$$\boxed{u_t = (|u_x|^\sigma u_x)_x,}$$

with $\sigma > 0$, and other equations with lower-order terms representing absorption, reaction or convection effects. Mainly, we are interested in the concavity or convexity properties of the solutions of the above equations with respect to the spatial variable. Such properties have to be stated after a convenient change of dependent variable. For the PME and the fast diffusion one (2.2), such variable is the *pressure* (a term from filtration theory) defined as

$$v = \tfrac{m}{m-1}\, u^{m-1} \quad (m \neq 1). \tag{2.3}$$

Then v satisfies the following parabolic equation with quadratic nonlinearities:

$$v_t = (m-1)vv_{xx} + (v_x)^2. \tag{2.4}$$

Let us begin with some extra motivation of our geometric concavity study. A well-known result by D.G. Aronson and Ph. Bénilan (1979) [31] states that all nonnegative solutions of the PME defined in $S = I\!R \times I\!R_+$ satisfy the semiconvexity estimate

$$v_{xx} \geq -\tfrac{1}{m+1}\, \frac{1}{t+\tau}, \tag{2.5}$$

where the constant $\tau \geq 0$ depends on the initial data v_0,

$$\tau = -[(m+1)\,(\inf\, v_0'')]^{-1}.$$

This estimate plays a fundamental role in the general regularity theory of the PME as explained in the books by E. DiBenedetto [99] and A. Friedman [122], as well as in A.S. Kalashnikov's survey [213]. The proof in [31], as in the classical Bernstein technique, is based on differential and algebraic manipulations with the pressure equation (2.4) by using the MP.

Estimate (2.5) has a simple geometric interpretation establishing that (here $\tau = 0$ means that the estimate holds for any data v_0)

$$v(x,t) + \tfrac{1}{2(m+1)}\, \frac{x^2}{t} \quad \text{is convex in } I\!R \text{ for all } t > 0. \tag{2.6}$$

The result is also true for $m = 1$, i.e., for the heat equation with the logarithmic pressure $v = \log u$ satisfying

$$v_t = v_{xx} + (v_x)^2. \tag{2.7}$$

More precise results for (2.4) depend on the initial data $v_0(x)$. Thus, if v_0 is convex, it easily follows by a direct application of the MP that $v(\cdot, t)$ is convex for every $t > 0$. A more difficult situation occurs when we want to study concavity since we cannot have a concave and nonnegative function in the whole line unless it is constant. However, a strong concavity result holds in the following sense: if the support of v_0 is an interval and $v_0'' \le 0$ in the positivity domain $\{v_0 > 0\} = \{x \in I\!R : \; v_0(x) > 0\}$, then

$$v_{xx}(x,t) \le 0 \quad \text{in } \{v(x,t) > 0\} \quad \text{ for } t > 0. \tag{2.8}$$

The main idea of the approach to prove (2.8) is that the geometric positive concavity property in question can be formulated in terms of intersection comparison with particular solutions from a functional subset B, and the intersection comparison theory allows us to transmit the relevant information from $t = 0$ to $t > 0$. Indeed, deriving properties of general classes of solutions from the properties of a suitable set of particular solutions (hopefully, a small subset of explicit solutions) is a basic idea in nonlinear partial differential equations. More specifically, it is the basis of much of the present knowledge of the large-time behaviour of nonlinear parabolic equations, where explicit self-similar solutions play a fundamental role.

In Section 2.2, for the general filtration equation (2.1), we begin with a simple property of transversality, giving the well-known *Bernstein estimate* on the first-order derivative v_x, which is thus shown to have a purely geometric nature. When dealing with the issue of transversality and concavity or convexity for the solutions of the PME or the filtration equation, the set of particular solutions B will be chosen to consist of solutions that are piecewise *linear* in space. We need to impose special restrictions on the set B, which are summarized by saying that it must be complete, continuous and monotone. We call a set B of solutions with these properties a *proper* one. Section 2.3 establishes *eventual concavity* for solutions of (2.1) whose initial data have a suitable decay at infinity or even compact support. In later Sections 2.4 and 2.5 we show how the method applies to a number of equations. We consider equations having a diffusion term such as (2.1) together with additional convection, reaction or absorption terms, and also equations with the p-Laplacian operator.

The crucial point will always be the existence of good enough sets of linear solutions as above. Typically, such exact solutions belong to a certain linear subspace *invariant* under the quasilinear stationary operator of the parabolic equation under consideration. We next propose a different kind of use of the above technique. Thus, while in the applications mentioned above the set B is chosen so as to reproduce the classical geometric properties we are looking for, there is another way of looking at the same setting. It says that, given a set of solutions B with the structure mentioned above, application of the comparison technique allows us to derive certain geometric properties of the solutions that we term as B-*concavity* or B-*convexity*, notions, which are thus formulated in terms of the set B. In some cases the geometric result can be translated into the existence of *a priori* estimates or differential inequalities on the solutions, of use in the qualitative theory of the equation. This motivates a careful choice of B. Sections 2.6 to 2.10 show different instances of application of these concepts. In Section 2.11 we extend our concavity analysis to a number of radial N-dimensional quasilinear equations.

Summing up, the general idea of the geometric interpretation can be stated as follows. Every proper set of particular solutions B of a 1D nonlinear parabolic equation defines a property of B-concavity/convexity that is preserved in time for a wide class of general solutions and can also occur eventually in time. For all equations studied in this chapter, the generalized concavity/convexity properties are generated by functional subsets B belonging to *linear invariant subspaces* of the corresponding stationary operators. A general discussion in this direction is available in the final Section 2.12. Extensions of such a geometric approach to parabolic equations with arbitrary nonlinearities not having the above linear invariant property, will be an important part of the future analysis in the next chapters.

2.2 Proofs of transversality and concavity estimates by intersection comparison with travelling waves

We consider the general filtration equation (2.1), where φ is a nondecreasing real function and $\varphi(0) = 0$ for convenience. For such φ, existence, uniqueness and comparison results for continuous weak solutions are classical, see Remarks.

Transversality and concavity: finite propagation

Though the convexity and concavity results will be proved for general φ, we will consider in greater detail those φ for which the property of *finite speed of propagation* (of disturbances from 0) holds. Such property is equivalent for the solutions of (2.1) to the condition of convergence of the integral

$$\Phi(u) = \int_0^u \frac{d\varphi(s)}{s} < \infty \quad \text{for } u > 0. \tag{2.9}$$

If (2.9) holds, the solution of the Cauchy problem to (2.1) with compactly supported data will be compactly supported for all times $t > 0$. Such property is typically found in filtration problems. As explained above, we may assume by approximation that φ has some regularity, $\varphi \in C^2(I\!R_+) \cap C^1([0,\infty))$ and $\varphi'(u) > 0$ for $u > 0$. Under condition (2.9) we may introduce the pressure

$$v = \Phi(u) \tag{2.10}$$

that is nonnegative ($v = 0$ for $u = 0$) and satisfies the equation

$$\boxed{v_t = \mathbf{P}_*(v) \equiv F(v)\, v_{xx} + (v_x)^2 \quad \text{in } S,} \tag{2.11}$$

where $F(v) = \varphi'(\Phi^{-1}(v))$. For the choice $\varphi(u) = u^m$, condition (2.9) is equivalent to $m > 1$ (i.e., the PME) and v is given by formula (2.3).

We take initial data

$$v(x, 0) = v_0(x) \geq 0 \quad \text{in } I\!R, \tag{2.12}$$

where v_0 is continuous and smooth whenever positive. Furthermore, we impose on

$v_0(x)$ a certain assumption on the growth as $|x| \to \infty$ (for instance, for the PME, $v_0(x)$ has less than quadratic growth as $|x| \to \infty$). Then the problem is known to have a unique solution globally in time. This always holds for the case of bounded v_0 to be studied in detail. Under these assumptions it is natural to suppose that a standard comparison argument can be used on any bounded or unbounded subset of S with a smooth parabolic boundary.

Optimal Bernstein estimates as a transversality condition. First we state a simple Bernstein gradient bound on general solutions of the filtration equation. This first result serves as an introduction to the future general regularity theory via intersection comparison.

Theorem 2.1 *Let* $0 \le v_0 \le 1$ *and* $|v_0'| \le 1$ *in* $\{v_0 > 0\}$. *Then* $|v_x| \le 1$ *in* $S \cap \{v > 0\}$.

A proof is postponed until the concavity analysis.

Concavity. Next, we study the concavity of the pressure of the solutions of equation (2.1) under condition (2.9). We choose this case because it involves, in its simple formulation, a certain generality and some mathematical difficulties, which make it suitable to show the method. Actually, we are interested in studying concavity and convexity. Both can be treated in a similar fashion. The concavity result has an interesting extension, namely eventual concavity, which will be dealt with in the next section. We finish the analysis with the cases having infinite speed of propagation.

We will need a modified notion of concavity. Thus we say that a nonnegative and continuous real function f is *positively concave* if, for any x_0 from the positivity domain of f denoted by $\{f > 0\}$, there exists a straight line $L(x)$ such that

$$L(x_0) = f(x_0) \quad \text{and} \quad f(x) \le (L(x))_+ \quad \text{in } I\!R,$$

where, as usual, $(\cdot)_+$ means $\max\{\cdot, 0\}$. For functions that are C^2 where they are positive, this is equivalent to the condition

$$\operatorname{supp} f \quad \text{is connected and} \quad f'' < 0 \quad \text{in } \{f > 0\}. \tag{2.13}$$

Theorem 2.2 *Under the above hypotheses, if* v_0 *is positively concave, then, for every* $t > 0$, *the function* $v(\cdot, t)$ *is positively concave, i.e.,*

$$v_{xx}(x,t) \le 0 \quad \textit{in the interval } \{v(x,t) > 0\}. \tag{2.14}$$

Presenting a proof of this result, we will explain the main ideas of the method to be used later on in various more delicate settings. The proof consists of putting together the ingredients mentioned above: defining a convenient set of particular solutions, intersection comparison techniques and application.

Proper set of explicit solutions. Properties. Hodograph plane. Equation (2.11) admits the following elementary set B of piecewise linear solutions of the travelling wave (TW) type:

$$V(x,t) = (\lambda^2 t + \lambda x + a)_+, \tag{2.15}$$

where λ, $a \in I\!R$ are arbitrary constants. Functions $V(x,t)$ are Lipschitz continuous and $u = \Phi^{-1}(V)$ are weak solutions of the filtration equation (2.1) according to the classical definition based on integration by parts. Functions (2.15) form a two-parameter set. The following properties will be needed.

(i) *Completeness* of B as a set of tangent solutions in the hodograph plane $I\!R^2$. Fix an arbitrary $(x_0, t_0) \in \bar{S}$ such that $v(x_0, t_0) > 0$. Then a function $V \in B$ is a *tangent solution* to $v(x, t)$ at the point (x_0, t_0) if

$$V(x_0, t_0) = v(x_0, t_0) = \mu > 0, \quad V_x(x_0, t_0) = v_x(x_0, t_0) = \nu, \tag{2.16}$$

and $V(x, t)$ exists for all $t \in [0, t_0]$.

Proposition 2.3 *The set B is complete in the sense that, for any $(x_0, t_0) \in \bar{S} \cap \{v > 0\}$, there exists a unique tangent solution $V \in B$.*

Proof. At a given tangency point (x_0, t_0), (2.15) and (2.16) become $\lambda^2 t_0 + \lambda x_0 + a = \mu$ and $\lambda = \nu$, whence the unique values of the parameters λ, a. □

If necessary, we will denote the tangent solution by $V(x, t; x_0, t_0, \mu, \nu)$. The interest of the complete set B for our problem lies in the fact that the transversality and positive concavity of a solution $v(x, t)$ can be checked by comparison with the TWs of B. Two further properties of the set B will play a role in the proof. Both properties are trivially satisfied in the current consideration, but they are not so straightforward for more complicated subsets B to be considered later on.

(ii) *Continuity*. The functions V depend continuously with respect to the parameters in the C^1-norm on compact subsets of the domain of positivity.

(iii) *Monotonicity*. The functions V are monotone with respect to the parameter a, and furthermore, for a fixed $t_0 \ge 0$, we have $V \to \infty$ as $a \to \infty$, $V \to 0$ as $a \to -\infty$ uniformly on compact subsets of $\{V(x, t_0) > 0\}$. We also have monotonicity with respect to λ.

A set B of solutions satisfying the above three properties is called *proper*. The transformation

$$w(v, t) = v_x(x, t)$$

is the *hodograph transformation*, and the $\{v, w\}$-plane is the hodograph plane $I\!R^2$. Given a smooth function $v(x, t)$ with a fixed $t \ge 0$ (a curve in the $\{x, v\}$-plane), we obtain a curve in the hodograph plane and *vice versa*. The completeness of B means that, for any point of the solution curve $\Gamma(t, v) = \{v, v_x\}$ in the hodograph plane, there exists a unique curve $\{V, V_x\}$ passing through it. In the present case of piecewise linear solutions V, this is obvious since in the hodograph plane each curve $\{V, V_x\}$ with $V_x \equiv \lambda$ is a horizontal straight line. Let $\mathcal{T}_B(t, v)$ be the corresponding subset of tangent solutions, i.e., of all V-curves crossing the v-curve. Then $\mathcal{T}_B(t, v)$ is a dense subset of horizontal lines.

In general, completeness of a given solution set B means that all mutually non-intersecting V-curves uniquely cover the hodograph plane where the set B is *ordered*. The intersection comparison argument we are going to use now is based on the time evolution analysis of the tangential subset $\mathcal{T}_B(t, v)$. In particular, the positive concavity means that any tangential horizontal V-line intersects $\Gamma(t, v)$ exactly once in the hodograph plane. We will prove that, due to the Sturm Theorem, such intersection properties are preserved in time.

Number of intersections $\mathrm{Int}(t, V)$ **of weak solutions**. The basic tool in the proof of the transversality estimate and (2.14) is based on intersection comparison of $v(x,t)$ with the complete set $B = \{V(x,t)\}$. Given solutions $v(x,t)$ and $V(x,t) \in B$ and a fixed $t \geq 0$, we denote by $\mathrm{Int}(t,v,V)$ the number of intersections in x of these profiles or the number of sign changes in x of the difference $w(x,t) = v(x,t) - V(x,t)$. Since v will be usually fixed, we will write $\mathrm{Int}(t,V)$ instead of $\mathrm{Int}(t,v,V)$ if no confusion is likely. Then the main comparison argument can be stated as follows: assuming that $\mathrm{Int}(0,V) < \infty$, we have

$$\mathrm{Int}(t,V) \quad \text{does not increase with time.} \tag{2.17}$$

In the domain of positivity of both solutions v and V, the difference $w(x,t)$ satisfies a linear parabolic equation

$$w_t = a(x,t)w_{xx} + b(x,t)w_x + c(x,t)w, \tag{2.18}$$

where $a = F(v) > 0$, $b = v_x + V_x$ and

$$c = \tfrac{F(v)-F(V)}{v-V} V_{xx}.$$

For parabolic equations (2.18) with bounded continuous coefficients and $a \geq \mu > 0$, (2.17) follows from the first Sturm Theorem.

For quasilinear degenerate equations of the type (2.11), the conclusion is proved by regularization and passage to the limit. Given initial data $v_0(x)$ and $V_0(x)$ with finite number of intersections $\mathrm{Int}(0,V)$, one can construct the corresponding monotone sequences of the regularized strictly positive data $v_{0\varepsilon}(x)$ and $V_{0\varepsilon}(x)$ for $\varepsilon > 0$ with the same number of intersections $\mathrm{Int}_\varepsilon(0,V)$. For the filtration equation we just put $v_{0\varepsilon} = v_0 + \varepsilon$ and $V_{0\varepsilon}(x) = V(x,0) + \varepsilon$. By the MP, the corresponding smooth solutions are strictly positive and $v_\varepsilon(x,t)$, $V_\varepsilon(x,t) \geq \varepsilon$ in S. Then, by the first Sturm Theorem applied to classical strictly positive solutions, we have

$$\mathrm{Int}(t,V_\varepsilon) \leq \mathrm{Int}(0,V_\varepsilon) = \mathrm{Int}(0,V) \quad \text{for} \quad t > 0$$

and

$$\mathrm{Int}(t,V_\varepsilon) \quad \text{does not increase with time.}$$

Passing to the limit $\varepsilon \to 0^+$ and using the fact that the approximating sequences $\{v_\varepsilon\}$ and $\{V_\varepsilon\}$ converge monotonically to the continuous weak solutions v and V, respectively, uniformly on compact subsets, we arrive at (2.17). Indeed, one can see that, if this were false at some $t = t_0 \geq 0$ for the weak solutions, then it would be false also for the corresponding smooth approximations with arbitrarily small $\varepsilon > 0$.

Property (2.17) is true for the solutions $v(x,t)$ and $V(x,t)$ in an arbitrary domain $\Omega \subseteq I\!R \times I\!R_+$ with a smooth boundary provided that the difference $w(x,t)$ does not change sign on the lateral boundary of Ω. In the case of the Cauchy problem this means that new intersections cannot appear at $x = \pm\infty$. For the problem under consideration with initial data $v_0(x)$ and $V(x,0)$, which are unbounded as $|x| \to \infty$ (but do not grow very fast), this follows from comparison in any unbounded domain of the form $(\ell,\infty) \times (0,T)$ or $(-\infty,\ell) \times (0,T)$ with $\ell \in I\!R$.

Bernstein estimates by transversality: proof of Theorem 2.1. It literally repeats the proof of Proposition 1.9 in Chapter 1. Let $B_1 = \{V, |\lambda| > 1\}$ and then

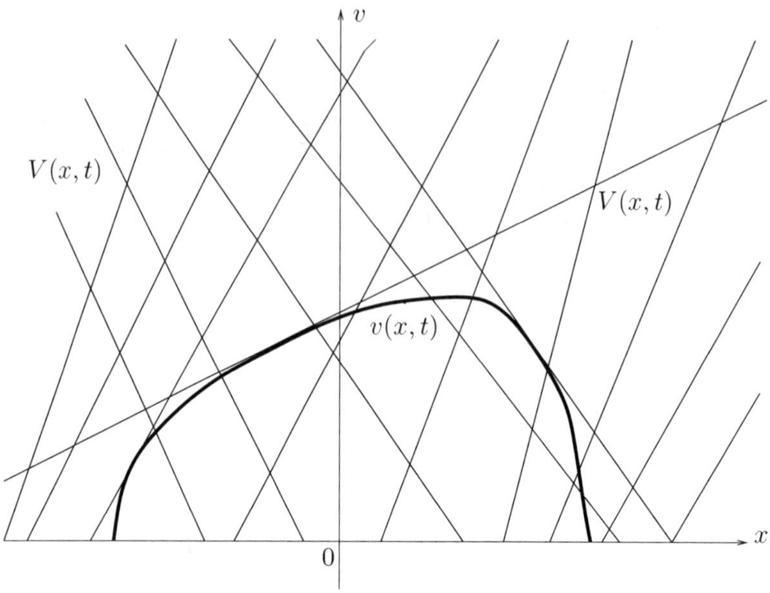

Figure 2.1 *Positive concavity: under condition* (2.20), *only tangency points can occur with TWs.*

v_0 intersects each $V \in B_1$ at most once and transversally: $\mathrm{Int}(0,V) \le 1$ for any $V \in B_1$. Hence, the inequality $\mathrm{Int}(t,V) \le 1$ for $V \in B_1$ means that the same transversality remains valid for any $t > 0$. The proof applies to unbounded solutions in the class $\{v(x,t) \le C(1+|x|)\}$. $\square$

Proof of concavity: intersection comparison with B. First, assumption (2.13) with $f = v_0$ implies that the initial number of intersections is

$$\mathrm{Int}(0,V) \le 2 \quad \text{for any } V \in B. \tag{2.19}$$

This is clear in view of the formula (2.15) for the TWs V. It then follows from (2.17) that

$$\mathrm{Int}(t,V) \le 2 \quad \text{for } t > 0 \text{ and any } V \in B. \tag{2.20}$$

Let us prove that $v(x,t)$ is positively concave. Figure 2.1 outlines our intersection strategy. Take an arbitrary $(x_0,t_0) \in S$ such that $\mu = v(x_0,t_0) > 0$ and set $\nu = v_x(x_0,t_0)$. Let $V(x,t) \in B$ be the unique tangent solution at the point (x_0,t_0) as defined in (2.16). We will analyze the behaviour of the difference $w(x,t_0) = v(x,t_0) - V(x,t_0)$ near the point $x = x_0$. An intersection point is called an *inflection point* if $w(x_0,t_0) = w_x(x_0,t_0) = 0$, i.e., the solutions are *tangent* to each other at this point. Our first step is to eliminate such a possibility.

Lemma 2.4 *The tangency point $x = x_0$ is not an inflection point for $w(x,t_0)$.*

Proof. Of course, this follows from the second Sturm Theorem since an inflection, as a multiple zero of the difference, can only occur from at least three intersections, which must be available at $t = 0$ contradicting (2.19) (i.e., the positive convexity

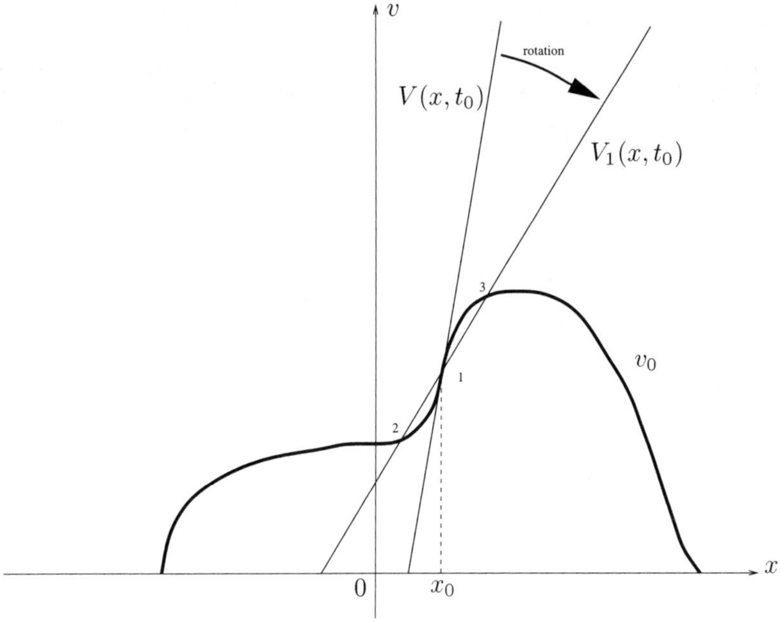

Figure 2.2 *Illustration of the proof of Lemma* 2.4*: a small rotation of* $V(x,t_0)$ *around the inflection point produces at least three intersections with a TW* $V_1(x,t_0)$.

of initial data). Such an application assumes additional regularity of the solution in the positivity domain, $u \in C^{4,2}_{x,t}$. In order to use the internal structure of the set B in more detail, we present a direct purely geometric proof of the nonexistence of an inflection, which assumes no extra regularity and shows that the first Sturm Theorem suffices to conclude.

We argue by contradiction. Assume that $x = x_0$ is an inflection point for $w(x,t_0)$, so $w(x_0,t_0) = w_x(x_0,t_0) = 0$ and furthermore, $w(x,t_0)$ changes sign in a neighbourhood $(x_0-\delta, x_0+\delta)$, $\delta > 0$. Assume, to be specific, that $\nu = v_x(x_0,t_0) > 0$ and that

$$w(x,t_0) > 0 \quad \text{on } (x_0-\delta, x_0), \quad w(x,t_0) < 0 \quad \text{on } (x_0, x_0-\delta) \tag{2.21}$$

(recall that $\text{Int}(t_0,V) \le 2$).

We now use the continuity of the set B to make a small perturbation of the tangent solution $V = V(x,t;x_0,t_0,\mu,\nu)$, as shown in Figure 2.2. Since we have an inflection point, one can see from (2.21) that, for any small $\varepsilon > 0$, there holds

$$\text{Int}(t_0,V_1) \ge 3 \quad \text{with the solution } V_1 = V(x,t;x_0,t_0,\mu,\nu-\varepsilon) \tag{2.22}$$

obtained by a small clockwise rotation of $V(x,t)$. This contradicts (2.20). The proof for $\nu \le 0$ is similar. Note that $\nu = 0$ gives $\lambda = 0$ in (2.15), i.e., $V \equiv \mu$. $\square$

We have proved that $w(x,t_0)$ does not change sign in $(x_0-\delta, x_0+\delta)$ for small $\delta > 0$. In fact, we can control the sign of w and thus complete the proof of Theorem 2.2. We have

Lemma 2.5 $w(x, t_0) \le 0$ *in* $I\!R$.

Proof. We will do this in two steps. (i) First we obtain the sign result in a small neighbourhood of x_0. Assume for contradiction that $w(x, t_0) \ge 0$ in $I = (x_0 - \delta, x_0 + \delta)$. Since, by regularity of the solutions in the positivity domains, the tangency points are isolated, we may assume that $w(x, t_0) > 0$ in $\{0 < |x - x_0| < \delta\}$. Set for a fixed small $\varepsilon > 0$

$$V_2(x, t) = V(x, t; x_0, t_0, \mu + \varepsilon, \nu). \tag{2.23}$$

Using the continuity of the set B in μ and ν, one can see that $\mathrm{Int}(t_0, V_2) \ge 2$. Since $\mathrm{Int}(t_0, V_2) \le 2$ in general, we conclude that $\mathrm{Int}(t_0, V_2) = 2$, and by (2.17)

$$\mathrm{Int}(t, V_2) = 2 \quad \text{for all } t \in [0, t_0]. \tag{2.24}$$

This contradicts the initial concavity. This is seen as follows: let $\{x = \xi_i(t),\ t \ge 0\}$, $i = 1, 2$, $\xi_1 < \xi_2$, be two continuous intersection curves of the solutions $v(x, t)$ and $V_2(x, t)$. Then they cannot meet for $t \in (0, t_0]$ since any meeting will mean the disappearance of both intersections thus contradicting (2.24). Therefore, we may assume that $\xi_1(t) < \xi_2(t)$ for all $t \in [0, t_0]$. Now the assumption (2.13) on v_0 and (2.24) imply that the difference $w_2 = v - V_2$ has to satisfy

$$w_2(x, 0) > 0 \quad \text{in } (\xi_1(0), \xi_2(0)). \tag{2.25}$$

By construction of the perturbed tangent solution $V_2(x, t)$, we have that

$$w_2(x, t_0) < 0 \quad \text{in } (\xi_1(t_0), \xi_2(t_0)). \tag{2.26}$$

From (2.25), by the fact that the curves do not meet in the time interval $0 \le t \le t_0$ and that $w = 0$ on them, the MP implies a contradiction with (2.26).

(ii) In order to finish the proof we need to transform this local sign control into a global sign control. If the inequality $w(x, t_0) \le 0$ is not valid everywhere, then, since the sign is correct near x_0, we conclude that necessarily there is another intersection, i.e., $\mathrm{Int}(t_0, V) \ge 1$. Hence, by the same small perturbation, now shifting downwards with $\varepsilon < 0$ in (2.23), we obtain at least three intersections, $\mathrm{Int}(t_0, V_2) \ge 3$. This contradicts the basic estimate (2.20) and ends the proof of Lemma 2.5 and Theorem 2.2. □

Proofs by approximation. Often, as we have just seen, transversality and concavity proofs rely on known regularity properties of the weak solutions near the singularity set and characterize some properties of the interfaces. Using the particular filtration equation, we next present an alternative proof based on suitable approximations. In general, we will use the fact that proper (weak or maximal, see below) solutions of singular parabolic equations can be obtained by monotone approximations, and, actually, it is easier to deal with the approximations rather than with the limit proper solutions exhibiting special, and possibly unknown in detail, regularity properties.

Namely, we perform a simultaneous ε-approximation of v and of TWs $V \in B$ as above, with initial data $v_{0\varepsilon} = v_0 + \varepsilon$ and $V_{0\varepsilon} = V_0 + \varepsilon$, respectively. By the initial concavity, the solutions $v_\varepsilon(x, t)$, $V_\varepsilon(x, t)$ satisfy $\mathrm{Int}(0, v_\varepsilon, V_\varepsilon) \le 2$, whence

$$\mathrm{Int}(t, v_\varepsilon, V_\varepsilon) \le 2 \quad \text{for } t > 0.$$

Here $V_\varepsilon(x,t)$ is not a simple piecewise linear TW solution any longer, but the set $B_\varepsilon = \{V_\varepsilon(x,t)\}$ satisfies the main property.

Proposition 2.6 *For any fixed $\varepsilon > 0$, the set B_ε is complete in $\{v > \varepsilon\}$.*

Proof. This follows from the transversality of intersections of pairs of different initial functions $V_{0\varepsilon}(x)$ corresponding to distinct speeds $\lambda_1 \neq \lambda_2$. Indeed, either there exists a single transversal intersection or no intersections at all. Hence, the same transversality holds for $t > 0$ and there exists at most one curve of transversal intersections for $t > 0$. This is a general property: for a sufficiently regular parabolic flow, a complete set of initial data $\{V(0,x)\}$ generates a set of solutions $\{V(x,t)\}$, which remains complete for all $t > 0$. Continuity and monotonicity of B_ε follow from standard results from parabolic theory. □

We thus obtain that v_ε is concave relative to the set B_ε (a concept to be used later on in Section 2.6). Passing to the limit $\varepsilon \to 0^+$ (then $v_\varepsilon \to v$ and $V_\varepsilon \to V$ uniformly on compact subsets), we obtain that $v(x,t)$ is concave relative to B, which in the present case means the standard positive concavity.

Convexity. Exactly the same analysis is true when we consider the convexity instead of the concavity of the solution.

Theorem 2.7 *Let v be a nonnegative solution of* (2.11), (2.12). *If v_0 is positively convex, $v(\cdot,t)$ is positively convex for every $t > 0$.*

Concavity with infinite propagation

The analysis of the filtration equation when (2.9) is not satisfied, i.e.,

$$\int_0 \frac{d\varphi(s)}{s} = \infty \tag{2.27}$$

(the so-called infinite propagation condition), is easier though it has been less considered in the literature. Then we take as pressure

$$v = \Phi(u) \equiv \int_a^u \frac{d\varphi(s)}{s}, \tag{2.28}$$

where $a > 0$ is a reference level, say $a = 1$ (we recall that typically pressure is defined but for a constant). Then $\Phi(0) = -\infty$ and $-\infty < v < \Phi(\infty)$ when $0 < u < \infty$. The maximum pressure, $\Phi(\infty)$, can be finite or infinite. The former case implies some non-essential technical difficulties. The latter case includes the heat equation $u_t = u_{xx}$ and then $v = \log u$. The log-convexity of the solutions of the heat equation is a standard question discussed in the literature; see Remarks. With respect to the finite-propagation case, the main difference in the theory is that nonnegative nontrivial solutions are automatically positive in S.

After the change of variables (2.28) we arrive at the same equation (2.11). We may assume that $u_0 > 0$, so $v_0 > -\infty$. We can formulate and prove a convexity result such as Theorem 2.2 but in discussing concavity we have to bear in mind that now there is no free boundary since $v > -\infty$ everywhere in S. Hence, the proof of concavity is much easier and unconditional.

Theorem 2.8 *Let v be a solution of* (2.11) *under condition* (2.27)*. If v_0 is concave, then $v(\cdot, t)$ is concave for $t > 0$.*

Remark: on other estimates via intersections. Theorem 2.2 establishes an estimate on the second derivative v_{xx}. This strongly relies on the existence of at most two intersections of v_0 with an arbitrary $V \in B$. Indeed, in the case of compactly supported $v_0(x)$ the result can be stated as follows:

$$\text{if } \mathrm{Int}(0, V) \le 2 \quad \text{for any } V \in B, \quad \text{then} \tag{2.29}$$

$$v_{xx} < 0 \quad \text{in the interval } \{v > 0\} \quad \text{for } t > 0. \tag{2.30}$$

The strict inequality in (2.30) follows from (2.14) by using the strong MP for the linear parabolic equation satisfied by v_{xx} (see (2.35) below). Let us detect some consequences of comparison with families having 0, 1, or 3 intersections.

(i) It is easy to state comparison with the subset

$$B_0 = \{V \in B : \ \mathrm{Int}(0, V) = 0\},$$

which yields an estimate on the solution

$$v(x, t) \le \inf_{V \in B_0} V(x, t) \quad \text{in } S.$$

(ii) By Theorem 2.1, the transversality analysis with the subset

$$B_1 = \{V \in B : \ \mathrm{Int}(0, V) \le 1\} \tag{2.31}$$

gives Bernstein-type estimates on the first derivative v_x. Recall the proof by using the subset $B_1^+ = \{V \in B_1 : \ V_x > 0\}$. If for some $(x_0, t_0) \in S$, $v(x_0, t_0) = V(x_0, t_0) > 0$ and $V \in B_1^+$, then

$$v_x(x_0, t_0) \le V_x(x_0, t_0) \quad \text{or} \quad v_x \le V_x|_{v=V}. \tag{2.32}$$

Obviously, if (2.32) is false and $v_x > V_x$ at the intersection point, then, since v_0 is compactly supported, we have

$$\mathrm{Int}(t_0, V) \ge 2$$

contradicting (2.31) and (2.17).

(iii) As a next generalization, consider the case

$$\mathrm{Int}(0, V) \le 3 \quad \text{for any } V \in B. \tag{2.33}$$

This yields an estimate of the third derivative v_{xxx}: if for some $(x_0, t_0) \in S$, $v(x_0, t_0) > 0$ and $v_{xx}(x_0, t_0) = 0$, then

$$v_x(x_0, t_0) v_{xxx}(x_0, t_0) \le 0.$$

Indeed, if $v_x v_{xxx} > 0$ at (x_0, t_0), there exists a solution $V_1 \in B$ of the form given in (2.22) with $\varepsilon < 0$ for $v_x(x_0, t_0) \ge 0$, and $\varepsilon > 0$ for $v_x(x_0, t_0) < 0$ that is a small perturbation of the tangent one $V(x, t)$ at the point (x_0, t_0). Then

$$\mathrm{Int}(t_0, V_1) \ge 4$$

contradicting (2.33) and (2.17).

One can continue the analysis with a greater maximal number of intersections, which controls a type of multiple tangency or inflection points occurring between v and tangent solutions $V \in B$.

2.3 Eventual concavity for the filtration equation

We now present a different type of concavity result for the filtration equation (2.1). The concavity condition on the initial data is weakened, and the concavity of the solution happens eventually in time.

Case of finite propagation

As in the previous section, we study in greater detail the case of finite propagation, where φ satisfies (2.9) and the same regularity assumptions. We consider the pressure function $v = \Phi(u)$ and take bounded, continuous and nonnegative initial data $v_0 \in C^2(\{v_0 > 0\})$, $v_0(x) \not\equiv 0$. For simplicity we suppose that the second derivative v_0'' has a finite number of zeros in $\{v_0 > 0\}$.

Theorem 2.9 *Let $v_0(x)$ be compactly supported. Then there exists $t_* \geq 0$ such that*

$$v_{xx} < 0 \quad \text{in} \ \ \{v(x,t) > 0\} \ \ \text{for all} \ \ t \geq t_*. \tag{2.34}$$

The proof is based on intersection comparison with the set B of TW solutions (2.15) plus an argument using the compactness of the set B in C.

We perform our analysis dealing directly with weak solutions $v(x,t)$ and $V(x,t) \in B$. Unlike the previous section, it includes rather delicate information concerning the behaviour of solutions near finite interfaces, which, fortunately, is well-known for the filtration equation. On the other hand, we will show later on that there is a proof based on approximations, when we deal with regularizing sequences $\{v_\varepsilon\}$ and $\{V_\varepsilon\}$ only. For the filtration equations, we prefer to conduct a direct proof describing interesting properties of interfaces.

Lemma 2.10 *No new sign changes of the second derivative $v_{xx}(x,t)$ are generated at the interfaces.*

Proof. Since, for large times, the support of compactly supported solutions is connected, we may assume that the support is a connected interval from the beginning, i.e., $\operatorname{supp} v(x,t) = [\zeta_-(t), \zeta_+(t)]$ for $t \geq 0$. Then $\zeta_\pm(t)$ are continuous functions, $\zeta_-(t)$ is nonincreasing and $\zeta_+(t)$ is nondecreasing, which is proved by comparison with TWs; see Remarks.

Under the given assumptions on v_0, we may suppose that, in the positivity domain, the number of zeros of the second derivative v_{xx} is finite for any $t > 0$. Indeed, differentiating equation (2.11) twice yields that $z = v_{xx}$ solves the following parabolic equation:

$$z_t = F z_{xx} + 2v_x(F' + 1)z_x + [(F' + 2)z + F''(v_x)^2]z \quad \text{in} \ \{v > 0\}. \tag{2.35}$$

Therefore, the above conclusion follows from the results for linear uniformly parabolic equations (Chapter 1) by monotone approximation for equation (2.11).

In a similar way, we conclude that the number of zeros of the first derivative $z = v_x$ is finite for all $t > 0$ since it satisfies, in $\{v > 0\}$, the parabolic equation

$$z_t = F z_{xx} + (F' + 2) z z_x$$

and z cannot change sign of both interfaces.

Suppose now for contradiction that, for some $t = t_1 \geq 0$, the second derivative v_{xx} changes sign at the left-hand interface $x = \zeta_-(t)$. In view of the above conclusions, there holds: (i) there exists $\delta > 0$ such that, for instance,

$$v_{xx}(x, t_1) \leq 0 \ \text{ on } (\zeta_-(t_1), \zeta_-(t_1) + \delta], \tag{2.36}$$

and (ii) for any $\varepsilon \in (0, \varepsilon_0]$, where $\varepsilon_0 > 0$ is a small constant, there exists $x_\varepsilon \in (\zeta_-(t_1 + \varepsilon), \zeta_-(t_1) + \frac{\delta}{2})$ such that

$$v_{xx}(x_\varepsilon, t_1 + \varepsilon) > 0. \tag{2.37}$$

By the strong MP applied to equation (2.35) in $\{v > 0\}$, we may assume from (2.36) that $v_{xx}(x, t_1) < 0$ for $x = \zeta_-(t_1) + \delta$, hence by continuity

$$v_{xx}(x, t_1 + \varepsilon) < 0 \quad \text{for } x = \zeta_-(t_1) + \delta \ \text{ and any } \varepsilon \in (0, \varepsilon_0]. \tag{2.38}$$

Let $\text{Int}_*(t, V)$ be the number of intersections in the domain $\{x \leq \zeta_-(t_1) + \delta\}$ of the solution $v(x, t)$ and a given function $V \in B$. It follows from (2.36) that $\text{Int}_*(t_1, V) \leq 2$ for all $V \in B$. Taking δ small, we conclude in view of (2.38) that the difference $w = v - V$ does not change sign for $x = \zeta_-(t_1) + \delta$, $t \in (t_1, t_1 + \varepsilon_0]$, so

$$\text{Int}_*(t, V) \leq 2 \quad \text{for all } t \in (t_1, t_1 + \varepsilon_0] \ \text{ and } V \in B. \tag{2.39}$$

It follows from (2.37) and (2.38) that, for a fixed small $\varepsilon \in (0, \varepsilon_0)$, there exists $V' \in B$ such that $\text{Int}_*(t + \varepsilon, V') \geq 3$, thus contradicting (2.39) and completing the proof of Lemma 2.10. □

Denote now by $I(t)$ for $t \geq 0$ the number of sign changes of $z = v_{xx}(x, t)$ satisfying equation (2.35). It follows from Lemma 2.10 that, in the case of solutions with compact supports, $I(t)$ is bounded for $t > 0$ and satisfies the standard intersection comparison property

$$I(t) \quad \text{does not increase for } t > 0. \tag{2.40}$$

Proof of Theorem 2.9. We argue by contradiction assuming that, for every $t \gg 1$, there exists an interval where $v_{xx}(x, t) > 0$. We begin our construction using the compactness of the TW set B.

Compactness of B: the case of two inflection curves. Assume that, for all $t > t_0 \gg 1$, there exist two inflection curves $x_-(t) < x_+(t)$ of the solution $v(x, t)$, both lying inside the positivity set of v and such that, by the strong MP applied to equation (2.35),

$$v_{xx} > 0 \quad \text{for } x \in (x_-(t), x_+(t)) \quad \text{and all } t > t_0, \tag{2.41}$$

and $v_{xx} = 0$ for $x = x_\pm(t)$. Set $x_0 = \frac{1}{2}(x_-(t_0) + x_+(t_0))$. Let $V_0(x, t) \in B$ be the unique solution that is tangent to $v(x, t)$ at the point (x_0, t_0), as shown on Figure 2.3.

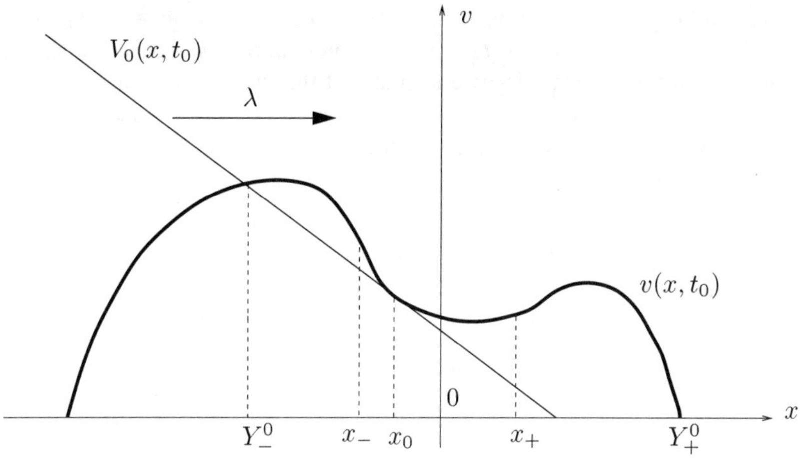

Figure 2.3 *The first moving tangent TW $V_0(x, t_0)$ makes $v(x, t)$ locally more concave in a neighbourhood of the tangency point at $x = x_0$.*

By $x = Y_-^0(t)$ and $x = Y_+^0(t)$ we denote two continuous zero curves of the difference $w = v - V_0$ such that $Y_-^0(t) < Y_+^0(t)$ and $w > 0$ in $I_0(t) = (Y_-^0(t), Y_+^0(t))$ for $t > t_0$. Let T_1 be the first point where they meet, i.e.,

$$T_1 = \sup\{t > t_0 : \ I_0(t) \neq \emptyset\}. \tag{2.42}$$

Proposition 2.11 $T_1 < \infty$.

Proof. Indeed, if V_0 is constant in x, i.e., $(V_0)_x = 0$, this follows from the fact that $v(x, t) \to 0$ as $t \to \infty$ uniformly in $I\!R$. If $(V_0)_x \neq 0$, we arrive at the same result by comparison of the supports of solutions. By comparison with explicit solutions of the form $f(x/\sqrt{t})$, we have that

$$|\zeta_\pm(t)| \le c\sqrt{1+t} \quad \text{for } t > 1. \tag{2.43}$$

Since the unique interface of a non-constant solution $V_0(x, t)$ moves like λt with $\lambda \neq 0$, we have that $V_0(x, t) \to \infty$ as $t \to \infty$ uniformly in $\operatorname{supp} v(\cdot, t)$, whence $T_1 < \infty$. □

To proceed further, we introduce the time

$$t_1 = \sup\{t \in (t_0, T_1) : \ (x_-(t), x_+(t)) \subseteq (Y_-^0(t), Y_+^0(t))\}.$$

Proposition 2.12 $t_1 < T_1$.

Proof. It is immediate that $t_1 = T_1$ would imply the disappearance of the interval $(x_-(t), x_+(t))$ at time T_1 contradicting our assumption. □

We now proceed at t_1 as above at t_0. We take $x_1 = \frac{1}{2}(x_-(t_1) + x_+(t_1))$ and introduce the tangent solution $V_1(x, t)$ at the point (x_1, t_1), the interval $I_1(t) = (Y_-^1(t), Y_+^1(t))$ for $t \ge t_1$, and the values T_2 and t_2. We have $t_1 < t_2 < T_2$. Iteratively, we derive a monotone sequence $\{t_k\}$ and its corresponding sequence of tangent TW solutions $\{V_k(x, t)\} \subset B$ such that each function $V_k(x, t)$ is tangent to $v(x,t)$ at the point (x_k, t_k), $x_k = \frac{1}{2}(x_-(t_k)+x_+(t_k))$, and $w \equiv v - V_k > 0$ on $I_k(t) = (Y_-^k(t), Y_+^k(t))$ for $t - t_k > 0$ small enough. By a standard compactness argument, using the Bernstein-type estimate of the uniformly bounded solution $|v_x| \le C$ in $I\!R \times [t_0, \infty)$ with $t_0 > 0$ (proved above by transversality), we conclude that there exists an increasing subsequence that we again label by $\{k\}$ such that

$$V_k \to \bar{V} \quad \text{as } k \to \infty \tag{2.44}$$

uniformly on any compact subset of S. Let $\{t_k\} \to T \le \infty$.

Proposition 2.13 $\bar{V} \not\equiv 0$ *and* $\text{Int}(t_0, \bar{V}) \ge 3$.

Proof. If $\bar{V} \equiv 0$, then, by (2.44), $V_k(x, t_0)$ is small enough on $\text{supp}\, v(\cdot, t_0)$ for all $k \gg 1$. Of course, $\bar{V} \not\equiv \infty$ since then $\text{Int}(0, V_k) = 0$ for $k \gg 1$. Hence, since the numbers of zeros of $v_{xx}(x, t_0)$ and $v_x(x, t_0)$ are finite, we deduce that $\text{Int}(t_0, V_k) \le 2$ for any fixed $k \gg 1$. This contradicts the fact that $V_k(x,t)$ is the tangent solution to $v(x,t)$ at the point (x_k, t_k) with $t_k > t_0$, since, by construction of $\{V_k\}$, we have that $\text{Int}(t_k^-, V_k) \ge 3$ by the strong MP.

It is also easily seen from (2.44) that $\text{Int}(t_0, \bar{V}) \ge 3$. Indeed, if $\text{Int}(t_0, \bar{V}) \le 2$, then (2.44) implies that $\text{Int}(t_0^+, V_k) \le 2$ for $k \gg 1$ again contradicting the construction of $\{V_k\}$. □

Since $\bar{V} \not\equiv 0$, using (2.43) as above, we conclude that

$$\tilde{T} = \inf\{t > t_0 : \; v(x,t) \le \bar{V}(x,t) \;\text{ in } I\!R\} < \infty. \tag{2.45}$$

Proposition 2.14 *T is finite, and moreover $T \le \tilde{T}$.*

Proof. Obviously, if $T > \tilde{T}$, then, since $V_k \approx \bar{V}$ for $k \gg 1$ on $(\text{supp}\, v(\cdot, \bar{t})) \times (0, \bar{t})$ with $\bar{t} = \frac{1}{2}(T + \tilde{T})$, by the strong MP, we have from (2.45) that $\bar{V}(x, \bar{t}) > v(x, \bar{t})$ in $\text{supp}\, v(\cdot, \bar{t})$. Hence, $\text{Int}(\bar{t}, V_k) = 0$ contradicting the fact that V_k is tangent to $v(x, t_k)$ at the point $x = x_k$ with $t_k > \bar{t}$. □

Proposition 2.15 *The curves $x_\pm(t)$ meet at T, i.e.,*

$$x_-(T) = x_+(T). \tag{2.46}$$

Proof. If $x_-(T) < x_+(T)$, then, by the strong MP for equation (2.35), it follows from (2.41) that we can continue our construction of the sequence $\{V_k\}$ with $t_k > T$ contradicting Proposition 2.14. □

Thus the existence for all $t \gg 1$ of two inflection curves $\{x = x_\pm(t)\}$ such that (2.41) holds is impossible. Condition (2.46) implies that, at $t = T$, these curves meet inside the support, and for small $t - T > 0$ these inflection points in a small neighbourhood of $x = x_\pm(T)$ do not exist.

The case of a single inflection curve. The second possibility of existence for all $t > t_0 \gg 1$ of an inflection curve $\{x = x_+(t)\}$ is as follows:

$$v_{xx} > 0 \quad \text{for } x \in (\zeta_-(t), x_+(t)), \; t \ge t_0, \tag{2.47}$$

i.e., the solution is convex near the interface, for definiteness the left-hand one. Then we can use the same approach as above setting $x_-(t) \equiv \xi_-(t)$. Therefore, (2.47) will imply that $\zeta_-(T) = x_+(T)$, i.e., in this case the inflection curve must disappear in a finite time on the corresponding interface.

End of proof of Theorem 2.9. Since $I(t_0) < \infty$, according to the above results we have that a finite number of inflection curves existing for small $t - t_0 > 0$ either annul each other inside the support of $v(x,t)$ or disappear on the interface. Hence, (2.34) is valid for $t \gg 1$ completing the proof. □

On the proof by approximation. As in the previous section, the proof of the eventual concavity can be done by intersection comparison of regular approximations v_ε and the regularized solutions $B_\varepsilon = \{V_\varepsilon\}$, where $V_\varepsilon(x,t)$ with initial data $V_\varepsilon(x,0) = V(x,0) + \varepsilon$ are not TWs but are classical strictly positive solutions. The completeness and compactness of B_ε are proved by the same standard arguments. Then, in order to establish the eventual concavity of a weak (proper) continuous solution v we do not need extra regularity results in the positivity domain. On the other hand, as we have already seen, the above proof has revealed some special properties of singular interfaces.

On eventual concavity with infinite propagation

As in the previous section, eventual concavity also occurs for equations with infinite propagation, when condition (2.27) holds. For simplicity of the statement we consider the cases with power-like nonlinearities, namely the heat equation and the fast diffusion equation.

In the case of the heat equation, $u_t = u_{xx}$, the pressure (2.28), which can be taken in the form $v = \log u$, solves equation (2.7). Using known analyticity properties of solutions, we can suppose that the initial function $u_0(x) > 0$ is C^2, and zeros of v_0'' are not accumulated at $x = \pm\infty$.

Theorem 2.16 *Consider the heat equations in $I\!R \times I\!R_+$. Let the initial data $u_0(x)$ satisfy*

$$0 < u_0(x) \le e^{-x^2/4} \quad \text{in } I\!R. \tag{2.48}$$

Then there exists $t_* \ge 0$ such that

$$v_{xx} < 0 \quad \text{in } I\!R \ \text{ for all } \ t \ge t_*. \tag{2.49}$$

Note that an assumption like (2.48) is essential for the eventual log-concavity of the solution $u(x,t)$, and it plays a similar role as the first hypothesis in Theorem 2.9, stated in the case of the finite propagation.

Proof. This is similar to the proof of Theorem 2.9. The fact that $T_1 < \infty$ (and also $T_k < \infty$) follows from (2.48). Indeed, by comparison (2.48) implies that

$$u(x,t) \le \frac{1}{\sqrt{1+t}}\, e^{-x^2/4(1+t)} \quad \text{in } S,$$

or in terms of v,

$$v(x,t) \le -\tfrac{1}{2}\log(1+t) - \tfrac{1}{4}\tfrac{x^2}{1+t}.$$

Therefore, for the TW $V_0(x,t) = \lambda^2 t + \lambda x + a \in B$, we conclude that T_1 given by (2.42) is finite. The rest of the proof is the same. The only difference is that we show that the limit function $\bar V$ exists, i.e., $\bar V \not\equiv -\infty$ (in fact, it is easier to see that $\bar U = e^{\bar V} \not\equiv 0$). We then again argue by contradiction. □

Finally, consider the fast diffusion equation $u_t = (u^m)_{xx}$ with $0 < m < 1$. According to (2.28) we take the pressure

$$v = -\tfrac{m}{1-m}\, u^{m-1} < 0$$

satisfying equation (2.4). As above, we assume that $u_0(x) > 0$ is C^2 and v_0'' has a finite number of zeros.

Theorem 2.17 *Consider the Cauchy problem for the fast diffusion equation with initial data satisfying*

$$0 < u_0(x) \le [c_*(1+x^2)]^{-1/(1-m)} \quad \text{in } I\!R, \tag{2.50}$$

where $c_* = \frac{1-m}{2m(m+1)}$. *Then* (2.49) *holds for some* $t_* \ge 0$.

Proof. We again follow the proof of Theorem 2.9. Then (2.50) implies that, by comparison with the explicit solution, there holds

$$v(x,t) \le -\tfrac{1}{2(m+1)} \left[(1+t)^{\frac{1-m}{m+1}} + \frac{x^2}{1+t} \right] < 0 \quad \text{in } S.$$

Therefore, $T_k < \infty$ for all $k = 1, 2, \ldots$. The end of the proof is similar. Indeed, assuming for contradiction that $\bar{V} \equiv -\infty$, we have that $\mathrm{Int}(t_0, V_k) = 2$ for $k \gg 1$ contradicting the fact that $\mathrm{Int}(T_k^-, V_k) \ge 4$ by construction. □

2.4 Concavity for filtration equations with lower-order terms

Equations with absorption and source terms

Let us show that the results of Section 2.2 are valid for more general quasilinear equations

$$\boxed{u_t = (\varphi(u))_{xx} - \psi(u),} \tag{2.51}$$

where $\varphi \in C^2((0,\infty)) \cap C^1([0,\infty))$ as above and $\psi \in C^1((0,\infty)) \cap C([0,\infty))$ satisfies certain assumptions to be specified later. In this case the pressure (2.10) solves the following equation (cf. (2.11)):

$$v_t = \mathbf{P}(v) \equiv F(v)v_{xx} + (v_x)^2 - q(v), \tag{2.52}$$

where $F(v) = \varphi'(u)$ and $q(v) = \frac{1}{u}\psi(u)\varphi'(u)$ with $u = \Phi^{-1}(v)$. According to the general philosophy, the result about the preservation in time of the initial concavity or convexity of the solution is expected to be valid if equation (2.52) admits a good enough set B of explicit solutions $\{V(x,t)\}$ that are piecewise linear in x. It is convenient to express this fact by introducing a certain invariant property of the nonlinear ordinary differential (elliptic) operator $\mathbf{P}$ in (2.52). Namely, we require $\mathbf{P}$ to admit the following two-dimensional *linear invariant subspace*:

$$W_2 = \mathrm{Span}\{1, x\}, \tag{2.53}$$

where $\mathrm{Span}\{\cdot\}$ denotes the linear span of the given functions, i.e., $\mathbf{P}(W_2) \subseteq W_2$ and $\mathbf{P}(V) \in W_2$ for any $V = C_0 + C_1 x \in W_2$. This has a strong implication for the set of admissible functions: $q(v)$ has to be a linear function of v for $v > 0$, i.e., for all $v \geq 0$, $q(v) = \gamma - \alpha v$, where $\gamma \geq 0$ and α are fixed constants. For the original equation (2.51), this implies a rather cumbersome form

$$\psi(u) = u \frac{\gamma - \alpha\Phi(u)}{\varphi'(u)} \quad \text{for } u > 0. \tag{2.54}$$

For simplicity we assume that ψ given by (2.54) is smooth enough at $u = 0$, namely that

$$u \frac{\Phi(u)}{\varphi'(u)} \in C^1([0,\infty)), \quad \frac{u}{\varphi'(u)} \to 0 \text{ as } u \to 0, \quad \psi(0) = 0. \tag{2.55}$$

These assumptions are convenient and guarantee the existence and uniqueness of weak solutions of (2.51). On the other hand, if $\psi(u)$ is singular at $u = 0$, i.e., $\psi(u) \to \infty$ as $u \to 0$, we can then deal with unique maximal solutions constructed by regular monotone approximations. It is curious that the existence of such nontrivial maximal solutions $u(x,t) \not\equiv 0$ for small $t > 0$ is directly related (moreover, is equivalent) to existence of the corresponding singular TWs. This is established in Chapter 7 in a maximal generality.

In the case of the porous medium operator, equation (2.51), (2.54) with $\varphi(u) = u^m$ has the form

$$\boxed{u_t = (u^m)_{xx} - \tfrac{\gamma}{m}\, u^{2-m} + \tfrac{\alpha}{m-1}\, u.} \tag{2.56}$$

Then (2.55) means that $m \in (1,2)$. Since, for such values of m, the lower order term in (2.56) is the concave function of u, we have to assume that $\gamma \geq 0$. This condition is also necessary for maximal solutions that can be constructed for any $m > 1$. Using the usual comparison with TW solutions, we have that under assumptions (2.55) equation (2.51), (2.54) (and hence (2.52)) describe processes with finite speed of propagation. Thus we consider the following quasilinear equation:

$$v_t = \mathbf{P}_*(v) \equiv F(v)v_{xx} + (v_x)^2 - \gamma + \alpha v \quad \text{in } S \cap \{v > 0\}, \tag{2.57}$$

with nonnegative continuous initial data v_0 smooth enough in the positivity domain. Since in general we can deal with proper maximal solutions (or weak ones if any), which are constructed by regular strictly positive monotone decreasing approximations, we do not need to carry over any extension of (2.57) into the singularity level $\{v = 0\}$. For weak solutions defined via integration by parts, this is done by replacing the constant γ by $\gamma H(v)$, where $H(v)$ is the Heaviside function. In general, integration by parts is not well-defined for equations (2.51) with essentially singular absorption terms satisfying $\psi(0) = \infty$. As usual, we assume that v_0 satisfies natural assumptions in order to have a unique, global, proper (weak) solution.

Let us state the main result on concavity.

Theorem 2.18 *Theorem* 2.2 *is true for equation* (2.57) *with any fixed* $\gamma \geq 0$ *and* $\alpha \in \mathbb{R}$.

Proof. According to the proof of Theorem 2.2 we have to check some properties of the set of explicit solutions to equation (2.57).

Proper set of explicit solutions. The operator $\mathbf{P}_*(v)$ in (2.57) was constructed to admit invariant subspace (2.53). Therefore, equation (2.57) has the set $B = \{V(x,t)\}$ of explicit solutions of the form

$$V(x,t) = [C_0(t) + C_1(t)x]_+. \tag{2.58}$$

Substituting V into (2.57) yields the dynamical system (DS) for the coefficients

$$\begin{cases} C_0' = C_1^2 + \alpha C_0 - \gamma, \\ C_1' = \alpha C_1. \end{cases} \tag{2.59}$$

If $\alpha = 0$, the set B consists of the TW solutions

$$V(x,t) = (B\xi)_+, \quad \xi = x - \lambda t + a, \tag{2.60}$$

where $\lambda,\ a \in I\!R$ are arbitrary constants and $B = -\frac{\lambda}{2} \pm \sqrt{\frac{1}{4}\lambda^2 + \gamma}$. If $\alpha \neq 0$, then integrating (2.59) yields

$$C_0(t) = Ae^{\alpha t} + \tfrac{1}{\alpha}B^2 e^{2\alpha t} + \tfrac{1}{\alpha}\gamma, \quad C_1(t) = Be^{\alpha t}, \tag{2.61}$$

where A and B are free constants. We now show that B is a proper set of solutions. The first property we need is *completeness.*

Proposition 2.19 *For the above set B, Proposition* 2.3 *is valid.*

Proof. In the case (2.60) the result is easier and has been proved. Consider now solution (2.58) with coefficients (2.61). Then solving the tangential system (2.16) yields the unique values of the parameters

$$A = (\mu - \tfrac{1}{\alpha}\nu^2 - \tfrac{1}{\alpha}\gamma - \nu x_0)e^{-\alpha t_0}, \quad B = \nu e^{-\alpha t_0},$$

whence the result. □

Continuity with respect to parameters in (2.60) and (2.61) is straightforward. Clearly, we also have *monotonicity* relative to a in (2.60) and A in (2.61).

As in Section 2.2, the application of the intersection comparison argument uses a regularization argument. For equation (2.56) with strong absorption, a positive regularization of the initial data is not enough (a finite-time extinction singularity will occur), and we need to regularize the equation as well. It is a well-known procedure for such degenerate equations [213]. Therefore, using the same proof as in Section 2.2 we arrive at Theorem 2.18. For such subsets $B \subset W_2$, the transversality condition gives the same Bernstein estimate $|v_x| \le C_0$.

Equations with convection terms

The first-order nonlinear operator

$$\mathbf{K}(v) = (v + d)v_x,$$

where $d \in I\!R$ is a constant, satisfies $\mathbf{K} : W_2 \to W_2$. Hence, the operator $\mathbf{P}_* + \mathbf{K}$, where $\mathbf{P}_*$ is as given in (2.57), also admits the invariant subspace (2.53). Thus the quasilinear equation with convection terms

$$v_t = \mathbf{P}_*(v) + \mathbf{K}(v) \tag{2.62}$$

admits the set B of explicit solutions of the form (2.58), where the expansion coefficients $\{C_0, C_1\}$ satisfy the following dynamical system:

$$\begin{cases} C_0' = C_1^2 + C_0 C_1 + \alpha C_0 + dC_1 - \gamma, \\ C_1' = C_1^2 + \alpha C_1. \end{cases} \tag{2.63}$$

In terms of the original variable $u(x,t)$ the corresponding filtration equation with convection takes the form

$$u_t = (\varphi(u))_{xx} - u\,\frac{\gamma - \alpha\Phi(u)}{\varphi'(u)} + (\Phi(u) + d)u_x \quad \text{in } \{u > 0\}.$$

If $\gamma \geq 0$, then under hypotheses (2.55), by the comparison argument, we conclude that the equation describes processes with finite speed of propagation.

The above set B of piecewise linear in x explicit solutions is proper. Indeed, (2.16) implies for $t = t_0$ that $C_0 + C_1 x_0 = \mu,\ C_1 = \nu$, whence the unique initial values

$$C_0(t_0) = \mu - \nu x_0, \quad C_1(t_0) = \nu. \tag{2.64}$$

The dynamical system (2.63) has a unique local (in time) solution $\{C_0(t),\ C_1(t)\}$ satisfying the above conditions at $t = t_0$. This solution is continued on the interval $[0, t_0]$, whence the uniqueness of the solution $V(x,t)$ tangent at (x_0, t_0) and the resulting completeness of B. Therefore, under a suitable hypotheses on the growth of $v_0(x)$ as $|x| \to \infty$, such that equation (2.62) has a unique weak solution in $S_T = I\!R \times (0,T)$ with $T = T(v_0) \leq \infty$, we have the following:

Theorem 2.20 *Theorem* 2.2 *is valid for equation* (2.62).

Notice that $W_2 = \text{Span}\{1, x\}$ is invariant under more general operators of the quasilinear parabolic equation

$$v_t = F(v)v_{xx} + vR_1(v_x) + R_2(v_x), \tag{2.65}$$

where $R_1(s)$ and $R_2(s)$ are smooth functions satisfying certain assumptions on the behaviour as $|s| \to \infty$. Substituting (2.58) into (2.65), we arrive at the DS

$$\begin{cases} C_0' = C_0 R_1(C_1) + R_2(C_1), \\ C_1' = C_1 R_1(C_1). \end{cases}$$

Assuming that the ODE problem (2.64) has a unique solution, we deduce the completeness of the corresponding set B of explicit solutions. Continuity and monotonicity of B can also be easily checked. Therefore, under some hypotheses on the coefficients of equation (2.65) and on the initial data to have existence and uniqueness of a compactly supported solution $v(x,t)$ to the well-posed Cauchy problem, we conclude that concavity (convexity) of the solutions is preserved in time.

2.5 Singular equations with the p-Laplacian operator preserving concavity

Let us apply the geometric technique to prove transversality, concavity and convexity results for equations with gradient-dependent diffusivity. We start with the p-Laplacian equation,

$$u_t = (|u_x|^\sigma u_x)_x \quad \text{in } S, \tag{2.66}$$

where $\sigma > 0$ is a fixed constant. The important point in the study of concavity and convexity is the introduction of a convenient "pressure" function, namely

$$v = \tfrac{1+\sigma}{\sigma}\, u^{\frac{\sigma}{\sigma+1}}. \tag{2.67}$$

Then equation (2.66) is transformed into

$$v_t = \mathbf{P}_1(v) \equiv |v_x|^\sigma [\sigma v v_{xx} + (v_x)^2]. \tag{2.68}$$

As in Section 2.2, we assume that the initial data $u_0 \geq 0$ satisfy suitable hypotheses, to ensure global existence of a unique weak solution $u(x,t)$. By known regularity results we may suppose that u_x is continuous for $t > 0$. Since $u_x = u^{1/(1+\sigma)} v_x$, we then deduce that, for $t > 0$, the derivative v_x exists in $\{v > 0\}$. In fact, v_x is bounded for $t \geq t_0 > 0$.

Equation (2.68) has a proper set B of piecewise linear in x TW solutions

$$V(x,t) = (B\xi)_+, \quad \xi = x - \lambda t + a, \quad \text{where } |B|^\sigma B = -\lambda. \tag{2.69}$$

Intersection comparison for the degenerate equation (2.66) is established by a standard approximation by using uniformly parabolic equations.

As above for equations of filtration type, the transversality with a subset $B_1 = \{V,\, |\lambda| \gg 1\} \subset B$ implies the geometric Bernstein bound $|v_x| \leq C_0$. As in Section 2.2, we arrive at the following concavity result.

Theorem 2.21 *Under the above hypothesis, if v_0 is positively concave (resp. convex), then $v(\cdot,t)$ is positively concave (resp. convex) for $t > 0$.*

Eventual concavity can be studied as in Section 2.3. We may repeat the proof of Theorem 2.9 with straightforward adaptation. Let us only remark that, in order to prove Proposition 2.11, we compare v with explicit self-similar solutions, which now take the form $f(x/t^{1/(2+\sigma)})$ and give the interface estimate $|\zeta_\pm(t)| \leq O(t^{1/(2+\sigma)})$. Thus we arrive at the following result.

Theorem 2.22 *Let, moreover, v_0 be compactly supported. Then there exists a time $t_* \geq 0$ such that $v(\cdot,t)$ is positively concave for all $t \geq t_*$.*

The results of Theorem 2.21 are true for equations with absorption

$$\boxed{u_t = (|u_x|^\sigma u_x)_x - \gamma u^{1/(1+\sigma)} + \alpha u, \quad \gamma \geq 0,\ \alpha \in I\!R.} \tag{2.70}$$

With the same definition, the pressure v satisfies

$$v_t = \mathbf{P}_2(v) \equiv \mathbf{P}_1(v) - \gamma + \delta v \quad \text{in } \{v > 0\}, \quad \delta = \tfrac{\alpha\sigma}{1+\sigma}, \tag{2.71}$$

where operator $\mathbf{P}_2$ possesses the invariant subspace (2.53). We then obtain a set B of explicit solutions (2.58) with coefficients satisfying the dynamical system

$$\begin{cases} C_0' = |C_1|^\sigma C_1^2 + \delta C_0 - \gamma, \\ C_1' = \delta C_1. \end{cases}$$

It is again a practical computation to check that B has the necessary properties. In particular, if $\alpha = 0$ in (2.70), then B consists of TW solutions such as (2.60), where the constant B satisfies, instead of (2.69), the algebraic equation

$$|B|^{2+\sigma} + \lambda B - \gamma = 0.$$

A more general equation

$$u_t = (\varphi(u_x))_x$$

with monotone increasing φ also admits a good set of TW solutions $B = \{V = f(\xi)\}$ obtained from the quadrature

$$-\int_0^{-\lambda f(\xi)} \frac{\mathrm{d}z}{\varphi^{-1}(z)} = -\lambda\xi.$$

There is no way of finding, in this generality, a change of variables like (2.67) (independent of λ) so as to obtain a suitable set B of piecewise linear solutions. This situation leads to the need of a generalization of geometric concepts, i.e., to the B-concavity, which we begin to study in the next section. On the other hand, the transversality with $B_1 = \{V, |\lambda| \gg 1\}$ gives a Bernstein-type bound on $|u_x|$ in terms of u via the so-called gradient function (a general derivation is given in Chapter 7).

2.6 Concepts of B-concavity and B-convexity. First example of sign-invariants

We will now enlarge our point of view and derive properties of the solutions by using sets B, which enjoy all above listed properties but do not consist of piecewise linear functions. We prove a strict concavity result with precise estimates for a more general (than the PME) diffusion-absorption equation written down in terms of the pressure

$$\boxed{v_t = \mathbf{P}_0(v) \equiv (m-1)vv_{xx} + (v_x)^2 - \gamma \quad \text{in } S \cap \{v > 0\},} \tag{2.72}$$

where $\gamma \geq 0$. We assume that the initial function $v_0 \not\equiv 0$ is nonnegative, bounded, compactly supported and is C^1 in $\{v_0 > 0\}$. If $\gamma > 0$, then $v(x,t)$ vanishes in a finite time $T = T(v_0) > 0$. In this case we consider the Cauchy problem in $S_T = I\!R \times (0,T)$. If $\gamma = 0$, we can set $T = \infty$.

Three-dimensional set of explicit solutions on W_3

To begin our study, we remark that operator $\mathbf{P}_0$ in (2.72) possesses the three-dimensional invariant subspace

$$W_3 = \mathrm{Span}\{1, x, x^2\}.$$

Then setting

$$V(x,t) = [C_0(t) + C_1(t)x + C_2(t)x^2]_+ \tag{2.73}$$

yields a third-order nonlinear dynamical system for the expansion coefficients $\{C_0, C_1, C_2\}$. We denote this set of explicit solutions (2.73) by D_3. Using it, we derive new generalized concavity or convexity properties of solutions. The set D_3 consists of the solutions

$$V(x,t) = A(t+\tau)\left[B(t+\tau) - D(t+\tau)(x-a)^2\right]_+, \tag{2.74}$$

where

$$A(s) = \tfrac{1}{2}(m+1)\, s^{-\frac{m-1}{m+1}}, \quad B(s) = \lambda^2 - \gamma \tfrac{(m+1)^2}{m}\, s^{\frac{2m}{m+1}} \tag{2.75}$$

and $D(s) = s^{-\frac{2}{m+1}}$, τ, λ, a being arbitrary constants. Since D_3 is three-dimensional, it is too wide and *is not complete* because a tangent solution defined by system (2.16) is not unique.

For $\gamma = 0$, (2.75) is the famous Ya.B. Zel'dovich–A.S. Kompaneetz–G.I. Barenblatt similarity solution (the ZKB-solution) of the PME. Note that, for $\gamma \neq 0$, (2.73)–(2.75) is not self-similar or group-invariant and belongs to the class of exact solutions on linear subspaces invariant under nonlinear operators.

B-concavity

The idea is to choose a particular proper *two-dimensional* subset $B \subset D_3$. The application of the methods of Section 2.2 shows that certain transversality and concavity properties of the solutions are preserved in time, which is proved by comparison with the subset B. A basic property that such a set must satisfy is completeness (as a set of tangent solutions; see Section 2.2). Then we can define generalized concavity with respect to B.

Definition 2.1 For a given $t_0 \in [0, T)$, a solution $v(x,t) \in C^1(\{v(x,t) > 0\})$ is said to be *concave with respect to the subset* B, or B*-concave*, if, for any $x_0 \in \{v(x, t_0) > 0\}$, there holds

$$v(x, t_0) \le V(x, t_0) \quad \text{in } I\!R, \tag{2.76}$$

where $V \in B$ is the unique tangent solution at the point (x_0, t_0). The sign $\ge$ in (2.76) defines the convexity relative to B, or the B*-convexity*.

The choice of proper subsets of solutions

There are in principle many possible choices of B. The particular selection depends on the properties we are looking for. Let us consider a simple choice of a proper set B_τ (other possibilities will be discussed in Section 2.7). Fix any constant $\tau > 0$. We define B_τ as follows:

$$B \equiv B_\tau = \{V = V(x,t;\lambda,a) \ \text{ in (2.74) with fixed } \tau > 0 \ \text{ and } \lambda, a \in I\!R\}.$$

Then $\dim B_\tau = 2$, and we have all properties listed in Section 2.2.

(i) *Completeness.* This time it is not immediate from the form of the solutions. We have to check that, for a fixed $(x_0, t_0) \in \bar{S}$ and arbitrary $\mu > 0, \nu \in I\!R$, there exist unique $\lambda > 0,\ a \in I\!R$ such that

$$V(x_0, t_0) = \mu, \quad V_x(x_0, t_0) = \nu. \tag{2.77}$$

In view of (2.74) this yields

$$A[B - D(x_0 - a)^2] = \mu, \quad -2AD(x_0 - a) = \nu.$$

Substituting $x_0 - a = -\nu/2AD$ from the second equation into the first one, we arrive at the unique value of $\lambda > 0$

$$\lambda^2 = \gamma(m+1)^2 \tfrac{1}{m} s^{\frac{2m}{m+1}} + \frac{1}{A(s)} \left[\mu + \frac{\nu^2}{4A(s)D(s)} \right], \quad s = t_0 + \tau,$$

and also of $a \in I\!R$. Thus the tangent solution exists and it is unique. If $\tau > 0$ in (2.74) is not fixed, system (2.77) admits infinitely many values $\lambda = \lambda(\tau)$ meaning that the whole set D_3 is overdetermined.

(ii) *Continuity.* By (2.74) the subset B_τ is *continuous*, i.e., V and V_x depend continuously on the parameters λ, a on compact subsets of $\{V > 0\}$.

(iii) *Monotonicity.* The next property plays an important role in the intersection comparison with the subset B_τ. In the case $B \subseteq W_2 = \text{Span}\{1, x\}$, studied in Sections 2.2–2.5, this property was trivial and we did not pay serious attention to it. Namely, set B_τ is *monotone* in the following sense. For any $V_0 \in B_\tau$, there exists a one-parameter family of solutions $\{V(\cdot; s),\ 0 < s < \infty\} \subset B_\tau$ with $V(\cdot; s_0) = V_0(\cdot)$ for some $s_0 > 0$, such that V, V_x are continuous in s in $\{V > 0\}$, and for any fixed $t \geq 0$,

$$\begin{aligned} &V(x,t;s) \uparrow \ \text{ in } s, \\ &V(x,t;s) \to 0, \ \text{ as } s \to 0, \\ &V(x,t;s) \to \infty, \ \text{ as } s \to \infty \end{aligned} \tag{2.78}$$

uniformly on any compact subsets in x from $\{V_0(x,t) > 0\}$. Indeed, for B_τ, this property is true, and the role of s is played by λ in (2.74), (2.75).

B-concavity (convexity) is preserved in time

We are now ready to prove the main result.

Theorem 2.23 *With the above notation, we have that v_0 is B-concave (B-convex) implies that $v(\cdot, t)$ is B-concave (B-convex) for all $t \in (0, T)$.*

Proof. We argue as in Section 2.2. Consider, for instance, the case of B-concavity. We need only to check that (2.19) is valid for any $V \in B_\tau$. We argue by contradiction. Assume that, for some $V_0(x,t) = V(x,t;\lambda_0, a_0)$,

$$\text{Int}(0, V_0) \geq 3.$$

Then there exist intersection points $x_1 < x_2$ such that

$$v_0(x) < V_0(x, 0) \quad \text{on} \quad (x_1,\, x_2),$$

and $v_0(x) = V_0(x, 0)$ for $x = x_{1,2}$. Then using the monotonicity property (2.78) with respect to the parameter $s = \lambda$, one can see by continuity that there exist $\lambda_* \in (0, \lambda_0)$ and a point $x_* \in (x_1, x_2)$ such that

$$V(x, 0; \lambda_*, a_0) < v_0(x) \quad \text{on} \quad (x_1,\, x_2), \quad x \neq x_*,$$

and these profiles are tangent at $x = x_*$. This contradicts the assumption that v_0 is B-concave; see (2.76). The rest of the proof is similar. □

Using the simple structure of the explicit solutions from B_τ, the notion of B-concavity can be expressed as a certain estimate on the second space derivative of the solution.

Proposition 2.24 *If $v(\cdot,t)$ is B-concave for $t \in [0,T)$ relative to B_τ, then*

$$v_{xx} \le -2A(t+\tau)D(t+\tau) \equiv -\tfrac{1}{(m+1)}\,\frac{1}{t+\tau} \tag{2.79}$$

in $S \cap \{v > 0\}$. B-convexity corresponds to the opposite inequality sign $\ge$ in (2.79).

Proof. Since $v = V$ and $v_x = V_x$ at any tangency point, we conclude from (2.76) that $v_{xx} \le V_{xx}$ there. Differentiating (2.74) twice yields (2.79). □

Thus (2.79) is the desired *semiconcavity estimate* complementary to the semiconvexity one (2.5), which can be treated as follows.

Proposition 2.25 *The Aronson-Bénilan estimate* (2.5) *is the B-convexity estimate relative to B_τ where any fixed $\tau > 0$ suits for a class of initial data v_0.*

The B-convexity analysis can be performed directly with B_0, i.e., we can set $\tau = 0$. Indeed, initial functions for such solutions are singular, $V(x,0) = C\delta(x-a)$, and any δ-function intersects v_0 (after approximation or, if necessary, a small shifting in time, when the solution becomes sufficiently smooth) at, at most, two points, so

$$\mathrm{Int}(0,V) \le 2 \quad \text{for all} \quad V \in B_0,$$

a crucial condition, leading to the B-convexity of arbitrary solutions.

If we consider the property of B-concavity relative to D_3 (in this case the tangent solution V in (2.77) is not unique and we perform intersection comparison with all of the tangent solutions), then $\tau > 0$ in (2.79) is arbitrary. Letting $\tau \to \infty$ yields $v_{xx} \le 0$. The B-concavity with respect to D_3 (defined in a natural way) is equivalent to the usual concavity property obtained by intersection comparison with piecewise linear solutions on $W_2 = \mathrm{Span}\{1, x\}$.

Sign-invariants

Let us rewrite (2.79) in the equivalent form

$$H_\tau(v) \equiv v_{xx} + \tfrac{1}{m+1}\,\frac{1}{t+\tau} \le 0, \tag{2.80}$$

and $\ge$ for the B-convex solutions. We then obtain the second-order differential operator $H_\tau(v)$ preserving its signs on suitable classes of solutions of the parabolic equation (2.72). This is a simple example of a *sign-invariant* as a differential operator preserving its signs on evolution orbits of the parabolic flow. It is important that this operator has been derived as the result of B-concavity (convexity) with respect to the fixed proper subset B_τ of particular solutions. Though (2.80) stays close to the obvious one $H_\infty(v) = v_{xx}$ obtained by comparison with piecewise linear solutions as in the previous sections (of course, H_0 is also a sign-invariant that needs no extra explanations), it gives certain new information about general solutions of the PME-type equations.

Such a simple structure of the sign-invariant H_τ followed from the above simple choice of the proper subset B_τ. Other choices of B will lead to different B-concavity properties and to more complicated sign-invariants as we show in the next section.

B-concavity for gradient-dependent diffusion equations

As expected, the above considerations suitably adapted apply to such equations. Consider for instance equation (2.71). The operator $\mathbf{P}_2$ has also the invariant subspace

$$\bar{W}_2 = \text{Span}\{1, |x|^\rho\}, \quad \rho = \tfrac{\sigma+2}{\sigma+1}. \tag{2.81}$$

In view of the translational invariance of the equation, this means that there exists the set B of explicit solutions

$$V(x,t) = [C_0(t) + C_1(t)|x-a|^\rho]_+ \, ,$$

where the coefficients satisfy the dynamical system

$$\begin{cases} C_0' = \frac{\sigma}{1+\sigma}\,(\rho^{\sigma+1}|C_1|^\sigma C_1 + \alpha)C_0 - \gamma, \\ C_1' = 2\rho^{1+\sigma}|C_1|^{2+\sigma} + \delta C_1 . \end{cases}$$

The set of such solutions D_3 is again three-dimensional. By choosing a suitable proper subset $B \subset D_3$, we can introduce the invariant property of B-concavity for the solutions of (2.71) in the sense of Definition 2.1 and derive the corresponding estimates.

2.7 Various B-concavity properties for the porous medium equation and sign-invariants

Consider the PME in terms of the pressure variable

$$v_t = (m-1)vv_{xx} + (v_x)^2 \quad (m > 1). \tag{2.82}$$

It admits the set D_3 of explicit solutions (2.74) with $\gamma = 0$,

$$V(x,t) = \tfrac{1}{2}\,\frac{m+1}{(t+\tau)^{\frac{m-1}{m+1}}}\left[\lambda^2 - \frac{(x-a)^2}{(t+\tau)^{\frac{2}{m+1}}}\right]_+ , \tag{2.83}$$

This set depends on three parameters τ, λ and a. In the previous section we considered the case $\tau = \text{const.} > 0$. We now assume that

$$\tau = \tau(\lambda) : I\!R_+ \to I\!R_+ \quad \text{is a smooth positive function,} \quad \tau'(\lambda) \le 0, \tag{2.84}$$

thus choosing the two-dimensional continuous subset $B(\tau) \subset D_3$. Then $B(\tau)$ is contained in the linear invariant subspace W_3 and forms a smooth two-dimensional manifold on W_3.

Proposition 2.26 *For any $\tau(\lambda)$ satisfying* (2.84), *the subset $B(\tau)$ is complete.*

Proof. The second equation of the tangential system (2.77) with $t = t_0$ and $\gamma = 0$ now has the form

$$-\frac{1}{t+\tau(\lambda)}\,\frac{x_0 - a}{m+1} = \nu. \tag{2.85}$$

Substituting $(x_0 - a)$ from (2.85) into the first equation (2.77) and using (2.83), we deduce the following algebraic equation for the parameter λ:

$$\Phi(\lambda; \tau, t) \equiv a_*(t+\tau(\lambda))^{-\frac{m-1}{m+1}}\lambda^2 - \tfrac{1}{2}\,(m+1)\nu^2(t+\tau(\lambda)) = \mu, \tag{2.86}$$

where $\Phi(\lambda)$ is smooth,

$$\Phi'_\lambda(\lambda) > 0 \quad \text{for} \quad \lambda > 0,$$
$$\Phi(\infty) = \infty,$$
$$\Phi(0^+) \le 0$$

provided that (2.84) holds. Hence, the equation (2.86) admits a unique root $\lambda > 0$ for any fixed $t \ge 0,\ \mu > 0,\ \nu \in I\!R$, completing the proof. □

Given a fixed function $\tau(\lambda)$, we denote the root of equation (2.86) by $\lambda = \lambda(\mu, \nu, t)$. Clearly, subset $B(\tau)$ is continuous, and we need to check the last necessary property of $B(\tau)$.

Proposition 2.27 *Under assumption* (2.84), *the subset* $B(\tau)$ *is monotone with respect to* λ *if the function* $\tau(\lambda)$ *satisfies*

$$\lambda\tau'_\lambda + (m+1)\tau \ge 0 \quad \textit{for} \ \lambda > 0. \tag{2.87}$$

Proof. We have to check properties (2.78) with s replaced by λ. It follows from (2.83) and (2.84) that, in the positivity domain of V, where $(x-a)^2 \le \lambda^2(t+\tau)^{\frac{2}{m+1}}$, there holds

$$(m+1)^2\, V'_\lambda \ge \lambda\, \frac{\lambda\tau'_\lambda + (m+1)\tau}{(t+\tau)^{\frac{2m}{m+1}}},$$

whence by (2.87) the first monotonicity hypothesis in (2.78). The rest follows from (2.83). □

A typical function $\tau(\lambda)$ satisfying (2.84) and (2.87) is

$$\tau(\lambda) = \lambda^{-\alpha} \quad \text{for} \quad \lambda > 0,$$

where $\alpha \in [0, m+1]$ is a fixed exponent. According to Definition 2.1, we introduce the notion of B-concavity (convexity) of the solution $v(x,t)$ relative to the given proper subset $B(\tau)$. In this case, instead of Proposition 2.24, we have the following characterization, which is a straightforward consequence of (2.76) and (2.83).

Proposition 2.28 *Let* (2.84) *hold. If* $v(x,t)$ *is* B*-concave, then*

$$H(v) \equiv v_{xx} + \tfrac{1}{m+1}\, \frac{1}{t + \tau(\lambda(v, v_x, t))} \le 0 \tag{2.88}$$

in $S \cap \{v > 0\}$, *where* $\lambda(\mu,\nu,t)$ *is the unique root of the algebraic equation* (2.86). B*-convexity corresponds to the opposite inequality sign* $\ge$ *in* (2.88).

Theorem 2.29 *Let* (2.84) *and* (2.87) *hold. Then Theorem* 2.23 *is valid for* $B(\tau)$.

The sign-invariant $H(v)$ defined in (2.88) gives a typical estimate of the second-order derivative v_{xx} in terms of the lower-order ones v_x and v and t. Second-order estimates are known to play a key role in the general regularity theory of nonlinear parabolic equations. We will not spend much effort in trying to derive essentially new estimates of solutions of the PME or other related equations via B-concavity analysis (though some of them will be presented). Observe that the function $\lambda(\mu,\nu,t)$ in (2.88) depending on a choice of a non-constant function $\tau(\lambda)$

cannot be expressed in an explicit form in general. In particular, this means that the general filtration equation in principle does not admit any *simpler* concavity estimates similar to the semiconvexity one for the PME. In the next few chapters we present other applications of such B-concavity techniques of studying evolution properties of reaction-diffusion equations.

2.8 B-concavity and sign-invariants for the heat equation

We apply our technique to describe further geometric properties of the heat equation written down in terms of the logarithmic pressure $v = \log u$

$$v_t = v_{xx} + (v_x)^2 \quad \text{in } S. \tag{2.89}$$

The initial data $v_0 \not\equiv 0$ are assumed to be locally bounded and smooth enough.

As it was shown above, well-known concavity or convexity results for (2.89) (or, which is the same, log-concavity and convexity in the original variable u) are a consequence of intersection comparison with the proper set of *linear* TW solutions

$$B = \{V(x,t) = \lambda^2 t + \lambda x + a\}.$$

We now consider some other particular examples of B-concavity/convexity for the heat equation (2.89).

B-concavity with respect to fundamental solutions

We first consider a simple subset B_τ of fundamental solutions to the heat equation that, in terms of the new variable v, have the form

$$V(x,t) = \lambda - \tfrac{1}{2}\log(t+\tau) - \tfrac{1}{4}\,\frac{(x-a)^2}{t+\tau}, \qquad \lambda,\, a \in I\!R, \tag{2.90}$$

where $\tau > 0$ is a fixed constant.

Proposition 2.30 *B_τ is a proper set.*

Proof. In order to prove completeness, we consider the corresponding tangential system (2.77) taking the form

$$\mu = \lambda - \tfrac{1}{2}\log(t_0+\tau) - \tfrac{1}{4}\,\frac{(x_0-a)^2}{t_0+\tau}, \qquad \nu = -\tfrac{1}{2}\,\frac{x_0-a}{t_0+\tau}.$$

Therefore for any $\mu,\ \nu \in I\!R$, there exist unique values of parameters λ and a given by

$$\lambda = \mu + \tfrac{1}{2}\log(t_0+\tau) + \nu^2(t_0+\tau), \quad a = x_0 + 2\nu(t_0+\tau).$$

Continuity and monotonicity (2.78) with $s = \lambda$ are obvious. □

Thus Theorem 2.23 is valid. Finally, from (2.90) we deduce that

$$V_{xx} = -\tfrac{1}{2}\,\frac{1}{t+\tau}. \tag{2.91}$$

Since $\mu = v = V$ and $\nu = v_x = V_x$ at tangency points, it follows from (2.91) that B-concavity with respect to B_τ of the initial function $v_0(x)$ implies that the solution $v(x,t)$ satisfies

$$H_\tau(v) \equiv v_{xx} + \tfrac{1}{2}\,\frac{1}{t+\tau} \le 0.$$

B-convexity corresponds to the sign $\ge$. Hence, H_τ is the corresponding sign-invariant.

B-convexity to the subset of $\log \mathrm{Span}\{1, x, x^2\}$

Equation (2.89) has explicit solutions $V(x,t)$ such that e^V belongs to the linear subspace $\mathrm{Span}\{1, x, x^2\}$. Setting

$$u = C_1(t) + C_2(t)(x-a)^2$$

yields a simple dynamical system

$$\begin{cases} C_1' = 2C_2, \\ C_2' = 0. \end{cases}$$

Solving it, we obtain the following subset $B(\tau)$ of explicit solutions:

$$V(x,t) = \log\left[\tau(\lambda) - \lambda t - \tfrac{1}{2}\,\lambda(x-a)^2\right]_+ \equiv \log[\,\cdot\,], \tag{2.92}$$

where $\lambda > 0$, $a \in I\!R$ and $\tau(\lambda) \ge 0$ is a smooth function satisfying

$$\lambda\tau' \ge \tau \ \text{ for } \lambda > 0, \quad \tau(0) = 0, \quad \frac{\tau(\lambda)}{\lambda} \to +\infty \ \text{ as } \lambda \to \infty. \tag{2.93}$$

Since V is defined in a bounded domain with $V = -\infty$ on the lateral boundary and v is finite everywhere in S, we can study B-convexity with respect to $B(\tau)$. Indeed, no sign changes of the difference $v - V$ can appear on the lateral boundary of $\{V > -\infty\}$. One can see that, in this case, no classical B-concave solutions exist.

Proposition 2.30 holds. In order to prove monotonicity, we calculate from (2.92) that, by (2.93),

$$V_\lambda' = e^{-V}(\tau' - t - \tfrac{1}{2}\,(x-a)^2) > e^{-V}\left(\tau' - \frac{\tau}{\lambda}\right) \ge 0$$

in the domain where $[\cdot] > 0$. To prove completeness, we have from (2.77) and (2.92) that

$$\mu = \log[\,\cdot\,], \quad \nu = -\lambda(x_0 - a)[\,\cdot\,]^{-1}, \tag{2.94}$$

whence the following algebraic equation for the parameter λ:

$$F(\lambda) \equiv \frac{\tau(\lambda)}{\lambda} - t_0 - \frac{e^\mu}{\lambda} - \frac{\nu^2 e^{2\mu}}{2\lambda^2} = 0. \tag{2.95}$$

It follows from (2.93) that $F(0^+) = -\infty$, $F(\infty) = \infty$ and $F'(\lambda) > 0$ for $\lambda > 0$, and hence there exists a unique root $\lambda = \lambda(\mu, \nu, t_0)$. Then a is uniquely determined from (2.94). Finally, since

$$V_{xx} = -(V_x)^2 - \lambda e^{-V},$$

we conclude that B-convexity of v_0 yields the corresponding sign-invariant and the following inequality on the solution:

$$H(v) \equiv v_{xx} + (v_x)^2 + \lambda(v, v_x, t)e^{-v} \geq 0 \quad \text{in } S. \tag{2.96}$$

In particular, if $\tau(\lambda) = \lambda^2$, then (2.95) becomes the cubic equation

$$\lambda^3 - t_0\lambda^2 - e^{\mu}\lambda - \tfrac{1}{2}\nu^2 e^{2\mu} = 0.$$

In the case $\tau(\lambda) = \lambda$ (one can see from (2.93)–(2.95) that completeness, continuity and monotonicity of $B(\tau)$ exist locally in time, for all $t \in [0, 1)$), (2.95) is the quadratic equation

$$(1 - t_0)\lambda^2 - e^{\mu}\lambda - \tfrac{1}{2}\nu^2 e^{2\mu} = 0,$$

whence the following estimate in $I\!R \times (0, 1)$ on the B-convex solutions and the sign-invariant:

$$H(v) \equiv v_{xx} + (v_x)^2 + \tfrac{1}{2}\,\frac{1}{1-t}\left[\sqrt{1 + 2(1-t)(v_x)^2} + 1\right] \geq 0.$$

2.9 B-concavity and transversality for the porous medium equation with source

Let us apply the technique to reaction-diffusion equations admitting blow-up solutions. The study of blow-up singularity phenomena will play an important part in the rest of the book. We consider the PME with source

$$\boxed{u_t = (u^m)_{xx} + u^m, \quad m > 1,} \tag{2.97}$$

describing combustion processes in a nonlinear medium. Consider for (2.97) the Cauchy problem with compactly supported bounded initial data $u_0 \geq 0$. The problem has a unique local in time weak continuous solution that blows up in a finite time $T = T(u_0) < \infty$ and blow-up is *regional*, i.e., occurs on a bounded interval in x; see Remarks. The pressure $v = \frac{m}{m-1}u^{m-1}$ satisfies the equation with quadratic nonlinearities

$$v_t = \mathbf{P}_*(v) \equiv (m-1)vv_{xx} + (v_x)^2 + \mu\, v^2, \quad \text{where } \mu = \tfrac{(m-1)^2}{m}. \tag{2.98}$$

1D invariant subspace and B-concavity. Equation (2.98) admits the following set B of explicit weak blow-up solutions in separate variables

$$V(x, t) = \frac{A}{\lambda - t}\,\theta(x - a), \quad \text{where } A = \tfrac{2m^2}{(m+1)(m-1)^2}. \tag{2.99}$$

Here $\lambda > 0$, $a \in I\!R$ are parameters and $\theta(y)$ is compactly supported,

$$\theta(y) = \cos^2(\alpha y) \quad \text{if } |y| \leq \tfrac{\pi}{2\alpha}, \quad \text{where } \alpha = \tfrac{m-1}{2m},$$

and we set $\theta(y) = 0$ if $|y| \geq \frac{\pi}{2\alpha}$. The value of λ in (2.99) is the blow-up time of the solution $V(x, t)$, which is localized in the positivity domain $\{|x - a| < \frac{\pi}{2\alpha}\}$ for all $t < \lambda$. This solution belongs to the 1D linear subspace $W_1 = \text{Span}\{\theta\}$ invariant under the quadratic operator $\mathbf{P}_*$ in (2.98).

Let us show that the set B is complete. The tangential algebraic system (2.77) takes the form

$$As_0 \cos^2(\alpha(x_0 - a)) = \mu, \quad -A\alpha s_0 \sin(2\alpha(x_0 - a)) = \nu, \tag{2.100}$$

where $s_0 = (\lambda - t_0)^{-1}$. This yields unique values of the parameters

$$\lambda = \lambda(\mu, \nu, t_0) \equiv t_0 + 4\mu A \left(4\mu^2 + \tfrac{\nu^2}{\alpha^2}\right)^{-1} \tag{2.101}$$

and a calculated from (2.100). Obviously, the set (2.99) is continuous and monotone with respect to $s = (\lambda - t)^{-1}$. Hence, we can define the B-concavity (convexity) of the solution $v(x,t)$ relative to the proper set B and Theorem 2.23 is true.

As the last step, we show that the B-concavity is equivalent to an estimate of the second derivative and derive the corresponding sign-invariant.

Proposition 2.31 *If $v(x,t)$ is B-concave for $t \in [0,T)$, then*

$$H(v) \equiv 2vv_{xx} - (v_x)^2 + \tfrac{(m-1)^2}{m^2}\, v^2 \le 0 \quad in \ \{v > 0\}. \tag{2.102}$$

B-convexity corresponds to the sign $\ge$ in (2.102).

Proof. By Definition 2.1, B-concavity implies that $v_{xx} \le V_{xx}$ at a given tangency point. By (2.99),

$$V_{xx} = -2A\alpha^2(\lambda - t)^{-1} \cos(2\alpha(x-a)),$$

and hence by (2.100)

$$V_{xx} \equiv -2\alpha^2 V + \tfrac{1}{2V}\,(V_x)^2,$$

so (2.102) follows. □

Using equation (2.98), estimate (2.102) can be rewritten as follows:

$$H(v) \equiv 2v_t - (m+1)(v_x)^2 - \tfrac{(m+1)(m-1)^2}{m^2}\, v^2 \le 0,$$

which is a nonlinear partial differential inequality of Hamilton-Jacobi type on solutions of quasilinear parabolic equation (2.98).

2D invariant subspace and transversality-type estimates. The 1D subspace W_1 has an invariant extension to the 2D invariant subspace

$$W_2 = \mathrm{Span}\{1, \cos(\lambda x)\}, \quad \text{where } \lambda = 2\alpha.$$

Indeed, for any $V = C_0 + C_1 \cos(\lambda x) \in W_2$, there holds

$$\mathbf{P}_*(V) = \mu C_0^2 + \lambda^2 C_1^2 + [2\mu - \lambda^2(m-1)]C_0C_1\cos(\lambda x) \in W_2.$$

Hence, the parabolic equation (2.98) restricted to the invariant subspace W_2 is a dynamical system for the Fourier coefficients $\{C_0(t),\, C_1(t)\}$,

$$\begin{cases} C_0' = \frac{(m-1)^2}{m^2}\,(mC_0^2 + C_1^2), \\ C_1' = \frac{(m-1)^2(m+1)}{m^2}\, C_0C_1. \end{cases}$$

The 1D subspace W_1 used above is a straight line $\{C_0 = C_1\}$ on the 2D plane W_2. Bearing in mind translations in x, we obtain a three-dimensional set of exact weak solutions $B = \{(V)_+\}$. As in the case of the PME, choosing a proper 2D subset $B' \subset B$, we obtain B-convexity (concavity) relative to B' preserved in time.

We now present a new application of intersection comparison. It is convenient to compare a general solution $v(x,t)$ blowing up at a finite T with exact solutions $B = \{V_*\} \subset B$ having the same blow-up time T. Integrating the DS once and setting for convenience $\varphi = C_1$ and $\psi = C_0/C_1$, we obtain the following one-dimensional family of solutions $B = \{V_*(x-a,t),\ a \in I\!R\}$:

$$V_*(x,t) = \varphi(t)[\psi(t) + \cos(\lambda x)]_+, \tag{2.103}$$

where ψ solves the first-order ODE

$$\psi' = \tfrac{m-1}{m}\, D_0(1-\psi^2)^{-(m-1)/2}, \ \ t \in (0,T); \quad \psi(0) = 1,\ \psi(T) = 1,$$

with the constant

$$D_0 = \tfrac{m}{m-1}\, \tfrac{1}{T}\, B\left(\tfrac{m+1}{2}, \tfrac{1}{2}\right),$$

$B(p,q)$ being Euler's Beta function and

$$\varphi(t) = \tfrac{m}{m-1}\, D_0(1-\psi^2(t)).$$

This is a periodic in x solution. Taking its central one hump profile, we obtain the localized blow-up solution with moving uniformly bounded support

$$\operatorname{supp} V_*(\cdot,t) = \left\{|x| \le g(t) \equiv \tfrac{m}{m-1}\left(\frac{\pi}{2} + \arcsin\psi(t)\right)\right\},$$

so, at the blow-up instance,

$$\operatorname{meas}\, \operatorname{supp} V_*(\cdot,T^-) = 2\pi \tfrac{m}{m-1} \equiv L_s,$$

where L_s is called the *fundamental length* of this blow-up S-regime. Moreover, the initial function for V_* is Dirac's measure, i.e.,

$$V_*(x,0) = E_0\, \delta(x), \quad E_0 = \text{const.} > 0,$$

and the asymptotic behaviour as $t \to 0^+$ is given by the ZKB-solution of the PME. Let us prove the main estimates on the L^∞-norm and the interface propagation following from the transversality analysis.

Proposition 2.32 *For any smooth compactly supported initial data v_0 there holds:*

(i) $\sup_x v(x,t) < \sup_x V_*(x,t) \equiv \varphi(t)[\psi(t)+1]$ *on* $(0,T)$, *and*

(ii) the right-hand interface $s(t) = \sup \operatorname{supp} v(\cdot,t)$ *satisfies*

$$s(t) \le s(0) + g(t) < s(0) + \tfrac{1}{2}\, L_s \quad \textit{for all } \ t \in (0,T).$$

Proof. (i) We work with two intersections as in the B-concavity analysis. The singular initial behaviour of V_* implies that, for arbitrarily small $\varepsilon > 0$, any intersection of $V_*(x-a,\varepsilon)$ and a sufficiently smooth $v(x,\varepsilon)$ in the positivity domain is transversal and

$$\operatorname{Int}(\varepsilon, V_*) \le 2 \implies \operatorname{Int}(t, V_*) \le 2 \ \text{ for all } \ t \in (\varepsilon, T).$$

Since the subset B is composed of solutions on W_2 having the same blow-up time T as $v(x,t)$ (the crucial assumption), the transversality of intersections must remain for all $t < T$. Indeed, if a tangency situation occurs at some $t = t_0 \in (0,T)$, where two intersections disappear, then, by a slight shifting in space and time, we reduce the mutual geometric location of such v and V_* to the ordered one

$$\frac{V_*(x-a,t_0) \le v(x,t_0) \ \text{ in } I\!\!R,}{\operatorname{supp} V_*(x-a,t_0) \subset \operatorname{supp} v(x,t_0).} \tag{2.104}$$

Obviously, by the MP, this contradicts the fact that both solutions v and V have the same blow-up time T. Such a transversality analysis implies the above ordering of L^∞-norms of solutions involved.

(ii) In order to derive an estimate of the interface, we compare $v(x,t)$ with a single solution $V_*(x-a)$, where $a = s(0)$, so that the initial function $V_*(x,0)$ is localized at $x = s(0)$. This implies that (after an arbitrarily small shifting in x and t)

$$\operatorname{Int}(0,V_*) = 1 \implies \operatorname{Int}(t,V_*) \le 1 \quad \text{for } t \in (0,T).$$

This leads to the above order of the interfaces. Otherwise, if this order is wrong at some $t = t_0 \in (0,T)$, we arrive at (2.104), whence the contradiction. □

2.10 B-convexity for equations with exponential nonlinearities

We consider in S the following quasilinear parabolic equations of the heat conduction type:

$$u_t = (e^u)_{xx} \quad \text{and} \quad u_t = (e^u)_{xx} - e^u \tag{2.105}$$

with a given bounded initial function u_0 that is not necessarily nonnegative. By setting $v = e^u$, we arrive at equations with quadratic nonlinearities

$$v_t = vv_{xx} \quad \text{and} \quad v_t = vv_{xx} - v^2. \tag{2.106}$$

We derive generalized B-convexity properties of solutions to both equations (2.106) with respect to proper sets B of compactly supported functions $V(x,t)$. By the parabolic regularity theory, we suppose that $v > 0$ is smooth in S. We prove the B-convexity only (by comparison "from below"). The B-concavity is not true in general. It is important that the exact solutions B belong to different linear subspaces, and this provides us with sign-invariants and estimates of the higher-order derivatives of different structures.

B-convexity on subspace of power functions

For the first equation in (2.106), we construct solutions $B \subset \operatorname{Span}\{1, x^2\}$ of the form

$$V(x,t) = \frac{1}{t+\tau}\,[\lambda - \tfrac{1}{2}\,(x-a)^2]_+, \tag{2.107}$$

where $\tau > 0$ is a fixed constant and $\lambda > 0$, $a \in I\!\!R$, as usual, are arbitrary parameters. This set is proper: solving the corresponding tangential system (2.77) for the parameters of the tangent solution $V \in B$, we arrive at a unique profile (2.107) with the parameters

$$\lambda = \mu(t_0+\tau) + \tfrac{1}{2}\,\nu^2(t_0+\tau)^2, \tag{2.108}$$

and $a = x_0 + \nu(t_0 + \tau)$. Then B-convexity of $v_0(x)$ implies that (cf. (2.79) with the opposite sign)

$$v_{xx}(x,t) \ge V_{xx}(x,t) \equiv -\frac{1}{t+\tau} \quad \text{in } S, \tag{2.109}$$

where $V(x,t)$ is the tangent solution at (x,t). This gives the sign invariant of already known simple structure

$$H(v) = v_{xx} + \frac{1}{t+\tau} \ge 0$$

for $t > 0$ if it holds at $t = 0$.

Consider other B-convexity properties. Assume that $\tau = \tau(\lambda) > 0$ in (2.107) is a smooth function satisfying

$$\tau(0^+) = +\infty, \quad \tau'(\lambda) < 0 \quad \text{for } \lambda > 0. \tag{2.110}$$

Then equation (2.108) has the unique root $\lambda = \lambda(\mu, \nu, t) > 0$, whence the completeness of $B(\tau)$. Continuity and monotonicity follow from (2.107). B-convexity means (2.109) with $\tau = \tau(\lambda(v, v_x, t))$.

In particular, if $\tau(\lambda) = \frac{1}{\lambda}$, we obtain the sign-invariant

$$H(v) \equiv v_{xx} + \frac{\lambda(v, v_x, t)}{1 + t\lambda(v, v_x, t)} \ge 0, \tag{2.111}$$

where $\lambda(\mu, \nu, t)$ is the unique root of the cubic equation

$$\lambda^3 - t(\mu + \tfrac{1}{2}\nu^2 t)\lambda^2 - (\mu + \nu^2 t)\lambda - \tfrac{1}{2}\nu^2 = 0.$$

One can see that this is precisely (2.108) with $\tau(\lambda) = \frac{1}{\lambda}$. Using the first equation (2.106), estimate (2.111) can be rewritten as a Hamilton-Jacobi inequality with the equivalent sign-invariant

$$H(v) = v_t + v\,\frac{\lambda(v, v_x, t)}{1 + t\lambda} \ge 0$$

in S, which is valid on any B-convex solution $v(x,t)$

B-convexity on subspace of hyperbolic functions

For the second equation in (2.106), we take a subset

$$B \subset W_2 = \text{Span}\{1, \cosh x\}$$

that is invariant under the quadratic operator $vv_{xx} - (v_x)^2$ on the right-hand side. B is composed of separate variable solutions

$$V(x,t) = \frac{1}{t+\lambda}\theta(x-a), \tag{2.112}$$

where $\lambda,\ a \in I\!R$ are parameters and $\theta(y)$ satisfies the ODE $\theta'' - \theta + 1 = 0$. Hence,

$$\theta(y) = (1 - A_0 \cosh y)_+\,, \tag{2.113}$$

where $A_0 \in (0,1)$ is fixed, so that function (2.113) is compactly supported. Completeness of B follows from the tangential system (2.77), which for given $t_0 \ge 0$

yields the unique tangent solution $V(x,t)$ with the parameter $\lambda = \lambda(\mu,\nu,t_0)$ satisfying

$$\lambda = \begin{cases} -t_0 + \frac{\mu - \sqrt{\mu^2 - (1-A_0^2)(\mu^2-\nu^2)}}{\mu^2-\nu^2}, \ \mu^2 \neq \nu^2, \\ -t_0 + \frac{1}{2\mu}(1-A_0^2), \ \mu^2 = \nu^2. \end{cases} \tag{2.114}$$

Continuity and monotonicity in λ are obvious. Notice that, if $\lambda < 0$ in (2.114), the corresponding solution (2.112) exists for $t > -\lambda > 0$ and cannot be continued to $t = 0$. Nevertheless, applying the intersection comparison argument, instead of the inequality like (2.19), we have $\text{Int}(-\lambda^+, V) \leq 2$, which immediately follows from (2.112) with $t = -\lambda^+$ and the regularity of the profile $v(x,-\lambda)$. This makes it possible to use a similar technique on the time interval $t \in (-\lambda, t_0]$.

Therefore, B-convexity of the solution implies that

$$v_{xx} \geq V_{xx} = -\frac{A_0}{t+\tau}\cosh(x-a).$$

Hence, by (2.113) and (2.114) we obtain the sign-invariant and the following estimate on the second-order derivative in S:

$$H(v) \equiv v_{xx} - \frac{(v_x)^2 - v\sqrt{v^2-(1-A_0^2)(v^2-(v_x)^2)}}{v - \sqrt{v^2-(1-A_0^2)(v^2-(v_x)^2)}} \geq 0.$$

Note that this awkward looking estimate comes from a very simple subset of separable solutions, so we do not expect simpler estimates to exist.

2.11 Singular parabolic diffusion equations in the radial N-dimensional geometry

In this section we consider radial solutions $u = u(r,t)$ with $r = |x|$ of the N-dimensional quasilinear heat equations and derive transversality and B-concavity (convexity) properties generated by various proper sets of explicit solutions. The main goal is to justify that the same 1D intersection comparison techniques apply in the radial geometry and provide us with similar results on B-concavity and sign-invariants. As usual, we begin with the PME admitting explicit radial ZKB solutions of the instantaneous point-source type.

The porous medium equation in $I\!R^N$

Let $u = u(r,t) \geq 0$ be a weak, continuous, radial solution of the PME in $I\!R^N$

$$u_t = \Delta u^m \equiv \frac{1}{r^{N-1}}(r^{N-1}(u^m)_r)_r \quad \text{in } S = I\!R_+ \times I\!R_+, \tag{2.115}$$

where $m > 1$ and $N > 1$. We assume that radial initial data $u(r,0) = u_0(r) \geq 0$ in $I\!R_+$, $u_0 \not\equiv 0$, are bounded, continuous and compactly supported. Then the solution is smooth in the positivity domain. The pressure function $v = \frac{m}{m-1}u^{m-1}$ satisfies the following equation with quadratic nonlinearities:

$$v_t = \mathbf{A}(v) = (m-1)v\Delta v + |\nabla v|^2 \equiv (m-1)v\frac{1}{r^{N-1}}(r^{N-1}v_r)_r + (v_r)^2 \tag{2.116}$$

with bounded compactly supported v_0 smooth in $\{v_0 > 0\}$. By symmetry, at the origin we have $v_r(0,t) = 0$ provided that $v(0,t) > 0$. We consider a class of nonnegative solutions $\mathcal{M}_- = \{v(r,t)\}$ that are strictly monotone with $r > 0$ in the positivity domain. Namely, we assume that

$$v_0'(r) < 0 \quad \text{in } \{v_0 > 0\}.$$

Then, by the strong MP applied to a parabolic equation for the derivative v_r in $S \cap \{v > 0\}$, for each $v \in \mathcal{M}_-$, there holds

$$v_r(r,t) < 0 \quad \text{in } \{v(r,t) > 0\} \quad \text{for } t > 0.$$

The set B of ZKB solutions of the PME in $I\!R^N$ in terms of the pressure variable has the form

$$V(r,t) = \frac{k}{2(t+\tau)^{N(m-1)k}} \left[\lambda^2 - \frac{r^2}{(t+\tau)^{2k}}\right]_+, \tag{2.117}$$

where $k = \frac{1}{N(m-1)+2}$ and $\tau \in I\!R$, $\lambda > 0$ are arbitrary parameters. Since equations (2.115) and (2.116) with $N > 1$ are not autonomous in r, the translational invariance in this space variable is not available. Therefore in contrast to $N = 1$ in previous sections, the set $B = \{V\}$ of solutions (2.117) is two-dimensional. We will use the same notions and notation as in Section 2.6 and first prove the following crucial property of this set.

Lemma 2.33 *The set B is proper (complete, continuous and monotone).*

Proof. Fix an arbitrary $(r_0, t_0) \in \overline{S} = (0,\infty) \times [0,\infty)$ and $\mu > 0$, $\nu < 0$, and consider the tangential system

$$V(r_0,t_0) = \mu, \quad V_r(r_0,t_0) = \nu. \tag{2.118}$$

The profile $V(r,t)$ is tangent to the solutions $v(r,t)$ at the point (x_0,t_0) if $v(r_0,t_0) = \mu$ and $v_r(r_0,t_0) = \nu$. The completeness of B means that such a tangent solution is defined for all $t \in [0,t_0]$ and it is unique. In view of (2.117) we have from (2.118) the following algebraic equations for the parameters λ and τ:

$$\frac{k}{2(t_0+\tau)^{N(m-1)k}} \left[\lambda^2 - \frac{r_0^2}{(t_0+\tau)^{2k}}\right] = \mu, \quad \frac{kr_0}{t_0+\tau} = |\nu|.$$

Substituting

$$t_0 + \tau = \frac{kr_0}{|\nu|} > 0 \tag{2.119}$$

from the second equation into the first one, we obtain the unique value of the second parameter

$$\lambda^2 = \frac{2\mu}{k}\left(\frac{kr_0}{|\nu|}\right)^{N(m-1)k} + \left(\frac{|\nu|}{k}\right)^{2k} r_0^{2(1-k)}. \tag{2.120}$$

Thus (2.119) and (2.120) imply that a tangent solution exists and it is unique. The continuity of V with respect to the parameters in the C^1-norm on compact subsets of the domain of positivity is obvious. The monotonicity property of B according to Section 2.6 in this case reduces to checking that solutions (2.117) are monotone with respect to λ and $V \to \infty$ as $\lambda \to \infty$ and $V \to 0$ as $\lambda \to 0$ uniformly on any compact subset in (r, t).

Therefore, the proof of Lemma 2.33 will be completed if the above unique tangent solution V can be continued to $t = 0$.

Proposition 2.34 *Let $(r_0, t_0) \in \overline{S}$ and $V = V(r, t; \lambda, \tau) \in B$ be a unique tangent solution to $v(r, t) \in \mathcal{M}_-$ at (r_0, t_0). Then*

$$\tau = \tau(\nu, r_0, t_0) > 0. \tag{2.121}$$

Proof. We argue by contradiction. Assume that $\tau \leq 0$. Since, by construction, the explicit solution

$$U(x, t) \equiv [\tfrac{m-1}{m} V(r, t)]^{\frac{1}{m-1}}$$

satisfies, in the sense of distributions,

$$U(x, t) \to c\,\delta(x) \quad \text{as } t \to -\tau^+$$

and $u_0 \in \mathcal{M}_-$ is assumed to be smooth enough, we conclude that the number of intersections of the solutions $V(r, t)$ and $v(r, t)$ satisfies $\text{Int}(-\tau^+, V) = 1$. By the first Sturm Theorem

$$\text{Int}(t, V) \leq 1 \quad \text{for all } t > -\tau. \tag{2.122}$$

Since, by construction of the tangent solution, we have that $V(r, t_0)$ with $t_0 > -\tau$ is tangent to the profile $v(r, t_0)$ at the point $r = r_0$, by the strong MP, we conclude that there exists a small $\varepsilon > 0$ such that

$$\text{Int}(t_0 - \varepsilon, V) \geq 2$$

contradicting (2.122). □

It is curious that using (2.121) in (2.119) means that $t_0 < kr_0/|\nu|$ and since, by construction, $\nu = V_r = v_r$, we arrive at the following gradient estimate proved by the transversality comparison with solutions having Dirac's measure as initial data.

Corollary 2.35 *Let $v \in \mathcal{M}_-$. Then*

$$|v_r| < k\,\frac{r}{t} \quad \text{in } S. \tag{2.123}$$

By Lemma 2.33 we introduce B-concavity (convexity) with respect to the set B for solutions $v \in \mathcal{M}_-$ exactly as in 1D.

Definition 2.2 For a given $t_0 \geq 0$, a solution $v(r, t) \in C^1(\{v > 0\})$ is said to be concave relative to set B or *B-concave*, if, for any $r_0 > 0$ from $\{v(r, t_0) > 0\}$, there holds

$$v(r, t_0) \leq V(r, t_0) \quad \text{in } \mathbb{R}_+, \tag{2.124}$$

where $V \in B$ is a unique tangent solution at the point (r_0, t_0). The sign $\geq$ in (2.124) corresponds to the *B-convexity*.

We now state the main result meaning that under the above hypotheses, the B-concavity (convexity) is preserved in time. The proof is exactly the same as that of Theorem 2.23 in 1D.

Theorem 2.36 *If $v_0 \in \mathcal{M}_-$ is B-concave (resp. B-convex), then $v(\cdot, t) \in \mathcal{M}_-$ is B-concave (resp. B-convex) for $t > 0$.*

Finally we derive the corresponding B-concavity estimate and the sign-invariant.

Proposition 2.37 *If a solution $v(\cdot, t) \in \mathcal{M}_-$ is B-concave, then*

$$H(v) \equiv \Delta_0 v = r \left(\frac{1}{r} v_r \right)_r \equiv v_{rr} - \frac{1}{r} v_r \leq 0 \tag{2.125}$$

(Δ_0 is the radial Laplace operator in zero dimension). The B-convexity estimate corresponds to the sign $\geq$.

Proof. By the definition, B-concavity implies that, at any tangency point,

$$v_{rr} \leq V_{rr} = -\frac{k}{t+\tau}, \tag{2.126}$$

where V is the tangent solution. Substituting $t + \tau$ from (2.119) with $t_0 = t$ and $\nu = v_r$, we obtain (2.125). □

Plugging the gradient bound (2.123) $v_r > -kr/t$ into the B-convexity estimate $v_{rr} \geq v_r/r$, we arrive at

$$v_{rr} \geq \frac{1}{r} v_r > -\frac{k}{t} \quad \text{or} \quad \Delta_r v > -\frac{kN}{t},$$

showing once more that the Aronson-Bénilan semiconvexity estimate is of the geometric nature.

Thanks to (2.116), estimate (2.125) transforms into a nonlinear first-order inequality of Hamilton-Jacobi type

$$\tilde{H}(v) \equiv v_t - \tfrac{(m-1)N}{r} v v_r - (v_r)^2 \leq 0.$$

Since $v \in \mathcal{M}_-$ and hence $v_r \leq 0$ a.e., (2.125) implies a usual concavity in r

$$v_{rr} \leq \frac{1}{r} v_r \leq 0 \quad \text{in } \{v > 0\} \quad \text{for } t > 0. \tag{2.127}$$

However, in contrast with the 1D problem, where both concavity and convexity are known to be preserved in time (Section 2.2), inequality (2.127) is guaranteed by a stronger estimate on the initial function

$$v_0'' \leq \frac{1}{r} v_0' \quad \text{in } \{v_0 > 0\}.$$

We emphasize that with respect to the operator in (2.125) both signs have been proved to be preserved in time, so that this invariant B-concavity (convexity) is naturally generated by the quasilinear PDE. For $r \gg 1$, when the spatial operator in (2.116) is "almost" one-dimensional on regular solutions, inequality (2.125) is "almost" equivalent to the usual concavity $v_{rr} \leq 0$. Respectively, for the B-convexity, we have the opposite sign $\geq$ in (2.125), and this implies that the convexity preserved in time cannot be proved by comparison with a set B of particular explicit solutions that are strictly positively concave.

The heat equation in $I\!R^N$

Consider the linear case $m = 1$, i.e., the heat equation

$$u_t = \Delta u$$

with radial solution $u = u(r, t)$. The pressure $v = \log u$ satisfies

$$v_t = \Delta v + |\nabla v|^2$$

with smooth initial data v_0. It is easy to see that the corresponding set B of fundamental solutions

$$V(r,t) = \lambda - \tfrac{N}{2} \log(t + \tau) - \tfrac{1}{4} \frac{r^2}{t + \tau}, \quad (\lambda, \tau) \in I\!R^2,$$

satisfies Lemma 2.33 and that Proposition 2.34 is also valid (after, if necessary, a small shifting of the origin in time). Therefore, using B-concavity (convexity) relative to B, we arrive at Theorem 2.36 and at the same sign-invariant (2.125).

The fast diffusion equation in $I\!R^N$

Consider now equation (2.115) with $0 < m < 1$. The pressure $v = -\frac{1-m}{m} u^{m-1} < 0$ solves the equation

$$v_t = \mathbf{P}_*(v) \equiv -(1-m)v\Delta v + |\nabla v|^2 \tag{2.128}$$

with smooth negative v_0. In order to construct the set B of explicit solutions, we note that the quadratic operator $\mathbf{P}_*$ admits the invariant subspace $W_2 = \text{Span}\{1, r^2\}$ satisfying $\mathbf{P}_*(W_2) \subseteq W_2$. Substituting

$$V(r,t) = [-C_0(t) - C_1(t) r^2]_- \tag{2.129}$$

into (2.128) yields a nonlinear dynamical system for the nonnegative coefficients $\{C_0, C_1\}$:

$$\begin{cases} C_0' = 2N(1-m) C_0 C_1, \\ C_1' = 2(N - 2 - Nm) C_1^2. \end{cases} \tag{2.130}$$

(i) If $m > \frac{N-2}{N}$, then

$$\begin{aligned} C_0(t) &= \tfrac{1}{2} k\lambda (t+\tau)^{N(1-m)k}, \\ C_1(t) &= \tfrac{1}{2} \tfrac{k}{t+\tau}, \end{aligned} \tag{2.131}$$

where $k > 0$ is as given in (2.117). Structurally, this is essentially the same solution as the ZKB one (2.117) and can be found in the famous L.D. Landay–E.M. Lifschitz's book [238].

(ii) If $m = \frac{N-2}{N}$ $(N \geq 3)$, then (2.130) becomes simpler and

$$\begin{aligned} C_0(t) &= \lambda e^{4\tau t}, \\ C_1(t) &= \tau, \end{aligned} \tag{2.132}$$

where $\tau > 0$.

(iii) If $m < \frac{N-2}{N}$ $(N \geq 3)$, then (2.130) yields solutions V that blow-up in finite time $\tau > 0$,

$$\begin{aligned} C_0(t) &= \lambda \tfrac{1}{2} \, |k| \, (\tau - t)^{-N(1-m)|k|}, \\ C_1(t) &= \tfrac{1}{2} \tfrac{|k|}{\tau - t}. \end{aligned} \tag{2.133}$$

By $\mathcal{M}_+$ we denote the class of negative solutions satisfying $v_r < 0$ for $r > 0$, and, without loss of generality, we assume that

$$|v_r| < 2 \, \frac{|v|}{r}.$$

If this is not true, then explicit solutions $V(r,t)$ vanishing at some finite r should be taken into account (this implies blow-up for u). It is easy to see that Lemma 2.33 is valid for the above set B. Indeed, the tangential system (2.118) with $\mu < 0$, $\nu < 0$, for the tangent solution, (2.129) yields

$$\begin{aligned} C_0 + C_1 r_0^2 &= |\mu|, \\ 2C_1 r_0 &= |\nu|, \end{aligned}$$

whence the following initial conditions for the DS (2.130):

$$\begin{aligned} C_0(t_0) &= |\mu| - \tfrac{1}{2} \, |\nu| r_0 > 0, \\ C_1(t_0) &= \tfrac{1}{2} \, \tfrac{|\nu|}{r_0} > 0. \end{aligned}$$

Therefore, a unique V exists. In the cases (2.132) and (2.133) V is automatically continued to $t = 0$. In the case (2.131) the analysis of $\tau \le 0$ is the same as in the proof of Proposition 2.34. Continuity and monotonicity of B in λ are easily checked. Thus, in the given class of solutions, we arrive at Theorem 2.36 for the B-convexity property. Proposition 2.37 and estimate (2.125) are valid.

Inequality (2.125) for B-concave solutions depends only on the internal structure ("curvature") of the linear invariant subspace $W_2 = \text{Span}\{1, r^2\}$. Indeed, in view of the general representation of tangent solutions (2.129), we have $v_{rr} \le V_{rr} = -2C_1$. Since, at any tangency point, $v_r = -2C_1 r$, i.e., $C_1 = -v_r/2r$, substituting C_1 into the above inequality yields (2.125). We will consider other examples explaining sign-invariants via sets $B \subset W_2$.

Example 2.1: exact solutions on a 3D linear invariant subspace. A wider set D_3 of particular solutions of the fast diffusion equation exists in the case $m = \frac{N-2}{N+2} \in (0,1)$ for $N \ge 3$, i.e., $m = 1/p_S$, where p_S is the *critical Sobolev exponent* for the operator $\Delta u + u^p$. Then the operator $\mathbf{P}_0$ of the pressure equation

$$v_t = \mathbf{P}_0 \equiv -\gamma_0 v \Delta v + |\nabla v|^2, \quad \gamma_0 = \tfrac{4}{N+2}, \tag{2.134}$$

admits the three-dimensional invariant subspace $W_3 = \text{Span}\{1, r^2, r^4\}$. Therefore, substituting

$$V(r,t) = [-C_0(t) - C_1(t) r^2 - C_2(t) r^4]_- \tag{2.135}$$

into (2.134) yields the following dynamical system for the coefficients:

$$\begin{cases} C_0' = 2N\gamma_0 C_0 C_1, \\ C_1' = 16 C_0 C_2 + \gamma_0 (N-2) C_1^2, \\ C_2' = 2N\gamma_0 C_1 C_2, \end{cases}$$

which can be solved via quadratures. The set D_3 is three-dimensional. Similar to our analysis in Section 2.7, choosing suitable proper subsets $B \subset D_3$, one can define different invariant B-concavity properties and the corresponding sign-invariants.

PME with lower-order terms

Consider the PME with extra reaction-diffusion operators

$$\boxed{u_t = \Delta u^m - \tfrac{\gamma}{m}\, u^{2-m} + \tfrac{\alpha}{m-1}\, u, \quad m > 1,}$$

where $\gamma \geq 0$ and α are fixed constants. Then the pressure $v = \frac{m}{m-1}\, u^{m-1}$ satisfies

$$v_t = \mathbf{P}_*(v) \equiv (m-1)v\Delta v + |\nabla v|^2 - \gamma + \alpha v \quad \text{in } \{v > 0\}. \tag{2.136}$$

Operator $\mathbf{P}_*$ admits the invariant subspace $W_2 = \text{Span}\{1, r^2\}$ and hence there exists the set B of compactly supported solutions

$$V(r,t) = [C_0(t) - C_1(t) r^2]_+,$$

where the coefficients solve the dynamical system (cf. (2.130))

$$\begin{cases} C_0' = -2\rho C_0 C_1 + \alpha C_0 - \gamma, \\ C_1' = -2(\rho+2)C_1^2 + \alpha C_1, \end{cases}$$

with $\rho = N(m-1)$. Assuming that $v \in \mathcal{M}_-$, we have that Lemma 2.33 is valid. The tangential system (2.118) yields $C_0(t_0) = \mu - \nu r_0/2 > 0$ and $C_1(t_0) = -\nu/2r_0 > 0$, i.e., a unique trajectory of the above DS. Then Theorem 2.36 holds, so that such a subset $B \subseteq W_2$ generates the sign-invariant (2.125). The quasilinear equation (cf. (2.134))

$$v_t = -\gamma_0 v\Delta v + |\nabla v|^2 - \gamma + \alpha v$$

($\gamma_0 = \frac{4}{N+2}$) admits explicit solutions (2.135) on the subspace

$$W_3 = \text{Span}\{1, r^2, r^4\},$$

which can be used in the B-convexity (concavity) analysis.

Equation with the p-Laplace operator in $\mathbb{R}^N$

Consider the p-Laplacian equation with lower-order terms

$$\boxed{u_t = \text{div}(|\nabla u|^\sigma \nabla u) - \gamma u^{1/(\sigma+1)} + \alpha u,}$$

where $\sigma > 0$ and $\gamma \geq 0$ are fixed constants. The pressure is

$$v = \tfrac{1+\sigma}{\sigma}\, u^{\frac{\sigma}{1+\sigma}}.$$

Nonnegative radial solutions $v(r,t)$ satisfy, in $\{v > 0\}$, the equation

$$v_t = \mathbf{P}_*(v) \equiv |v_r|^2[\sigma v\, v_{rr} + (v_r)^2 + \tfrac{N-1}{r}\, \delta\, v v_r] - \gamma + \alpha\delta v, \tag{2.137}$$

where $\delta = \frac{\sigma}{\sigma+1}$. Operator $\mathbf{P}_*$ is known to admit the invariant subspace

$$W_2 = \text{Span}\{1, r^\rho\}, \quad \rho = \tfrac{2+\sigma}{1+\sigma}, \tag{2.138}$$

and the expansion coefficients of solutions on W_2

$$V(r,t) = [C_0(t) - C_1(t)r^\rho]_+ \tag{2.139}$$

satisfy the two-dimensional dynamical system

$$\begin{cases} C_0' = -N\delta\rho^{\sigma+1}|C_1|^\sigma C_1 C_0 + \alpha\delta C_0 - \gamma, \\ C_1' = -\rho^{\sigma+1}(\delta + N\rho)|C_1|^{\sigma+2} + \alpha\delta C_1. \end{cases}$$

It is the PDE (2.137) on W_2. Assume that $v_0(r) \in \mathcal{M}_-$ is nonnegative, C^1, and compactly supported. Let $u(r,t) \geq 0$ be a unique weak compactly supported solution of the Cauchy problem. By approximation and by the strong MP, we conclude that $u(r,t) \in \mathcal{M}_-$ for all $t \in (0,T)$, where $T \in (0,\infty]$ is the extinction time (if $\gamma = 0$ then $T = \infty$).

Therefore we introduce B-concavity (convexity) relative to the proper set B satisfying necessary properties on the class $\mathcal{M}_-$. Hence, Theorem 2.36 is valid. Finally, for the given invariant subspace W_2, instead of (2.125), we have the following B-concavity estimate.

Proposition 2.38 *If $v(\cdot,r) \in \mathcal{M}_-$ is B-concave relative to $B \subseteq W_2$, then*

$$H(v) \equiv v_{rr} - \tfrac{1}{(1+\sigma)r} v_r \leq 0 \quad in \;\; S \cap \{v > 0\}. \tag{2.140}$$

Proof. It follows from (2.139) that, at any tangency point,

$$v_{rr} \leq V_{rr} = -\rho(\rho-1)C_1 r^{\rho-2}$$

and $v_r = -r(\rho C_1 r^{\rho-2})$. □

B-convexity is true with the sign $\geq$ in (2.140). The sign-invariant (2.140) is naturally associated with the radial Laplacian $H = \Delta_\delta$ with dimension $\delta = \frac{\sigma}{1+\sigma}$.

Linear sign-invariant associated with invariant subspace

Let us describe a typical dependence of the B-convexity property relative to a proper subset B upon the structure of the corresponding invariant subspace W_2. Then we need quadratic operators possessing more arbitrary linear invariant subspaces. Consider reaction-diffusion equation with exponential nonlinearities

$$\boxed{u_t = \Delta e^u - \gamma e^u + \delta + \varepsilon e^{-u}, \quad \gamma > 0.}$$

Setting $e^u = v > 0$ yields the following equation with quadratic nonlinearities:

$$v_t = \mathbf{P}_0(v) \equiv v\Delta v - \gamma v^2 + \delta v + \varepsilon.$$

The quadratic operator $\mathbf{P}_0$ admits the invariant subspace

$$W_2 = \text{Span}\{1, \varphi(r)\}, \tag{2.141}$$

where $\varphi(r) > 0$ is a strictly increasing solution of the elliptic equation

$$\Delta\varphi - \gamma\varphi = 0 \quad \text{in } I\!R^N.$$

Finally, we arrive at a proper set B of compactly supported solutions

$$V(r,t) = [C_0(t) - C_1(t)\varphi(r)]_+,$$

where the coefficients solve the dynamical system

$$\begin{cases} C_0' = -\gamma C_0^2 + \delta C_0 + \varepsilon, \\ C_1' = -\gamma C_0 C_1 + \delta C_1. \end{cases}$$

Let us derive the sign-invariant corresponding to B-convex solutions $u \in \mathcal{M}_-$. Since, at any tangency point,

$$v_r = V_r = -C_1 \varphi'(r),$$

we have that

$$v_{rr} \geq V_{rr} = -C_1 \varphi''(r) \equiv v_r \, \frac{\varphi''(r)}{\varphi'(r)}.$$

Hence, the linear sign-invariant takes the form

$$H_\varphi(v) \equiv \varphi'(r) \left(\frac{v_r}{\varphi'(r)} \right)_r \geq 0 \quad \text{in } S. \tag{2.142}$$

This formula describes the structure of the above differential inequalities via B-convexity.

2.12 On general B-concavity via solutions on linear invariant subspaces

Finally, we discuss some general aspects of the method of tangent solutions, B-concavity and sign-invariants on invariant subspaces. We recall that the above characteristic subsets B were always composed of particular solutions V, belonging to finite-dimensional linear subspaces invariant under some quasilinear operators. Let us briefly list the main steps of such a geometric analysis.

(i) Let X be an infinite-dimensional space of smooth functions, e.g., $X = \{f \in C^1(I\!R)\}$, and $v = v(\cdot, t) \in X$ for $t > 0$ be a smooth solution of a nonlinear evolution equation

$$v_t = \mathbf{F}(v) \equiv F(x, t, v, v_x, v_{xx}) \quad \text{in } S, \quad v(0) = v_0 \in X. \tag{2.143}$$

(ii) Assume that the nonlinear operator $\mathbf{F}$ admits a certain $(s+1)$-dimensional $(s \geq 1)$ linear invariant subspace

$$W_{s+1} = \text{Span}\{f_0, ..., f_s\} \subset X,$$

so

$$\mathbf{F}(W_{s+1}) \subseteq W_{s+1}.$$

This implies that, for any

$$V = C_0(t) f_0 + ... + C_s(t) f_s \in W_{s+1}, \tag{2.144}$$

there holds

$$\mathbf{F}(V) = F_0(C_0, ..., C_s) f_0 + ... + F_s(C_0, ..., C_s) f_s \in W_{s+1}. \tag{2.145}$$

Equation (2.143) restricted to W_{s+1} is equivalent to the $(s+1)$-dimensional DS for $t > 0$

$$\begin{cases} C_0' = F_0(C_0, ..., C_s), \\ \quad \cdots \qquad \cdots \\ C_s' = F_s(C_0, ..., C_s). \end{cases}$$

(iii) We choose a proper (complete, continuous and monotone) two-dimensional subset

$$B = \{V_{\lambda,a}\} \subseteq W_{s+1}$$

indexed by two parameters $(\lambda, a) \in \mathbb{R}^2$.

(iv) Using the completeness of B, for a fixed $t_0 \geq 0$, we can define the *tangent set* with respect to the corresponding profile $\{v(x, t_0),\ x \in \mathbb{R}\}$ consisting of tangent solutions crossing the v-curve in the hodograph plane

$$\mathcal{T}_B(t_0, v) = \{V_{\lambda,a},\ (\lambda, a) \in \mathcal{M}(t_0, v)\}, \tag{2.146}$$

where the set $\mathcal{M}(t_0, v) \subseteq \mathbb{R}^2$ of possible values of the parameters (λ, a) is given by

$$V_{\lambda,a}(x_0, t_0) = v(x, t_0),\ (V_{\lambda,a})_x(x_0, t_0) = v_x(x, t_0), \quad x_0 \in \mathbb{R}. \tag{2.147}$$

A function $V_{\lambda,a} \in \mathcal{T}_B(t_0)$ is a tangent solution at a point (x_0, t_0), so $\mathcal{M}(t_0, v)$ includes all pairs (λ, a) corresponding to tangent V-curves from B in the hodograph plane of $v(x, t_0)$. The algebraic system (2.147) is the equation of the tangent set $\mathcal{T}_B(t_0, v)$. It can be treated as a nonlinear algebraic equation of the change of variables from the original *geometric hodograph coordinates* (v, v_x) of the solution v at a point (x_0, t_0), to the *parametric coordinates* (λ, a) of tangent solutions $V_{\lambda,a}$ at (x_0, t_0) on the complete set $B \subset W_{s+1}$. In the previous sections we introduced a number of examples, in which algebraic equations of the type (2.147) were solved explicitly. This is not possible in general; see nonlinear equations (2.86) or (2.95).

(v) Finally we apply the intersection comparison technique, which is assumed to be valid for the 1D equation (2.143). We assume that the number of intersections $\mathrm{Int}(t, v, V_{\lambda,a})$ is such that

$$\mathrm{Int}(0, v, V_{\lambda,a}) \leq 2 \quad \text{for all } V_{\lambda,a} \in B.$$

Then $\mathrm{Int}(t, v, V_{\lambda,a}) \leq 2$ for $t > 0$ by the Sturm Theorem. As we have seen, this implies that, in the hodograph plane, the tangential V-curves intersect the v-curve $\Gamma(t, v)$ exactly once. Then we arrive at the estimate on the solution for $t_0 > 0$ by means of the corresponding tangent solutions: for any $V \in \mathcal{T}_B(t_0, v)$,

$$\text{either} \quad v(x, t_0) \geq V(x, t_0) \quad \text{or} \quad v(x, t_0) \leq V(x, t_0) \ \text{ in } \mathbb{R}. \tag{2.148}$$

If the functions v and V are C^2 smooth, then (2.148) implies that

$$\text{either} \quad v_{xx} \geq (V_{\lambda,a})_{xx} \ \text{ in } \mathbb{R} \quad (B\text{-convexity}), \tag{2.149}$$

$$\text{or} \quad v_{xx} \leq (V_{\lambda,a})_{xx} \quad \text{in } \mathbb{R} \quad (B\text{-concavity}). \tag{2.150}$$

Finally, we calculate $(V_{\lambda,a})_{xx}$ in terms of the lower-order derivatives $(V_{\lambda,a})_x$ and $V_{\lambda,a}$, say,

$$(V_{\lambda,a})_{xx} = \psi_B(x, t, V_{\lambda,a}, (V_{\lambda,a})_x),$$

where ψ_B depends on the peculiar structure of the set B under consideration. Using the tangential system (2.147), we then obtain the corresponding sign-invariant

$$H_B(v) = v_{xx} - \psi_B(x, t, v, v_x),$$

which preserves its signs: $H_B(v) \geq 0$ for the B-convex solutions and $H_B(v) \leq 0$ for the B-concave ones. Calculating v_{xx} from the parabolic equation (2.143),

$$v_{xx} = F^{-1}(x, t, v, v_x, v_t),$$

the B-concavity reduces to the Hamilton-Jacobi inequality

$$F^{-1}(x, t, v, v_x, v_t) \geq \psi_B(x, t, v, v_x).$$

Finally, H_B is also a zero-invariant: $H_B(V) \equiv 0$ on B.

The B-concavity property of solutions of nonlinear parabolic equations and the corresponding sign-invariant follow from the geometric structure of the proper set B of particular solutions generating a dense ordered subset in the hodograph plane $\{v, v_x\}$. Moreover, in the intersection comparison analysis the only thing arising from the parabolic PDE is the fact that the number of intersections of two arbitrary solutions does not increase with time (the first Sturm Theorem). Postulating such a property, at this stage we can exclude the PDE and consider the evolution with time of a function $v(x,t)$ in a complete, continuous, monotone set of given functions $B = \{V(x,t)\}$. Then the same conclusions follow and the B-concavity is the invariant of such evolution (flow).

The choice of proper subsets B, which before was naturally restricted to some nice explicit profiles, is wide. Since the space of general solutions $D_{\mathbf{F}}$ of a parabolic PDE (2.143) is infinite-dimensional, there exists an infinite-dimensional family $\{B\}$ of different two-dimensional proper subsets $B \subset D_{\mathbf{F}}$ of particular solutions. It generates an infinite number of the corresponding invariant B-concavity/convexity properties and the sign and zero-invariants H_B.

Remarks and comments on the literature

Many results are taken from [164] and [137]. The questions of spatial concavity or of the logarithmic concavity of solutions of linear, semilinear or quasilinear elliptic and parabolic equations form an important part in the theory of nonlinear PDEs and are popular in the mathematical literature. We refer to papers [6], [62], [75], [77], [227], [218], [249], where further references can be found.

As we have seen, special aspects of the concavity/convexity analysis occur for the quasilinear singular equations with continuous weak solutions having finite interfaces. This is an essential feature of the present geometric analysis. For regular equations admitting stationary linear in x solutions, the result on concavity (convexity) $u_{xx} \leq 0$ ($u_{xx} \geq 0$) being preserved in time was proved by W. Walter, [344, p. 209]. We study a particular one-dimensional aspect of the general concavity problem and the main feature is that we use intersection comparison with some subsets of exact solutions B. This leads to optimal notions of concavity/convexity generated by the parabolic equations under consideration.

§ **2.1.** An estimate of Δv such as (2.5) is also true in $I\!R^N$ [31]. The concavity result (2.8) was first proved in [47] by using an iterative technique of semigroups corresponding to splitting the nonlinear operator on the right-hand side of (2.4) and using a Trotter-Lie formula. The eventual concavity result for the PME was first proved in [32]. There are a number of results on the asymptotic convexity (i.e., after taking $t \to \infty$) for nonnegative solutions of parabolic equations, but eventual concavity is a stronger property. An extended list of references on asymptotic behaviour for different types of parabolic reaction-diffusion equations can be found in [213], [317], [43], [306], [170].

§ **2.2.** Concerning the filtration equation, we should mention the celebrated paper [277], where a complete theory of weak solutions for such equations was established. In particular, finite speed of propagation of perturbations was first proved there by comparison with TW solutions with finite interfaces. For more general functions φ, the existence of a solution is proved by the method of implicit discretization in time using the theory of nonlinear semigroups. We obtain in this way a function $u \in C([0,\infty) : L^1(I\!R))$ that solves the problem in a *mild* sense. Such a mild solution is unique. Under certain conditions on φ the mild solution is a weak solution, and if φ is a C^2 function with $\varphi' > 0$, then u is a classical solution. Moreover, the solution depends continuously on φ in suitable topologies (u in $C([0,\infty) : L^1(I\!R))$, φ in the topology of graphs). This fact makes it possible to assume in the proofs that φ is smooth. We consider in all cases the Cauchy problem with nonnegative solutions and data. For details on the above theory we refer to [45]. Optimal conditions on the initial data that guarantee at least local solvability were obtained in [208] and [46]; see the survey [213]. We observe again that the proof of concavity in [47] for the PME case needed a delicate semigroup argument and cannot be applied to equation (2.11) with the rather arbitrary nonlinear coefficient F of the second-order derivative.

In the proof of Lemma 2.4 we use a typical idea of parameters perturbations of profiles in B, transforming an inflection into at least three points of intersections, which was used in [150] and [154]. It is important that such a perturbation analysis works for equations with coefficients that are not very smooth (but proper continuous solutions still can be constructed by monotone regular approximations), where general results on zero-set analysis [11], [234] are not straightforward. We observe that the second Sturm Theorem in finite regularity classes (unlike the C^∞ case) does not specify the number of zeros lost at a formation of a multiple zero in general. Recall that under necessary regularity assumptions (no analyticity is necessary, just those, which are necessary to apply the strong MP), the results in [11], [234] show that the intersections are isolated points for $t > 0$ even if the number of intersections is initially infinite. In the proof of Lemma 2.4 we do not need this since we deal with at most two intersections for $t > 0$ for any $V \in B$.

The idea of Bernstein estimates of the first-order derivative v_x in terms of v by auxiliary functions originates from the classical work of S.N. Bernstein (1938) [50]; see references in the survey by O.A. Oleinik and S.N. Kruzhkov [278]. For the PME, the Bernstein estimate on v_x was proved by D.G. Aronson (1969) [29]. This Bernstein technique was extended to a general filtration equation by A.S. Kalashnikov (1974) [212]. Proposition 2.1 shows that the first Bernstein gradient estimate has a purely geometric nature, and later on, in Chapter 7, we show that such a transversality estimate is available for general fully nonlinear singular parabolic PDEs.

§ **2.3.** The first result in this direction was proved for the PME in [32]. That proof uses detailed information on the large-time profile of the solution, which is known for the particular PME but not for the general filtration equation. An "eventual monotonicity" (single maximum) result [304] for the PME in $I\!R^N \times I\!R_+$ with compactly supported data also uses the convergence to a ZKB solution. The properties of interfaces we use in the proof of Lemma 2.10 are well established; see references in [213]. Self-similar solutions of the form $f(x/\sqrt{t})$ of general filtration equations have been studied in detail in the 1970s; see [35], though such a study began in the 1940s, [294]. In the proof of Proposition 2.12 we use Bernstein estimates first proved in [29] for the PME and in [212] for the general filtration equation; see references in [213].

We do not discuss another phenomenon of the *eventual spatial monotonicity* of solutions e.g., approaching blow-up or extinction time. The intersection comparison analysis is similar and easier than that for the eventual concavity. We refer to [273] (global solutions), [162] (finite-time extinction), [166] (blow-up), [305] (monotonicity near interfaces).

§ **2.4.** If $\gamma < 0$, the Cauchy problem for (2.56) is not well-posed, there exist non-unique TW solutions [126] and there is no uniqueness of nonnegative solutions [283] in general (nevertheless, there exists a unique proper maximal solution; see Chapter 7 for general results).

§ **2.5.** About typical regularity for equations with p-Laplacian operators; see [213], [99]. A justification of intersection comparison for gradient-dependent diffusion terms can be found in [153].

§ **2.6.** The concept of B-concavity relative to a given proper B was introduced in [164] and different extensions were performed in [137] and [139]. Sign-invariants were introduced in [134], where earlier related references are available. This concept summarized various ideas and techniques of deriving parabolic estimates by means of the MP in parabolic problems, especially those including singular blow-up phenomena. Equation $H(v) = 0$, where H is supposed to be a zero-invariant of the parabolic equations preserving its zero values on parabolic orbits, is a differential constraint to be in *involution* with the PDE. It turns out [134] that, for parabolic PDEs, each zero-invariant H is also a sign-invariant, a property proved by the MP. As usual for differential constraints, the problem of finding zero-invariant operators H reduces to a complicated PDE, which often is not easier than the original evolution equation. We show how to determine zero and sign-invariants following from the geometry of proper sets B. The backward problem of finding, firstly, sign-invariants and, secondly, determining the set B of solutions generating H via B-concavity, is also of importance and can lead to new solutions of classical quasilinear parabolic PDEs; see examples in [134], [157], [158]. The invariant subspace (2.81) and the corresponding exact solutions were found in [155] for a class of parabolic equations with gradient-dependent operators.

The ZKB-solution of the PME was first constructed by Ya.B. Zel'dovich and A.S. Kompaneetz (1950) [347] in dimensions $N = 1$ and 3 and extended to arbitrary N by G.I. Barenblatt (1952) [39].

In this chapter we do not consider the eventual B-convexity, which can be studied as in Section 2.3; see [137]. Such a property was first used in [150], [154] in order to prove the monotonicity with time of large nonnegative solutions of a general quasilinear equation with source,

$$\boxed{u_t = (\varphi(u))_{xx} + \psi(u) \quad \text{in } S,}$$

where $\psi(u) > 0$ for $u > 0$. The proof was based on intersection comparison with a set B of stationary solutions $\{U\}$ satisfying the ODE

$$(\varphi(U))_{xx} + \psi(U) = 0.$$

Introducing the concept of the *tangent stationary solution*, it was proved that, under a natural regularity of the initial data, B-convexity of solutions $u(x,t)$ with respect $B = \{U\}$ implies that

$$u_{xx} \geq U_{xx}$$

at any tangency point where $u \gg 1$, i.e., the B-convexity property occurs eventually when the solutions become sufficiently large. As a straightforward consequence, this yields that

$$u_t > 0$$

on any subset of S, where $u \gg 1$. A similar monotonicity of large solutions is true in $I\!\!R^N$ for radial solutions in the subcritical case of nonlinearities. For the PME with source

$$\boxed{u_t = \Delta u^m + u^p, \quad m \geq 1, \;\; p > 0,}$$

such eventual time monotonicity in $\{u \gg 1\}$ holds in the subcritical Sobolev range $p < p_S = m\,\frac{N+2}{N-2}$ (and this is not always true if $p > p_S$), [149].

§ **2.7.** It is easy to find other examples. For instance, equation (2.82) admits also the following simple set of blow-up separate-variables solutions:

$$V(x,t) = \tfrac{1}{2(m+1)}\,\frac{1}{\lambda - t}\{[\pm(x-a)]_+\}^2, \;\; \lambda > 0.$$

If we add the flat solutions $V \equiv \mu$ with arbitrary $\mu > 0$, we arrive at a complete, continuous and monotone set B (monotone in the sense of (2.78) with $s = \frac{1}{\lambda - t}$). Then B-concavity means that

$$v_{xx} \leq V_{xx} = \tfrac{1}{m+1}\,\frac{1}{\lambda - t}$$

and since, at any tangency point,

$$\frac{1}{\lambda - t} = \tfrac{m+1}{2}\,\frac{(v_x)^2}{v},$$

we obtain that the invariant B-concavity property means that

$$H(v) \equiv 2vv_{xx} - (v_x)^2 \leq 0$$

in the positivity domain, where H is the corresponding sign-invariant.

§ **2.8.** There are many papers on the log-concavity for the heat equation in $I\!\!R^N$; see references at the beginning of Remarks. We use a different approach and treat other concavity aspects.

§ **2.9, 2.10.** The localized blow-up solution (2.99) with steady interfaces was constructed in [307]. It was a cornerstone of a general theory of blow-up localization of heat and combustion process in nonlinear media; see [306, Chapters 3 and 4].

Explicit blow-up solutions with prescribed blow-up time T on the 2D invariant subspace W_2 give sharp estimates on localization domains and the behaviour of the L^∞ norm of general solutions [132], [133]. Proof of localization is easier by comparison with the separable standing-wave solution (2.99), $\lambda = T$, in W_1; see [306, Chapter 4]. Namely, any solution of (2.97) blowing up as $t \to T^- < \infty$ with compactly supported initial data u_0, is localized, and if $\operatorname{supp} u_0$ is connected, there holds

$$\operatorname{supp} u(\cdot, t) \le \operatorname{supp} u_0 + 2\pi \tfrac{m}{m-1} \quad \text{for all } \; t \in (0, T).$$

This localization result is rather general and is true for any autonomous equation

$$v_t = F(v, v_x, v_{xx})$$

with finite blow-up propagation. Namely, if there exists a single localized blow-up solution $V(x,t)$ such that

$$L_s = \sup_{t \in (0,T)} \; \text{meas supp}\, V(\cdot, t) < \infty,$$

then, for a bounded initial function u_0 compactly supported on an interval, the corresponding solution is localized and

$$\text{meas supp}\, v(\cdot, t) \le \text{meas supp}\, v_0 + 2L_s \quad \text{for all } t \in (0, T).$$

Other examples of invariant subspaces for nonlinear operators generating concavity and transversality properties can be found in [131], [136], [155].

§ **2.11.** Estimate (2.123) can be obtained from the general semiconvexity estimate by D.G. Aronson and Ph. Bénilan [31], and here we emphasize its geometric nature. An extended three-dimensional invariant subspace (2.135) and the corresponding exact solutions were found in [136]. This shows that three-dimensional families of exact solutions (which, as in Section 2.7, guarantee an infinite number of different B-concavity properties preserved in time) do exist in radial geometry. Explicit solutions of equations such as (2.136) on W_2 were considered in [131] and [136]. The invariant subspace (2.138) for (2.137) was found in [155]. Notice that the corresponding DS cannot be solved explicitly. Subspaces (2.141) for quadratic operators were considered in [131].

§ **2.12.** The concept of B-concavity (convexity) of general solutions crucially depends on the properties of a proper set of particular solutions $B = \{V\}$ under consideration generated by the PDE. In the examples considered, the B-concavity (convexity) in its turn depends on the structure of invariant subspaces admitted by nonlinear differential operators.

CHAPTER 3

B-Concavity and Transversality on Nonlinear Subsets for Quasilinear Heat Equations

This chapter extends the results of the previous one in the following principal direction. Instead of dealing with subsets of exact solutions on linear subspaces invariant under nonlinear operators of the PDEs (such exact solutions are easier), we consider "nonlinear" subsets B of particular solutions. We then loose some simple mathematics and structural aspects inherited from the linear subspaces. Though linear subspaces W_k for quadratic differential operators and the corresponding reductions of the parabolic PDEs on W_k to finite-dimensional dynamical systems occur rather often for the PME-type operators, as a rule, quasilinear operators with general nonlinearities do not admit linear invariant subspaces. Hence, TW solutions existing for arbitrary autonomous PDEs do not exhibit an easy piecewise linear structure, as it used to be in Chapter 2. Such TW sets B are essentially "nonlinear", as well as other subsets to be treated in this chapter.

Then we use the concepts of concavity, B-concavity and sign-invariants relative to such nonlinear sets, and show how to prove estimates of general solutions for a number of quasilinear heat equations. We also comment on transversality, leading to first-order Bernstein estimates to be considered in Chapter 7 in greater generality.

3.1 Introduction: Basic equations and concavity estimates

As in the previous chapter, we begin with the standard spatial concavity properties for compactly supported nonnegative solutions $u = u(x,t)$ of one-dimensional quasilinear degenerate parabolic equations. We establish that the concavity is preserved in time for general solutions provided that there exists a *complete* set of convex particular solutions with vanishing convexity at finite interfaces. As a basic example, we consider a general quasilinear heat equations with absorption of the form

$$\boxed{u_t = (\varphi(u))_{xx} - \psi(u) \quad \text{in } S = I\!R \times I\!R_+.} \tag{3.1}$$

Here $\varphi \in C^2(I\!R_+) \cap C^1([0,\infty))$, $\psi \in C^2(I\!R_+) \cap C([0,\infty))$ are given functions satisfying $\varphi'(u) > 0$, $\psi(u) > 0$ for $u > 0$, $\varphi'(0) = 0$. A typical example of (3.1) arising in mechanics and plasma physics is the PME with absorption

$$\boxed{u_t = (u^m)_{xx} - \gamma\, u^p,} \tag{3.2}$$

where $m > 1, p > 1$ and $\gamma \geq 0$. We also consider the gradient diffusivity equation (the p-Laplacian equation) with absorption

$$\boxed{u_t = (|u_x|^\sigma u_x)_x - \gamma\, u^p, \quad \sigma > 0, \;\; p > 1, \;\; \gamma \geq 0,} \tag{3.3}$$

which is known in the theory of turbulent diffusion, non-Newtonian, dilatant liquids and combustion of solid fuels.

We study concavity and B-concavity (convexity) properties in a general situation where the set B of exact solutions is not induced by a linear invariant subspace since (3.1) contains rather arbitrary nonlinearities. Then the operator in the pressure equation does not admit any invariant subspace of the type

$$W_2 = \text{Span}\{1, x\} \quad \text{or} \quad W_3 = \text{Span}\{1, x, x^2\},$$

unlike many equations in Chapter 2. Nevertheless, we show that existence of suitable nonlinear sets B of particular solutions can guarantee the usual concavity property. In Section 3.2 we prove that the positive concavity is preserved in time for the pressure corresponding to (3.1). Equation (3.3) is treated in Section 3.3. The proof is based on a local property of "small" convexity of the travelling wave (TW) solutions. On the other hand, in Section 3.4 we prove that the above set B of TWs is proper and derive estimates of B-concavity preserved in time. In Section 3.5 we consider the general filtration equation

$$u_t = (\varphi(u))_{xx} \tag{3.4}$$

with bounded compactly supported initial data $u_0(x)$. The pressure for the PME $v = \frac{m}{m-1}\, u^{m-1}$ is known to satisfy the semiconvexity Aronson-Bénilan estimate

$$v_{xx} \geq -\tfrac{1}{m+1}\, \frac{1}{t + \tau} \tag{3.5}$$

with the constant $\tau = -[(m+1)\,(\inf v_0'')]^{-1} \geq 0$. In Section 2.6 we showed that this estimate and a similar semiconvexity one are equivalent to the invariant B-concavity and B-convexity properties induced by a subset of explicit ZKB-solutions belonging to the three-dimensional invariant subspace W_3.

The general filtration equation (3.4) with arbitrary φ does not admit such exceptional explicit solutions, so that the only known result (Section 2.2) is the positive convexity (concavity) via intersection comparison with the piecewise linear TW solutions on the subspace W_2. This estimate is not optimal in the class of bounded compactly supported solutions since the set $B = \{V\}$ is composed of *unbounded* TW solutions $V(x,t)$. We show that for equation (3.4), there exists a better B-concavity (semiconcavity) estimate of the form (cf. (3.5) with the opposite sign)

$$v_{xx} \leq \frac{g(v, v_x, t)}{t + \tau}$$

derived by a subset B of standard exact self-similar solutions $V(x,t) = \theta(x/\sqrt{t+\tau})$. The coefficient $g(v, v_x, t)$ on the right-hand side is uniquely determined from the tangential system but cannot be expressed explicitly to derive a better and simpler estimate (it does not exist for general φ).

We also study the case where the set B is not complete and show that this yields *conditional* B-concavity (convexity) properties, Section 3.6. Here we use various subsets of stationary solutions of (3.2). Finally, in Section 3.7 we introduce the property of *partial eventual* B-concavity relative to complete sets B of stationary solutions.

Combining these results and those from Chapter 2 implies that a proper (complete, continuous and monotone) two-dimensional set B of particular solutions of a quasilinear parabolic equation (3.1) generates a certain *geometric evolution* with invariant properties of B-concavity (convexity) established by intersection comparison by the first Sturm Theorem. For classes of solutions that are not B-concave (convex) initially these properties can occur *eventually* in time and also *partially* in space. Various sets of explicit solutions can describe a sharp asymptotic geometry of wide classes of more general solutions. Since our geometric approach is based on intersection comparison with different subsets of explicit solutions, the results can be treated from the point of view of a certain "superposition principle" for nonlinear parabolic equations, where we translate some common properties of proper sets of particular (explicit) solutions to general solutions.

We treat the *forward* problem of determining invariant B-concavity properties via a given proper set B of particular solutions. These properties define a differential operator H_B preserving signs on the orbits of the PDE (3.1), which is a *sign-invariant* of the equation. The *backward* problem of reconstructing the set of particular solutions B via a given sign-invariant, is also a fruitful approach often leading to new exact solutions of quasilinear parabolic equations; see Remarks.

3.2 Local concavity analysis via travelling wave solutions

In this section we study the concavity property of nonnegative solutions of (3.1). We assume that $\varphi(u)$ satisfies the condition of finite speed of propagation

$$\Phi(u) = \int_0^u \frac{\mathrm{d}\varphi(s)}{s} < \infty \quad \text{for } u > 0. \tag{3.6}$$

Then the pressure

$$v = \Phi(u) \geq 0$$

satisfies the equation

$$\boxed{v_t = \mathbf{P}_*(v) \equiv F(v)v_{xx} + (v_x)^2 - q(v),} \tag{3.7}$$

where $F(v) = \varphi'(u)$, and

$$q(v) = \tfrac{1}{u}\,\psi(u)\varphi'(u) \quad \text{with } u = \Phi^{-1}(v).$$

Let us impose the necessary restriction on the coefficients reflecting properties of equation (3.2). We assume that

$$q'(s) \geq 0, \;\; q''(s) \geq 0 \text{ and } F''(s) \leq 0 \quad \text{for } s > 0, \tag{3.8}$$

$$\tfrac{q(s)}{F(s)} \to 0 \quad \text{as } s \to 0, \quad q(0) = F(0) = 0 \text{ and } q(\infty) = \infty, \tag{3.9}$$

$$\int^\infty \frac{q(s)}{F(s)}\,\mathrm{d}s = \infty. \tag{3.10}$$

For convenience, we also impose the condition, under which equation (3.1) does not admit nontrivial stationary solutions with a finite interface, so that the space localization of compactly supported solutions does not occur. This implies that the stationary equation

$$(\varphi(S))'' - \psi(S) = 0 \quad \text{for } x > 0, \;\; S(0) = (\varphi(S))'(0) = 0,$$

does not have a nontrivial solution $S(x) > 0$ for $x > 0$. Integrating this ODE, one obtains the criterion of nonexistence of space localization

$$\int_0 \left(\int_0^s \psi(\Phi^{-1}(\eta))\, \mathrm{d}\eta \right)^{-1/2} \mathrm{d}s = \infty. \tag{3.11}$$

All the above assumptions are valid for equation (3.2) with $p \geq m$. The case $1 < p < m$ will be discussed later.

The set of travelling wave solutions

Equation (3.7) admits the two-dimensional set $B = \{V(x,t)\}$ of TWs

$$V(x,t) = f(\xi), \quad \xi = x - \lambda t + a, \tag{3.12}$$

where $\lambda, a \in I\!R$ are the parameters and the function $f(\xi) \geq 0$ is assumed to have a finite interface, say, at $\xi = 0$. Substituting (3.12) into (3.7) yields the ODE

$$F(f) f'' + (f')^2 + \lambda f' - q(f) = 0 \tag{3.13}$$

with the conditions at the origin

$$f(0) = 0, \quad [\varphi(\Phi^{-1}(f))]'(0) = 0. \tag{3.14}$$

The second condition is the continuity of the heat flux $-(\varphi(V))_x$ of the solution $V(x,t)$ with the zero continuation either for $x > 0$ or $x < 0$. Setting $f' = P$ in (3.13) and hence

$$f'' = P \frac{\mathrm{d}P}{\mathrm{d}f}, \tag{3.15}$$

we obtain the first-order ODE

$$\frac{\mathrm{d}P}{\mathrm{d}f} = \frac{1}{PF(f)} [q(f) - P(P + \lambda)]. \tag{3.16}$$

Assuming that $\lambda > 0$, we consider equation (3.16) in the quarter plane $\{f > 0, \; P < 0\}$. It is easy to check that both conditions (3.14) are valid for the trajectory $P = P_*(f)$ satisfying

$$P_*(0) = -\lambda \, . \tag{3.17}$$

Notice that $P_*(f) \equiv -\lambda$ is a solution if $q \equiv 0$. Under hypothesis (3.11), the stationary problem with $\lambda = 0$ does not have a nontrivial solution.

Proposition 3.1 *Under the above hypotheses on the coefficients, for any $\lambda > 0$, problem* (3.16), (3.17) *has a unique solution $P_*(f)$ such that*

$$P_*(f) < -\lambda, \quad \frac{\mathrm{d}P_*}{\mathrm{d}f}(f) < 0 \quad \textit{for } f > 0, \tag{3.18}$$

$$P_*(f) \to -\infty \quad as \ f \to \infty, \tag{3.19}$$

$$P_*(f) \to 0 \quad as \ \lambda \to 0 \ uniformly \ on \ compact \ subsets. \tag{3.20}$$

Proof. It follows from (3.6) that

$$\int_0^1 \frac{ds}{F(s)} = \infty.$$

By the standard local analysis of the ODE (3.16), (3.17) in a small neighbourhood of the point $(0, -\lambda)$ by reducing it to the integral equation and applying Banach's Fixed Point Theorem, we have that a unique solution exists and for small $f > 0$ it has the expansion

$$P_*(f) = -\lambda - \frac{1}{\lambda} G(f) \int_0^f \frac{q(z)\,dz}{F(z)G(z)} + ..., \quad G(z) = \exp\left\{\int_z^1 \frac{ds}{F(s)}\right\}.$$

The trajectory P_* is the separatrix on the phase-plane. Monotonicity of $P_*(f)$ follows from the monotonicity of the function $q(s)$. Indeed, we have from (3.16) that the zero-cline

$$P_0(f) = -\tfrac{1}{2}\left[\lambda + \sqrt{\lambda^2 + 4q(f)}\right] < -\lambda \quad \text{for } f > 0$$

is monotone decreasing, and since by construction $P_*(f) > P_0(f)$ for $f > 0$ and $P_*(0) = P_0(0) = -\lambda$, we arrive at (3.18). Unboundedness (3.19) easily follows from (3.10), and (3.20) is a straightforward consequence of (3.11). □

In terms of the TW profile $f(\xi)$ we have the following properties.

Corollary 3.2 *For any* $\lambda > 0$, *problem* (3.13), (3.14) *has a unique solution* $f(\xi) > 0$ *for* $\xi < 0$ *that is a positively convex function satisfying*

$$f''_{\xi\xi}(\xi) \equiv V_{xx} \geq 0, \quad whenever \ f > 0, \tag{3.21}$$

$$f''_{\xi\xi} \to 0 \quad as \ f \to 0, \tag{3.22}$$

$$f'_\xi(0) = -\lambda, \ f'_\xi(\xi) \to -\infty \quad as \ f \to \infty. \tag{3.23}$$

Proof. (3.21) follows from (3.15) and (3.18). Since $P_*(f) < -\lambda$, (3.16) implies

$$f''_{\xi\xi} \equiv P\frac{dP}{df} \leq \frac{q(f)}{F(f)} \to 0 \quad \text{as } f \to 0 \tag{3.24}$$

by assumption (3.9). □

If $f(\xi)$ becomes unbounded and blows up as $\xi \to \bar{\xi}^+$, for convenience, we set $f = \infty$ for $\xi \leq \bar{\xi}$. Finally, we notice that for $\lambda < 0$ by means of the invariant transformation in (3.16) $\lambda \mapsto -\lambda$, $P \mapsto -P$, we obtain a similar trajectory $-P_*(f) > 0$ such that properties (3.21) and (3.22) hold.

Main result on concavity

We consider the Cauchy problem for equation (3.7) with initial data $v(x,0) = v_0(x) \geq 0$ in $I\!R$, where $v_0 \in C(I\!R)$ is bounded and smooth whenever positive. A global weak solution $v(x,t)$ exists, unique and smooth in $\{v > 0\}$.

Theorem 3.3 *Let the assumptions on the coefficients of equation* (3.7) *be valid. Let the support of* v_0 *be bounded, connected and* v_0 *be positively concave, i.e.,*

$$v_0'' \le 0 \quad in \ \{v_0 > 0\}. \tag{3.25}$$

Then for every $t > 0$, $v(\cdot, t)$ *is also positively concave,*

$$v_{xx}(x,t) \le 0 \quad in \ \{v(\cdot,t) > 0\}. \tag{3.26}$$

Below, we present a proof based on the direct intersection comparison of the weak solution $v(x,t)$ and the TWs $V(x,t) = f(\xi)$. Such an analysis reveals some interesting properties of interfaces. At the end of the section we show that the concavity result can be obtained by approximations where we deal with classical smooth solutions v_ε and $B_\varepsilon = \{V_\varepsilon\}$ of the regularized equations and can avoid a delicate interface analysis.

Proof. By (3.25) the initial function $v_0(x)$ is bell-shaped and does not have strict minimum points in the positivity domain. Then $v(x,t)$ is bell-shaped for all $t > 0$, which is proved by the MP applied to the linear parabolic equation for the derivative v_x plus a suitable approximation. Let $l(t) = (\zeta_-(t), \zeta_+(t))$ be the support of $v(x,t)$. The second derivative $z = v_{xx}$ satisfies in $\Omega = l(t) \times I\!R_+$ the following linear parabolic equation:

$$z_t = F\, z_{xx} + 2v_x(F'+1)z_x + [(F'+2)z + F''(v_x)^2 - q']z - q''(v_x)^2. \tag{3.27}$$

In view of (3.8), the parabolic inequality holds in $\Omega \cap \{z > 0\}$

$$z_t \le F z_{xx} + 2v_x(F'+1)z_x + (F'+2)z^2. \tag{3.28}$$

In what follows we do not study this inequality in the maximal generality. For instance, the hypothesis $F'' \le 0$ in (3.8) can be weakened if we take into account a typical Bernstein estimate on the first derivative $|v_x| \le C$ in order to estimate the term $F''(v_x)^2 z$ on the right-hand side of (3.27).

Fix an arbitrary $T > 0$ such that $v(x,t) \not\equiv 0$ for all $t \in (0,T)$. By the MP we have that $v \le M_0 = \sup v_0$ in $\Omega_T = \Omega \cap \{t < T\}$. Consider the main problem concerning the behaviour of $z(x,t) = v_{xx}(x,t)$ near interfaces. Fix a small $\varepsilon > 0$ and a $\delta = \delta(\varepsilon) > 0$ such that

$$v \le \varepsilon \quad \text{in } S_\varepsilon = \Omega_T \setminus \Omega_T^\delta, \ \Omega_T^\delta = (\zeta_-(t)+\delta, \zeta_+(t)-\delta) \times (0,T).$$

We first prove that v_{xx} cannot be large and positive near the interfaces. The analysis is based on a local study of the convexity of TWs.

Proposition 3.4 *There holds*

$$v_{xx} \le m_0(\varepsilon) = \sup_{s \in (0,\varepsilon)} \frac{q(s)}{F(s)} \quad in \ S_\varepsilon. \tag{3.29}$$

Proof. We use the same intersection technique as in the interface analysis in Section 2.3. For a fixed $t \in (0,T)$, we denote by $\mathrm{Int}(t,V)$ the number of intersections of the solution $v(x,t)$ and a given TW solution $V(x,t) = f(\xi) \in B$. It follows from (3.25) and (3.21) that

$$\mathrm{Int}(0,V) \le 2 \quad \text{for any } V \in B_*. \tag{3.30}$$

$\text{Int}(t, V)$ does not increase with time by the Sturm Theorem so that

$$\text{Int}(t, V) \le 2 \quad \text{for } t \ge 0 \text{ and any, } V \in B_* \,. \tag{3.31}$$

We prove (3.29) arguing by contradiction. Assume that there exists a point $(x_1, t_1) \in S_\varepsilon$ such that

$$v_{xx}(x_1, t_1) > m_0(\varepsilon). \tag{3.32}$$

Set $v(x_1, t_1) = \mu \in (0, \varepsilon)$ and, without loss of generality, assume that $\nu = v_x(x_1, t_1) < 0$. By the MP, in view of assumption (3.25), for any $t \in (0, T)$, the function $v(x, t)$ has a unique positive maximum where $v_x = 0$ and $v_{xx} \le 0$, so we may assume that $\nu \neq 0$, say, $\nu < 0$. Let $V(x, t) \in B$ be the *tangent solution* at the point (x_1, t_1), i.e.,

$$V(x_1, t_1) = \mu, \quad V_x(x_1, t_1) = \nu. \tag{3.33}$$

It follows from (3.20) and (3.23) that such a solution

$$V(x, t) = f(x - \lambda t + a)$$

exists. From (3.24), (3.30) and (3.32) we then conclude that at the point (x_1, t_1),

$$v_{xx} > V_{xx} \quad (\text{ and } v = V, \; v_x = V_x). \tag{3.34}$$

Therefore, since $v(x, t_1)$ is uniformly bounded and $V(x, t_1)$ is unbounded as $x \to -\infty$ by (3.23), we have that there exists at least one intersection in $\{x < x_1\}$ of the profiles $v(x, t_1)$ and $V(x, t_1)$ so that

$$\text{Int}(t_1, V) \ge 1. \tag{3.35}$$

We now introduce another solution $V'(x, t)$ that is $V(x, t)$ slightly shifted in the positive x-direction. Namely, we choose $a' < a, \; a - a' \ll 1$, and set

$$V'(x, t) = f(x - \lambda t + a').$$

One can see from (3.34) and (3.35) that

$$\text{Int}(t_1, V') \ge 3$$

by construction. This contradicts (3.31) completing the proof of Proposition 3.4.
□

End of proof of Theorem 3.3. The rest of the proof is based on a standard application of the MP. We introduce a supersolution $\bar{z}(t)$ of the equation (3.27) in Ω_T^δ. Set

$$b_F = \sup_{s \in (0, M_0)} F'(s),$$

and let $\bar{z}(t)$ be the unique solution of the ODE

$$\bar{z}' = (b_F + 2)\bar{z}^2 \quad \text{for } t > 0, \; \bar{z}(0) = m_0(\varepsilon).$$

From (3.9) we deduce that for any small $\varepsilon > 0$, the function $\bar{z}(t)$ exists on $[0, T]$ and $\bar{z}(t) \to 0$ as $\varepsilon \to 0$ uniformly on $[0, T]$. It follows from (3.28) and (3.8) that $\bar{z}(t)$ is a supersolution of equation (3.27), and (3.25) implies that $z(x, 0) \le \bar{z}(0)$

in the interval $(\zeta_-(0)+\delta, \zeta_+(0)-\delta)$. Moreover, $z \le \bar z$ on the lateral boundary of Ω_T^δ by (3.29). Therefore by comparison

$$z \equiv v_{xx} \le \bar z(t) \quad \text{in } \Omega_T^\delta.$$

Passing to the limit $\varepsilon \to 0$ and using the above estimates, we conclude that $z \le 0$ in Ω_T. Since T is arbitrary, we arrive at (3.26). □

Let us comment on some extensions. If (3.11) does not hold, then (3.20) is not valid. Therefore, a tangent solution $V \in B$ does not exist provided that $|\nu|$ is small enough. This implies that the subset B *is not complete*. In this case a complete subset $B^s = B \cup B_s$ is formed by adding to B a subset of *stationary* solutions $B_s = \{V, \ \lambda = 0\}$ including the stationary solution $V_0 = \Phi(S)$ and all others that are strictly positive. It is easily seen that on the $\{f, P\}$-plane with $\lambda = 0$, the subset B_s consists of the nontrivial separatrix P_* (corresponding to V_0) and all other orbits above the separatrix corresponding to convex stationary solutions. One can check that (3.33) defines a tangent solution $V \in B^s$ for all $\mu > 0, \nu \in I\!R$, so that B^s is a complete functional subset. The end of the proof of Theorem 3.3 remains the same. A more detailed analysis is presented in Section 3.7.

Combining these results with the MP applied to the parabolic differential inequality (3.28), under necessary natural assumptions we can state the following general conclusion about concavity of weak solutions with finite interfaces: *initially positively concave solutions preserve concavity if the equation admits a complete set* $B = \{V\}$ *of particular solutions that are convex in the positivity domain and* $V_{xx} = 0$ *on the free boundaries.*

Concavity result by approximation. As usual, the above concavity proof needs an extra analysis of the behaviour of weak solutions near finite singular interfaces. If we are not specially interested in such properties of interfaces, we can deal with positive smooth approximations of the solutions involved, $v_\varepsilon \to v$, $V_\varepsilon \to V$ (V_ε are not TWs any more) as $\varepsilon \to 0$ satisfying regularized uniformly parabolic equations. Following the lines of the approach explained in Section 2.2, we prove the invariance of the concavity property by passing to the limit. There are two main ingredients of this geometric analysis:

(i) Proposition 2.6 on completeness of the solution set $B_\varepsilon = \{V_\varepsilon\}$ approximating the TWs, and

(ii) a correct choice of approximation of initial data $v_{0\varepsilon}(x)$, under which the main intersection hypothesis $\text{Int}(0, v_\varepsilon, V_\varepsilon) \le 2$ holds for any $V_\varepsilon \in B_\varepsilon$.

Then passing to the limit $\varepsilon \to 0$ we obtain that the same is true for weak solutions v and Vs, and this is equivalent to the positive concavity.

On Bernstein estimates by transversality. As in the case of the filtration equation in Section 2.2, the transversality condition with a subset of "steep" TWs $B_1 = \{f_\lambda\} \subset B$ gives a gradient bound near the interface. For sufficiently smooth compactly supported initial data v_0 such that $\text{Int}(0, v, V) \le 1$ for any $V \in B_1$, the derivative satisfies

$$|v_x| \le G(v) \equiv |f'(\xi)|\,|_{f=v} \quad \text{for all } v \approx 0.$$

It is a sharp bound since equality is attained on the TWs $f(x - \lambda t)$. The *gradient function* $G(v)$ on the right-hand side is not a constant as for the filtration equation, and is obtained from the ODE for the TW profiles f. We postpone this general geometric Bernstein-type approach until Chapter 7, where we derive such bounds for fully nonlinear equations. Instead, we now turn our attention to more interesting tangential constructions, leading to B-concavity properties and finally to estimates of derivatives v_{xx}.

3.3 Concavity for the p-Laplacian equation with absorption

Let us show that a similar concavity result is valid for equation (3.3) with $p \geq 1 + \sigma$. The pressure function

$$v = u^{\frac{\sigma}{\sigma+1}}$$

satisfies the equation

$$\boxed{v_t = \mathbf{P}_1(v) \equiv \beta^\sigma |v_x|^\sigma [(\sigma+1) v\, v_{xx} + \beta (v_x)^2] - \delta v^\alpha,} \tag{3.36}$$

where $\beta = \frac{\sigma+1}{\sigma}$, $\delta = \frac{\gamma}{\beta}$ and $\alpha = \frac{p(\sigma+1)-1}{\sigma} > 1$. By the classical regularity results, we may assume that for $t > 0$ the derivative v_x exists in $\{v > 0\}$. Since in general v_{xx} does not exist at a positive maxima or minima, we use the notion of concavity and convexity for C^1-functions. Namely, a nonnegative function $v_0 \in C^1$ is positively *concave* (*convex*) if for any $x_0 \in \{v_0 > 0\}$ there holds

$$v_0(x) \leq [l(x)]_+ \quad (\text{resp. } \geq [l(x)]_+) \;\text{ in } I\!R,$$

where $l(x)$ is the tangent straight line to the function $v_0(x)$ at $x = x_0$. Then we can state the concavity result similar to Theorem 3.3.

Theorem 3.5 *Let $p \geq 1+\sigma$ and the initial data $v_0 \in C^1$ have bounded, connected support. If v_0 is positively concave, then the solution $v(x,t)$ of (3.36) is positively concave for $t > 0$.*

Proof. As in the proof of Theorem 3.3 we need to check certain properties of the set B of TWs (3.12). Here f satisfies the equation

$$\beta^\sigma |f'|^\sigma [(\sigma+1) f\, f'' + \beta (f')^2] + \lambda f' - \delta f^\alpha = 0$$

with the boundary conditions $f(0) = 0$, $(f^{1/\sigma} f')(0) = 0$. Setting $f' = P$ yields the equation (cf. (3.16))

$$\frac{\mathrm{d}P}{\mathrm{d}f} = \frac{1}{(\sigma+1) f P} \left[\frac{1}{\beta^\sigma |P|^\sigma} (\delta f^\alpha - \lambda P) - \beta P^2 \right]. \tag{3.37}$$

Let $\lambda > 0$. It is easy to check that in the quarter plane $\{f > 0, P < 0\}$, equation (3.37) admits a unique trajectory $P = P_*(f)$ such that

$$P_*(f) \to -\rho = -\beta^{-1} \lambda^{1/(\sigma+1)} \quad \text{as } f \to 0, \quad \frac{\mathrm{d}P_*}{\mathrm{d}f} < 0 \;\text{ for } f > 0, \tag{3.38}$$

$P_*(f) < -\rho$ and (3.19) holds. In addition, (3.20) is valid for $p \geq \sigma + 1$. In particular, by (3.15) all solutions $V(x,t) \equiv f(\xi)$ are convex functions and (3.21) holds. Using (3.38) in (3.37), we deduce that as $f \to 0$ (cf. (3.24)),

$$f''_{\xi\xi} \equiv P\frac{\mathrm{d}P}{\mathrm{d}f} \le \frac{\gamma}{(\sigma+1)\beta^{\sigma+1}}\frac{1}{|P|^{\sigma}}f^{\alpha-1} \le \frac{2\gamma}{(\sigma+1)\beta^{\sigma+1}\rho^{\sigma}}f^{\alpha-1} \to 0\,. \quad (3.39)$$

The end of the proof is based on the intersection comparison argument. The only difference is as follows. Since equation (3.36) is doubly degenerate, the second derivative $z = v_{xx}$ satisfies the equation, which is uniformly parabolic on any compact subset of $S \cap \{v_x \neq 0\}$, where $v \in C^{\infty}$. Therefore we first study the behaviour of z near the curves, where $v_x = 0$. Then as in the proof of Proposition 3.4, z is small positive via (3.39) in a δ-neighbourhood of the interfaces. Near other curves with $v_x = 0$ inside the support of $v(x,t)$, from the concavity of v_0 we easily deduce that z takes nonpositive values since $v(x,t)$ attains maximum in x on these curves. In fact, these are technical difficulties only, and we get rid of these applying intersection comparison with smooth approximations of solutions v_ε and $B_\varepsilon = \{V_\varepsilon\}$ and passing to the limit $\varepsilon \to 0$. □

In the case $1 < p < \sigma + 1$, Theorem 3.5 is still true. The proof uses comparison with the set B^s constructed in exactly the same way as explained at the end of the previous section.

3.4 B-concavity relative to travelling waves

We return to equation (3.1) in the pressure form

$$\boxed{v_t = \mathbf{P}_*(v) \equiv F(v)v_{xx} + (v_x)^2 - q(v) \quad \text{in } S,} \tag{3.40}$$

and assume that all hypotheses on the coefficients of (3.40) stated in Section 3.2 are valid. This makes it possible to introduce the set B of TWs that are convex in x functions with finite interfaces.

Denote

$$B_* = B \cup B_\infty,$$

where $B_\infty = \{V = V(t) \, : \, V' = -q(V),\, t > 0\}$ is the set of nontrivial solutions of (3.40) independent of x. B_* is a complete supplement of B, see below. Functions $V \in B_\infty$ can be treated as a limit case $\lambda = \infty$ of the TW solutions $V = f(\eta)$ with the variable $\eta = \frac{x}{\lambda} - t$ and the set B_* is the closure of B by continuity. Obviously, B_∞ is a one-dimensional family $\{\tilde V(t+\alpha),\ \ \alpha \in I\!R\}$, where $\tilde V(t)$ is any fixed solution of the ODE.

The functions $V \in B_*$ are not piecewise linear in x and are supposed to define the corresponding B-concavity of solutions $v(x,t)$ relative the functional set B_*. We assume that the initial function $v_0 \ge 0$ is bounded, compactly supported and smooth whenever positive.

Proper set of TW solutions

Let us recall that, in order to use the intersection comparison argument as in Section 2.6, the set B_* has to satisfy the following three properties.

(i) *Completeness*: for an arbitrary $(x_0, t_0) \in \bar S$ such that $v(x_0, t_0) > 0$, there exists a unique function $V \in B_*$ (the tangent solution to $v(x,t)$ at the point (x_0, t_0)) such that

$$V(x_0,t_0) = v(x_0,t_0) = \mu > 0, \quad V_x(x_0,t_0) = v_x(x_0,t_0) = \nu, \tag{3.41}$$

and $V(x,t)$ exists for all $t \in [0, t_0]$.

Proposition 3.6 *B_* is complete.*

Proof. Existence of such a solution $V \equiv f(\xi) \in B$ satisfying (3.41) with $\nu \neq 0$ follows from the properties of the function $f(\xi)$ with different $\lambda \neq 0$ stated in Section 3.2; see Proposition 3.1 and Corollary 3.2. $V(x,t)$ exists for all $t \in I\!R$. If $\nu = 0$, then the tangent solution $V(t) \in B_\infty$ is defined from the ODE

$$V' = -q(V) \quad \text{for} \quad t > t_0$$

with $V(t_0) = \mu$. By the strong MP in the positivity domain we have that the number of intersections satisfies

$$\text{Int}(t, V) \geq 2 \qquad \text{for all } t \in [0, t_0),$$

so that the tangent solutions $V(t)$ is bounded and can be extended on the interval $[0, t_0]$.

Let us prove uniqueness. For $\nu = 0$ it is obvious. Assume that $\nu < 0$, then $\lambda > 0$. Let us show that the derivative $V_x \equiv f'_\xi$ on a fixed level $V \equiv f = \mu$ is strictly increasing with $\lambda > 0$. Fix an arbitrary $\lambda_2 > \lambda_1$ and let $P_*^{(1)}(f)$ and $P_*^{(2)}(f)$ be the corresponding solutions of equation (3.16) with the boundary conditions

$$P_*^{(\nu)}(0) = -\lambda_\nu \quad \text{for} \quad \nu = 1, 2.$$

Then $P_*^{(2)}(f) < P_*^{(1)}(f)$ for small $f > 0$. Assume that the trajectory $P_*^{(1)}$ intersects for the first time $P_*^{(2)}$ at some $\bar{f} > 0$. Then at $f = \bar{f}$,

$$P_*^{(1)} = P_*^{(2)} < 0, \quad \frac{\mathrm{d}P_*^{(1)}}{\mathrm{d}f} \leq \frac{\mathrm{d}P_*^{(2)}}{\mathrm{d}f}. \tag{3.42}$$

It follows from equation (3.16) for both functions $P_*^{(\nu)}$ that at $f = \bar{f}$,

$$F(\bar{f})P_*^{(1)}(\bar{f}) \left(\frac{\mathrm{d}P_*^{(1)}}{\mathrm{d}f} - \frac{\mathrm{d}P_*^{(2)}}{\mathrm{d}f} \right) = P_*^{(1)}(\bar{f})(\lambda_2 - \lambda_1) < 0$$

contradicting (3.42). Therefore

$$P_*(f) \quad \text{is strictly decreasing with } \lambda > 0 \text{ for all } f \geq 0.$$

This yields uniqueness of the tangent solution and completes the proof. □

(ii) *Continuity:* any function $V \in B_*$ and the derivative V_x depend continuously on the parameters (λ, a) on compact subsets of the domain of positivity $\{V > 0\}$. Indeed, it is true in our case. Obviously, $\tilde{V}(t + a)$ depends continuously on a.

(iii) *Monotonicity:* the general statement is given in Section 2.6. In the case of TWs (3.12), where the second parameter a is translation in space, the monotonicity is guaranteed by the obvious properties that $V(x,t) \equiv f(x - \lambda t + a) \to \infty$ as $a \to -\infty$ and $V \to 0$ as $a \to \infty$ monotonically and uniformly on compact subsets of S. For the set B_∞, monotonicity with respect to a is trivial.

B-concavity and sign-invariants

Since B_* is proper, it generates the notion of B-concavity. Namely, given a $t_0 \geq 0$, a solution $v(x,t) \in C^1(\{v(x,t) > 0\})$ is B-concave at $t = t_0$ with respect to the set B_* if for any $x_0 \in \{v(x,t_0) > 0\}$ there holds

$$v(x,t_0) \leq V(x,t_0) \quad \text{in } I\!R,$$

where $V \in B_*$ is the unique tangent solution at the point (x_0,t_0). Due to the regularity in the positivity domain, B-concavity implies the inequality

$$v_{xx} \leq V_{xx}|_{V=v,V_x=v_x} \quad \text{in } \{v > 0\}. \tag{3.43}$$

In the case of piecewise linear functions V such that $V_{xx} \equiv 0$ a.e., (3.43) implies the usual positive concavity $v_{xx} \leq 0$ in $\{v > 0\}$. By Theorem 2.23 in Chapter 2 we deduce that the B-concavity is preserved in time.

Theorem 3.7 *Under the given assumptions, if $v_0(x)$ is B-concave, then $v(x,t)$ is B-concave for $t > 0$.*

As usual, the B-concavity of the solution yields a nonlinear differential inequality on the solutions. In order to rewrite (3.43) in terms of a differential inequality, we first translate the geometric hodograph coordinates $\{v, v_x\}$ of the solution $v(x,t)$ at a given point (x_0,t_0) into the parametric coordinates $\{\lambda, a\}$ (or α) of the corresponding tangent solution $V \in B_*$. Assume first that $\nu \neq 0$. In view of (3.12) and (3.41) this yields the following algebraic system of equations of the above change of variables:

$$f(\xi) = \mu > 0, \quad f'(\xi) = \nu \in I\!R \setminus \{0\}, \quad \text{where } \xi = x_0 - \lambda t_0 + a. \tag{3.44}$$

We have proved in Proposition 3.6 that these equations admit a unique solution $\lambda = \lambda(\mu,\nu)$ (one can see that λ is independent of t_0) and $a = a(\mu,\nu,x_0,t_0)$. In general, this solution cannot be written in an explicit form. If $\nu = 0$, then $V \in B_\infty$ with $V_x = V_{xx} \equiv 0$. We now formulate the desired estimate in terms of the corresponding sign-invariant H.

Proposition 3.8 *If $v(x,t)$ is B-concave for $t \geq 0$, then in $S \cap \{v > 0,\ v_x \neq 0\}$*

$$H(v) \equiv v_{xx} - \frac{1}{F(v)}\,\{q(v) - v_x[v_x + \lambda(v,v_x)]\} \leq 0. \tag{3.45}$$

Proof. Let $V \equiv f(\xi) \in B$ be the unique tangent solution at (x_0,t_0). Then by (3.13)

$$V_{xx} \equiv f'' = \frac{1}{F(f)}\,[q(f) - f'(f' + \lambda)]. \tag{3.46}$$

Since at any tangency point there holds

$$V \equiv f = v, \ V_x \equiv f' = v_x,$$

(3.45) follows from (3.43) and (3.46). □

If $v_x = 0$, then $V = V(t) \in B_\infty$ and (3.43) implies the obvious inequality $v_{xx} \leq 0$ at the maximum point. Using equation (3.40), estimate (3.45) can be rewritten in the form of a differential inequality of the Hamilton-Jacobi type

$$H(v) \equiv v_t + \lambda(v,v_x)v_x \leq 0 \quad \text{in } \{v > 0,\ v_x \neq 0\}.$$

The B-convexity property for another class of solutions is also invariant and is equivalent to

$$H(v) \geq 0.$$

The Hamilton-Jacobi operator H is the sign-invariant of the parabolic equation (3.7) and preserves both signs $\leq$ and $\geq$ in the positivity domains for classes of suitable solutions. Thus, due to the construction, H is generated by the proper set B of TWs.

3.5 B-concavity for the filtration equation

In this section we prove that the general filtration equation (3.7) with $q \equiv 0$ admits a *semiconcavity* estimate of a special form. In the case of the PME it is the counterpart of the Aronson-Bénilan semiconvexity one (3.5). We consider the pressure equation

$$v_t = F(v)v_{xx} + (v_x)^2 \tag{3.47}$$

in the case of finite propagation (3.6), with the same assumptions on compactly supported initial data.

Subset of similarity solutions is proper

We will study the B-concavity property relative to the subset B_s of self-similar solutions

$$V(x,t) = \theta(\xi), \quad \xi = x/\sqrt{t+\tau}, \tag{3.48}$$

where $\tau \geq 0$ is a fixed constant. Substituting (3.48) into (3.47) yields the ODE

$$F(\theta)\theta'' + (\theta')^2 + \tfrac{1}{2}\,\theta'\xi = 0. \tag{3.49}$$

We are interested in profiles $\theta = \theta(\xi; b)$ satisfying the interface conditions at a fixed finite $\xi = b > 0$:

$$\theta(b) = 0, \quad \theta'(b^-) = -\tfrac{1}{2}\,b, \tag{3.50}$$

and $\theta(\xi) \equiv 0$ for $\xi > b$. The second condition implies continuity of the heat flux $-\varphi'(V)V_x$ on the interface. It is easy to see that $\theta'(\xi) < 0$ for $\xi < b$. Let us impose the condition

$$\int_1^\infty \frac{\mathrm{d}\varphi(s)}{s} = \infty.$$

Then the solution of (3.49), (3.50) can be continued for all $\xi \in (\infty, b)$ and $\theta' < 0$ there. In fact, this plays no important role since we deal with bounded solutions $v(x,t) \leq \sup v_0$. One can see that $\theta(\xi; b) \to \infty$ as $b \to \infty$.

Fix a $\tau \geq 0$. We define the functional subset

$$B_s = \{V = \theta(\pm(x-a)/\sqrt{t+\tau}; b),\ a \in I\!R,\ b > 0\} \cup \{V \equiv c,\ c > 0\}. \tag{3.51}$$

Proposition 3.9 *Subset* (3.51) *is proper.*

Proof. Consider the tangential system (3.41). If $\nu = 0$, then $V \equiv \mu$ is the unique tangent solution. Let $\nu < 0$. It follows from (3.50) that by continuity, a tangent solution exists. In order to prove uniqueness, we need to establish monotonicity in the parameter b.

Assume for contradiction that it is not unique and there exist two profiles $\theta_s = \theta(\xi; b_s)$, $s = 1, 2$, such that $b_1 < b_2$ and

$$\theta_2'(\xi) < \theta_1'(\xi) \quad \text{on any level } \theta = \mu \in (0, \bar{\mu}), \tag{3.52}$$

and $\theta_2'(\xi) = \theta_1'(\xi)$ on the level $\theta = \bar{\mu}$. Let $\theta_1(\xi_1) = \theta_2(\xi_2) = \bar{\mu}$, where $\xi_1 < \xi_2$. A contradiction can be obtained from the ODE (3.49) by using the PDE (3.47). Consider the functions $f_s(\xi) = \Phi^{-1}(\theta_s(\xi))$ giving the self-similar solutions $u_s(x,t) = f_s(x/\sqrt{t+\tau})$ of the filtration equation (3.4). It follows from (3.52) that

$$\int_{\xi_1}^{b_1} f_1(\eta)\, d\eta > \int_{\xi_2}^{b_2} f_2(\eta)\, d\eta. \tag{3.53}$$

Integrating equation (3.4) with $u = u_s$ over $(\xi_s\sqrt{t+\tau}, b_s\sqrt{t+\tau})$ and taking into account (3.52), we arrive at a contradiction with (3.53) for $t \gg 1$. Thus, for any $b_1 < b_2$, we have that

$$\theta_2'(\xi) < \theta_1'(\xi) \quad \text{on any level } \theta = \mu \in (0, \sup \theta_2).$$

Continuity and monotonicity are straightforward as in Sections 2.2 and 2.6. □

Semiconcavity estimate and sign-invariant

It follows from Proposition 3.9 that we can directly apply Theorem 2.23 from Chapter 2. In view of (3.43) and (3.51) the corresponding B-convexity estimate has the form

$$v_{xx} \le V_{xx}(x,t) \equiv (t+\tau)^{-1}\theta''_{\xi\xi}(\xi; b) \quad \text{in } \{v > 0\}. \tag{3.54}$$

By Proposition 3.9 the second derivative $\theta''(\xi; b)$ at the tangency point ξ is uniquely determined from the tangential system (3.41) having the form

$$\theta(\xi; b) = \mu = v, \quad (t+\tau)^{-1/2}\,\theta_\xi'(\xi; b) = \nu = v_x, \tag{3.55}$$

where $\xi = (x-a)/\sqrt{t+\tau}$. Equations

$$\theta(\xi; b) = v, \quad \theta_\xi'(\xi; b) = v_x\sqrt{t+\tau}$$

determine $b = b(v, v_x\sqrt{t+\tau})$ and $\xi = \xi(v, v_x\sqrt{t+\tau})$, and next we obtain

$$a = a(x,t,v,v_x) = x - \xi(v, v_x\sqrt{t+\tau})\sqrt{t+\tau}.$$

We suppose that $\nu = v_x < 0$. If $\nu > 0$, then we fix the profiles $\theta(-\xi; b)$ and if $\nu = 0$, then by (3.51) $\theta(\xi) \equiv \mu$ and hence $\theta'' = 0$. From the ODE (3.49) we derive the following equivalent form of (3.54):

$$H(v) \equiv F(v)v_{xx} + (v_x)^2 + \tfrac{1}{2}\, v_x\xi/\sqrt{t+\tau} \le 0, \tag{3.56}$$

where H is the corresponding sign-invariant.

Theorem 3.10 *If for a fixed* $\tau \ge 0$ *inequality* (3.54), (3.55) *holds at* $t = 0$, *then it is true for all* $t > 0$.

Inequality (3.54) determines an upper bound of the second derivative v_{xx} on the interfaces, which together with the standard semiconvexity estimate gives control of v_{xx} in the positivity domain by the MP applied to a linear parabolic equation for v_{xx} obtained by differentiating. There exists the corresponding B-convexity estimate with the opposite inequality sign $\geq$ in (3.54). In view of the evident properties of the subset (3.51), this result is true for v_0, which is supported from one side.

Thus, (3.56) defines the first-order sign-invariant derived from equation (3.47),

$$H(v) = v_t + \tfrac{1}{2} v_x \xi / \sqrt{t + \tau}$$

preserving both signs on suitable solutions of the filtration equation (3.47).

There exist infinitely many other proper subsets B belonging to the set (3.48), each one corresponds to a suitable non-constant function $\tau = \tau(b)$. Therefore, solutions (3.48) define an infinite number of B-concavity (convexity) properties preserved in time or, under certain assumptions, occurring eventually in time similar to the analysis in Section 2.3. Each proper subset generates the corresponding sign-invariant. It is worth mentioning that concavity/convexity estimates of a simpler form for solutions of general filtration equations do not exist in principle.

3.6 B-concavity relative to incomplete functional subsets

We now discuss some possible extensions related to the B-concavity. We begin with B-concavity with respect to a subset B that *is not complete* according to Section 3.4. We show that if B is also continuous and monotone then the intersection comparison can be performed in a similar way but gives a slightly different result.

We explain the main difference by taking particular equations. Consider the PME with absorption

$$\boxed{u_t = (u^m)_{xx} - u^p \quad \text{in } S,} \tag{3.57}$$

where $m > 1$, $0 < p < m$ and $p \neq m - 2$.

Stationary solutions. Equation (3.57) admits a subset B_0 of the following stationary solutions having a single finite interface:

$$U(x) = \{c_* [\pm(x - a)]_+\}^{\frac{2}{m-p}}, \quad c_* = \frac{m-p}{\sqrt{2m(m+p)}}, \tag{3.58}$$

where $a \in I\!R$ is an arbitrary parameter.

B-concavity. The subset $B_0 = \{U\}$ is one-dimensional, and hence it is not complete. Nevertheless, we introduce a standard notion of B-concavity, and are going to use the intersection comparison developed for proper sets B. We assume that the initial data u_0 are bounded, compactly supported and smooth in $\{u_0 > 0\}$.

We say that for a given $(x_0, t_0) \in \bar{Q}$ such that $\mu = u(x_0, t_0) > 0$, $\nu = u_x(x_0, t_0) \in I\!R$, the point with the geometric coordinates (μ, ν) is *admissible* to the subset $B = \{U(x, t)\}$ if there exists a unique function $U \in B$ such that

$$U(x_0, t_0) = \mu, \quad U_x(x_0, t_0) = \nu, \tag{3.59}$$

and U exists for all $t \in [0, t_0]$.

Given $t_0 \geq 0$, the solution $u(x,t) \in C^2(\{u > 0\})$ is said to be B-concave relative to a subset B if

$$\mathrm{Int}(t_0, u, U) \leq 2 \quad \text{for any} \quad U \in B.$$

Hence, for any $x_0 \in \{u(x,t_0) > 0\}$ such that $(u(x_0,t_0), u_x(x_0,t_0))$ is admissible to B there holds $u(x,t_0) \leq U(x,t_0)$ in $I\!R$, where $U \in B_0$ is the unique tangent solution at (x_0, t_0). The proof of this B-concavity property stays the same. This implies the differential inequality $u_{xx}(x_0,t_0) \leq U_{xx}(x_0,t_0)$.

One can see that B_0 is continuous and also it is monotone with respect to the translational parameter $a \in I\!R$. Therefore, exactly as in Section 3.4, we conclude that B-concavity of solutions $u(x,t)$ is invariant.

Theorem 3.11 *u_0 is B-concave $\Longrightarrow$ $u(\cdot,t)$ is B-concave for $t > 0$.*

B-concavity of $u(x,t)$ implies that at any point $(u(x,t), u_x(x,t))$ admissible to B_0 there holds

$$H(u) \equiv u_{xx} - c_1 u^{1+p-m} \leq 0, \quad c_1 = \tfrac{2-m+p}{m(m+p)}. \tag{3.60}$$

Using (3.58) and the tangential system (3.59), inequality (3.60) becomes

$$u_{xx} \leq C|u_x|^{\frac{2(1+p-m)}{2-m+p}},$$

where $C = C(m,p)$ is a constant. In view of equation (3.57) it is equivalent to the following inequality of the Hamilton-Jacobi type:

$$\tilde{H}(u) \equiv u_t - m(m-1)u^{m-2}(u_x)^2 - c_2 u^p \leq 0, \quad c_2 = \tfrac{2(m-1)}{m+p}.$$

3.7 Eventual B-concavity

We consider next a special application of the eventual B-concavity to the study of positive solutions of the PME with absorption (3.57) in the parameter range

$$m > 1, \quad p \in I\!R. \tag{3.61}$$

Equation (3.57) with $p < 1$ is known to exhibit the phenomenon of finite-time extinction due to the *strong* absorption for $p \in [0,1)$ or the *singular* one for $p < 0$, so that any bounded solution vanishes identically in finite time. We fix a strictly positive and sufficiently smooth initial function u_0 satisfying

$$0 < c_0 \leq u_0(x) \leq C_0, \quad [(u_0^m)']^2 \leq C_1 \tag{3.62}$$

and let

$$T = \sup\{\tau > 0: \; u(x,\tau) > 0 \; \text{in} \; I\!R\}$$

be the extinction time. Then $T = \infty$ if $p \geq 1$ and $T < \infty$ if $p < 1$, which is proved by comparison with solutions of the ODE. We consider the solution $u(x,t)$ in $S_T = S \cap \{t < T\}$, so that by the regularity of the coefficients of the equation, $u \in C^\infty(S_{T'})$ with any $T' \in (0,T)$.

We show that under certain necessary assumptions on the initial data, the geometry of the complete set B of stationary solutions generates the property of *eventual* B-concavity having interesting applications. A simpler property of eventual concavity is true for general filtration equations as shown in Section 2.3 in Chapter 2.

Subsets of stationary solutions

The stationary equation

$$(U^m)_{xx} - U^p = 0 \tag{3.63}$$

admits the following subsets of solutions.

(i) If $|p| < m$, then (3.63) has the one-dimensional subset B_0 of solutions (3.58).

(ii) In the parameter range (3.61), there exist strictly positive solutions $U(x;\lambda)$ satisfying, at the minimum point $x = 0$, the conditions

$$U(0;\lambda) = \lambda > 0, \quad U_x'(0;\lambda) = 0. \tag{3.64}$$

Multiplying (3.63) by $(U^m)_x$ and integrating over $(0, x)$ with conditions (3.64) yields

$$[(U^m)_x]^2 = \begin{cases} a_*(U^{p+m} - \lambda^{p+m}) & \text{if } \; p \neq -m, \\ m\log(\frac{U}{\lambda}) & \text{if } \; p = -m, \end{cases} \tag{3.65}$$

where $a_* = \frac{2m}{p+m}$. Let $U_*(x)$ be the unique even solution of problem (3.63), (3.64) with $\lambda = 1$. Then by scaling, the full subset B_1 of such solutions shifted in space is given by

$$B_1 = \{U(x) = \lambda U_*((x-a)\lambda^{(p-m)/2}), \;\; \lambda > 0, \; a \in I\!R\}.$$

It follows that if $p \leq m$, then the functions U are well defined in $I\!R$. If $p > m$, then $U(x)$ blows up as $x \to a \pm x_0(\lambda)$, where $x_0(\lambda) = x_*\lambda^{-(p-m)/2}$ with $x_* > 0$ being the blow-up point of U_*. It follows from (3.65) that as $\lambda \to 0$, on any level set $\{x: \;\; U = c > 0\}$,

$$|(U^m)_x| \to \begin{cases} a_*\, c^{p+m} & \text{for } \; p > -m, \\ \infty & \text{for } \; p \leq -m. \end{cases} \tag{3.66}$$

(iii) The last subset B_2 exists for $p > -m$. Any solution $U \in B_2$ vanishes at some point, say, $x = 0$, where it satisfies the conditions $U(0) = 0$ and $(U^m)_x(0) = \lambda \neq 0$. Integrating (3.63) with these boundary conditions yields (cf. (3.65))

$$[(U^m)_x]^2 = \lambda^2 + a_* U^{p+m}. \tag{3.67}$$

Notice that similar to B_1, $B_2 = \{U(x+a), a, \lambda \in I\!R\}$ is two-dimensional.

Completeness and proper subsets

Continuity properties relative to parameters (λ, a) and monotonicity with respect to a or λ of the subsets B_0, B_1 and B_2 are straightforward. The property of completeness depends on m and p.

Proposition 3.12 *The following sets are proper:* $B = B_1 \cup B_2$ *if* $p \geq m$, $B = B_0 \cup B_1 \cup B_2$ *if* $|p| < m$, *and* $B = B_1$ *if* $p \leq -m$.

Proof. We prove completeness. Existence and uniqueness of a tangent solution $U(x)$ satisfying the tangential system (3.59) for arbitrary given $x_0 \in I\!R,\ \mu > 0,\ \nu \in I\!R$, follow from the above results for the ODE (3.63). Then using possible continuations of the function $U(x)$ into the domain $\{0 < U \ll 1\}$ (see (3.65) and (3.67)), we obtain the result. Observe that completeness of B_1 for $p \le -m$ is easily seen from (3.66). □

As usual, this implies that the B-convexity is preserved in time.

Eventual B-concavity

This property means that the solution eventually becomes B-convex in a domain where, in addition, it satisfies some extra requirements. For the present problem with absorption, it is natural to consider a domain, where the solution is sufficiently small, i.e., the B-concavity is *eventual* (and, possibly, partial) in S_T. Let us state the corresponding definition.

Definition 3.1 A solution $u(x,t) > 0$ in S_T is said to be *eventually* B-concave relative a proper set B given in Proposition 3.12, if there exists a constant $\delta_0 = \delta_0(u_0) > 0$ such that

$$u(\cdot,t) \quad \text{is } B\text{-concave in } \ S_T \cap \{0 < u < \delta_0\}. \tag{3.68}$$

Theorem 3.13 *Let* (3.61) *and* (3.62) *hold. Then:*

(i) If $p \le -m$, the solution $u(x,t)$ satisfies (3.68)*, where $\delta_0 = c_0 e^{-C_1/m}$ if $p = -m$ and $\delta_0^{p+m} = C_1/|a_*| + c_0^{p+m}$ if $p < -m$.*

(ii) If $p > -m$, (3.68) *is also true under the assumption $C_1 < a_* c_0^{p+m}$, where $\delta_0^{p+m} = c_0^{p+m} - C_1/a_*$.*

Proof. We prove that for any $(x_0,t_0) \in S_T \cap \{0 < u < \delta_0\}$, there holds

$$u(x,t_0) \le U(x) \quad \text{in } I\!R, \tag{3.69}$$

where $U \in B$ is the tangent solution at (x_0,t_0). See Figure 3.1 below. Indeed, (3.69) implies B-concavity in $\{0 < u < \delta_0\}$. According to the general scheme in Section 3.4, (3.69) is valid if for any $\mu \in (0,\delta_0)$ and $\nu \in I\!R$, the number of intersections $\mathrm{Int}(0,U)$ of the initial function u_0 and the tangent profile $U \in B$ satisfying (3.59) at $t_0 = 0$ is such that

$$\mathrm{Int}(0,U) \le 2 \quad \text{for any } x_0 \in I\!R. \tag{3.70}$$

Then since new intersections cannot appear at $x = \pm\infty$ ($u(x,t)$ is uniformly bounded unlike the stationary solutions), we have

$$\mathrm{Int}(t,U) \le 2 \quad \text{for all } \ t \in (0,T). \tag{3.71}$$

Therefore, if the solution u takes, at a point (x_0,t_0), the values $u(x_0,t_0) = \mu$ and $u_x(x_0,t_0) = \nu$, then (3.71) with $t = t_0$ gives (3.69). In view of assumptions (3.62), one can see that (3.70) is valid if $|(U^m)_x| \ge |(u_0^m)'|$ on any level set $\{x \in I\!R : u_0(x) = c\}$ with an arbitrary $c \in [c_0, C_0]$. With the choice of the values of δ_0 given in the statement of the theorem this follows from (3.65) or (3.67). □

Theorem 3.13 applies to equations (3.1) with general nonlinearities including the case of singular absorption where $\psi(u) \to \infty$ as $u \to 0$.

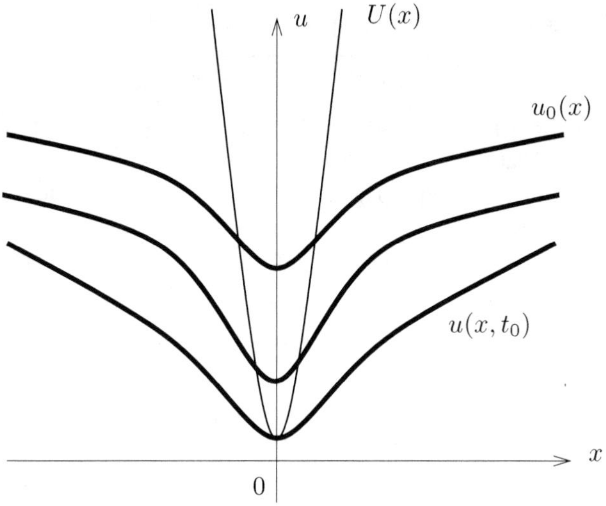

Figure 3.1 *Illustration of the proof of Theorem* 3.13*:* $u(x, t)$ *has two intersections with the stationary one* $U(x)$ *and hence, evolving with time,* $u(x, t_0) \le U(x)$ *for some* $t = t_0 > 0$.

Application: monotonicity with time of small solutions

Finally, we derive a simple differential inequality equivalent to (3.68).

Corollary 3.14 *If* $u(x, t)$ *is eventually* B*-concave, then*

$$H(u) \equiv u_t < 0 \quad \text{in } S_T \cap \{0 < u < \delta_0\}. \tag{3.72}$$

Proof. Inequality (3.69) at an arbitrary $(x_0, t_0) \in S_T \cap \{u < \delta_0\}$ implies that $u_{xx}(x_0, t_0) \le U_{xx}(x_0)$. In view of (3.57) and (3.63), since $u = U$ and $u_x = U_x$ at the tangency point (x_0, t_0), we have that $u_t \le 0$ in $S_T \cap \{u < \delta_0\}$. The strict inequality in (3.72) follows by the strong MP applied to the linear parabolic equation $z_t = m(u^{m-1}z)_{xx} - p\,u^{p-1}z$ satisfied by the derivative $z = u_t \not\equiv 0$. □

Inequality (3.72) shows that under the above hypotheses, near the extinction point any suitable solution always *decreases monotonically with time*. This is a general property of the nonlinear equation (3.57) establishing that no oscillations in time can appear near extinction.

Remarks and comments on the literature

The main results are contained in [137].

§ **3.1.** Regularity and other properties of weak solutions of degenerate equations (3.1) and (3.2), (3.3), as well as some physical background can be found in [213]. The backward problem of B-concavity (exact solutions via sign-invariants) made it possible to construct infinite-dimensional classes of quasilinear equations (3.1) admitting explicit solutions; see examples in [134].

§ **3.2–3.4.** First concavity result for equation

$$\boxed{v_t = (m-1)vv_{xx} + (v_x)^2 - v^p, \quad p \geq 1}$$

was proved in [336] by an iterative technique of semigroups corresponding to splitting the nonlinear operator into two $(m-1)vv_{xx}$ and $(v_x)^2 - v^p$ and using a Trotter-Lie formula. Finite propagation criterion (3.6) for the filtration equation with $q \equiv 0$ was invented in [277] as a sufficient condition and was proved to be necessary in [287]; see also earlier partial results in [209] and a survey [213].

§ **3.5.** Similarity solutions (3.48) have been understood in detail; see [34], [35] and [287]. The importance of the semiconvexity estimate (3.5) is explained in the beginning of Chapter 2.

§ **3.6–3.7.** Let $p \geq m$ and $u_0(x) \geq 0$ be compactly supported. Then $T = \infty$ and meas supp $u(\cdot,t) \to \infty$ as $t \to \infty$ [213]. In this case one can also prove the eventual B-concavity result for the solutions where the "partial" property includes not only the restriction $u < \delta_0$ as in (3.68) but also $x \in K$, where K is a compact subset from supp u_0. Actually, a suitable notion of eventual B-concavity depends on the properties of the solution under consideration.

Another example of the eventual B-convexity similar to that in Corollary 3.14 is presented in [150], [154], where for a reaction-diffusion equation

$$\boxed{u_t = (\varphi(u))_{xx} + \psi(u)}$$

it is proved that $u_t > 0$ in $\{u \gg 1\}$ (monotonicity in time of large solutions). The "envelope analysis" in the method of stationary states [306, p. 423] can be treated in such a way. The linear sign-invariant $H(u) = u_t$ is elementary and can be obtained by differentiating any autonomous parabolic equations and applying the MP. It admits a natural generalization of the form $H_\Psi(u) \equiv u_t - \Psi(u)$ with functions Ψ to be determined from a two-dimensional DS for the nonlinear coefficients φ, ψ of the equation [134]. Applications of the Ψ-criticality inequality $H_\Psi(u) \geq 0$ can be found in [306, p. 332, 353].

CHAPTER 4

Eventual B-convexity: a Criterion of Complete Blow-up and Extinction for Quasilinear Heat Equations

This chapter is devoted to a new important application of the geometric convexity approach to the study of blow-up and extinction phenomena for general quasilinear reaction-diffusion equations in one dimension. We derive a criterion of *complete/incomplete* finite-time singularities for this class of nonlinear PDEs. This is the first time that we face essentially *discontinuous* nonlinear semigroups induced by singular parabolic equations. This particular example of blow-up singularity propagation phenomena in quasilinear parabolic equations represents some typical aspects of the geometric theory of fully nonlinear singular parabolic equations to be developed in Chapter 7.

Here we consider the first questions concerning the existence or nonexistence of a nontrivial solution extension beyond the finite-time blow-up or extinction singularity. Complete blow-up at a finite time $t = T$ is a simple case of discontinuous semigroups generated by a singular quasilinear heat equations. In the case of blow-up the geometric constructions deal with the extension of solutions at the infinite singularity level $\{u = \infty\}$ instead of the universal zero-level $\{u = 0\}$ studied before. This emphasizes some special unusual features of formation and propagation of blow-up singularities.

4.1 Introduction: The blow-up problem

We consider quasilinear heat equations with rather arbitrary nonlinearities

$$\boxed{u_t = (\varphi(u))_{xx} + \psi(u) \quad \text{in } S = I\!R \times I\!R_+,} \tag{4.1}$$

where φ and ψ are real functions that are defined and positive for $u > 0$. The term $(\varphi(u))_{xx}$ represents nonlinear diffusion, possibly degenerate or singular meaning fast diffusion. As above, we assume that $\varphi \in C([0,\infty)) \cap C^1(I\!R_+)$, $\varphi'(u) > 0$ for $u > 0$ and $\varphi(0) = 0$. The term $\psi(u)$, which is assumed to be smooth for $u > 0$, denotes a reaction. We consider the Cauchy problem with bounded, continuous and nonnegative initial data

$$u(x,0) = u_0(x) \not\equiv \text{const.} \quad \text{in } I\!R. \tag{4.2}$$

Moreover, we assume that u_0 is bell-shaped in a natural sense. We are interested in the occurrence of *finite-time blow-up*, i.e., the existence of a time $T = T(u_0) < \infty$

such that

$$\sup_{x\in \mathbb{R}} u(x,t) \to \infty \quad \text{as } t \to T^-. \tag{4.3}$$

This singular formation phenomenon occurs whenever $\psi(u)$ is superlinear for large u, as is well-known. More precisely, ψ has to satisfy the *Osgood criterion*

$$\int^\infty \frac{\mathrm{d}s}{\psi(s)} < \infty, \tag{4.4}$$

which follows by comparison with the spatially flat solutions $u = u(t)$ satisfying the ODE

$$u_t = \psi(u) \quad \text{for } t > 0.$$

An important aspect of blow-up problems is the possibility of having a *nontrivial extension* of the solution for times $t > T$. If such a continuation exists, we say that the blow-up is *incomplete*; otherwise it is called *complete*. A natural way of obtaining a continuation consists of approximating the equation by a reaction-diffusion model with the reaction nonlinearity $\psi(u)$ replaced by a sequence of uniformly Lipschitz functions $\{\psi_n(u) \geq 0\}$ such that $\psi_n(u) \to \psi(u)$ uniformly on bounded intervals. Then we obtain a sequence of global solutions $\{u_n\}$ defined for all $t \geq 0$. If we also impose the convenient condition that $\{\psi_n\}$ is monotone nondecreasing with n, the sequence $\{u_n\}$ will be monotone, so that the limit

$$u(x,t) = \lim_{n\to\infty} u_n(x,t) \tag{4.5}$$

exists (finite or infinite) for all times.* For $t < T$, it is easy to see that we recover the standard solution of the problem. The question of complete/incomplete blow-up is then reduced to determining whether the limit (4.5) becomes identically infinite for $t > T$ or not. We call the limit (4.5) the *proper solution* of (4.1), (4.2). It is always a *minimal solution* of the Cauchy problem. Since the 1980s, the problem of complete blow-up has been studied mainly for semilinear heat equations (see Remarks)

$$u_t = \Delta u + \psi(u). \tag{4.6}$$

For general quasilinear heat equations, a natural problem is to find conditions for complete/incomplete blow-up in terms of the constitutive functions φ and ψ. In principle, the alternative will also depend on the initial data. It is clear that for flat initial data (u_0 constant) blow-up is always flat, hence complete. To avoid such trivial situations we make the typical assumption that the initial data are bell-shaped. We prove *necessary and sufficient conditions* for complete blow-up of the solutions of (4.1), (4.2) depending on the balance between φ and ψ, and more precisely on the behaviour for large u of the function

$$F(u) = \frac{1}{u^2}\int_1^u G(s)\,\mathrm{d}s, \quad \text{where} \quad G(s) = \varphi'(s)\psi(s). \tag{4.7}$$

We will prove that *complete blow-up occurs* if $F(u)$ is unbounded as $u \to \infty$, while *incomplete blow-up happens* if

$$F(u) \quad \text{is uniformly bounded as } u \to \infty, \tag{4.8}$$

* A detailed construction is presented in Chapters 6 and 7

together with

$$\int_1^\infty \frac{\mathrm{d}\varphi(s)}{s} < \infty. \tag{4.9}$$

The latter condition is necessary and sufficient for the *finite* speed of propagation of *blow-up interfaces* of travelling wave solutions for the corresponding quasilinear heat equation without source

$$u_t = (\varphi(u))_{xx}.$$

Conditions (4.4) and (4.8) "almost" imply (4.9). For instance, when we deal with power nonlinearities or even when ψ is merely monotone, this is always the case. However, we show that there exist "pathological" choices of φ and ψ for which (4.4) and (4.8) hold but (4.9) does not. We also outline some of the curious phenomena that may happen for such a class. In particular, we show that equations (4.1) from the pathological class exhibit the following very special property: *all solutions with flat initial data* $u_0 \equiv \mathrm{const.} > 0$ *blow-up in finite time, while no solution with bell-shaped compactly supported data does.*

Our results are true not only for solutions to the Cauchy problem, but can also be directly applied to initial-boundary value problems in bounded spatial domains with Dirichlet or Neumann boundary conditions. In fact, our analysis of complete/incomplete blow-up is local in the sense that the behaviour for $t > T$ depends only on the behaviour of the solution in a small neighbourhood of a given blow-up point, thus being independent of any standard boundary conditions.

The analysis is done in terms of the *eventual* B-concavity relative to a subset $B = \{f(\xi),\ \lambda \in I\!R\}$ of the TW solutions

$$V(x,t) = f(\xi), \quad \xi = x - \lambda t + a, \tag{4.10}$$

where λ is the speed parameter, $a \in I\!R$ is arbitrary, and the profile function f solves a nonlinear ODE. We carry out a complete study of the solutions of the form (4.10) in Sections 4.2 and 4.3. The crucial point is the behaviour of the envelope of the set B for large values of the parameter λ. Using the set of TW solutions for large speeds $|\lambda| \gg 1$ makes it possible to check if the speed of propagation of perturbations at the singular level $\{u = \infty\}$ is infinite or not. This implies complete or incomplete propagation of singularities after T. Our geometric analysis in Section 4.4 relies on an intersection comparison technique between the given solution $u(x,t)$ and the family B.

Precise statements of the conditions are found in Section 4.4. In order to give an intuitive idea of the result, let us consider the case of power nonlinearities

$$\boxed{u_t = (u^m)_{xx} + u^p} \tag{4.11}$$

with $m > 0$ and $p > 0$. Then it is well-known that blow-up happens only if $p > 1$. Moreover, for $1 < p \le m+2$, all solutions will blow-up, while for $p > m+2$ the specific occurrence of blow-up depends on the initial data. Under the assumption that the data are bell-shaped we have single-point blow-up if $p > m$, while for $p = m$, it is regional and for $p \in (1, m)$, blow-up is global; see details in [306, Chapter 4]. As a consequence of our results, it follows that, in the particular model

(4.11), *incomplete blow-up* occurs if and only if

$$p + m \le 2 \quad \text{and } p > 1, \tag{4.12}$$

which in particular implies $m < 1$. The description of evolution after the beginning of blow-up for equation (4.11) in the range (4.12) will be the object of the next chapter. For the p-Laplacian equation with source

$$\boxed{u_t = (|u_x|^\sigma u_x)_x + u^q, \quad \sigma > -1, \ q > 1,} \tag{4.13}$$

the criterion of incomplete blow-up is also easy to compute,

$$q \in (1, \tfrac{1}{1+\sigma}], \qquad \text{where } \; -1 < \sigma < 0. \tag{4.14}$$

Sections 4.5 and 4.6 are devoted to the study of complete/incomplete extinction for the quasilinear equation (4.1) with a singular *absorption* term. The precise statement of this problem is given in Section 4.5. Although it can be reduced to a blow-up problem for a different parabolic equation (so the results of Section 4.4 can be directly applied), it is convenient to study such singular behaviour separately; see Section 4.6.

4.2 Existence and nonexistence of singular blow-up travelling waves

The characterization of complete/incomplete blow-up for equation (4.1) will be done in terms of the existence of suitable TW solutions of the equation. For arbitrary constant wave speeds $\lambda > 0$ and $a \in I\!R$, we consider solutions of the form (4.10). Solutions with negative speed are given by symmetry $V(x,t;-\lambda) = V(-x,t;\lambda)$. Substituting this form into equation (4.1), we obtain the following ODE for the wave profile $f \ge 0$:

$$(\varphi(f))'' + \lambda f' + \psi(f) = 0. \tag{4.15}$$

To investigate (4.15) we carry out a phase-plane analysis. We set

$$\frac{\mathrm{d}\varphi(f)}{\mathrm{d}\xi} = -P$$

and multiply (4.15) by $\varphi'(f)$ to obtain the trajectory equation in the $\{f, P\}$-plane

$$P\frac{\mathrm{d}P}{\mathrm{d}f} - \lambda P + G(f) = 0 \quad \text{for } f > 0. \tag{4.16}$$

The analysis of blow-up behaviour depends on the existence of solutions $P = P(f)$ of equation (4.16) contained in the quadrant $\{f > 0, \ P \ge 0\}$, and more precisely, on the existence of a solution of the equation (4.16) such that

$$P(f) > 0 \quad \text{is well defined for all } \; f \gg 1.$$

This implies that the corresponding TW is defined for all large $f > 0$ and has a monotone profile. We call such solutions *monotone TWs* (observe that, for $\lambda > 0$, these are monotone decreasing with ξ). Moreover, if the TW $f = f(\xi)$ blows up at a finite ξ, we call such a solution a *singular TW*; see Figure 4.1.

The existence of monotone TWs is characterized as follows.

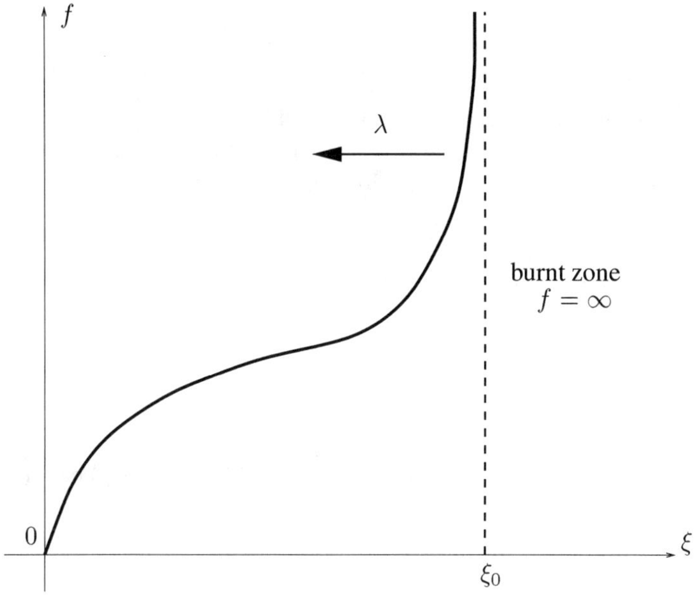

Figure 4.1 *Monotone singular TW $f(\xi)$ with the finite blow-up interface at $\xi = \xi_0$, i.e., at $x = \xi_0 - a + \lambda t$ with speed $\lambda < 0$.*

Theorem 4.1 *There exists a monotone TW for some $\lambda_0 > 0$ if and only if condition* (4.8) *is satisfied, and then such waves exist for all $\lambda \geq \lambda_0$.*

Proof. (i) Let us begin by remarking that the line $P = \frac{1}{\lambda} G(f)$ represents the zero-cline in the phase-plane of (4.16). On the other hand, the slope $\mathrm{d}P/\mathrm{d}f$ becomes infinite as $P \to 0$. Therefore, a solution starting in the first quadrant evolves according to two possibilities: either it reaches the $P = 0$ axis at a finite height f or it is defined and positive for all large f. Our question is precisely the existence of solutions in the second class. Of course, if such a class is non-empty, according to a standard ODE analysis there exists a minimal monotone solution called a *separatrix* of the flow picture.

On the other hand, every solution of (4.16) with $\lambda = \lambda_0 > 0$ is a subsolution for (4.16) with $\lambda > \lambda_0$. It easily follows that the existence of a monotone solution for λ_0 implies the existence for all $\lambda > \lambda_0$. Finally, let us remark that all solutions of (4.16) satisfy

$$\frac{\mathrm{d}P}{\mathrm{d}f} \leq \lambda, \tag{4.17}$$

so they grow at most linearly as $f \to \infty$, and hence $P \leq \lambda f + c$.

(ii) Let us now prove that whenever a monotone solution exists, then condition (4.8) holds. Indeed, integrating (4.16) from 1 to $f > 1$ yields

$$P^2(f) - P_0^2 = 2\lambda \int_1^f P(s)\,\mathrm{d}s - 2\int_1^f G(s)\,\mathrm{d}s, \quad P_0 = P(1). \tag{4.18}$$

In view of (4.17), using the definition (4.7), we have

$$f^2F(f) \le \tfrac{1}{2}\,\lambda^2 f^2 + \lambda c f + K,$$

where K depends on the initial data, whence (4.8).

(iii) Let us now prove conversely that (4.8) implies the existence of monotone TWs for large λ. From (4.8) we have for some $a > 0$ that

$$F(f) \le a \quad \text{for} \;\; f \ge 1. \tag{4.19}$$

Let $z = \int_1^f P(s)\,\mathrm{d}s$. Then (4.18) and (4.19) imply that

$$(z')^2 \ge 2\lambda z - 2af^2. \tag{4.20}$$

We want to show that, for a suitably chosen $\mu > 0$, the function

$$\underline{z}(f) = \mu f^2$$

is a subsolution of the equation corresponding to (4.20). This will be true if

$$4\mu^2 f^2 \le (2\lambda\mu - 2a) f^2,$$

i.e., if

$$2\mu^2 - \lambda\mu + a \le 0. \tag{4.21}$$

For that to take place, it is necessary that the discriminant be nonnegative,

$$\lambda^2 - 8a \ge 0. \tag{4.22}$$

Then we have to pick a μ between the two positive roots of the quadratic polynomial in (4.21). Therefore, we obtain an increasing subsolution of (4.20). It follows that, for every solution starting with a value larger than $\underline{z}(f)$ at, say, $f = 1$, we have

$$z(f) \ge \underline{z}(f) \quad \text{for} \;\; f \ge 1.$$

Using (4.20) yields the inequality

$$P^2 = (z')^2 \ge (\underline{z}')^2.$$

We conclude that the solution can be continued with positive values up to $f = \infty$. $\square$

The solution we have constructed is not necessarily the separatrix. For λ satisfying (4.22), we can show that the solutions grow linearly as $f \to \infty$. Indeed,

$$P^2 = (z')^2 \ge 2\lambda z - 2af^2 \ge (2\lambda\mu - 2a) f^2 \ge 4\mu^2 f^2, \tag{4.23}$$

cf. the upper bound obtained from (4.17).

We now investigate the existence of singular TWs. This depends on the calculation of the total ξ-range of the orbits we have constructed. From the definition of P we have

$$\mathrm{d}\xi = -\frac{\mathrm{d}\varphi(f)}{P}.$$

Hence, using the upper bound on the linear growth of the solutions, we have

$$\int_{\xi_\infty}^{\xi} \mathrm{d}\xi \ge c \int_f^\infty \frac{\mathrm{d}\varphi(s)}{s}, \quad \text{where} \;\; c = \tfrac{1}{\lambda}.$$

The converse inequality with $c = \frac{1}{2\mu}$ can be obtained for the monotone solutions constructed above using (4.23). In summary, we deduce the following characterization.

Theorem 4.2 *There exist singular TWs for some* $\lambda > 0$ *if and only if conditions* (4.8) *and* (4.9) *hold, and then such TWs exist for all large* λ.

If a monotone TW is not singular, it is defined for all $\xi \ll -1$ and reaches the singular level $f = \infty$ at $\xi = -\infty$. The existence of such waves is discussed in the next section.

4.3 Discussion of the blow-up conditions. Pathological equations

The discussion of blow-up relies on three conditions on the coefficients, namely:

(4.4) that controls the blow-up occurrence in the ODE in terms of ψ,

(4.8) that controls the balance between φ and ψ and characterizes the existence of monotone TW solutions, and

(4.9) that affects only φ and determines the property of finite speed of propagation of blow-up interfaces in a non-reactive problem.

We show here that under some mild extra non-oscillation conditions on $\psi(u)$ and $\varphi(u)$ for $u \gg 1$, (4.9) is a consequence of (4.4) and (4.8).

Theorem 4.3 *Let, besides* (4.4) *and* (4.8), *one of the following assumptions hold:*

(i) the function $\psi(u)$ *is monotone for* $u \gg 1$,

(ii) a stronger version of (4.8)*: the function* $\varphi'(u)\psi(u)/u$ *is bounded for* $u \gg 1$.

Then (4.9) *is valid.*

Proof. (i) Assume that ψ is monotone increasing. Take $a_n = 2^n$ and $I_n = (a_n, a_{n+1})$. Then, thanks to (4.8)

$$\psi(a_n)a_n \int_{I_n} \frac{d\varphi(s)}{s} \le \int_{I_n} \psi(s)\, d\varphi(s) \le C a_{n+1}^2,$$

so

$$\int_{I_n} \frac{d\varphi(s)}{s} \le 4C \frac{a_n}{\psi(a_n)} \le 4C \int_{I_{n-1}} \frac{ds}{\psi(s)},$$

where we have used again the monotonicity of ψ in the last step. Recalling now (4.4) and summing over n, (4.9) follows.

(ii) This case is very simple. The extra assumption means that

$$\frac{\varphi'(s)}{s} \le \frac{C}{\psi(s)}.$$

Therefore, (4.4) implies (4.9). □

However, this implication can fail for highly oscillatory functions.

Proposition 4.4 *There exist smooth choices of the functions* φ *and* ψ *such that* (4.4) *and* (4.8) *hold but* (4.9) *does not.*

Proof. Take $a_n = 2^n$, $\varepsilon_n = \frac{1}{n}$, $b_n = a_n + \varepsilon_n$ for $n = 1, 2, ...$, and let $I_n =$

(a_n, b_n). Consider a smooth and positive function that approximates (in a sense that will be obvious to the reader) the following step sum:

$$h(u) \sim \sum_{(n)} \frac{1}{n}\chi[I_n](u),$$

where $\chi[A]$ denotes the characteristic function of a measurable set $A \subset I\!R$. Further, let j be smooth, positive and

$$j(u) \sim \sum n\chi[I_n](u)\,.$$

Then we define

$$\psi(u) = \frac{1}{h(u)} \quad \text{and} \quad \varphi'(u) = j(u)h(u)u.$$

In this way we have

$$\int_1^\infty \frac{\mathrm{d}s}{\psi(s)} = \int_1^\infty h(s)\,\mathrm{d}s \sim \sum_{n=1}^\infty \frac{\varepsilon_n}{n} = \sum_{n=1}^\infty \frac{1}{n^2} < \infty,$$

i.e., (4.4) holds. Furthermore,

$$\int_1^\infty \frac{\mathrm{d}\varphi(s)}{s} = \int_1^\infty j(s)h(s)\,\mathrm{d}s \sim \sum \varepsilon_n = \sum \frac{1}{n} = \infty,$$

so (4.9) does not hold. Finally, if f lies in the interval $(2^N, 2^{N+1})$,

$$\int_1^f \varphi'(s)\psi(s)\,\mathrm{d}s = \int_1^s j(s)s\,\mathrm{d}s \sim \sum_{n=1}^N n2^n\varepsilon_n = \sum_{n=1}^N 2^n \sim 2^{N+1} \le 2f.$$

In this way we see that not only (4.8) holds, but even that it does so with a very slow growth rate

$$\int_1^f G(s)\,\mathrm{d}s = O(f) \quad \text{for } f \gg 1.$$

This completes the proof. □

We will call equations (4.1) with coefficients ψ and φ such that (4.4) and (4.8) hold, but (4.9) does not, *pathological ones*. Let us first consider a particular subclass with a special property.

Lemma 4.5 *There exists a class $\mathcal{P}$ of equations* (4.1) *with the functions φ and ψ satisfying the assumptions of Proposition* 4.4*, such that there exist global monotone TW solutions of the form* (4.10) *defined for all $\xi \in I\!R$ and satisfying*

$$\frac{\mathrm{d}\varphi(f)}{\mathrm{d}\xi} \to 0 \quad \textit{as } f \to 0. \tag{4.24}$$

Proof. Let us assume that $G(f) = o(f)$ as $f \to 0$. It then follows from equation (4.16) that, for all $\lambda \gg 1$, there exists a solution such that

$$P(f) \approx \lambda f, \quad P'(f) \approx \lambda \quad \text{for small } f > 0,$$

i.e., a solution satisfying (4.24). By continuity, for $\lambda \gg 1$, the values of P at $f \approx 1$

are so large that we can compare the corresponding solutions to (4.20) with the subsolution

$$\underline{z}(f) = \mu f^2$$

to conclude that, for $f > 2$, the function $P(f)$ satisfies (4.23) and hence f is monotone. Since (4.9) does not hold, the corresponding TW solution is global in ξ and satisfies (4.24). □

We can now establish the following *alternative* behaviour, which seems to be characteristic of the pathological class.

Theorem 4.6 *Given an arbitrary equation from the class $\mathcal{P}$, we have:*

(i) any solution with flat initial data $u_0 \equiv \text{constant} > 0$ blows up in finite time,

(ii) any solution with bounded, compactly supported data $u_0(x)$ is global in time. The same holds if u_0 is bounded and supported in a half-infinite interval, e.g., the Heaviside (step) function

$$u_0(x) = \begin{cases} 1 & \textit{for } x \le 0, \\ 0 & \textit{for } x > 0. \end{cases}$$

Proof. (i) Since (4.4) holds, this follows. (ii) Given suitable data $u_0(x)$, we take a global TW solution $V(x,t)$ from Lemma 4.5 shifted in x such that

$$u_0(x) \le V(x-a,0) \quad \text{in} \quad I\!R,$$

as Figure 4.2 shows. Then, since V is bounded and global in time and

$$u(x,t) \le V(x-a,t) \quad \text{in} \quad I\!R \times I\!R_+$$

by comparison, $u(x,t)$ is bounded for all $t > 0$. □

4.4 Proof of complete and incomplete blow-up

Let us consider now the behaviour after blow-up of the solutions of the Cauchy problem for (4.1). The functions $\varphi(u)$ and $\psi(u)$ satisfy standard conditions listed in Section 4.1 plus condition (4.4), to ensure the existence of blow-up.

We take a proper minimal solution $u(x,t)$ with bounded initial data $0 \le u_0(x) \le C$, smooth in the sense that $|\mathrm{d}\varphi(u_0)/\mathrm{d}x| \le C$, and such that u has blow-up time $T = T(u_0) < \infty$. We define the blow-up set as

$$\mathcal{B}[u_0] = \{x \in I\!R : \exists \ \{x_k\} \to x, \ \{t_k\} \to T^- \text{ with } u(x_k,t_k) \to \infty\}. \quad (4.25)$$

Typical situations are $\mathcal{B}[u_0] = I\!R$, called *global blow-up*, and $\mathcal{B}[u_0]$ is a point (*single-point blow-up*). Less prevalent is the case where $\mathcal{B}[u_0]$ is an interval (*regional blow-up*). Other configurations may also happen depending on the equation and the data. In particular, $\mathcal{B}[u_0]$ may be empty in some exceptional situation. It is easy to see that under standard assumptions on the initial data, like the function u_0 being bell-shaped, the blow-up set is always non-empty.

Nonexistence of nontrivial continuation

We now state the main results about a possible extension of the solution for $t > T$.

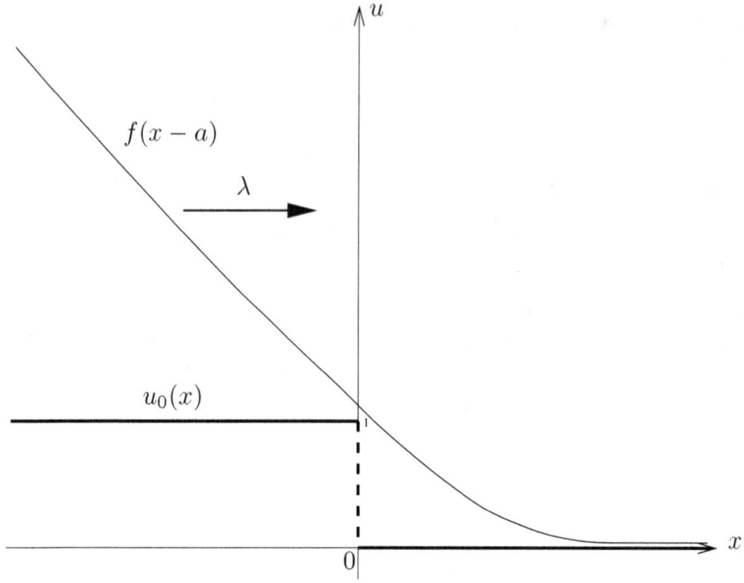

Figure 4.2 *Comparison from above with a pathological TW implies global existence.*

Theorem 4.7 (Complete blow-up) *Assume that condition* (4.8) *does not hold and that the blow-up set is not empty. Then* $u(x,t) \equiv \infty$ *for* $t > T$.

Proof. The proof depends on the nonexistence of monotone TWs plus an intersection comparison analysis. Assume without loss of generality that $0 \in \mathcal{B}[u_0]$. Fix a monotone sequence $\{\varepsilon = \varepsilon_k\} \to \infty$ such that function (4.7) satisfies

$$F(\varepsilon_k) \to \infty \quad \text{as} \quad k \to \infty.$$

Next, fix a $\lambda \gg 1$. Consider the TW solution (4.10) with a fixed $a \in I\!\!R$ and $f(\xi)$ satisfying (4.15) and having a maximum at $\xi = 0$, with the conditions

$$f(0) = \varepsilon_k \gg 1, \quad f'(0) = 0. \tag{4.26}$$

By the properties of solutions of the ODE (4.16), under the above hypothesis on G, given a constant $C > 0$, the function f satisfies

$$\left| \frac{\mathrm{d}\varphi(f)}{\mathrm{d}\xi} \right| \gg C \quad \text{on any level} \quad \{f = m \in [0, C]\}, \tag{4.27}$$

provided that $\varepsilon \gg 1$ and λ is not very large, say $1 \ll \lambda \leq \lambda_\varepsilon$, with $\lambda_\varepsilon \to \infty$ as $\varepsilon = \varepsilon_k \to \infty$. More precisely, it follows from inequality (4.20) and from the upper bound $P \leq \lambda f + c$ that we may set

$$\lambda_\varepsilon^2 \sim F(\varepsilon) \to \infty \quad \text{as} \quad \varepsilon \to \infty.$$

Therefore, the corresponding initial function $V(x,0)$ intersects sufficiently smooth initial data $u_0(x)$ at, at most, two points. Hence, under the given hypotheses on ε and λ, the number of intersections $\mathrm{Int}(t, V)$ of the solutions $u(x,t)$ and

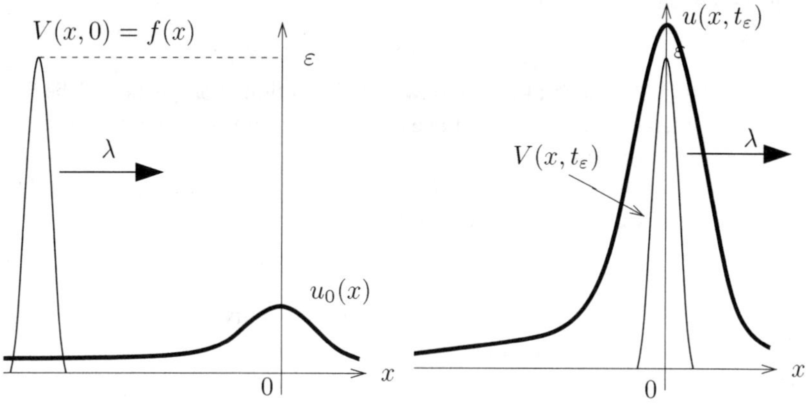

(a) Initial situation (b) TW V is now below u

Figure 4.3 *Illustration of the proof of Theorem 4.7: mutual location of $u(x,t)$ and the TW $V(x,t)$ (a) at $t=0$ (two intersection), and (b) at $t=t_\varepsilon$ (no intersections), when $u(x,t_\varepsilon)$ becomes large and $V(x,t_\varepsilon)$ stays below.*

$V(x,t)$ in the positivity domain $\{V(x,t)>0\}$ satisfies

$$\text{Int}(0,V) \le 2 \quad \text{for all } 1 \ll \lambda \le \lambda_\varepsilon \quad \text{and } a \in I\!R; \tag{4.28}$$

see Figure 4.3 (a). Since $u(x,t) \ge 0$ and $V(x,t)=0$ at the end points of $\{V > 0\}$, by the first Sturm Theorem from Chapter 1, we have that

$$\text{Int}(t,V) \le 2 \quad \text{for all } t \in (0,T) \text{ and for any } 1 \ll \lambda \le \lambda_\varepsilon,\ a \in I\!R. \tag{4.29}$$

In view of the assumption that $0 \in \mathcal{B}[u_0]$, there exist sequences $\{x_k\} \to 0$ and $\{t_k\} \to T^-$ such that $u(x_k,t_k) \to \infty$ monotonically as $k \to \infty$. Therefore by continuity, for any ε large enough, there exist x_ε and t_ε such that

$$u(x_\varepsilon,t_\varepsilon) > \varepsilon, \quad \text{where } x_\varepsilon \to 0 \quad \text{and } t_\varepsilon \to T^- \quad \text{as } \varepsilon \to \infty. \tag{4.30}$$

Fix an $\varepsilon = \varepsilon_k > 0$ large and put $\lambda = \lambda_\varepsilon,\ a = \lambda_\varepsilon t_\varepsilon$ in (4.10). Then $V(x,t_\varepsilon) = f(x)$, and by (4.29)

$$\text{Int}(t_\varepsilon, V) \le 2.$$

Due to the spatial forms of the profiles $u(x,t_\varepsilon)$ and $V(x,t_\varepsilon)$, the number of intersections can be equal 0, 1, or 2. Assume that $\text{Int}(t_\varepsilon,V) > 0$. (The case $\text{Int}(t_\varepsilon,V)=0$ is easier, see below.) Inspection of the values of $u(x,t_\varepsilon) - V(x,t_\varepsilon)$ at $x = x_\varepsilon$ and for x close to endpoints of the domain of positivity $\{V(x,t_\varepsilon)>0\}$ allows us to conclude that all intersections of $u(x,t_\varepsilon)$ and $V(x,t_\varepsilon)$ lie on the same side with respect to $x = x_\varepsilon$, say, to the left. We then translate the profile $V(x,t_\varepsilon)$ to the right. One can see that by construction there exists a finite $b_\varepsilon < 0$ such that $\tilde V(x,t_\varepsilon) \equiv f(x+b_\varepsilon)$ does not intersect $u(x,t_\varepsilon)$. Indeed, if it is not the case, there exists another $b_\varepsilon < 0$ such that

$$\mathrm{Int}(t_\varepsilon, \tilde{V}) \geq 3$$

contradicting (4.29). Recall that, in view of the mentioned properties of the function f (it is steep near $x = x_\varepsilon$), we have that $|b_\varepsilon|$ is uniformly bounded for $\varepsilon \gg 1$, and moreover $b_\varepsilon \to 0$ as $\varepsilon \to \infty$.

Thus we have that $\mathrm{Int}(t_\varepsilon, \tilde{V}) = 0$; see Figure 4.3 (b) (where for simplicity $\tilde{V} = V$). Then, by comparison, we deduce that

$$u(x,t) \geq \tilde{V}(x,t) \equiv f(x - \lambda_\varepsilon(t - t_\varepsilon) + b_\varepsilon) \tag{4.31}$$

for all $t \in (t_\varepsilon, T)$. By comparison, the same inequality (4.31) is valid for the proper (minimal) solution for all $t \geq T$.

Let us now take a $\delta > 0$ independent of ε and put $t = T + \delta > T$ in (4.31). This means that, for any small $\delta > 0$,

$$u(x, T+\delta) \geq \tilde{V}(x, T+\delta) \equiv f(x - \lambda_\varepsilon(T - t_\varepsilon + \delta) + b_\varepsilon), \tag{4.32}$$

and by construction this estimate is valid for all $\varepsilon = \varepsilon_k \gg 1$ (recall that $\lambda_\varepsilon \to \infty$, $t_\varepsilon \to T^-$ and $b_\varepsilon \to 0$ as $\varepsilon \to \infty$). Taking the envelope of the set of functions

$$\{V = f(x - \lambda_\varepsilon(T - t_\varepsilon + \delta) + b_\varepsilon)\}$$

with respect to the parameter $\varepsilon = \varepsilon_k \gg 1$, or equivalently with respect to $\lambda = \lambda_\varepsilon \gg 1$, we easily calculate from inequality (4.32) that, since it is valid for all small $\delta > 0$, there holds

$$u(x, T+\delta) \geq L(x) = \sup\nolimits_{\varepsilon \gg 1} f(x - \lambda_\varepsilon(T - t_\varepsilon + \delta) + b_\varepsilon). \tag{4.33}$$

Hence,

$$L(x) \equiv \sup_{\lambda \gg 1} f(x - \lambda\delta) = \infty \quad \text{for all} \quad x > 0.$$

This inequality means that after the blow-up time $t = T$ the ∞-level propagates with infinite speed to the right, away from the blow-up point $x = 0 \in \mathcal{B}[u_0]$.

The proof of infinite propagation of the ∞-level to the left is the same after a symmetry, which is equivalent to taking negative $\lambda \ll -1$ in our construction. □

Existence of nontrivial continuation

In the case of incomplete blow-up we can easily derive a condition of *global continuation in time*, when the singular TW ends nicely at $f = 0$ and the initial data are located below it.

Theorem 4.8 (Global continuation) *Assume that* (4.8) *and* (4.9) *hold, and there exists a global singular TW solution with* $f(\xi)$ *satisfying* (4.24). *Then, for any compactly supported* $u_0(x)$, *blow-up is incomplete and the proper solution satisfies* $u(x,t) \not\equiv \infty$ *for all* $t > T$.

It follows that finite *blow-up interfaces* occur, on which $u = \infty$. The proof is based on a standard comparison as in the proof of Theorem 4.6 as shown on Figure 4.4. Observe that, on the contrary, blow-up is always global and complete for flat

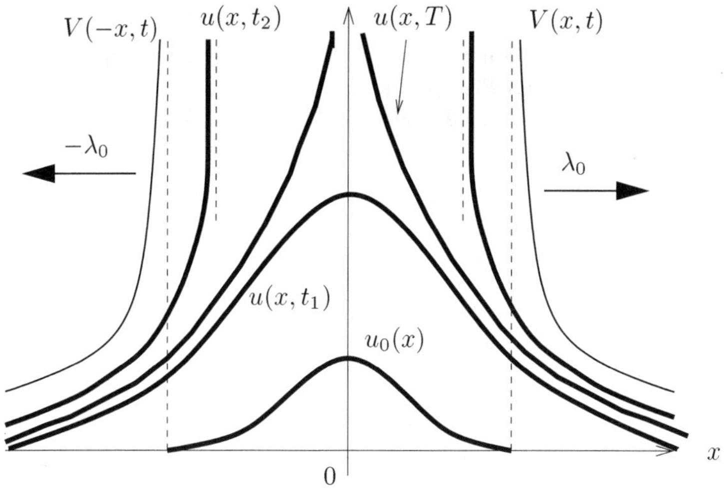

Figure 4.4 *The idea of incomplete blow-up is comparison: $u(x,t)$ cannot overtake the singular TWs $V(x,t)$ and $V(-x,t)$ for all $t>0$. Here $0<t_1<T<t_2$.*

solutions. For more general singular TWs, we can still prove incomplete blow-up, but continuation of the solution is only asserted for a certain time beyond T.

Theorem 4.9 (Local incomplete blow-up) *Let* (4.8) *and* (4.9) *be valid. Then, if the blow-up at $t=T$ is not global, i.e., $\mathcal{B}[u_0] \neq I\!R$, the proper solution continues as a finite function in some rectangle $R=\{(x,t): x_1<x<x_2, T<t<T+\delta\}$.*

Proof. In this more general case, we need a slight modification to the direct comparison proof illustrated in Figure 4.4. Since, by assumption, the closed blow-up set $\mathcal{B}[u_0]$ is not the whole line $I\!R$, there exists an x interval disjoint with $\mathcal{B}[u_0]$. Moreover, we may assume that there is an interval, say, $I=(0,2l)$, and a constant $M>0$ such that $\limsup_{t\uparrow T} u(x,t) \leq M$ for all $x \in I$. We will need to take l small, so that a singular TW fits above the level $u=M$. This will be explained below. Recall also that the approximations u_n are uniformly bounded for all fixed $t>0$. In particular, at $t=T$, we have $u_n(x,T) \leq c_n < \infty$. Now we simplify the geometry by replacing the u_n by the functions $\overline{u}_n(x,t)$ defined for $t \geq T$ as the solutions of the approximate problems

$$u_t = (\varphi(u))_{xx} + \psi_n(u)$$

with initial data

$$\overline{u}_n(x,T) = \begin{cases} M & \text{if } x \in I, \\ c_n & \text{otherwise.} \end{cases} \tag{4.34}$$

In this way it is clear that:

(i) $u_n(x,t) \leq \overline{u}_n(x,t)$ for all $t \geq T$ and $x \in I\!R$, hence our continuation problem is solved if we prove local finite continuation for $\overline{u} = \lim_{n\to\infty} \overline{u}_n$.

(ii) For all $t \geq T$, the functions $\overline{u}_n(x,t)$ are monotone decreasing for $x<l$ and increasing for $x>l$.

Our further analysis is based on comparison. The only influence of the parabolic equation under consideration is that solutions u_n and V satisfy the strong MP and comparison in any domain where both are uniformly bounded.

By the assumptions (4.8) and (4.9), there exists a singular TW solution $f(\xi)$ that blows up at a finite ξ, say 0, thus having an ∞-level interface that propagates with the finite speed $\lambda_0 > 0$. Our analysis is based on local comparison applied for $t \geq T$ to the singular TW

$$V(x,t) = f(x - \lambda_0(t-T)),$$

and the bounded positive approximations $\overline{u}_n(x,t)$, $n \gg 1$, which we write as u_n for convenience and without fear of confusion. We need to take l small so that $f(l) \geq 4M$. Then, by (4.34), we have

$$u_n(x,T) \leq V(x,T) \quad \text{for} \quad 0 \leq x \leq l, \tag{4.35}$$

$$V(x,t) \geq 2M \quad \text{in} \quad S_\delta = (0,l) \times (T, T+\delta)$$

provided that $\delta > 0$ is sufficiently small. The rest of the proof is based on the local comparison. We introduce another TW

$$\tilde{V}(x,t) = f((2l - x) + \lambda_0(t-T)),$$

which is the reflection of V about $x = l$. Arguing by contradiction, we assume that

$$u_n(x,t) \to \infty \quad \text{as} \quad n \to \infty \quad \text{on} \quad I \tag{4.36}$$

for any $t \in (T, T+\delta)$. This means that the inequality

$$u_n(x,t) \leq \max\{V(x,t),\, \tilde{V}(x,t)\} \tag{4.37}$$

must be first violated near the TW blow-up points, i.e., at $x \approx 0$ and $x \approx 2l$; see Figure 4.5. But this is impossible by the local comparison with these singular TWs, exactly as shown on Figure 4.4. Indeed, if, for all $t \in (T, T+\delta)$, $u_n(x,t)$ stays below $V(x,t)$ for $x \approx 0$ and below $\tilde{V}(x,t)$ for $x \approx 2l$, then we deal with the uniformly bounded solutions u_n (the L^∞-bound of u_n depends on V and $\tilde{V}$ only and is independent of n) of the uniformly parabolic equations for which (4.36) cannot happen for $t \in (T, T+\delta)$ arbitrarily close to T.

We thus obtain a small rectangle $R = [\frac{l}{2}, \frac{3l}{2}] \times [T, T+\delta]$, where the monotone sequence $\{u_n\}$ is uniformly bounded and hence the proper solution $u(x,t)$ is bounded. By interior regularity results we conclude that u is a weak or a classical solution in R if the coefficients are sufficiently smooth in the range of u. $\square$

The analysis of the particular equation (4.11) with power nonlinearities is easier. We arrive finally at the criterion (4.12) for incomplete blow-up. Similarly, for equation (4.13) with the p-Laplacian operator, the same analysis gives the criterion (4.14). We postpone these rather straightforward computations until Section 7.11 devoted to more general models with doubly nonlinear operators.

A global continuation theorem cannot be obtained in the situation of Theorem 4.9 without some additional information on the behaviour of ψ and φ or the data u_0 near the level $u = 0$, as is done in Theorem 4.8. Let us mention some situations, leading to non-global results. Thus, if we assume that $\psi(u) \geq c > 0$

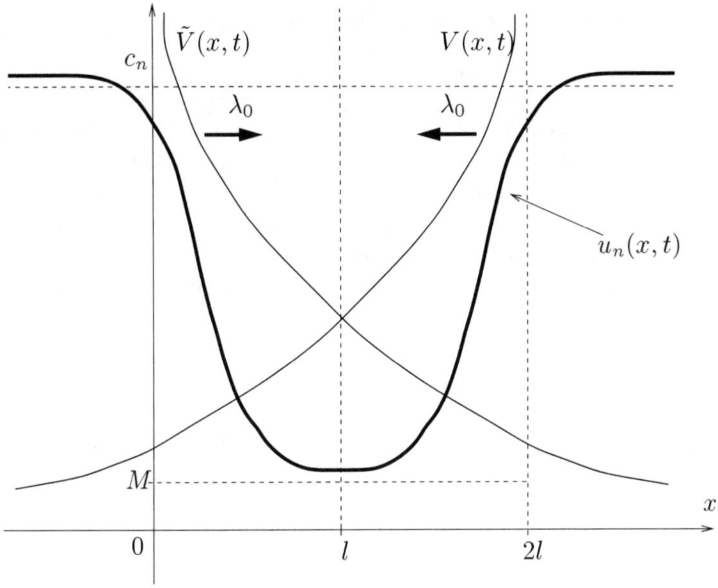

Figure 4.5 *Illustration of the proof of Theorem* 4.9*: for* $t > T$ *the solution* $u_n(x,t)$ *cannot grow fast as* $n \to \infty$ *without overtaking singular TWs* $V(x,t)$ *and* $\tilde V(x,t)$ *near blow-up interfaces, which is forbidden by local comparison.*

for all $u \geq 0$, it is immediately clear from comparison with the solution of the ODE $u_t = \psi(u)$ with $u(0) = 0$ that all solutions will eventually blow up in the whole space. The same happens if $u_0 \geq c_1 > 0$ and $\psi(u) \geq c > 0$ for $u \geq c_1$.

Finally, we present a monotonicity property of the blow-up set of the proper minimal solution

$$\mathcal{B}[u](t) = \{x \in I\!R : \exists \ \{x_k\} \to x, \ \{t_k\} \to T^- \text{ such that } u(x_k,t_k) \to \infty\},$$

which is defined for $t \geq T(u_0)$.

Proposition 4.10 *Assume that*

$$\int_1^\infty G(s)\,\mathrm{d}s = \infty. \tag{4.38}$$

Then $\mathcal{B}[u](t_1) \subseteq \mathcal{B}[u](t_2)$ *for all* $t_2 \geq t_1 \geq T$.

The proof relies on a general time-monotonicity property of large solutions; see Remarks. This obviously implies that *global blow-up is always complete.*

4.5 The extinction problem

The techniques developed for the blow-up problem allow us to understand the problem of extinction in nonlinear diffusion-absorption equations, in principle a different problem, which is in fact closely related mathematically. Furthermore, the results have a strong resemblance. Consider the equation

$$\boxed{u_t = (\varphi(u))_{xx} - \psi(u) \quad \text{in } S.} \tag{4.39}$$

We make the same assumptions on φ as in Section 4.1. The absorption term is $-\psi(u)$, where $\psi : I\!R_+ \to I\!R_+$ is a positive C^1-function. Our $\psi(u)$ will not be continuous at $u = 0$ in general. Different variants of this model, mainly with linear diffusion $\varphi(u) = cu$, are used in the chemical literature and have been investigated mathematically; see references in Remarks. In these applications u represents the concentration of a reactant. We consider the Cauchy problem with initial data u_0 that are C^1-smooth, $|\mathrm{d}\varphi(u_0)/\mathrm{d}x| \le C$, and bounded above and below uniformly away from zero, i.e., $0 < c \le u_0(x) \le C$. We also assume an inverse bell-shaped form of u_0.

It is well-known that under these conditions when $\psi(u)$ is continuous at 0 but the absorption rate $\frac{1}{u}\,\psi(u)$ diverges as $u \to 0$ (so-called *strong absorption* in the literature), the solution is strictly positive up to a finite time, $0 < t < T$, but as $t \to T^-$ the solution reaches the 0-level, i.e.,

$$\inf_{x \in I\!R} u(x,t) \to 0 \quad \text{as } t \to T^-. \tag{4.40}$$

The solution can be continued in the weak sense for $t > T$ (or, in general, as the maximal solution) and a non-empty region appears, called the *dead core* or *depleted zone*, where u vanishes,

$$\mathcal{D}(t) = \{x : u(x,t) = 0\}. \tag{4.41}$$

The properties of $\mathcal{D}(t)$ are an important object of research. When, as in this case, $\mathcal{D}(t)$ is not the whole space, we say that *incomplete extinction* occurs. *Complete extinction* at time t means that $\mathcal{D}(t) = I\!R$. The words *quenching* or *vanishing in finite time* are also used to denote these phenomena.

In this section we assume that the absorption is singular at the level $u = 0$, i.e.,

$$\psi(u) \to \infty \quad \text{as } u \to 0. \tag{4.42}$$

Due to the assumption on ψ, a singularity develops in the equation as $t \to T^-$, which is reflected in the behaviour of the solution at the extinction time. Again, there is a question of continuation after extinction. The natural way to investigate the possible continuation consists of replacing f by smooth approximations ψ_n that converge monotonically to ψ from below. Then we obtain global positive solutions u_n and a proper solution of our original problem is still given by formula (4.5). Complete extinction means that $u \equiv 0$ for $t > T$.

We discuss in some detail the methods developed above to deal with the blow-up problem to study the existence of complete/incomplete extinction after $t = T$ in terms of the behaviour at $u = 0$ of the function $G(u) = \varphi'(u)\psi(u)$ given in (4.7). Let us introduce the basic condition for incomplete extinction

$$\int_0^1 G(s)\,\mathrm{d}s < \infty. \tag{4.43}$$

Before we proceed with the proofs we want to illustrate the results with the simplest and more popular instance of equation (4.39), namely, the one-dimensional semilinear heat equation with singular absorption

$$\boxed{u_t = u_{xx} - u^p \quad \text{in } S, \quad p < 0,} \tag{4.44}$$

with data as explained above. We prove that *complete extinction* for (4.44) occurs *if and only if* $p \leq -1$. On the other hand, if $p > -1$, the maximal solution $u(x,t)$ is nontrivial for $t > T$. The connection between extinction and blow-up is established in this case by the change of variables $v = \frac{1}{u}$, which translates the extinction behaviour (4.40) for equation (4.44) into the equivalent study of blow-up for the nonlinear parabolic equation

$$v_t = v_{xx} - \tfrac{2}{v}\,(v_x)^2 + v^{2-p}.$$

This shows that both singular phenomena can be studied in a unified manner as a singularity occurring at the zero-level, but for rather general (not necessarily divergent) quasilinear parabolic equations. This will be done in Chapter 7 in the most general setting.

4.6 Complete and incomplete extinction via singular travelling waves

We consider the Cauchy problem for the quasilinear heat equation (4.39) with the conditions on φ and ψ already stated. In particular, ψ is singular at 0 and (4.42) holds. We also take initial data (4.2), which are assumed to be bounded and uniformly bounded away from zero. As usual, we introduce the extinction set

$$\mathcal{E}[u_0] = \{x \in I\!R : \exists\, \{x_k\} \to x, \ \{t_k\} \to T^- \ \text{with } u(x_k, t_k) \to 0\}. \tag{4.45}$$

We consider the possible continuation after extinction of the proper (maximal) solution of the problem. Actually, as we have mentioned, there are direct similarities between blow-up and extinction. Such a comparison with blow-up properties is not formal since, by the transformation $u = \frac{1}{v}$, the extinction problem for equation (4.39) reduces to the blow-up one for a different nonlinear parabolic equation

$$v_t = \varphi'(\tfrac{1}{v})v_{xx} + v^2(\varphi'(\tfrac{1}{v})\tfrac{1}{v^2})'(v_x)^2 + v^2 f(\tfrac{1}{v}),$$

and $v(x,t)$ blows up in the sense of (4.3). Recall that (4.42) implies a superlinear growth of the source term $v^2 f(\frac{1}{v})$ as $v \to \infty$. Therefore, the results of complete/incomplete extinction can be directly derived by using similar blow-up ones from Section 4.4. However, it is convenient to study some specific extinction properties in terms of the original equation (4.39) with singular absorption, but we consider the extinction behaviour in less detail. As in Section 4.4, the proof depends on the analysis of the nonexistence/existence of suitable singular TWs plus intersection comparison arguments.

Analysis of travelling wave solutions

Introducing the TW solution (4.10), after necessary transformations we arrive at the equation (cf. (4.16))

$$P\tfrac{\mathrm{d}P}{\mathrm{d}f} + \lambda P - G(f) = 0 \quad \text{for } f > 0, \quad \text{where } P = \tfrac{\mathrm{d}\varphi(f)}{\mathrm{d}\xi}, \tag{4.46}$$

where G is prescribed in (4.7). We are interested in the solutions $P = P(f)$ of equation (4.46) living in the first quadrant $\{f > 0, P \geq 0\}$. The important point is the existence of an orbit starting from the P-axis, i.e.,

$$P(f) > 0 \quad \text{exists for any small } f > 0. \tag{4.47}$$

Equivalently, what we want to find is a monotone and nonnegative TW with a finite speed of propagation λ that starts from the level $f = 0$. As above, we want it to start at a finite ξ and we call such a solution a *singular TW*. The study of the orbits P of (4.46) is similar to that given in Section 4.2. Let us comment on some details.

We note that when $P(0) \neq 0$, we obtain a TW that is defined only for some $\xi \geq \xi_0$ and $f(\xi_0) = 0$ without continuation to the region $\xi < \xi_0$. On the contrary, when $P(0) = 0$, the corresponding TW solution of (4.39) could be defined for all $(x,t) \in I\!R \times I\!R_+$ (a global TW). Finally, we remark that such a monotone TW can be zero at either finite or infinite ξ. In the former case, it exhibits an infinite tail where $f(\xi) \to 0^+$ as $\xi \to -\infty$. In the latter one, there is a ξ_0 (we can take $\xi_0 = 0$ in view of the translational invariance) such that $f(\xi) > 0$ for $\xi > \xi_0$ and $f(\xi) = 0$ for $\xi \leq \xi_0$. However, these details will not enter our present consideration.

Lemma 4.11 *Let* (4.42) *hold. Then:*

(i) If (4.43) *does not hold, there are no singular TWs for any* $\lambda > 0$.

(ii) If (4.43) *holds, then, for any* $\lambda > 0$ *and every* $P_0 \geq 0$, *there exists a singular TW with* $P(0) \geq P_0$.

Proof. First of all, we rewrite (4.46) as

$$\tfrac{\mathrm{d}P}{\mathrm{d}f} = -\lambda + \tfrac{G(f)}{P}. \tag{4.48}$$

Since the right-hand side is larger or equal than $-\lambda$, it is clear that any solution of (4.48) defined and positive in a small interval $(0, f_0)$ must have a finite limit $P(0) \geq 0$ as $f \to 0$. Therefore we may suppose that $P(f) \leq c$ for small $f > 0$.

(i) In the domain $\{P \leq c\}$ we have from (4.48) that

$$\tfrac{\mathrm{d}P}{\mathrm{d}f} \geq -\lambda + \tfrac{1}{c}\, G(f). \tag{4.49}$$

Integrating this inequality over $(0, f)$ with the assumption

$$\textstyle\int_0 G(s)\,\mathrm{d}s = \infty$$

implies the nonexistence of a solution satisfying (4.47).

(ii) We will find an orbit of (4.48) satisfying (4.47) by the analysis of the flow in a suitable region. Assuming that (4.43) holds, for a fixed constant $Y_0 > 0$, we define

$$Y(f) = Y_0 + \tfrac{1}{Y_0} \textstyle\int_0^f G(s)\,\mathrm{d}s.$$

One can check that, for small $f > 0$,

$$\tfrac{\mathrm{d}P}{\mathrm{d}f}\big|_{P=Y(f)} = \tfrac{G(f)}{Y(f)} - \lambda < \tfrac{G(f)}{Y_0} \equiv Y'(f).$$

The region we need is defined as $R = \{(f,P) : P > Y(f),\ 0 < f < c\}$ for small $c > 0$. The flow points inwards at the wall $P = Y(f)$, and outwards through

$f = c$. By a simple topological argument we conclude that there exist infinitely many solutions inside R each one satisfying (4.47). Moreover, we also have that

$$P(f) \geq Y_0 > 0 \quad \text{for all small } f > 0.$$

Integrating this inequality and using the fact that $\varphi(0) = 0$, we have that such $f(\xi)$ represents a singular TW. □

Extinction analysis

Theorem 4.12 *Assume that* (4.42) *holds. Then:*

(i) (Complete extinction) *Let* $\mathcal{E} \neq \emptyset$ *and the integral in* (4.43) *diverge. Then* $u \equiv 0$ *for* $t > T$.

(ii) (Incomplete extinction) *On the contrary, let* $\mathcal{E} \neq I\!\!R$ *and* (4.43) *hold. Then the solution has a nontrivial continuation after the extinction time* T.

The above criterion of complete/incomplete extinction is simpler than that of blow-up in Section 4.4, where the three conditions (4.4)–(4.9) were involved in the analysis, since we have already imposed a strong hypothesis on ψ, (4.42), and since φ is assumed to be continuous at 0. If we weaken such requirements in line with minimal conditions for extinction (as in the blow-up case), then we need to take into account all three conditions.

Proof of Theorem 4.12. (i) It is quite similar to that in Section 4.4 for complete blow-up. We take $\varepsilon \to 0$ in (4.26). Then using the same construction, since $u(x,t) > 0$ for $t < T$ is uniformly bounded and $V(x,0) \gg 1$ for $|x| \gg 1$, we conclude that

$$\text{Int}(0, V) = 2 \quad \text{for all } 1 \ll \lambda \leq \lambda_\varepsilon \text{ and } a \in I\!\!R.$$

Indeed, it follows from inequality (4.49) that under assumption (4.47) the function $V(x,0)$ is steep provided that $\lambda \leq \lambda_\varepsilon \to \infty$ as $\varepsilon \to \infty$. Therefore (4.29) is valid. Due to the form of the functions for large x the number of intersections is now either 2 or 0. The rest of the proof is the same.

(ii) Assume now that we have a singular TW solution $V(x,t)$ vanishing at a finite x and thus having a 0-level interface that propagates with finite speed. Then, as above, we use a local comparison of the proper solution $u(x,t)$ with the solution $V(x,t)$ in the domain where both are small enough to ensure a non-steep behavior of u and the corresponding integral estimate (see Section 4.4). □

It is clear from the method of the proof of complete extinction that our construction of the envelope of the set $\{V\}$ for $\lambda \gg 1$ does not depend on the fact that we consider the proper solution defined for all $x \in I\!\!R$ (the Cauchy problem). Indeed, consider e.g. the initial-boundary value problem on a bounded interval $x \in (a,b)$ with, say, Dirichlet boundary conditions $u = 1$ for $x = a$, b for $t \geq 0$. Then, assuming that $0 \in (a,b) \cap \mathcal{E}$, we perform the same analysis as above to show that $u(x, T + \delta) \equiv 0$ in (a,b). Indeed, one can see that TWs with $\varepsilon \ll 1$ and $\lambda_\varepsilon \gg 1$ are finally so steep that, in our intersection comparison analysis, they do not interact with any standard regular boundary conditions at the points $x = a$, b that stay away from the singular point $0 \in \mathcal{E}$. Roughly speaking, we conclude that, independently of boundary conditions, the envelope satisfies $L(x) \equiv 0$ in (a, b), whence complete extinction.

Remarks and comments on the literature

Main results are presented in [165].

§ **4.1.** W.F. Osgood's criterion (4.4) was established in 1898 [281]; see more details in E.L. Ince [203, Part I.3].

A construction of a minimal continuation beyond blow-up at a finite $t = T$ for the semilinear heat equation (4.6) by a monotone increasing truncation $\psi \mapsto \psi_n$ was done by P. Baras and L. Cohen [37]. See also other results on this subject in [236] and [235]. It was also proved in [37] that, for the power nonlinearity $\psi(u) = u^p$ for $p > 1$, blow-up is always complete in the subcritical Sobolev range (i.e., below the *critical Sobolev exponent*) $p < p_S = \frac{N+2}{N-2}$. In the critical case $p = p_S$ blow-up is also complete for at least radial solutions. See [167], where a construction of a unique proper (minimal) solution and the corresponding extended limit semigroup were performed for the general quasilinear heat equation $u_t = \Delta u^m + u^p$. Complete and incomplete blow-up for such equations in $I\!R^N$ is studied in Chapter 6.

§ **4.2.** First results on nonuniqueness of TW solutions for a family of quasilinear heat equations (4.1) were obtained in [126], where a description of other related properties of TWs is available. For (4.11) nonuniqueness of TW solutions for $m + p \leq 2$ and determining a unique maximal one by a smooth positive approximation were established in [285].

§ **4.3.** We have found no evidence of the pathological class of heat equations in the existing literature published earlier.

§ **4.4.** The idea of "envelopes" in the intersection comparison with a family of particular solutions was introduced in [127] and has been used in [148] for a wide class of quasilinear reaction-diffusion equations as the *method of stationary states* (intersection comparison with a family of stationary solutions).

A more general proof of Theorem 4.9 is given in Section 7.3. In the proof of Proposition 4.10 we use a general result on the monotonicity with time of any sufficiently large solution, [150], [154].

§ **4.5.** The quasilinear heat equations with absorption describe important physical processes, cf. references in the papers [213], [243] and in [170, Chapter 4].

§ **4.6.** The fact that under the above hypotheses on nonlinearities the solution with initial data having an inverse bell-shaped form has *single-point extinction* is proved by a slight modification to the method proposed in [123] for the semilinear equation (4.6). See a similar analysis of blow-up for general quasilinear equations in [151], [153], [161], [166] and [170, Section 10.4].

§ **4.7.** Different types of application of the method of stationary states to quasilinear parabolic problems can be found in [127], [148], [153], [161]; see also [306, Chapter 7] and [170, Chapter 10].

CHAPTER 5

Blow-up Interfaces for Quasilinear Heat Equations

In this chapter we choose a special porous medium equation (PME) with power source admitting *incomplete* blow-up. We then obtain the unusual *blow-up interfaces* describing singular propagation on the infinite level $\{u = \infty\}$. Using this particular equation, we explain basic intersection comparison techniques of studying some evolution and regularity properties of such remarkable interfaces that do not obey the classical Darcy law for the PME.

It is of key importance that blow-up interfaces cannot be very smooth and are at most $C^{1,1}$ functions (the first derivative is Lipschitz continuous) and are not C^2 in principle. This is in striking contrast with interfaces for the 1D PME in $I\!R \times I\!R_+$ or other similar quasilinear heat equations, where interfaces are known to become C^∞ or even analytic functions after waiting time. Such a special finite regularity property of singular blow-up interfaces is associated with the non-monotonicity of the corresponding Rankine–Hugoniot diagram describing the dependence of the interface velocity upon the interface spatial slope.

5.1 Introduction: First properties of incomplete blow-up

Nonnegative solutions $u = u(x,t)$ of the quasilinear heat equation

$$\boxed{u_t = (u^m)_{xx} + u^p \quad \text{in } S = I\!R \times I\!R_+, \quad m > 0, \ p > 1} \tag{5.1}$$

may blow up in finite time for some initial data $u(x,0) = u_0(x) \geq 0$, $u_0 \not\equiv 0$. For $1 < p \leq m+2$, solutions blow-up for arbitrarily small initial data u_0, while for $p > m+2$ blow-up always occurs for sufficiently large initial functions.

Let $u(x,t)$ be the unique global proper (minimal) solution constructed by monotone increasing approximations (see Sections 4.1 and 6.2 for more details), and $T = T(u_0)$ be its finite blow-up time. If the continuation of the solution beyond blow-up is trivial, i.e., $u(x,t) \equiv \infty$ for $t > T$, we say that the blow-up is *complete*, otherwise, if $u(x,t) \not\equiv \infty$ for $t > T$, it is *incomplete*.

The blow-up set $\mathcal{B}[u](t)$ is defined for every $t \geq T$ by the formula

$$\mathcal{B}[u](t) = \{x \in I\!R : \exists \ \{x_k\} \to x, \ \{t_k\} \to t^- \ \text{with} \ u(x_k,t_k) \to \infty\}, \tag{5.2}$$

and in the case of incomplete blow-up $\mathcal{B}[u](t) \not\equiv I\!R$ at least for $t \approx T^+$. This corresponds to the idea of *burnt zone* in the theory of flame propagation, while the boundary of this set $\partial\mathcal{B}[u]$ corresponds to the *flame front*. Let us summarize the known results on incomplete blow-up already established in Section 4.4.

Lemma 5.1 *Let u be the proper solution of* (5.1) *under the assumptions $m \in (0,1)$ and $1 < p \le 2 - m$ and with initial data $u_0(x) \to 0$ as $|x| \to \infty$.*

Then u can be continued in a nontrivial way for all times $t > 0$ (i.e., $u(\cdot, t) \not\equiv \infty$ for all $t > 0$) even if u blows up at a time $T < \infty$. For every $t > T$, the burned zone $\mathcal{B}[u](t)$ is a bounded subset of $I\!R$.

We perform a closer investigation of the behaviour of the solutions that undergo incomplete blow-up by studying especially the *blow-up interfaces* that separate the *burnt region* $\{u = \infty\}$ from the *unburnt zone*, where the solution remains finite.

From now on, we will concentrate on equation (5.1) for the parameter choice $0 < m < 1$ and the *critical* value $p = 2 - m$, i.e.,

$$\boxed{u_t = (u^m)_{xx} + u^{2-m},} \tag{5.3}$$

which leads to simpler and sometimes explicit mathematics but, nevertheless, describes the main distinguished properties of blow-up interfaces.

We consider the Cauchy problem posed for $x \in I\!R$ and time $t > 0$. We take initial data

$$u(x, 0) = u_0(x) \ge 0 \quad \text{in } I\!R, \ \sup u_0 < \infty, \ u_0(x) \to 0 \ \text{ as } |x| \to \infty, \tag{5.4}$$

so, by Lemma 5.1, a nontrivial proper solution exists globally in time. We establish the optimal linear expansion growth of the burnt zone with a precise estimate.

Theorem 5.2 *Let*

$$u_0(x) = o(|x|^{-1/(1-m)}) \quad \text{as} \ |x| \to \infty. \tag{5.5}$$

(i) Then the blow-up set $\mathcal{B}[u](t)$ of the proper solution satisfies

$$\mathcal{B}[u](t) \subseteq \{|x| \le 2\sqrt{m}(t + c_0)\} \quad \text{for } t \ge T, \tag{5.6}$$

where $c_0 = c_0(u_0)$ is a constant.

(ii) This estimate is asymptotically sharp and

$$\text{meas}\, \mathcal{B}[u](t) = 4\sqrt{m}t(1 + o(1)) \quad \text{as } t \to \infty. \tag{5.7}$$

Our investigation is based on the construction and intersection comparison with two important classes of explicit solutions exhibiting singular interfaces. They are the travelling wave (TW) solutions as in previous chapters and the *parabolic shaped* solutions.

As usual, the TWs are given by

$$U(x, t) = f(x - \lambda t),$$

where λ is the wave speed. Typically, we consider speeds $\lambda > 0$, and then solutions with negative λ are obtained by means of the symmetry $x \mapsto -x$. It is easy to show that such solutions exist for the speeds

$$\lambda \ge \lambda_m = 2\sqrt{m},$$

and then there are two branches of solutions $U_\pm$ satisfying

$$U_+(x, t; \lambda) \le U_-(x, t; \lambda) \quad \text{and} \quad U_+ \equiv U_- \ \text{ if } \lambda = \lambda_m.$$

Both $U_\pm(x,t)$ are defined in the same region $\{x > \lambda t\}$ with the singular interface $\{x = \lambda t\}$, at which $u = +\infty$ (flame front). It is remarkable that only the minimal branch $U_-(x,t)$ is a branch of proper (minimal) solutions of the corresponding Cauchy problem.

We continue the study with the class of explicit solutions $\tilde{v}(x,t)$ that, in terms of the pressure variable, have parabolic profiles in x for every t, i.e.,

$$\tilde{v}(x,t) \in \mathrm{Span}\{1, x^2\} \quad \text{for any } t \geq 0.$$

These solutions have the interest of starting with bounded data, blowing up at a finite time $t = T$, and then exhibiting finite blow-up fronts for $t > T$. An unusual phenomenon occurs for these explicit solutions, which are represented by analytic functions: they are proper solutions on some interval $[T, T_1]$ after the blow-up time $T < T_1$, and then they are no longer proper for $t > T_1$. It turns out that, for $t > T_1$, the interface propagation becomes *exactly linear* with time (exact solutions $\tilde{v}$ do not exhibit such a propagation) with eventual convergence to a TW with the minimal speed $\lambda_m = 2\sqrt{m}$.

In subsequent sections, by intersection comparison, we generalize these phenomena to wide classes of initial data. Thus we show that the singular interface $x = s(t)$ obeys the minimal speed law,

$$s'(t) \geq \lambda_m,$$

and moreover it starts with infinite speed (in particular, no waiting-time occurs). We investigate closely the dynamical equation at the interface for the proper solutions and prove that singular interfaces are essentially non-analytic for $t > T$. Moreover, they are $C^{1,1}$ and not C^2 functions. We also prove that, as $t \to \infty$, the general minimal solutions converge to the TW with the minimal speed.

In the last sections we describe some properties of blow-up interfaces for the p-Laplacian equation with source, for some quasilinear heat equations with general nonlinearities and in the N-dimensional geometry.

5.2 Explicit proper blow-up travelling waves and first estimates of blow-up propagation

In order to describe the properties of the general proper solutions of (5.3), we begin with a preliminary analysis of some explicit solutions. First of all, we introduce the pressure variable $v = \frac{m}{1-m} u^{m-1} > 0$ for $u > 0$ satisfying the quasilinear equation

$$v_t = \mathbf{A}(v) \equiv (1-m) v v_{xx} - (v_x)^2 - m \quad \text{in } S \cap \{v > 0\}. \tag{5.8}$$

It admits different types of explicit solutions due to the invariant properties of the quadratic operator $\mathbf{A}$. The simplest explicit solutions are the TW solutions $V(x,t) = f(x - \lambda t)$. The profile f is easy to compute as the solution of an ODE.

It is of interest, for the problems considered here, to view the construction in the following way: the operator $\mathbf{A}$ admits the invariant subspace

$$W_2 = \mathrm{Span}\{1, x\}, \quad \text{i.e.,} \quad \mathbf{A}(W_2) \subseteq W_2.$$

Therefore, there exist explicit solutions of the spatially linear form

$$V(x,t) = C_0(t) + C_1(t)\,x \in W_1. \tag{5.9}$$

Substituting (5.9) into (5.8) yields a simple dynamical system (DS) for the coefficients $\{C_0(t),\ C_1(t)\}$,

$$\begin{cases} C_0' = -m - C_1^2, \\ C_1' = 0, \end{cases}$$

which is the PDE (5.8) restricted to W_2. Solving this system yields the nonnegative solution

$$V(x,t;\lambda) = S(x - \lambda t)_+, \tag{5.10}$$

where the *wave or interface slope* $S = V_x$ and the *wave speed* λ are related by the formula

$$\lambda = S + \frac{m}{S} \equiv H(S). \tag{5.11}$$

Then V is positive in the domain $D_\lambda = \{(x,t) : x > \lambda t\}$ with the straight line interface at

$$\bar{s}(t) = \lambda t.$$

Formula (5.11) is the *Rankine–Hugoniot* condition of the TW interface propagation, which gives the relevant information about the dynamics of the interface, and will be basic in subsequent sections. The graph of H contains two monotone branches (Figure 5.1), one decreasing branch in the interval $0 < S < S_m$, where

$$S_m = \sqrt{m}$$

is the slope corresponding to the *minimal speed*

$$\lambda_m = 2\sqrt{m},$$

and an increasing branch for $S > S_m$, where λ goes from λ_m to infinity. As a first step, we will prove that, as usual, proper (maximal) solutions stay on the decreasing branch of H. This is true not only for the TWs but also for general classes of solutions.

Thus we have that all TWs have speed $\lambda \ge 2\sqrt{m}$ and that for every $\lambda > \lambda_m$, there exist two different profiles (5.10) with slopes $S_\pm = S_\pm(\lambda)$ given by

$$S_\pm = \tfrac{1}{2}\,(\lambda \pm \sqrt{\lambda^2 - 4m}), \tag{5.12}$$

so $V_-(x,t;\lambda) < V_+(x,t;\lambda)$ everywhere in their common positivity domain D_λ. For the minimal speed λ_m, both solutions coincide. In terms of u, the two solutions are given by the formulae

$$U_\pm(x,t;\lambda) = \left[\tfrac{1-m}{m}\, S_\pm(x - \lambda t)_+\right]^{-\frac{1}{1-m}} \tag{5.13}$$

in D_λ, where these are solutions of equation (5.3). They have finite blow-up interfaces at $x = \lambda t$ and are naturally extended by putting $u = +\infty$ in the remaining space-time domain $\{x \le \lambda t\}$. Let us recall for later use that, for a fixed λ satisfying the strict inequality $\lambda > 2\sqrt{m}$, we have the opposite inequality

$$U_+ < U_- \quad \text{in} \quad D_\lambda.$$

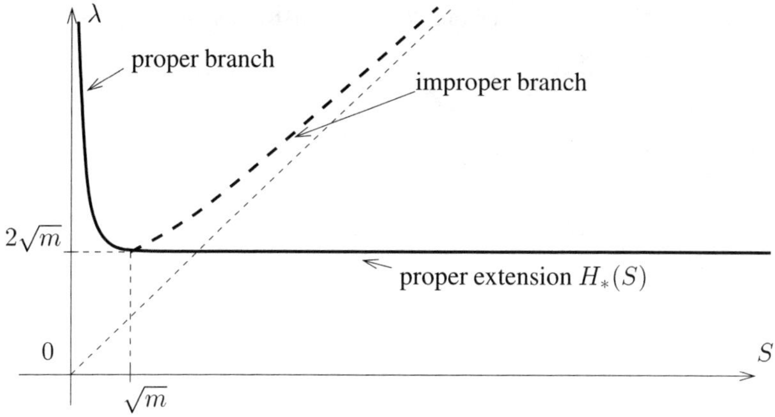

Figure 5.1 *Rankine–Hugoniot diagram* (5.11) *consisting of the decreasing branch of proper solutions and the increasing improper one.*

We now show that the solutions $V_-(x,t;\lambda)$ are proper maximal solutions of the Cauchy problem with the corresponding initial data. In particular, this implies that V_- can be used in comparison with other general proper solutions. Note that maximal for v is equivalent to minimal for u.

Lemma 5.3 *(i) For all* $\lambda \geq 2\sqrt{m}$, *the function* $V_-(x,t)$ *is a proper solution of the Cauchy problem for* (5.8).

(ii) For $\lambda > 2\sqrt{m}$, $V_+(x,t;\lambda)$ *is not a proper solution.*

Proof. (i) We will use a rare opportunity to perform some explicit computations establishing the result. Truncating (approximating) both the equation and the data, we construct U_- as the monotone limit as $n \to \infty$ of the sequence $\{u_n(x,t)\}$ of global smooth solutions to the problems

$$u_t - (u^m)_{xx} + \psi_n(u), \quad \psi_n(u) = \min\{u^{2-m}, n^{2-m}\}, \quad n = 1, 2, ..., \tag{5.14}$$

with the initial data $u_{0n}(x) = \min\{U_-(x,0), n\}$. By standard comparison, we have the upper bound

$$u_n(x,t) \leq U_-(x,t). \tag{5.15}$$

On the other hand, we deduce that the following lower bound holds:

$$u_n(x,t) \geq U_n(x,t), \tag{5.16}$$

where $U_n = f_n(\xi)$ is the corresponding TW solution of (5.14) satisfying, in the positivity domain, the ODE problem

$$(f^m)'' + \lambda f' + \psi_n(f) = 0 \quad \text{in } I\!R, \quad f(0) = n, \; f'(0) = 0. \tag{5.17}$$

One can see that $f_n(\xi) \leq n$, and therefore we may put $\psi_n(f) = f^{2-m}$ in this range. Then equation (5.17) becomes

$$(f^m)'' + \lambda f' + f^{2-m} = 0, \;\; \xi \in I\!R \quad (0 \leq f \leq n) \tag{5.18}$$

with the same boundary conditions. It is easily integrated by the transformation

$P = -(f^m)' = fF(f)$, which reduces it to the first-order ODE

$$f\frac{dF}{df} = -\frac{1}{F}(F^2 - \lambda F + m) \equiv -\frac{1}{F}(F - S_+)(F - S_-), \tag{5.19}$$

with $S_\pm$ given in (5.12). Then the solution $F_n(f)$ satisfying $F_n(0) = 0$ has the form

$$\frac{1}{S_+ - S_-}(S_+ \log|F_n - S_+| - S_- \log|F_n - S_-|) = -\log f + C_n, \tag{5.20}$$

where

$$C_n = \log n + \frac{S_+ \log S_+ - S_- \log S_-}{S_+ - S_-}$$

for $\lambda > \sqrt{m}$, while for $\lambda = 2\sqrt{m}$ we have

$$\log|F_n - \sqrt{m}| - \frac{\sqrt{m}}{F_n - \sqrt{m}} = -\log f + C_n, \tag{5.21}$$

where

$$C_n = \log n + \tfrac{1}{2} \log m + 1.$$

As we have shown above, in the domain $\{0 < f < n, \ P > 0\}$ equation (5.18) admits two explicit solutions (cf. (5.12))

$$P_\pm(f) = S_\pm f.$$

It is easy to see from the phase-plane of equation (5.19) that $F_n(f) < S_-$ for $f \in (0, n)$. It follows from (5.20) or (5.21) that $F_n(f) \to S_-$ as $n \to \infty$ pointwise for any $f > 0$; see Figure 5.2. Moreover, the convergence is uniform on compact subsets $[\delta, \frac{1}{\delta}]$ for any small $\delta > 0$. This implies that, as $n \to \infty$, $P_n(f) = fF_n(f) \to S_- f$ uniformly on subsets $\{0 \le f \le c\}$. Finally, passing to the limit in (5.15) and (5.16), we obtain that $u_n \to U_-$, which completes the proof of (i).

(ii) If $\lambda > \lambda_m = 2\sqrt{m}$, we fix $\lambda_1 \in (\lambda_m, \lambda)$. Then still $V_-(x, 0; \lambda_1) \le V_+(x, 0; \lambda)$ but since the propagation speeds verify $\lambda_1 < \lambda$, we have that the inequality $V_-(x, t; \lambda_1) \le V_+(x, t; \lambda)$ is not true in $I\!R$ for any $t > 0$ small, i.e., a new intersection between solutions occur at the interface as Figure 5.3 illustrates. Since V_- is the proper solution, we conclude that V_+ is not a proper one since it does not satisfy the MP (no comparison with this solution holds). $\square$

Proof of Theorem 5.2 *(i).* Estimate (5.6) follows by comparison with proper TWs $U_-(x \pm a, t; \lambda_m)$. $\square$

5.3 Explicit blow-up solutions on an invariant subspace

The operator $\mathbf{A}$ in (5.8) admits another 2D invariant subspace

$$\tilde{W}_2 = \mathrm{Span}\{1, x^2\}.$$

In fact, it admits a 3D invariant subspace $W_3 = \mathrm{Span}\{1, x, x^2\}$, where the 1D subspace $\mathrm{Span}\{x\}$ is not essential because of the translational invariance of the

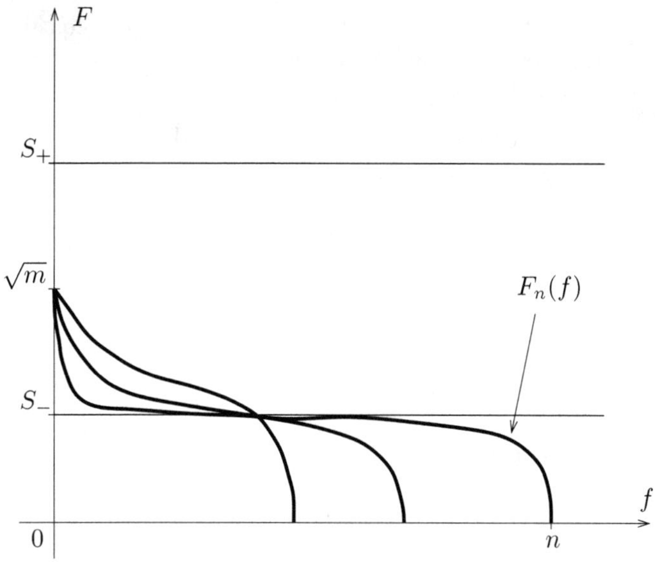

Figure 5.2 *On the* $\{f, F\}$*-plane, the approximating sequence* $\{F_n(f)\}$ *converges as* $n \to \infty$ *to the minimal slope* S_- *and cannot converge to the maximal one* S_+.

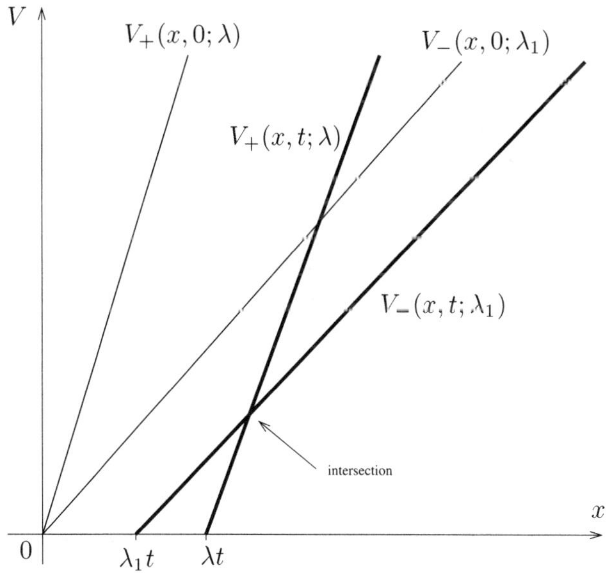

Figure 5.3 $V_+(x, t; \lambda)$ *is not a proper solution since, for* $t > 0$*, it violates the comparison principle: a new intersection with the proper TW* $V_-(x, t; \lambda_1)$*,* $\lambda_1 < \lambda$*, occurs at the interface.*

autonomous PDE. Since $\tilde{W}_2$ is invariant, the nonlinear parabolic equation (5.8) restricted to $\tilde{W}_2$ reduces to a dynamical system. Hence, there exist explicit solutions of the *parabolic* spatial shape

$$\tilde{v}(x,t) = D_0(t) + D_1(t)\,x^2 \in \tilde{W}_2, \tag{5.22}$$

where the coefficients $\{D_0, D_1\}$ solve the dynamical system

$$\begin{cases} D_0' = (1-m)D_0D_1 - m, \\ D_1' = -2(1+m)\,D_1^2. \end{cases} \tag{5.23}$$

This system is again solved explicitly,

$$\tilde{v}(x,t) = d_0 t^{\frac{1-m}{1+m}} - \tfrac{1+m}{2}\,t + \tfrac{1}{2(1+m)}\,\frac{x^2}{t}, \tag{5.24}$$

where d_0 is arbitrary constant of integration, and is well defined in the positivity domain $\tilde{D} = \{|x| > \tilde{s}(t)\}$. Here

$$\tilde{s}(t) = (1+m)t\sqrt{1 - c_* t^{-\frac{2m}{1+m}}} \quad \text{with } c_* = \tfrac{2d_0}{1+m}, \tag{5.25}$$

is the interface, at which (5.24) vanishes and $\tilde{u} = \infty$. In terms of the solutions of (5.3), we obtain the explicit solutions

$$\tilde{u}(x,t) = \left\{ \tfrac{1-m}{m} \left[d_0 t^{\frac{1-m}{1+m}} - \tfrac{1+m}{2}\,t + \tfrac{1}{2(1+m)}\,\frac{x^2}{t} \right] \right\}^{-\frac{1}{1-m}}$$

in $\tilde{D}$, extended by $\tilde{u} = +\infty$ otherwise. There are two cases depending on the value of d_0.

(i) If $d_0 > 0$, the spatially symmetric solution $\tilde{u}$ starts with regular analytic data at any small moment $t = \delta > 0$, and blows up at

$$T = \left(\tfrac{2d_0}{1+m}\right)^{\frac{1+m}{2m}}. \tag{5.26}$$

If we take $t = 0$ as the initial time, the initial function is a Dirac mass. For $t > T$, the solution creates a blow-up interface at

$$\tilde{s}(t) = (1+m)\,t\sqrt{1 - \left(\tfrac{T}{t}\right)^{\frac{2m}{1+m}}}. \tag{5.27}$$

Note that $\tilde{s}(t)$ is an analytic function for $t > T$ and

$$\tilde{s}(t) = (1+m)\,t\,(1+o(1)) \quad \text{as} \quad t \to \infty,$$

while $\tilde{s}$ is not Lipschitz continuous at the blow-up instance,

$$\tilde{s}(t) = 2\sqrt{md_0}\,\sqrt{t-T}\,(1+o(1)) \quad \text{as} \quad t \to T^+.$$

(ii) For $d_0 \le 0$, we have a solution that starts from a profile that is already singular and has a positive interface given by (5.25), thus an analytic function. For $d_0 = 0$, we obtain a linear interface

$$\tilde{s}(t) = (m+1)t \quad (d_0 = 0).$$

These are less interesting cases than (i), where the singularity develops after a finite time.

In the case $d_0 \le 0$ the interface velocity $\tilde{s}'(t)$ is equal or larger than $1+m > 2\sqrt{m}$. For $d_0 > 0$, we have the same velocity estimate for large times,

$$s'(t) \ge 1 + m > 2\sqrt{m} \quad \text{for} \quad t \gg 1. \tag{5.28}$$

Then, by a standard comparison with the proper TW having the minimal speed $\lambda = \lambda_m = 2\sqrt{m}$, exactly as in the proof of Lemma 5.3, we deduce from (5.28) the following conclusion.

Lemma 5.4 *Explicit solutions* (5.24) *of equation* (5.8) *are not proper for* $t \gg 1$.

We want to investigate the possibility that the explicit solutions (5.22) are proper for some time for $d_0 > 0$. We then need some elementary properties of the interface function $\tilde{s}(t)$ in (5.27). It has a unique inflection point at

$$T_1 = T(1-m)^{-\frac{1+m}{2m}} > T, \tag{5.29}$$

where $\tilde{s}''(T_1) = 0$ and $\tilde{s}'(T_1) = 2\sqrt{m}$, so that $\tilde{s}'(t) > 2\sqrt{m}$ for all $t > T$, $t \ne T_1$. The following key property of $\tilde{v}$ will be proved later on.

Theorem 5.5 *Let* $d_0 > 0$. *Then* $\tilde{v}(x,t)$ *is a proper maximal solution of equation* (5.8) *for* $t \le T_1$ *and is not a proper one for* $t > T_1$.

5.4 Lower speed estimate of blow-up interfaces

We now begin the study of general solutions of (5.3) with blow-up interfaces. Let $s(t)$ be the right-hand blow-up interface of the solution given by $s(t) = \sup \mathcal{B}[u](t)$ for $t \ge T$. The same considerations apply to the left-hand interface. Then $s(t)$ is continuous, nondecreasing (proved by the direct local comparison with TWs or parabolic solutions) and moreover, by intersection comparison, we easily establish the following new speed estimate from below.

Theorem 5.6 *For all* $t > T$,

$$\underline{D}^+ s(t) \equiv \liminf_{\Delta t \to 0^+} \frac{s(t+\Delta t) - s(t)}{\Delta t} \ge 2\sqrt{m}. \tag{5.30}$$

Proof. Fix an arbitrary $t_0 > T$. We apply the intersection comparison technique as in the proof of Theorem 4.7 in Section 4.4. This technique was used there to prove complete blow-up, a different type of blow-up behaviour. Nevertheless, we show that, by the same approach, we can derive a sharp bound of the incomplete blow-up.

Thus, fixing an arbitrarily small $\varepsilon > 0$, we take the speed $\lambda = 2\sqrt{m} - \varepsilon$, and for a fixed $n \gg 1$, let

$$U_n(x,t) = f_n(\xi), \quad \xi = x - s(t_0) - \lambda(t - t_0),$$

be the TW solution of (5.3) satisfying (5.18) and the boundary conditions in (5.17). Using the same phase-plane transformation, we arrive at the equation (cf. (5.19))

$$f\frac{\mathrm{d}F}{\mathrm{d}f} = -\frac{1}{F}\left[\left(F - \tfrac{1}{2}\lambda\right)^2 + a^2\right], \quad \text{where } a^2 = m - \tfrac{1}{4}\lambda^2 > 0.$$

The solution $F_n(f)$ is explicitly given by

$$\tfrac{1}{2}\log(F_n^2 - \lambda F_n + m) + \frac{\lambda}{2a}\arctan\left(\frac{2F_n-\lambda}{2a}\right) = -\log f + C_n,$$

where

$$C_n = \log n + \tfrac{1}{2}\log m - \frac{\lambda}{2a}\arctan\frac{\lambda}{2a}.$$

It follows that $F_n(f) = O(n) \to \infty$ as $n \to \infty$ uniformly on subsets $[\delta, \frac{1}{\delta}]$ for arbitrarily small $\delta > 0$. In view of the flux transformation, this implies that the corresponding profile $f_n(\xi)$ is steep on the level sets in the sense that, for $n \gg 1$,

$$|(f_n^m(\xi))'| \gg 1 \quad \text{on the level set } \{f_n(\xi) = c\}$$

for any constant $0 \le c \le \sup u_0$. This property is enough to apply the intersections comparison technique in exactly the same way as in the analysis of complete blow-up in Section 4.4. Therefore, using if necessary a small shifting of the solution U_n in x, as in the proof of Theorem 4.7, we conclude that there exists a sequence $\{b_n > 0\} \to 0$ such that

$$u_n(x,t_0) \ge U_n(x+b_n,t_0) \equiv f(x+b_n-s(t_0)) \quad \text{for } x \in I\!R.$$

By the usual comparison, we deduce that $u_n(x,t) \ge U_n(x+b_n,t)$ for all $t > t_0$. It is clear from the construction that the same inequality holds for all $t_0 - t > 0$ small (this follows from the fact that $t_0 > T$ and hence $u_n(x,t) \gg 1$ for all $s(t) - x > 0$ small and $t \approx t_0$). Passing to the limit $n \to \infty$, we obtain that the interface at $x = s(t)$ satisfies

$$s(t) - s(t_0) \ge (2\sqrt{m} - \varepsilon)(t - t_0) \quad \text{for all } t \approx t_0.$$

Since $\varepsilon > 0$ and $t_0 > T$ are arbitrary, (5.30) follows. □

Combining the results of Theorems 5.2 (i) (following from comparison with TWs) and 5.6, we arrive at our first sharp asymptotic result.

Corollary 5.7 *If* $u_0(x) = o(x^{-1/(1-m)})$ *as* $x \to +\infty$, *then*

$$s(t) = 2\sqrt{m}\,t\,(1+o(1)) \quad \text{as} \quad t \to \infty.$$

5.5 Dynamical equation of blow-up interfaces

We now establish the dynamic equation for the blow-up interface. Starting from the TW-analysis of Section 5.2, we have to introduce a modification to the Rankine–Hugoniot dependence (5.11) of the speed on the wave slope. We introduce the function

$$H_*(S) = \begin{cases} S + \frac{m}{S}, & \text{if } 0 < S \le \sqrt{m}, \\ 2\sqrt{m}, & \text{if } S > \sqrt{m}, \end{cases} \tag{5.31}$$

i.e., $H_*(S)$ consists of the descending branch for $0 < S < \sqrt{m}$ continued after the minimum with the horizontal line $H_* \equiv 2\sqrt{m}$, as shown on Figure 5.1. Therefore, $H_*(S)$ is indeed from $C^1(I\!R_+)$ but is not a C^2-function and the second derivative does not exist at $S = \sqrt{m}$.

Theorem 5.8 *Let $u_0^{m-1}(x)$ be convex and $(u_0^{m-1})'' \le C_0$. Then the interface equation in terms of the pressure $v(x,t) = u^{m-1}(x,t)$ has the form*

$$s'(t) = H_*(v_x(s(t),t)) \quad \textit{for} \quad t > T. \tag{5.32}$$

Observe that now (5.30) is a straightforward consequence of (5.32) and (5.31). The proof consists of several steps.

L^∞-bounds

Consider the Cauchy problem with bounded initial data $u_0(x)$ satisfying (5.5). Then, by comparison from above with the proper TW solution, we have that on any bounded intervals of $t > 0$,

$$u(x,t) \le C_1|x|^{-1/(1-m)} \quad \text{for } |x| \gg 1. \tag{5.33}$$

In order to detect the "minimal" behaviour of $u(x,t)$ as $x \to \infty$, we compare $u(x,t)$ from below with the standard self-similar solution

$$\overline{u}(x,t) = \theta(\eta), \quad \eta = x/\sqrt{t},$$

of the fast diffusion equation without the source term

$$u_t = (u^m)_{xx}.$$

Then θ solves the ODE

$$(\theta^m)'' + \tfrac{1}{2}\,\theta'\eta = 0$$

and hence $\theta(\eta) \sim \eta^{-2/(1-m)}$ as $\eta \to \infty$. We then deduce that $u(x,t) \ge \overline{u}(x,t)$ for $t > 0$, so

$$u(x,t) \ge C_2(t)|x|^{-2/(1-m)} \quad \text{for } |x| \gg 1, \tag{5.34}$$

where

$$C_2(t) = C\,t^{\frac{1}{1-m}}.$$

Finally, in terms of the pressure variable $v = \frac{m}{1-m}u^{m-1}$, these mean that

$$\tilde{C}_1(t)|x| \le v(x,t) \le \tilde{C}_2(t)x^2 \quad \text{for } |x| \gg 1,\ t > 0, \tag{5.35}$$

where $\tilde{C}_1,\ \tilde{C}_2$ are positive functions.

Convexity

We now restrict our attention to the proper solutions $v(x,t)$ of (5.8) having positive, symmetric, convex initial pressure

$$v(x,0) = v_0(x) > 0 \quad \text{and} \quad 0 \le v_0''(x) \le C_0 \quad \text{in } I\!R. \tag{5.36}$$

In view of the independence of the proper solution upon the type of monotone approximations (Section 6.2), we treat $v(x,t)$ as the limit as $\varepsilon \to 0$ of the monotone sequence of classical strictly positive solutions $\{v_\varepsilon(x,t)\}$ of the regularized equation

$$v_t = (1-m)vv_{xx} - (v_x)^2 - H_\varepsilon(v) \tag{5.37}$$

with the same initial data. The function $H_\varepsilon \in C^\infty([0,\infty))$ is a smooth concave approximation of the Heaviside function $m\,H(v)$, which we have to put formally into equation (5.8) if we want it to be valid at $v=0$, satisfying

$$H_\varepsilon(v) = \begin{cases} m & \text{for} \quad v \ge 3\varepsilon, \\ \frac{1}{2\varepsilon} mv & \text{for} \quad 0 \le v \le \varepsilon, \end{cases}$$

and $H_\varepsilon''(v) \le 0$ for all $v \ge 0$.

Proposition 5.9 *The proper solution is strictly positively convex, i.e.,*

$$v_{xx}(x,t) > 0 \quad \textit{in the positivity domain} \ \{v > 0\}. \tag{5.38}$$

Proof. The second derivative of the approximating solution $z = (v_\varepsilon)_{xx}$ satisfies the equation

$$z_t = (1-m)v_\varepsilon z_{xx} - 2m\,(v_\varepsilon)_x z_x - H_\varepsilon'(v_\varepsilon)z - (1+m)z^2 - H_\varepsilon''(v_\varepsilon)(v_\varepsilon)_x^2.$$

Since $H_\varepsilon'' \le 0$, using the "convex" behaviour (5.35) as $|x| \to \infty$, by the strong MP, we conclude that

$$(v_\varepsilon)_{xx} > 0 \quad \text{everywhere for } t > 0. \tag{5.39}$$

Passing to the limit $\varepsilon \to 0^+$ and using the standard C^∞-regularity of solutions in the positivity domain, we obtain (5.38). The strict inequality sign follows from the strong MP applied in $\{v > 0\}$. □

As a straightforward consequence of (5.35) and (5.38), we obtain a gradient estimate. As we have seen in the previous chapters (a general transversality principle is formulated in Section 1.4), such an estimate also follows as a transversality estimate by comparison with steep linear TWs.

Corollary 5.10 *There is a gradient bound of the form*

$$|v_x| \le C \quad \textit{on compact subsets.} \tag{5.40}$$

Estimate of v_{xx} from above

We now prove a second-order estimate by a geometric tangential construction using the fact that the subset $\tilde{B}$ of parabolic solutions is complete in the tangential space $I\!R^3$ in the sense of Section 1.5 (unlike the one composed of TW solutions that is complete in the hodograph plane $I\!R^2$ only).

Proposition 5.11 *Let $0 \le v_0'' \le C_0$. Then*

$$0 < v_{xx} \le C \quad \textit{on bounded subsets of} \ \{v > 0\} \cap \{t \ge T\}. \tag{5.41}$$

From equation (5.8) we then obtain that

$$|v_t| \le C \quad \text{on bounded subsets from } \{v > 0\} \cap \{t \ge T\}. \tag{5.42}$$

Passing to the limit $x \to s(t)^+$ in equation (5.8), we arrive at the following continuity result.

Corollary 5.12 *For any $t \ge T$, there exists the limit*

$$v_t(s(t),t) = -[v_x(s(t),t)]^2 - m. \tag{5.43}$$

Proof of Proposition 5.11. The proof is based on intersection comparison with the subset $\tilde{B} = \{\tilde{v}(x,t)\}$ of explicit parabolic solutions. Using the known blow-up results, we have that the second derivative is bounded at the blow-up time, $v_{xx}(x,T) \le C$; see Remarks. We follow the idea of the convexity construction from Section 1.4 similar to that illustrated on Figure 1.3.

Consider the three-dimensional family of the particular solutions from Section 5.3, $\tilde{v}(x,t) = \tilde{v}(x-a, t+t_0; d_0)$ that, in the positivity domain, have the form

$$\tilde{v}(x,t) = d_0(t+t_0)^{\frac{1-m}{1+m}} - \tfrac{1+m}{2}(t+t_0) + \tfrac{1}{2(1+m)}\frac{(x-a)^2}{t+t_0}, \tag{5.44}$$

where $a, t_0 \in I\!R$ and $d_0 > 0$ are parameters. Assume that there exists a point, say, $x = 0,\ t = t_1 > T$, such that

$$C_1 = v_{xx}(0,t_1) \gg 1,\ v(0,t_1) = \mu,\ v_x(0,t_1) = \nu \in (0, C]. \tag{5.45}$$

At the point $(0, t_1)$, we now construct the *tangent solution* $\tilde{v}$ satisfying

$$\tilde{v}(0,t_1) = \mu,\ \tilde{v}_x(0,t_1) = \nu. \tag{5.46}$$

Substituting (5.44) into (5.46), we obtain the following equations on the parameters with τ denoting $t_1 + t_0 > 0$:

$$d_0\tau^{\frac{1-m}{1+m}} - \tfrac{1+m}{2}\tau + \tfrac{1}{2(1+m)}\frac{a^2}{\tau} = \mu, \quad -\tfrac{1}{2(1+m)}\frac{a}{\tau} = \nu, \tag{5.47}$$

or, finally, substituting $(-a)$ from the last equation into the first one, the equation

$$d_0\tau^{\frac{1-m}{1+m}} + \tfrac{1+m}{2}(4\nu^2 - 1)\tau = \mu. \tag{5.48}$$

We now assume that

$$v_{xx}(0,t_1) = \tfrac{1}{1+m}\tau \gg 1, \quad \text{but} \ \ll C_1, \tag{5.49}$$

i.e., $C_1^{-1} \ll \tau \ll 1$. We put $t_0 = -t_1 + \delta$ with $0 < \delta \ll 1$. It follows from (5.47) and (5.48) that if $\tau \ll 1$, then $\tilde{v}(x,t)$ is strictly positive for all $-t_0 < t \le t_1$. Therefore, for any small $\varepsilon \in (0, \varepsilon_0)$, the function $\tilde{v}(x+\varepsilon, t_1)$ intersects $v(x,t_1)$ at least two times in a small neighbourhood of the origin $x = 0$. Moreover, in view of (5.49) and known behaviour of $v(x,t_1)$ as $|x| \to \infty$, the total number of intersections in $\{x > s(t_1)\}$ is not less than 3.

Let $\mathrm{Int}_\varepsilon(t,\tilde{v})$ be the number of intersections in $\{x > s(t)\}$ between the solutions $\tilde{v}(x+\varepsilon,t)$ and $v(x,t)$ for $t_1 - \delta = -t_0 < t \le t_1$. Since, by (5.49),

$$\tilde{v}_{xx}(0,t) = \tfrac{1}{1+m}\frac{1}{t+t_0} \to \infty \quad \text{as } t \to -t_0^+,$$

it follows from the convexity assumption on the initial data and from the local regularity results for $v(x,t)$ that there exists an $\varepsilon \in (0,\varepsilon_0)$ such that

$$\mathrm{Int}_\varepsilon(t,\tilde{v}) = 2 \quad \text{for } t = t_2 \approx -t_0^+.$$

By the Sturm Theorem, taking into account the well-defined standard comparison for $|x| \gg 1$, we deduce that

$$\mathrm{Int}_\varepsilon(t,\tilde{v}) \le 2 \quad \text{for all } t > t_2.$$

Recall that, as follows from (5.44) for $t = t_2 \approx -t_0^+$, the function $\tilde{v}(x - \varepsilon, t_2)$ is strictly positive. Finally, we obtain that (5.45) and (5.49) contradict our construction since $\mathrm{Int}_\varepsilon(t_1, \tilde{v}) \geq 3$. $\square$

Interface slope is finite and nondecreasing

First, it follows from (5.8) that, for any $t \geq T$, there exists a nonnegative limit

$$v_x(s(t), t) = \lim v_x(x, t) \quad \text{as } x \to s(t)^+.$$

Moreover, it follows from the asymptotic blow-up analysis as $t \to T^-$ (see Remarks) that $v_x(s(T), T) = 0$, and a standard comparison with the proper TWs gives that $s'(T^+) = +\infty$. Since $s(t)$ is nondecreasing and continuous, by the same comparison, we conclude that, for any $\delta > 0$, there exists $t_\delta \in (T, T + \delta)$ such that

$$v_x(s(t_\delta), t_\delta) > 0. \tag{5.50}$$

Proposition 5.13 *Under the convexity assumption* (5.36), *the interface slope of the pressure $S_v(t) = v_x(s(t), t)$ is positive for all $t > T$ and nondecreasing.*

Proof. (i) Take $t_\delta - T > 0$ small, to ensure that (5.50) holds, and assume that

$$S_v(t_\delta) < \sqrt{m}. \tag{5.51}$$

Since $v(x, t_\delta)$ is strictly positively convex, we have that any proper TW solution $V_-(x - a, t; \lambda)$ of the form (5.10) for any speed $\lambda \geq 2\sqrt{m}$ such that the TW-interface slope S_- satisfies

$$S_v = S_-(\lambda) \leq S_v(t_\delta), \tag{5.52}$$

intersects the profile $v(x, t_\delta)$ at, at most, one point for any $a \in I\!R$. We denote by $\mathrm{Int}(t, \lambda, a)$ the number of intersections between those v and V_-. We have from (5.51) that

$$\mathrm{Int}(t_\delta, \lambda, a) \leq 1 \quad \text{for all } \lambda \text{ satisfying (5.52) and } a \in I\!R. \tag{5.53}$$

By regular approximations, the Sturm Theorem applies to the free-boundary problem, to yield

$$\mathrm{Int}(t, \lambda, a) \quad \text{is not increasing in } t. \tag{5.54}$$

Indeed, in view of (5.38) new intersections with linear profiles $V_-(x, t)$ cannot appear at $x = \infty$. By the strong MP for uniformly parabolic equations, intersections cannot appear in the positivity domain. Finally, a careful analysis via the local comparison of the proper solutions shows that no new intersections appear on the free boundary (this is not necessary if a monotone approximation of both solutions is considered). Thus (5.54) is valid and then (5.53) implies that

$$\mathrm{Int}(t, \lambda, a) \leq 1 \quad \text{for all} \quad t > t_\delta, \tag{5.55}$$

with λ from (5.52) and $a \in I\!R$. Assume that there exists a moment $t_1 > t_\delta$, at which $S_v(t_1) < S_v(t_\delta)$. One can see that, in this case, there exists a TW of the form $V_-(x - a_1, t; \lambda_1)$ with the interface slope

$$S_v(t_1) < S_- < S_v(t_\delta)$$

such that

$$\mathrm{Int}(t_1, \lambda_1, a_1) \geq 2.$$

This contradicts (5.55).

(ii) Assume now that (5.51) is not valid. This situation is much easier, and the following result completes the proof of Proposition 5.13. □

Proposition 5.14 *If* $S_v(t_0) = \sqrt{m}$ *for some* $t = t_0 > T$*, then*

$$s'(t) = 2\sqrt{m} \quad \textit{for all} \quad t > t_0. \tag{5.56}$$

Proof. First we have (5.30). Second, via the convexity, the tangent proper TW $V_-(x - s(t_0), t - t_0)$ with the same interface slope $S_m = \sqrt{m}$ and the interface $\bar{s}(t) = 2\sqrt{m}(t - t_0)$ satisfies

$$v(x, t_0) \geq V_-(x - s(t_0), 0) \quad \text{in } I\!R.$$

Hence, by comparison $v(x,t) \geq V_-(x - s(t_0), t - t_0)$ for all $x \in I\!R$, $t > t_0$, and therefore

$$s(t) \leq \bar{s}(t) + s(t_0) = s(t_0) + 2\sqrt{m}(t - t_0) \quad \text{for } t > t_0. \tag{5.57}$$

It follows from (5.30) and (5.57) that $D^+s(t_0^+) = 2\sqrt{m}$. Since, by the same proof as in (i), Proposition 5.13, the interface slope of v does not decrease, we apply the same comparison at any point $t > t_0$ to conclude that $D^+s(t) \equiv 2\sqrt{m}$. Using Lagrange's formula of finite increments

$$s(t) - s(t_0) = D^+s(\bar{t})(t - t_0) = 2\sqrt{m}(t - t_0) \quad \text{for } \bar{t} \in (t, t_0),$$

we arrive at (5.56). □

Interface equation

Proposition 5.15 *If* (5.36) *is valid, then* $s'(t)$ *is continuous for* $t > T$ *and* (5.32) *holds.*

Proof. Fix a $t_\delta > T$ and let (5.51) hold. Denote by $V_1(x,t) = V_-(x - s(t_0), t - t_0; \lambda_1)$ the proper TW with the same interface slope $S_-(\lambda_1) = S_v(t_\delta)$ and $V_2(x,t) = V_-(x - s(t_0), t - t_0; \lambda_2)$ with a slightly larger slope $S_-(\lambda_2) = S_v(t_\delta) + \varepsilon$, where $\varepsilon > 0$ is small. Then

$$V_1(x, t_0) \leq v(x, t_0) \leq V_2(x, t_0)$$

in a small neighbourhood of the common interface at $x = s(t_0)$. Moreover, using the uniform estimate of the second derivative and Taylor's formula with the remainder in the integral form, we have that for $x > s(t)$,

$$|v(x,t) - v_x(s(t), t)(x - s(t))| \leq \tfrac{1}{2} C (x - s(t))^2. \tag{5.58}$$

Therefore, by local comparison, there exists a small $\tau > 0$, which by (5.58) and (5.41) is independent of t_0, such that, for all $t \in (t_0, t_0 + \tau)$,

$$V_1(x,t) \leq v(x,t) \leq V_2(x,t) \tag{5.59}$$

in a neighbourhood of $x = s(t_0)$. Comparing the corresponding interfaces, we have that

$$\lambda_2(t-t_0) \le s(t) - s(t_0) \le \lambda_1(t-t_0) \quad \text{for all} \quad t \in (t_0, t_0+\tau),$$

and in particular,

$$\lambda_2 \le \underline{D}^+ s(t_0) \le \overline{D}^+ s(t_0) \le \lambda_1.$$

Passing to the limit $\varepsilon \to 0^+$, when $\lambda_2 \to \lambda_1$, we conclude that there exists $D^+ s(t_0) = H_*(S_v(t_0))$. One can see from estimate (5.58) that the interface slope $v_x(s(t),t)$ is continuous at $t = t_0$. Hence, it follows from (5.43) that the time derivative $v_t(s(t),t)$ is also continuous. Therefore $s(t)$ is differentiable at $t = t_0$ and

$$s'(t_0) = H_*(S_v(t_0)).$$

If $S_v(t_0) = S_m$, then (5.32) follows from Proposition 5.14. □

Thus, beyond the blow-up time, the proper solution satisfies the following free-boundary problem for the quasilinear degenerate parabolic equation:

$$v_t = \mathbf{A}(v) \quad \text{for } x > s(t), \quad s'(t) = H_*(v_x(s(t),t)) \quad \text{for } t > T, \tag{5.60}$$

with given initial data $v(x,T)$. We have proved that the interface is a C^1-function.

5.6 Blow-up interfaces are not C^2 functions

We now show that $s(t)$ is not an analytic function in general, and moreover is not a C^2-function even if it is an analytic for small $t > 0$ unlike the interfaces for the porous medium equation, which was proved by S. Angenent [12]. In order to make the idea of the construction clear, we first turn our attention to a particular class of initial data that generates (locally in time) an explicit proper solution.

Theorem 5.16 *Let*

$$v_0(x) \equiv \tfrac{m}{1-m}\, u_0^{m-1}(x) = a_0 + a_1 x^2, \tag{5.61}$$

where a_0 and a_1 are arbitrary positive constants. Then there exists a finite time $T_1 = T_1(u_0) > T(u_0)$ such that $s(t) \in C^{1,1}([T+\delta,\infty))$ for any small $\delta > 0$, but $s(t) \notin C^2([T+\delta,\infty))$, and $s''(T_1^-) \ne s''(T_1^+)$. Moreover,

$$s'(t) > 2\sqrt{m} \quad \textit{for } t \in (T,T_1) \quad \textit{and} \quad s'(t) \equiv 2\sqrt{m} \quad \textit{for } t \ge T_1. \tag{5.62}$$

Analytic continuation up to the blow-up time

We begin the proof of Theorem 5.16 by constructing an explicit parabolic local in time solution having the initial data corresponding to (5.61). In the pressure form the solution $\tilde{v}(x,t)$ has the form (5.22) with the coefficients $\{D_0(t), D_1(t)\}$ satisfying the DS (5.23) with the initial conditions $D_0(0) = a_0$, $D_1(0) = a_1$. Therefore, in the positivity domain, $\tilde{v}(x,t)$ has the form (5.24) with $t \mapsto t + t_0$

$$\tilde{v}(x,t) = d_0(t+t_0)^{\frac{1-m}{1+m}} - \tfrac{1+m}{2}(t+t_0) + \tfrac{1}{2(1+m)}\, \frac{x^2}{t+t_0}, \tag{5.63}$$

where

$$t_0 = \frac{1}{2(1+m)a_1} \quad \text{and} \quad d_0 = t_0^{-\frac{1-m}{1+m}} (a_0 + \tfrac{1}{4}\, a_1).$$

This is a unique, strictly positive classical and moreover analytic, in both x and t, solution up to the blow-up time $T - t_0$ calculated explicitly by (5.26).

Analytic continuation up to the inflection point

For $t + t_0 - T > 0$ small, the explicit solution $\tilde{v}(x,t)$ satisfies (5.60) with the monotone increasing interface $\tilde{s}(t)$ given by

$$\tilde{s}(t) = (1+m)\,(t+t_0)\sqrt{1 - \left(\tfrac{T}{t+t_0}\right)^{\frac{2m}{1+m}}}. \tag{5.64}$$

Let $T_1 - t_0$ be the unique inflection point of function (5.64).

For convenience of computations and in view of the translational invariance of the PDE, we put $t_0 = 0$ in (5.63) and (5.64) recovering expression (5.27). We begin with the local uniqueness for the free-boundary problem (5.60). Recall that the uniqueness of the maximal solution follows from the monotonicity of the approximation (Section 6.2), and now we show that the free-boundary problem (5.60) also uniquely determines proper solutions.

Proposition 5.17 *Problem* (5.60) *with the given initial data* $\tilde{v}(x,T)$ *has a unique solution for* $t \in (T, T_1)$*, which is the parabolic one* (5.63) *(*$t_0 = 0$*).*

Proof. It is based on a standard application of the MP. Assume that there exist two different solutions v and $\bar{v}$. Moreover, since the unique proper solution is maximal in v, we may suppose that

$$\bar{v} \le v.$$

Then, by the MP applied in the positivity domains, we deduce that they have different interfaces. Namely, we suppose that there exists $T_2 \in [T, T_1)$ such that $s(t) = \bar{s}(t)$ for $t \in [T, T_2]$ and $s(t) \le s(t)$, $s(t) \not\equiv s(t)$ for small $t - T_2 > 0$. We now choose $t_1 \approx T_2^+$ such that

$$\Delta s_1 \equiv \bar{s}(t_1) - s(t_1) > \bar{s}(t) - s(t) \quad \text{for all } t \in [T_2, t_1). \tag{5.65}$$

Obviously, we may also assume that

$$(\bar{s} - s)'(t_1) > 0. \tag{5.66}$$

In view of the interface equation (5.32) and since, by assumption $S_v(t_1) < \sqrt{m}$, from (5.66) we have that

$$\bar{v}_x(\bar{s}(t_1), t_1) < v_x(s(t_1), t_1). \tag{5.67}$$

We now consider the solution shifted in x, $\bar{v}_1(x,t) = \bar{v}(x + \Delta s_1, t)$. Then $\bar{s}_1(t_1) = s(t_1)$, i.e., by (5.67)

$$\bar{v}_1(s(t_1), t_1) = v(s(t_1), t_1) = 0 \quad \text{and} \quad \bar{v}_1(x, t_1) < v(x, t_1) \tag{5.68}$$

in a small right-hand neighbourhood of $x = s(t_1)$. Recall that the initial data for $\bar{v}_1$ satisfy

$$\bar{v}_1(x, T_1) \equiv \bar{v}(x + \Delta s_1, T_1) = v(x + \Delta s_1, T_1) \ge v(x, T_1).$$

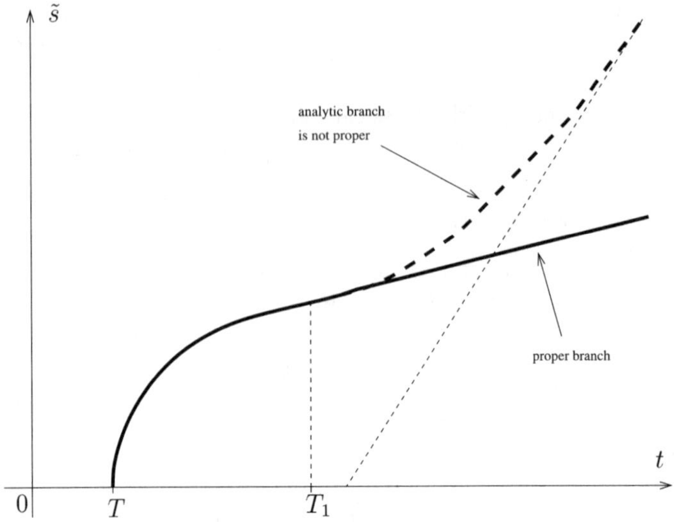

Figure 5.4 *The analytic interface* (5.64) *(*$t_0 = 0$*) is not proper after the inflection at* $t = T_1$*. The proper continuation for* $t > T_1$ *is the straight line.*

Hence, by comparison of classical solutions in the positivity domains, we have that $\bar{v}_1(x,t) \ge v(x,t)$ unless the interface $x = \bar{s}_1(t) - \Delta s_1$ intersects $x = s(t)$. Since, by assumptions (5.65) and (5.66), this occurs for the first time at $t = t_1$, we have that $\bar{v}_1(x,t_1) \ge v(x,t_1)$ in a neighbourhood of the common interface point $x = \bar{s}_1(t_1) = s(t_1)$, contradicting (5.68). Clearly, $\tilde{v}(x,t)$ is a solution to (5.60). Hence, $v \equiv \tilde{v}$ for $t \in (T, T_1)$. As a direct consequence, we have proved the first part of Theorem 5.5. $\square$

Breakdown of C^2-regularity at inflection

Consider now the proper solution $v(x,t)$ at the inflection point $t = T_1$ of the interface given by (5.29). Thus $v(x,t)$ satisfies equation (5.8) for $x \in I\!R$, $t > T_1$, with the initial function

$$v(x,T_1) = [D_0(T_1) + D_1(T_1)x^2]_+, \quad D_0(T_1) < 0, \;\; D_1(T_1) > 0$$

that is convex in the positivity domain. Then

$$v_x(s(T_1),T_1) = \sqrt{m}.$$

Therefore, by Theorem 5.8, we conclude that

$$s'(t) = 2\sqrt{m} \quad \text{for } t > T_1,$$

i.e., the proper solution has a straight line propagation for $t > T_1$, as Figure 5.4 illustrates. Hence, $s'(t) < \tilde{s}'(t)$ for all $t > T_1$.

This completes the proof of Theorems 5.16 and 5.5. $\square$

Extension to general solutions

We have proved that, due to (5.62), the explicit analytic continuation of the solution for $t > T_1$ *is not* the proper maximal one. Using the above properties of the explicit solutions with data (5.61), we next easily establish a similar result for a wide class of initial functions.

Corollary 5.18 *Let* $u_0 = u_0(|x|)$ *be nonincreasing,* $0 \le (u_0^{m-1})'' \le C_0$, *and*

$$\lim_{x \to \infty} \inf \tfrac{1}{x} \left[u_0^{m-1}(x)\right]' > 0. \tag{5.69}$$

Then there exists a finite $T_1 > T(u_0)$ *such that* (5.62) *holds.*

Proof. We first compare for $t < T$ the solution $v(x,t)$ with the family of the explicit parabolic solutions $\{\tilde{v}(x,t;t_0)\}$ with the same blow-up time $T = T(u_0)$. Then (5.63) yields that

$$d_0 = \tfrac{1}{2}\,(1+m)(T+t_0)^{\frac{2m}{1+m}}.$$

Hence,

$$\tilde{v}(x,0;t_0) \to mT \quad \text{and} \quad \tilde{v}_x(x,0;t_0) \to 0 \quad \text{as} \quad t_0 \to \infty$$

uniformly on compact subsets in x. Moreover,

$$\frac{\tilde{v}(x,0;t_0)}{x^2} = O\left(\frac{1}{t_0}\right) \to 0 \quad \text{as} \quad x \to \infty \quad \text{for} \quad t_0 \gg 1.$$

Therefore, after (if necessary) a small shifting of the origin in time, assumption (5.69) guarantees that the number of intersections $\text{Int}(t,t_0,a)$ between $v(x,t)$ and $\tilde{v}(x-a,t;t_0)$ satisfies

$$\text{Int}(0,t_0,a) \le 2 \quad \text{for all} \quad t_0 \gg 1 \ \text{and every} \quad a \in I\!R.$$

By the Sturm Theorem, we then have

$$\text{Int}(t,t_0,a) \le 2 \quad \text{for } t > 0 \text{ and } a \in I\!R.$$

Fix a $t_0 \gg 1$, and let us now count for $t > T$ the intersections in the domain $\{x > l(t) = \max\{a + \tilde{s}(t), s(t)\}\}$ only. Recall that new sign changes cannot appear on the lateral boundary $\{x = l(t),\ t > T\}$; see the proof of Proposition 5.13. Since both solutions $v(x,t)$ and $\tilde{v}(x-a,t;t_0)$ have the same blow-up time T, the corresponding number of intersections denoted by $\text{Int}_+(t,t_0,a)$ satisfies

$$\text{Int}_+(t,t_0,a) \le 1 \quad \text{for } t > T,\ a \in I\!R. \tag{5.70}$$

Indeed, this follows from the easy fact that, for $t = T$, the second intersection (if any) is always situated in $\{x < l(T)\}$. We now fix $t_0 \gg 1$ so that (5.70) is valid. It follows from Theorem 5.16 that there exists $T_1 > T$ such that $\tilde{s}'(t) = 2\sqrt{m}$ for all $t \ge T_1$.

Let us show that

$$s'(T_1) = 2\sqrt{m}$$

and hence $S_v(T_1) = \sqrt{m}$. We argue by contradiction. Assume that the above equalities are not true and $S_v(T_1) < \sqrt{m}$. Then there exists $a \in I\!R$ such that $v(x,T_1)$ intersects the profile $\tilde{v}(x-a,T_1;t_0)$ in $\{x > s(T_1)\}$ at least two times. This contradicts (5.70). □

5.7 Large time behaviour of proper blow-up solutions

Consider for $t > T_1$ the proper solution $v(x,t)$ such that $v \equiv \tilde{v}$ for $t \le T_1$ as explained in the previous section. Assume for convenience that $s(T_1) = 0$. We already know that after inflection at $t = T_1$, there holds

$$s(t) \equiv 2\sqrt{m}\,(t - T_1) \quad \text{for } t > T_1.$$

The corresponding TW solution with the same interface has the form

$$V_-(x,t) = \sqrt{m}(\xi)_+, \quad \xi = x - 2\sqrt{m}(t - T_1).$$

Let us show that the large-time behaviour of $v(x,t)$ is described by this TW. One can see that the corresponding rescaled function

$$f(\xi,t) = v(\xi + 2\sqrt{m}(t - T_1), t)$$

satisfies the following rescaled equation:

$$f_t = (1-m) f f_{\xi\xi} - (f_\xi)^2 + 2\sqrt{m} f_\xi - m \quad \text{in } I\!\!R_+ \times \{t > T_1\}.$$

Theorem 5.19 *As $t \to \infty$, uniformly on compact subsets,*

$$|f(\xi,t) - \sqrt{m}(\xi)_+| \le \frac{C}{\log t} \to 0. \tag{5.71}$$

Proof. The difference $w(\xi,t) = f(\xi,t) - \sqrt{m}\,(\xi)_+$ satisfies

$$w_t = \sqrt{m}(1-m)\xi\, w_{\xi\xi} + (1-m) w\, w_{\xi\xi} - (w_\xi)^2 \tag{5.72}$$

with the boundary conditions $w = w_\xi = 0$ at $\xi = 0$ for $t > T_1$. Our analysis is based on the construction of a suitable self-similar solution of (5.72) of the similarity form

$$w_*(\xi,t) = t\, g(\eta), \quad \text{where } \eta = \tfrac{\xi}{t}.$$

Then $g(\eta) \ge 0$ with $g(0) = g'(0) = 0$ satisfies the ODE

$$\sqrt{m}(1-m)\eta g'' + (1-m) g g'' - (g')^2 + \eta g' - g = 0 \quad \text{for } \eta > 0. \tag{5.73}$$

Proposition 5.20 *Problem* (5.73) *admits a nontrivial positive solution satisfying*

$$g(\eta) = \sqrt{m}(1-m)\frac{\eta}{|\log \eta|}(1 + o(1)) \quad \text{as } \eta \to 0, \tag{5.74}$$

$$g(\eta) = c_1 \eta^2 (1 + o(1)) \quad \text{as } \eta \to \infty, \quad c_1 = \tfrac{1}{2}(1+m). \tag{5.75}$$

Proof. The local solvability for small $\eta > 0$ is proved by applying the Banach Contraction Principle to the equivalent integral equation. Then the solution is continued for all $\eta > 0$ in a strictly monotone way by the MP (any solution of the ODE (5.73) does not admit a point of positive maximum where $g > 0$, $g' = 0$ and $g'' \le 0$). The asymptotic expansion (5.75) follows from a local analysis at $\eta = \infty$; see comments and references to similar ODE results in Remarks. □

We now compare the initial function $w(\xi, T_1) = \tilde{v}(x, T_1) - A(\xi)_+$ with the corresponding self-similar profile

$$w_*(\xi,\delta) = \delta g(\tfrac{\xi}{\delta}), \quad \text{where } 0 < \delta \ll 1.$$

It follows from (5.75) that, as $\delta \to 0^+$,

$$w_*(\xi, \delta) \sim c_1 \tfrac{\xi^2}{\delta} \to \infty$$

uniformly on subsets $\{\xi \geq c > 0\}$. Taking into account (5.74) and the unboundedness of $g''(0)$, by using the known convex behaviour of $\tilde{v}(x, T_1)$ as $x \to \infty$, we conclude that there exists a small $\delta > 0$ such that $w(\xi, T_1) \leq w_*(\xi, \delta)$ for all $\xi \geq 0$. Therefore, by comparison

$$w(\xi, t) \leq w_*(\xi, t - T_1 + \delta) \quad \text{for } \xi > 0 \text{ and } t > T_1.$$

Hence, on the subsets $\{0 \leq \xi \leq c\}$, by the monotonicity of $g(\eta)$, we have

$$0 \leq w(\xi, t) \leq (t - T_1 + \delta) g\big(\tfrac{c}{t - T_1 + \delta}\big) \approx t g\big(\tfrac{c}{t}\big)$$

for $t \gg 1$, and (5.71) follows from (5.74). □

The self-similar solution in Proposition 5.20 is expected to give a sharp estimate of the boundary layer occurring at the singular interface for $t \gg 1$. A similar asymptotic result holds for a more general class of solutions specified in Corollary 5.18.

5.8 Blow-up interfaces for the p-Laplacian equation with source

We now extend the main results on blow-up interfaces to other quasilinear parabolic equations. Consider the quasilinear equation with the p-Laplacian operator and a superlinear source term

$$\boxed{u_t = (|u_x|^\sigma u_x)_x + u^{1/(1+\sigma)}, \quad -1 < \sigma < 0.} \tag{5.76}$$

The exponent $q = \frac{1}{1+\sigma}$ in the source term u^q corresponds to the limit case of incomplete blow-up. We present below explicit blow-up TW solutions of (5.76), which by comparison in $\{u \gg 1\}$ prove that the blow up is incomplete for any $q \subset (1, \frac{1}{1+\sigma}]$. On the other hand, in view of the nonexistence of singular TWs for $q > \frac{1}{1+\sigma}$, the geometric proof of Theorem 4.7 gives complete blow-up in this supercritical range. For more general equations with power nonlinearities, criteria of complete blow-up will be computed in Section 7.11.

If $q > \frac{1}{1+\sigma}$, as we have proved in the previous chapter, blow-up is complete and no nontrivial continuation beyond finite blow-up time exists.

Let us state known properties of the right-hand interface $s(t)$ for $t \geq T$ of the global proper solution to (5.76) corresponding to, say, bounded, compactly supported, bell-shaped initial data $u_0(x) \geq 0$ satisfying

$$u_0(x) = o(x^{-(1+\sigma)/|\sigma|}) \quad \text{as } x \to +\infty.$$

This result follows from the usual comparison with proper blow-up TWs given below.

Theorem 5.21 *The right-hand blow-up interface $x = s(t)$ is nondecreasing, continuous and satisfies*

$$s(t) \leq \lambda_m (t + c_0) \quad \textit{for } t \geq T, \tag{5.77}$$

where

$$\lambda_m = \lambda_m(\sigma) = (2+\sigma)(1+\sigma)^{-\frac{1+\sigma}{2+\sigma}}.$$

As above, the analysis of (5.76) relies on the properties of proper explicit solutions of two types. The pressure function is now

$$v = \tfrac{1+\sigma}{|\sigma|}\, u^{\frac{\sigma}{1+\sigma}} \ge 0$$

satisfying the equation

$$v_t = \mathbf{B}(v) = |v_x|^\sigma[|\sigma|\, v v_{xx} - (v_x)^2] - 1. \tag{5.78}$$

The nonlinear operator $\mathbf{B}$ has remarkable invariant properties.

Linear TW solutions

First of all, $\mathbf{B}$ admits the invariant subspace $W_2 = \text{Span}\{1, x\}$, whence the explicit TW solutions

$$V_\pm(x,t) = B_\pm(\xi)_+, \quad \xi = x - \lambda t, \tag{5.79}$$

where $B_- \le B_+$ satisfy the algebraic equation

$$B^{2+\sigma} - \lambda B + 1 = 0.$$

It follows that solutions (5.79) exist under the assumption $\lambda \ge \lambda_m$. Thus $B_- < B_+$ if $\lambda > \lambda_m$ and $B_- = B_+$ at $\lambda = \lambda_m$. General TW solutions $V = f(\xi)$ satisfy the ODE

$$|f'|^\sigma[|\sigma|\, f f'' - (f')^2] + \lambda f' - 1 = 0,$$

which by setting $P = f' \ge 0$ reduces to the first-order one

$$\frac{\mathrm{d}P}{\mathrm{d}f} = \frac{P^{2+\sigma} - \lambda P + 1}{|\sigma| f P^{1+\sigma}}.$$

Using techniques similar to those in Section 5.2, we prove that, for $\lambda \ge \lambda_m$, the TW V_- is a proper solution, and V_+ is not proper if $\lambda > \lambda_m$ since it does not verify the comparison principle.

Transversality with a subset of steep TWs $B_1 = \{V, |\lambda| \gg 1\}$ implies a standard Bernstein bound $|v_x| \le C_0$ for general bounded solutions. In order to get more regularity, we derive estimates of v_{xx} by a tangential construction with a subset of *"parabolic"* solutions that is complete in $\mathbb{R}^3$ and is constructed below.

Explicit parabolic solutions

Secondly, $\mathbf{B}$ possesses another two-dimensional invariant subspace

$$\tilde W_2 = \text{Span}\{1, |x|^\gamma\}, \quad \text{where} \quad \gamma = \tfrac{2+\sigma}{1+\sigma} > 0.$$

Therefore, equation (5.78) restricted to $\tilde W_2$,

$$\tilde v(x,t) = [D_0(t) + D_1(t)|x|^\gamma]_+, \tag{5.80}$$

reduces to the dynamical system of the expansion coefficients

$$\begin{cases} D_0' = \mu\gamma^{1+\sigma} D_0 |D_1|^\sigma D_1 - 1, \\ D_1' = -2\gamma^{1+\sigma} |D_1|^\sigma D_1^2, \end{cases} \tag{5.81}$$

where $\mu = \frac{|\sigma|}{1+\sigma} > 0$. Assuming that $D_1(t) > 0$, we have

$$D_1(t) = d_1 t^{-\frac{1}{1+\sigma}},$$

where

$$d_1 = [2\gamma^{1+\sigma}(1+\sigma)]^{-\frac{1}{1+\sigma}},$$

and then the first equation in (5.81) reduces to the first-order ODE

$$D_0' = \rho D_0 \tfrac{1}{t} - 1,$$

with

$$\rho = \frac{|\sigma|}{2(1+\sigma)^2} > 0.$$

Finally, we obtain the following solutions on $\tilde{W}_2$:

$$\tilde{v}(x,t) = \begin{cases} d_0 t^\rho + \frac{t}{\rho - 1} + d_1 \frac{|x|^\gamma}{t^{1/(1+\sigma)}}, & \rho \neq 0, \\ \tilde{d}_0 t - t \log t + \frac{|x|^3}{3t^2}, & \rho = 1. \end{cases} \tag{5.82}$$

Here d_0 is a free constant. It follows that, if $\rho < 1$, i.e., $\sigma \in (-\frac{1}{2}, 0)$, then finite-time blow-up for $\tilde{u}$ occurs if $d_0 > 0$. For $\rho > 1$ ($\sigma \in (-1, -\frac{1}{2})$), we need $d_0 < 0$ for blow-up. If $\rho = 1$ ($\sigma = -\frac{1}{2}$), then $\tilde{u}$ blows up at $T = e^{d_0}$.

The interface equation is derived from equation (5.78),

$$\tilde{s}'(t) = F(\tilde{v}_x(\tilde{s}(t), t)), \quad \text{where } F(S) = |S|^\sigma S + \frac{1}{S}.$$

It follows from (5.82) that, if $\rho > 1$, then

$$\tilde{s}(t) \sim t^{\frac{1}{2(1+\sigma)}} \gg t \quad \text{for } t \gg 1.$$

Hence, by Theorem 5.21, the function $\tilde{v}(x,t)$ is not a proper maximal solution for large times.

In the critical case $\sigma = -\frac{1}{2}$, it follows from (5.82) that the right-hand interface is given by

$$\tilde{s}(t) = 3^{1/3}\, t \left(\log(\tfrac{t}{T})\right)^{1/3},$$

where $T = e^{d_0} > 0$ is the corresponding blow-up time. Again, by Theorem 5.21, this explicit solution cannot be a proper one for $t \gg 1$. On the other hand, exactly as in Section 5.2, we have that

$$\tilde{s}''(t) = 0 \quad \text{at } t = T_1 = Te^{2/3}$$

and moreover,

$$\tilde{s}'(T_1) = 3 \cdot 2^{-2/3} \equiv \lambda_m(-\tfrac{1}{2}).$$

This implies that for all $t \leq T_1$, the solution $\tilde{v}(x,t)$ is proper maximal, while it is not for $t > T_1$; cf. Theorem 5.5.

Finally, for $\rho < 1$, the blow-up interface

$$\tilde{s}(t) = 2(1+\sigma)(1+2\sigma)^{-\frac{1+\sigma}{2+\sigma}}\, t\, \Big[1 - \big(\tfrac{t}{T}\big)^{\rho-1}\Big]^{\frac{1+\sigma}{2+\sigma}}$$

has its unique inflection point at

$$T_1 = T\mu^{\frac{2(1+\sigma)^2}{(2+\sigma)(21+\sigma)}},$$

where the corresponding interface speed is minimal,

$$\tilde{s}'(T_1) = \lambda_m.$$

Therefore, $\tilde{v}$ is proper maximal for $t \le T_1$ and is not proper for $t > T_1$, exactly as in Theorem 5.5. Using the same intersection comparison techniques, most of the results from preceding sections can be translated to equation (5.76).

5.9 Blow-up interfaces for equations with general nonlinearities

Let us discuss singular blow-up interfaces for some quasilinear heat equations with more general nonlinearities

$$\boxed{u_t = (\varphi(u))_{xx} + \psi(u),} \tag{5.83}$$

where $\varphi \in C^1([0,\infty)) \cap C^2(I\!R_+)$, $\psi(u) \in C^1([0,\infty))$ are given functions satisfying

$$\varphi'(u) > 0,\;\; u > 0 \text{ (parabolicity)}, \qquad \int^{\infty} \frac{\mathrm{d}s}{\psi(s)} < \infty \text{ (blow-up)}, \tag{5.84}$$

$$\int^{\infty} \frac{\mathrm{d}\varphi(s)}{s} < \infty \quad \text{(finite speed of propagation on } \{u = \infty\}). \tag{5.85}$$

Our goal is to demonstrate that the results of the previous sections describe typical properties of blow-up interfaces for more general equations. We concentrate on the construction of families of exact solutions. The proof of several results similar to those in Sections 5.2–5.7 is based on comparison and intersection comparison and is straightforward.

Linear TW solutions

Introducing the pressure variable

$$v = \Phi(u) \equiv \int_u^{\infty} \frac{\mathrm{d}\varphi(s)}{s}, \tag{5.86}$$

we obtain the equation

$$v_t = \mathbf{A}(v) \equiv F(v)v_{xx} - (v_x)^2 - q(v), \tag{5.87}$$

with the coefficients

$$F(v) = \varphi'(u) \quad \text{and} \quad q(v) = \frac{\psi(u)\varphi'(u)}{u}.$$

Assuming that

$$\psi(u) = \frac{u}{\varphi'(u)},$$

we have $q(v) \equiv 1$ for $v > 0$ and hence the equation has the form

$$v_t = \mathbf{A}(v) \equiv F(v)v_{xx} - (v_x)^2 - 1 \quad \text{in } S \cap \{v > 0\}.$$

Operator $\mathbf{A}$ admits the invariant subspace $W_2 = \text{Span}\{1, x\}$, on which there exist explicit TW solutions

$$V_\pm(x,t) = S_\pm(x - \lambda t)_+, \quad S_\pm = \tfrac{1}{2}\,(\lambda \pm \sqrt{\lambda^2 - 4}), \tag{5.88}$$

with the speed restriction $\lambda \geq 2$. Similarly, we have that V_- is the proper maximal solution, and V_+ is not if $\lambda > 2$. By comparison, we deduce that the right-hand blow-up interface of the general solution $u(x,t)$ with compactly supported $u_0 \geq 0$ satisfies

$$s(t) \leq 2(t + c_0) \quad \text{for all } t \geq T$$

(cf. Theorem 5.21). As in Theorem 5.6, under a natural condition on $\varphi(u)$, which implies that the TW solutions with $\lambda = 2 - \varepsilon$ are steep enough, by passing to the limit $\varepsilon \to 0^+$, we prove that (cf. (5.30))

$$\underline{D}^+ s(t)(t) \geq 2 \quad \text{for} \quad t > T.$$

Linear explicit solutions

Let us describe some properties of other explicit solutions (not necessarily TWs) on the invariant subspace $W_2 = \text{Span}\{1, x\}$, which exist for some equations (5.87). Now let $q(v) = 1 + \alpha v$ for $v > 0$, where $\alpha > 0$ is a constant, i.e.,

$$\psi(u) = u\,\frac{\alpha\Phi(u) + 1}{\varphi'(u)}.$$

Then (5.87) takes the form

$$v_t = F(v)v_{xx} - (v_x)^2 - 1 - \alpha v. \tag{5.89}$$

Since $\mathbf{A}(W_2) \subseteq W_2$ still, there exist exact solutions

$$\overline{v}(x,t) = [C_0(t) + C_1(t)x]_+ \in W_2,$$

where the expansion coefficients on W_2 satisfy the dynamical system

$$\begin{cases} C_0' = -\alpha C_0 - C - 1^2 - 1, \\ C_1' = \alpha C_1. \end{cases}$$

Integrating it yields

$$\overline{v}(x,t) = \left(d_0 e^{-\alpha t} + \tfrac{1}{\alpha}\, e^{-2\alpha t} - \tfrac{1}{\alpha} + e^{-\alpha t} x\right)_+, \quad d_0 \in I\!R. \tag{5.90}$$

The corresponding singular interface moves exponentially fast

$$\overline{s}(t) = \tfrac{1}{\alpha}\, e^{\alpha t} - d_0 - \tfrac{1}{\alpha}\, e^{-\alpha t} = \frac{1}{\alpha}\, e^{\alpha t}(1 + o(1)) \quad \text{for } t \gg 1.$$

Observe that, in this case, $\overline{v}_x(\overline{s}(t),t) = e^{-\alpha t}$ and

$$\overline{s}'(t) = e^{-\alpha t} + e^{\alpha t} \equiv \left(\overline{v}_x + \frac{1}{\overline{v}_x}\right)(\overline{s}(t),t).$$

It follows that this free boundary equation corresponding to equation (5.89) is valid provided that $\overline{v}_x \le 1$, i.e., on the monotone decreasing branch that is defined for interface slopes $\bar{v}_x \in (0,1)$. The other increasing branch corresponds to improper solutions violating the comparison at the interface.

Thus, for all $t \ge 0$, when $\overline{v}_x(\overline{s}(t),t) = e^{-\alpha t} \le 1$, the linear explicit solution $\overline{v}(x,t)$ is proper maximal. If $t < 0$, then $\overline{v}_x > 1$ on the interface, no comparison is valid, and hence the solution $\overline{v}(x,t)$ is not proper. By comparison this implies the upper bound $s(t) \le \overline{s}(t) + c_0$ of the right-hand interfaces of all proper solutions $v(x,t)$ with, say, compactly supported initial data. For instance, setting $\varphi(u) = u^m$, we obtain that the blow-up interface of any proper minimal solution of

$$\boxed{u_t = (u^m)_{xx} + \tfrac{1}{m} u^{2-m} + \tfrac{\alpha}{1-m}\, u, \quad 0 < m < 1, \;\; \alpha > 0,}$$

with compactly supported u_0 satisfies

$$s(t) \le \tfrac{1}{\alpha}\, e^{\alpha t} + c_0 \quad \text{for all } t > T.$$

Solutions (5.90) make sense if $\alpha < 0$, and then $\overline{s}(t) \sim e^{|\alpha| t} \to \infty$ as $t \to \infty$. Hence,

$$\overline{v}_x(\overline{s}(t),t) = e^{|\alpha| t} \le 1 \quad \text{if} \quad t \le 0,$$

i.e., $\overline{v}(x,t)$ is proper for $t \le 0$ and is not proper for $t > 0$.

Explicit blow-up solutions on an invariant set

Consider a new class of explicit solutions describing infinite time collapse of two blow-up interfaces. Under assumptions (5.84) and (5.85) consider the equation

$$u_t = (\varphi(u))_{xx} + u\left[\frac{2}{\varphi'(u)} \int_u^{\infty} \frac{\varphi'(s)}{s}\, \mathrm{d}s - 1\right]. \tag{5.91}$$

Then the pressure (5.86) satisfies

$$v_t = \mathbf{B}(v) \equiv F(v)(v_{xx} + 1) - [(v_x)^2 + 2v]. \tag{5.92}$$

Since $F(v)$ is a rather arbitrary function, the operator $\mathbf{B}$ does not admit the invariant subspace $W_2 = \mathrm{Span}\{1, x^2\}$. Nevertheless, $\mathbf{B}$ is known to admit an *invariant set* (an affine subspace) of the form

$$\mathcal{M} = \{-\tfrac{1}{2}\, x^2 - C_0, \; C_0 \in I\!R\},$$

which is invariant on the linear subspace W_2 in the sense that

$$\mathbf{B}(\mathcal{M}) \subseteq W_2.$$

This means that equation (5.92) on $\mathcal{M}$ is equivalent to an *overdetermined* dynamical system. Therefore, substituting

$$\tilde{v}(x,t) = -\tfrac{1}{2}\, x^2 + C_0(t)$$

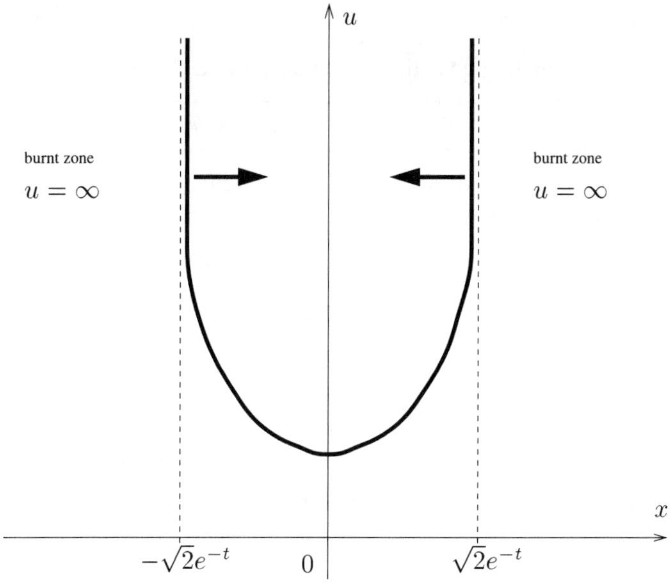

Figure 5.5 *The explicit solution $\tilde u(x,t)$ given by* (5.86), (5.93) *describes an exponentially slow collapse as* $t \to \infty$ *of two blow-up interfaces.*

into (5.92), we arrive at the ODE

$$C_0' = -2C_0,$$

whence the following exact solution on $\mathcal{M}$:

$$\tilde v(x,t) = \left(e^{-2t} - \tfrac{1}{2}\,x^2\right)_+ . \tag{5.93}$$

It has two symmetric blow-up interfaces

$$\tilde s(t) = \pm\sqrt{2}\,e^{-t},$$

on which

$$\tilde v_x(\tilde s(t),t) = -\tilde s(t),$$

so the free boundary equation $\tilde s'(t) = \tilde v_x(\tilde s(t),t)$ holds. In terms of u, solution (5.93) describes the large-time evolution of an exponentially thin layer of finite temperature $\{|x| < \sqrt{2}\,e^{-t}\}$ in the burnt $\{u = \infty\}$-zone; see Figure 5.5.

5.10 Examples of blow-up surfaces in $I\!R^N$

We finish our discussion with examples of singular blow-up interfaces (surfaces) described by the N-dimensional quasilinear heat equations, where a complete mathematical theory is still not available. The regularity properties of such blow-up interfaces generate a number of open challenging problems.

Nonsymmetric blow-up surfaces

Consider first the following quasilinear heat equation with power nonlinearities:

$$\boxed{u_t = \nabla \cdot (u^{-\mu} \nabla u) + u^{1+\mu} \quad \text{in } S = I\!R^N \times I\!R_+,} \tag{5.94}$$

where $\mu > 0$ is a fixed constant. The pressure function $v = u^{-\mu}$ satisfies a parabolic equation with the quadratic nonlinear operator

$$v_t = \mathbf{A}(v) \equiv v\Delta v - \mu\,|\nabla v|^2 - \mu \quad \text{in } \{v > 0\}. \tag{5.95}$$

The operator $\mathbf{A}$ admits the $(N+1)$-dimensional invariant subspace in $I\!R^N$

$$W_{N+1} = \text{Span}\{1, x_1^2, ..., x_N^2\}.$$

Therefore there exist exact solutions of the form

$$\tilde{v}(x,t) = [C_0(t) + C_1(t)x_1^2 + ... + C_N(t)x_N^2]_+, \tag{5.96}$$

where the expansion coefficients $\{C_k(t)\}$ satisfy the dynamical system

$$\begin{cases} C_0' = 2C_0 M - \mu, \\ C_k' = 2C_k M - \frac{4}{\mu}\, C_k^2, \end{cases} \tag{5.97}$$

for all $k = 1, ..., N$, where we use the notation

$$M = \sum_{k=1}^{N} C_k.$$

Assuming that $C_k^0 = C_k(0) > 0$ for all $k = 0, 1, ..., N$, we obtain the following representation of the solutions of (5.97):

$$C_0(t) = \frac{1}{C_0^0 - \frac{\mu}{F(G)} \int_0^G F^2(z)\,\mathrm{d}z}, \quad C_k(t) = \frac{C_k^0}{F(G)(1 + 4\frac{1}{\mu}C_k^0\, G)} \tag{5.98}$$

for $k = 1, ..., N$, where $G = G(t)$ is now the new time variable given by

$$t = \int_0^G F(z)\,\mathrm{d}z, \quad F(z) = \prod_{k=1}^{N} (1 + 4\tfrac{1}{\mu} C_k^0 z)^{-\frac{\mu}{2}}. \tag{5.99}$$

Note that besides the blow-up phenomenon for equation (5.94), there exists another singular in time behaviour of the initially positive solution (5.96) due to the *extinction* property driven by the fast diffusion. This can affect the behaviour of the exact solutions. Indeed, it follows from (5.99) that $G(t)$ is defined on a finite interval $(0, T_*)$ provided that

$$T_* = \int_0^\infty F(z)\,\mathrm{d}z < \infty.$$

Since

$$F(z) \sim z^{-\mu N/2} \quad \text{as } z \to \infty,$$

this gives the well-known condition of extinction in finite time via the fast diffusion $-\mu \frac{N}{2} < -1$, i.e., if $\mu > \frac{2}{N}$.

Suppose that $\mu \le \frac{2}{N}$, so no extinction in finite time occurs, i.e., $T_* = \infty$. Then we deduce that $\tilde{u} = \infty$ when $C_0(T) = 0$ and $C_0(t) < 0$ for $t > T$. This gives the singular *blow-up surface* (we continue to call it the interface) $\tilde{s}(t) = \partial\{\tilde{v}(\cdot, t) > 0\}$ in the form of the ellipsoid

$$\sum_{i=1}^{N} C_k(t) x_k^2 = |C_0(t)| \quad \text{for } t > T.$$

According to equation (5.95), where $\tilde{v}\Delta\tilde{v} = 0$ on $\{\tilde{v} = 0\}$, the normal velocity $v^\perp(x,t)$ at any $x \in \tilde{s}(t)$ of the propagation of the singular interface (in the direction of the outward normal n to $\tilde{s}(t)$, which has the same direction as $\nabla\tilde{v}$) satisfies the interface equation

$$v^\perp = \frac{1}{\mu} |\nabla \tilde{v}| + \frac{\mu}{|\nabla \tilde{v}|} \equiv H(|\nabla \tilde{v}|) \quad \text{for } x \in \tilde{s}(t) \tag{5.100}$$

provided that $|\nabla\tilde{v}| \le \mu$. If $|\nabla\tilde{v}(x,t)| > \mu$, the solution $\tilde{v}$ cannot be a proper maximal one since it does not verify the comparison principle locally, in a neighbourhood of any such point $x \in \tilde{s}(t)$. We observe a constant C^1-continuation for $S = |\nabla\tilde{v}| \ge \mu$ of the function H on the right-hand side of (5.100) similar to the 1D case in Section 5.5. In particular, we have the following property of the general solution.

Lemma 5.22 *Let $t_0 > T$ and $x_0 \in s(t_0)$. Assume that $s(t_0)$ is a C^1-surface in a neighbourhood of $x = x_0$. Then the normal velocity satisfies*

$$\underline{v}_+^\perp(x_0, t_0) \ge 2, \tag{5.101}$$

where the right-hand derivative $\underline{v}_+^\perp$ is understood in the $\liminf_{t \to t_0^+}$ *sense.*

Proof. Fix a point $a_0 \in \{v(\cdot, t_0) > 0\}$ with $|a_0 - x_0| \ll 1$. We construct a radially symmetric TW subsolution of equation (5.94) of the form

$$U(x,t) = f(y), \quad y = |x - a_0| - \lambda(t - t_0), \quad \lambda = 2 - \varepsilon,$$

with arbitrarily small $\varepsilon > 0$. There exists $\rho > 0$ such that $\bar{B}_\rho(a_0) \subset \{v(\cdot, t_0) > 0\}$. The subsolution U satisfies, for $r = |x - a_0| > \frac{1}{2}\rho$, the 1D equation

$$u_t = (u^{-\mu} u_r)_r + d_* u^{-\mu} u_r + u^{1+\mu}, \quad \text{where } d_* = \tfrac{2(N-1)}{\rho}. \tag{5.102}$$

We assume that $f(0) = n \gg 1$ and $f'(0) = 0$. It follows from the analysis of the ODE for the function f, that $f(y)$ is steep enough on any fixed finite level, so, for any small $\varepsilon > 0$, there exists $n = n_\varepsilon \gg 1$ such that $U(x,0) \le u(x,t_0)$ in $\{U(x,0) > 0\}$. In fact, one can see that the second term on the right-hand side of (5.102) plays no role as $n \to \infty$ if $\lambda = 2 - \varepsilon$. By comparison, passing to the limit $\varepsilon \to 0^+$, we obtain that, for $t > t_0$,

$$\{|x - a_0| < 2(t - t_0)\} \subset \{v(\cdot, t) > 0\} \quad \text{for any } a_0 \in s(t_0),$$

which implies (5.101). □

Thus we expect that the interface equation has a form similar to that, in the 1D geometry,

$$v^{\perp} = H_*(|\nabla v|), \quad \text{where} \quad H_*(S) = \begin{cases} H(S) & \text{for } S \le \mu, \\ 2 & \text{for } S > \mu. \end{cases}$$

Let us show that the explicit solutions $\tilde{v}$ for large times enter the region with the minimal speed of propagation

$$v^{\perp} = H_*(\mu) = 2,$$

i.e., $\tilde{v}$ is not a proper one for $t \gg 1$. This also proves that the interface of proper solutions cannot be analytic or even sufficiently smooth for all times and shows possible types of its irregularity. This is easily seen by comparison from below with simpler radially symmetric solutions $\tilde{v}(x,t) \in W_2 = \text{Span}\{1, |x|^2\}$, which obviously reach the region $\{|\nabla \tilde{v}| > \mu\}$.

For instance, let $N = 2$, where

$$\tilde{s}(t) = \{C_1(t)x_1^2 + C_2(t)x_2^2 = |C_0(t)|\} \quad \text{for } t > T,$$

and on $\tilde{s}(t)$

$$|\nabla \tilde{v}| = 2\sqrt{C_1^2 x_1^2 + C_2^2 x_2^2} \equiv 2\sqrt{C_1(C_1 - C_2)x_1^2 + C_2|C_0|}. \tag{5.103}$$

Therefore, assuming for definiteness that

$$C_1^0 < C_2^0$$

(and hence $C_1(t) < C_2(t)$), we have from (5.103) that the equality $|\nabla \tilde{v}| = \mu$ holds for the first time for $t = T_1$ at the point

$$Q_0 = \{0, \sqrt{|C_0(T_1)|/C_2(T_1)}\}, \tag{5.104}$$

where

$$|\nabla \tilde{v}| = 2\sqrt{C_2(T_1)|C_0(T_1)|} = \mu.$$

Let us prove that T_1 is finite for all $\mu \in (0,1)$. It follows from (5.98), (5.99) that, as $t \to \infty$ (or equivalently as $G \to \infty$),

$$C_2(t)|C_0(t)| \sim G^{2\mu-1} \to \infty \quad \text{if} \quad \mu > \tfrac{1}{2},$$

$$C_2(t)|C_0(t)| \sim \tfrac{1}{4}\,\mu^2 \log G \to \infty \quad \text{if} \quad \mu = \tfrac{1}{2},$$

$$C_2(t)|C_0(t)| \to \tfrac{\mu^2}{4(1-2\mu)} > \tfrac{1}{4}\,\mu^2 \quad \text{if } \mu < \tfrac{1}{2}.$$

Therefore, equality (5.104) holds at finite T_1, and hence $\tilde{v}(x,t)$ is not proper for $t > T_1$. As in the 1D case, we have that the analyticity of the solution and its blow-up surface break down at finite $t = T_1$.

Explicit blow-up solutions on $W_3 = \text{Span}\{1, |x|^2, |x|^4\}$

Finally, consider a different type of singular blow-up surfaces in the symmetric N-dimensional geometry. Consider the following special quasilinear heat equation:

$$u_t = \nabla \cdot (u^{-4/(N+2)} \nabla u) + u^{(N+6)/(N+2)}. \tag{5.105}$$

Then the pressure change $v = u^{-4/(N+2)}$ transforms (5.105) into the equation

$$v_t = \mathbf{A}(v) \equiv v\Delta v - \gamma|\nabla v|^2 - \tfrac{1}{\gamma}, \quad \text{where } \gamma = \tfrac{1}{4}(N+2). \tag{5.106}$$

The quadratic operator $\mathbf{A}$ admits the 3D invariant subspace

$$W_3 = \text{Span}\{1, |x|^2, |x|^4\}.$$

Hence, (5.106) has the exact solutions

$$v(x,t) = [C_1(t) + C_2(t)|x|^2 + C_3(t)|x|^4]_+, \tag{5.107}$$

where the coefficients $\{C_1, C_2, C_3\}$ satisfy the dynamical system

$$\begin{cases} C_1' = 2NC_1C_2 - \frac{4}{N+2}, \\ C_2' = 4(N+2)C_1C_3 + (N-2)C_2^2, \\ C_3' = 2NC_2C_3. \end{cases}$$

It follows from (5.107) (or directly from equation (5.106)) that the right-hand interface equation has the form

$$s' = \gamma v_r + \tfrac{1}{\gamma} v_r \quad \text{if} \quad v_r(s(t),t) \le \tfrac{1}{\gamma}.$$

By comparison with the radial TW-subsolutions, as in Lemma 5.22, we have that $\underline{D}^+ s(t) \ge 2$. We then deduce that these solutions are not proper if

$$v_r > \tfrac{4}{N+2}$$

on the interface. As in the 1D case, such exact solutions make it possible to establish some optimal regularity of singular interfaces.

Remarks and comments on the literature

Main results are presented in [168].

§ **5.1.** The critical Fujita exponent $p_F = m + 2$ for the 1D quasilinear equation (5.1) was calculated in [145], where $p_F = m + \frac{2}{N}$ for the equations in $I\!R^N$; see more references in survey papers [244], [97] and [169]. The critical exponent $p = p_F$ always belongs to the blow-up case [135]. An extended list of references on blow-up in nonlinear parabolic equations can be found in the books [306, Chapter 4], [267, Part II] and [170, Chapters 9, 10]. The terminology from Combustion Theory is explained in [71]. Estimate (5.6) can be found in [167, Section 4].

The existence of a minimal wave speed is to be compared with the classical KPP result [226] and the extensions to quasilinear equations done in [30], [126] and [285].

§ **5.2.** Examples of such explicit TW solutions are available in [126], [165], [285]. Using invariant subspaces for quadratic operators is motivated by [131], [136], where new types of exact solutions are constructed by such nonlinear invariance properties.

Comparison with a proper blow-up TW solution was used in the proof of Theorem 4.1 in [167]. A construction of a unique proper minimal solution by truncations of both the equation and (singular) initial data is available in [167, Section 2] and is explained in detail in Section 6.2 below.

The existence of two branches of TWs was detected in [126] for equations of the type (5.3) with general nonlinearities (see Section 5.9), where the increasing branch was shown to be unstable. In [285] equation (5.3) is studied in the non blow-up range $m > 1, p = 2 - m$, and the good branch is shown to be a branch of minimal solutions for both u and v. In that case the Cauchy problem is shown to have infinitely many non-minimal solutions with finite interfaces and also a maximal solution that is positive everywhere, [284].

§ **5.3.** As far as we know, these explicit non self-similar (with no group invariance) solutions were first constructed by R. Kersner (1976) [222]. Such solutions can also be constructed either by a nonlinear separation method [131], or by using linear subspaces invariant under quadratic operators [136].

§ **5.4.** $s(t)$ is nondecreasing due to a general monotonicity result [154] saying that, for general 1D quasilinear heat equations, any large solution becomes strictly monotone with time, i.e., this is true for any blow-up solutions. Continuity follows by an elementary local comparison; see [165, Theorem 4.3], where other intersection comparison ideas were introduced.

§ **5.5.** Comparison with proper TWs is presented in [167, Section 4], where the uniqueness of the blow-up proper solution (independence of a monotone regularization of the problem) was also proved; see Chapter 6.

The proof of the crucial upper bound on the second derivative $v_{xx} \le C$ is based on intersection comparison with a family of parabolic solutions. It uses some geometric ideas from [164]. Boundedness of v_{xx} at the blow-up time $t = T$ follows from the final-time profile $v(x, T^-) = c_0 x^2(1 + o(1))$ for $x \approx 0$, which is proved exactly as in [163]. This asymptotic estimate is important for the further analysis of the interface slope in this section. Similar ideas of intersection comparison were used in Sections 2 and 3 in [164].

§ **5.6.** Regularity properties of interfaces for the PME with different lower-order operators are studied in a number of papers; see extended lists of references in the book [122], [170, Chapters 2, 4] and papers [159] and [160].

§ **5.7.** Solvability of ODEs such as (5.73) and asymptotic properties of solutions were quite popular questions of the qualitative theory of quasilinear parabolic equations in the 1970s and 80s. See references in [306, Chapters 4–7].

§ **5.8.** Comparison with proper TWs for gradient diffusivity equations can be found in [165, Section 7] and [167, Section 17] . Explicit solutions (5.80) on the invariant subspace and more complicated examples are given in [155].

§ **5.9.** Two families of TWs (5.88) were available in [126]; see a more general analysis in [134] and in [164, Section 4]. The exact solution (5.93) belonging to the invariant set was constructed in [134, Section 3].

§ **5.10.** Exact solutions (5.96) on the invariant subspace W_{N+1} were studied in [156]. The proof of estimate (5.101) is based on the construction from [167, Section 5]. More general exact solutions of the type (5.107) can be found in [136]. Spatially non-monotone (in the linearized setting, spanned by Hermite polynomials) asymptotic structures of such blow-up solutions were described in [156].

CHAPTER 6

Complete and Incomplete Blow-up in Several Space Dimensions

In this chapter we apply intersection comparison techniques to the study of complete and incomplete blow-up phenomena for quasilinear heat equations in $\mathbb{R}^N$.

First we present a detailed construction of limit semigroups of minimal solutions to quasilinear heat equations admitting finite-time blow-up. Second we show that, in the complete-incomplete blow-up phenomena in $\mathbb{R}^N$, a crucial role is played by the *critical Sobolev exponent* for the elliptic operator of the reaction-diffusion equation under consideration. The results are extended to another singular phenomenon of finite-time extinction for equations with absorption.

6.1 Introduction: The blow-up problem in $\mathbb{R}^N$ and critical exponents

As in the previous chapter, our basic model is the quasilinear heat equation with source

$$\boxed{u_t = \Delta u^m + u^p \quad \text{in } S = \mathbb{R}^N \times \mathbb{R}_+, \quad m > 0, \;\; p > 1.} \tag{6.1}$$

We deal with nonnegative solutions $u(x, t)$ defined in $\mathbb{R}^N$ for some time interval $0 < t < T$. We now work in dimension $N > 1$. We assume that $m > m_c = \frac{(N-2)_+}{N}$, a well-known critical value, below which the very fast diffusion implies finite-time extinction at the zero-level $\{u = 0\}$. This falls out of our present interest. Otherwise, we take $u_0(x) \geq \delta > 0$ in $\mathbb{R}^N$ thus excluding this extra singularity.

The initial data $u_0(x)$ are assumed to be nonnegative and bounded. By standard theory we may also assume that u_0 is smooth in the positivity domain (after waiting a small amount of time). Let us again mention that we work with the Cauchy problem for convenience; our methods on complete and incomplete blow-up apply to the initial-boundary value problem in a bounded ball, to yield a similar classification.

Due to the superlinear heat source $Q(u) = u^p$ for $u \gg 1$, the Cauchy problem is known to admit a local (in time) solution that may blow-up in a finite time. The precise result depends on the relative values of m and p. Thus it is known that, if $1 < p \leq p_F = m + \frac{2}{N}$, p_F is the *critical Fujita exponent*, any solution $u \not\equiv 0$ blows up in finite time: there exists a blow-up time $T = T(u_0) < \infty$ such that the solution is well defined for all $x \in \mathbb{R}^N$ and $0 < t < T$, while

$$\sup_{x \in \mathbb{R}^N} u(x, t) \to \infty \quad \text{as } t \to T^-.$$

On the other hand, when $p > m + \frac{2}{N}$ blow-up occurs if u_0 is large enough, but there also exist global in time small solutions. Moreover, the possible modes of blow-up depend on m and p. Thus, for bell-shaped symmetric data, *single point blow-up* occurs if $p > m$. If $p = m$, there are solutions exhibiting *regional blow-up* (i.e., in a ball), and blow-up is *global* (on all of $I\!R^N$) if $1 < p < m$. As in the previous chapters, our concern here is the possible continuation of the solution for $t > T$. In Section 6.2 we present a more detailed concept of proper minimal solutions in the general framework of extending order-preserving semigroups. For such global proper solutions, we can define the blow-up set $\mathcal{B}[u](t)$ for $t \geq T$

$$\mathcal{B}[u](t) = \{x \in I\!R^N : \exists \ \{x_k\} \to x, \ \{t_k\} \to t^- \text{ with } u(x_k, t_k) \to \infty\}. \quad (6.2)$$

Then complete blow-up means that $\mathcal{B}[u](t) = I\!R^N$ for all $t > T$ (though possibly meas $\mathcal{B}[u](T) = 0$), while for the incomplete blow-up $\mathcal{B}[u](t) \neq I\!R^N$ at least for all small $t - T > 0$.

In Section 6.3 we deal with the phenomenon of incomplete blow-up for $m+p \leq 2$ with the appearance of finite blow-up interfaces. Section 6.4 gives a more precise idea of the behaviour of solutions undergoing incomplete blow-up in the limit case $p+m = 2$ by means of explicit solutions. Sections 6.5 to 6.9 cover the analysis of complete blow-up. The main difficulties occur in the case where the blow-up set is localized at the centre of symmetry $x = 0$ and in that respect there appears an important exponent, the *critical Sobolev exponent*, p_S. In Section 6.6 we perform the analysis of complete blow-up for focused solutions in the subcritical range $2 - m < p < p_S$, while Section 6.7 covers the limit case $p = p_S$. The unfocused cases are dealt with by a modification to 1D techniques in Section 6.8. Section 6.9 establishes complete blow-up in the supercritical range $p > p_S$, $N \geq 3$, under specific conditions on initial data. Let us remark that, since $p_S > p_F$, "small" solutions never blow-up (they are global in time and bounded throughout), while "large" data lead to blow-up.

6.2 Construction of the proper blow-up solution: extension of monotone semigroups

The construction is rather general, and deals with the possibility of extending ordered semigroups (in the application, the ones associated to the nonlinear PDEs we deal with), so they apply to more general data and reaction functions, under suitable monotonicity assumptions.

Order-preserving semigroups

We begin with an ordered topological space X of functions, $\Omega \mapsto \overline{I\!R}_+$, where Ω is an open subset of $I\!R^N$ and $\overline{I\!R}_+ = [0, \infty) \cup \{\infty\}$. We also have a semigroup $\mathcal{S}(t)$ acting in a space Y that is a subspace of X and approximates X as explained below. We want to extend $\mathcal{S}(t)$ to act on X. We assume that

(S1) $\mathcal{S}(t)$ is order-preserving (i.e., the solutions satisfy the usual comparison), and

(S2) $\mathcal{S}(t)$ is continuous and X is closed with respect to monotone increasing convergence (m.i.c. for short).

Next we will consider a family of “approximation” operators $\{P_n : X \to Y,\ n = 1, 2, ...\}$ such that:

(P1) the family $\{P_n\}$ is ordered in the sense that, for every $u \in X$ and $n > m$, we have $P_n u \geq P_m u$,

(P2) it is also continuous under m.i.c., and

(P3) as $n \to \infty$, we have $P_n u \to u$ in X.

Definition 6.1 We define the *extension* of $\mathcal{S}(t)$ as follows. For every $u \in X$ and $t > 0$, we put

$$\mathcal{T}(t)u = \lim_{n\to\infty} \mathcal{S}(t)P_n u. \tag{6.3}$$

Theorem 6.1 *$\mathcal{T}(t)$ is a semigroup in X that extends $\mathcal{S}(t)$ and is continuous under m.i.c. The limit is independent of the approximation sequence P_n satisfying $(P1)$–$(P3)$.*

Proof. (i) Take $u \in B$. Then, since $P_n u \to u$ and $\mathcal{S}_t$ is continuous, there holds

$$\mathcal{S}(t)P_n u \to \mathcal{S}(t)u.$$

In other words, $\mathcal{T}(t)u = \mathcal{S}(t)u$, so $\mathcal{T}(t)$ extends $\mathcal{S}(t)$.

(ii) Now put $\mathcal{T}^n(t) = \mathcal{S}(t)P_n$. For $t, s > 0$ and arbitrary $k \geq 1$, we have

$$\begin{aligned}\mathcal{T}^n(t+s)u &= \mathcal{S}(t+s)P_n u = \mathcal{S}(s)(\mathcal{S}(t)P_n u)\\ &\geq \mathcal{S}(s)P_k(\mathcal{S}(t)P_n u) = \mathcal{T}^k(s)(\mathcal{T}^n(t)u).\end{aligned}$$

In the limit $n \to \infty$ we obtain $\mathcal{T}(t+s)u \geq \mathcal{T}^k(s)(\mathcal{T}(t)u)$. Letting now $k \to \infty$, we have

$$\mathcal{T}(t+s) \geq \mathcal{T}(s) \circ \mathcal{T}(t).$$

(iii) We now prove the converse inequality, so that $\mathcal{T}(t)$ is shown to be a semigroup,

$$\begin{aligned}\mathcal{T}^n(t+s)u = \mathcal{S}(t+s)P_n u = \mathcal{S}(s)(\mathcal{S}(t)P_n u) &= \lim_{k\to\infty} \mathcal{S}(s)P_k(\mathcal{S}(t)P_n u)\\ = \lim_{k\to\infty} \mathcal{T}^k(s)\mathcal{S}(t)P_n u \leq \lim_{k\to\infty} \mathcal{T}^k(s)(\mathcal{T}(t)u) &= \mathcal{T}(s)\mathcal{T}(t)u.\end{aligned}$$

(iv) *Continuity*. Let $u_j \to u$ with m.i.c. in X. Then $\mathcal{T}(t)u_j \leq \mathcal{T}(t)u$ since $u_j \leq u$, and hence

$$\lim_{j\to\infty} \mathcal{T}(t)u_j \leq \mathcal{T}(t)u.$$

Next, we have $P_n u_j \to P_n u$. Hence, $\mathcal{T}(t)u_j \geq \mathcal{T}^n(t)u_j$ implies upon passing to the limit $j \to \infty$ that

$$\lim_{j\to\infty} \mathcal{T}(t)u_j \geq \lim_{j\to\infty} \mathcal{T}^n(t)u_j = \mathcal{T}^n(t)u,$$

whence the converse inequality by letting $n \to \infty$, completing the proof.

(v) *Independence*. Assume that Q_n is another approximation sequence. We have for every $u \in X$ that $Q_n u \to u$, hence, if we call $\hat{\mathcal{T}}(t) = \lim_{n\to\infty} \mathcal{S}(t)Q_n u$, we have

$$\mathcal{S}(t)P_n u = \lim_{j\to\infty} \mathcal{S}(t)P_n Q_j u \leq \lim_{j\to\infty} \mathcal{S}(t)Q_j u = \hat{\mathcal{T}}(t)u.$$

In the limit $n \to \infty$ we obtain $\mathcal{T}(t)u \le \hat{\mathcal{T}}(t)u$. The same argument can be done by interchanging the roles of P_n and Q_n. □

In our application, X will be the space of nonnegative, measurable functions $\mathbb{R}^N \to \overline{\mathbb{R}}_+$. There are a number of choices for Y and P_n. Thus Y has to be chosen so that the equation

$$u_t = \Delta u^m + \psi(u), \tag{6.4}$$

with $m > 0$ generates a semigroup $\mathcal{S}(t)$ in Y with the properties (S1) and (S2). For instance, we can take $Y = L^1(\mathbb{R}^N) \cap L^\infty(\mathbb{R}^N)$ or some other smaller domain, like nonnegative functions in $L^\infty(\mathbb{R}^N)$ having compact support (if $m > 1$). Monotone convergence is understood a.e.. We then assume that ψ is uniformly Lipschitz continuous at this stage, so that $\mathcal{S}(t)$ will be well defined in Y. We can also take the more general equation

$$\boxed{u_t = \Delta\varphi(u) + \psi(u)}$$

under suitable assumptions on φ; see Section 6.3. It generates a semigroup in $L^1(\mathbb{R}^N)$, ordered and continuous with respect to m.i.c., [92]. Finally, the operator P_n can be any of the usual cut-off operators producing bounded functions (possibly also with compact support).

Extension of the semigroup

As a second step, we want to pass from the semigroup for equation (6.4) corresponding to a uniformly Lipschitz continuous reaction function ψ to a semigroup having more arbitrary $\psi \ge 0$. We have two possible options. One of them is to assume that the previous stage is complete. Then, given a general $\psi \ge 0$ we perform an approximation process by choosing a monotone increasing sequence of nonnegative Lipschitz-continuous functions ψ_n such that

$$\psi_n(s) \to \psi(s) \quad \text{uniformly on bounded intervals in } s \in [0,\infty),$$

and then pass to the limit. Let us call $\mathcal{S}^n(t)$ the semigroup for equation (6.4) with reaction term ψ_n acting on Y, and its extension to X as constructed before by $\mathcal{T}_t^n$. Thus, for every $u \in X$, we define

$$\mathcal{T}(t)u = \lim_{n\to\infty} \mathcal{T}^n(t)u = \lim_{n\to\infty}\lim_{k\to\infty} \mathcal{S}^n(t)(P_k u).$$

Another natural definition does the two approximation processes at the same time

$$\mathcal{U}(t)u = \lim_{n\to\infty} \mathcal{S}^n(t)P_n u. \tag{6.5}$$

Theorem 6.2 *Both definitions are equivalent and provide us with a semigroup in X that is continuous under the m.i.c. The result is independent of the approximating sequences $\{P_n\}$ and $\{\psi_n\}$. The equivalent general definition is*

$$\mathcal{T}(t)u = \lim_{n,k\to\infty} \mathcal{S}^n(t)P_k u. \tag{6.6}$$

Proof. (i) The first observation is easy: $\mathcal{T}(t) \ge \mathcal{U}(t)$. But,

$$\mathcal{T}^n(t)u = \lim_{k\to\infty} \mathcal{S}^n(t)P_k u \le \lim_{k\to\infty} \mathcal{S}^k(t)P_k u = \mathcal{U}(t)u.$$

It means that $\mathcal{T}(t) \le \mathcal{U}(t)$, hence both are equal.

(ii) Now we present half of the semigroup property: if $\mathcal{U}^k(t) = \mathcal{S}^k(t)P_k$, we have $\mathcal{T}^k(t) \ge \mathcal{U}^k(t)$, and for $n > k$,

$$\mathcal{U}^n(t+s)u = \mathcal{S}^n(t+s)P_n u = \mathcal{S}^n(s)\mathcal{S}^n(t)P_n u$$

$$\ge \mathcal{S}^k(s)P_k\mathcal{S}^n(t)P_n u = \mathcal{U}^k(s)(\mathcal{U}^n(t)u).$$

Using the continuity of $\mathcal{U}^k(s)$, we have then in the limit $n \to \infty$,

$$\mathcal{U}(t+s)u \ge \mathcal{U}^k(s)(\mathcal{U}(t)u) \text{ and } \mathcal{U}(t+s) \ge \mathcal{U}(s) \circ \mathcal{U}(t).$$

(iii) Now the converse inequality using $\mathcal{T}(t)$:

$$\mathcal{T}^n(t+s)u = \mathcal{T}^n(s)(\mathcal{T}^n(t)u) \le \mathcal{T}^n(s)(\mathcal{T}(t)u).$$

Hence, in the limit

$$\mathcal{T}(t+s) \le \mathcal{T}(s) \circ \mathcal{S}(t).$$

(iv) Continuity under the m.i.c. is proved as before.

(v) The independence of P_n is also proved as before. As for the independence of ψ_n, only notation changes are needed.

(vi) The existence of the double limit in (6.6) and the validity of the formula comes from the inequalities

$$\mathcal{S}^n(t)P_n u \le \mathcal{S}^n(t)P_k u \le \mathcal{S}^k(t)P_k u,$$

which are valid when $k \ge n$. We obtain the converse formula for $n \ge k$ after passing to the limit $n, k \to \infty$ and using (6.5). $\sqcap$

Definition 6.2 We will call this extended semigroup the *limit semigroup*. For every $u_0 \in X$, the function $u : \Omega \times [0, \infty) \to \overline{I\!R}_+$ defined by

$$u(x,t) = \mathcal{T}(t)u_0(x)$$

is called the *proper solution* of the corresponding initial-value problem.

These solutions are *minimal ones* in a natural way. There are ordering properties that serve as basis for possible denomination that we state next. The proof is immediate.

Theorem 6.3 *The proper solutions satisfy the standard comparison theorem with respect to the data. In other words, the limit semigroup is order-preserving. Moreover, the proper solution is minimal with respect to any kind of weak solutions of the problem satisfying the Maximum Principle relative to bounded weak solutions.*

For parabolic equations, it is well known that the globally defined approximate solutions are continuous. Therefore, the proper solution, as a monotone limit, is defined everywhere, and not just almost everywhere (it is a lower semicontinuous function). This gives a pointwise sense to the definition of the blow-up sets $\mathcal{B}[u](t)$.

6.3 Global continuation of nontrivial proper solutions

We prove a general result on nontrivial global in time continuation of blow-up proper solutions of the general equation

$$\boxed{u_t = \Delta\varphi(u) + \psi(u),} \tag{6.7}$$

which includes equation (6.1) as a particular case. We assume that the coefficients of the equation $\varphi \in C^1([0,\infty)) \cap C^2((0,\infty))$ and $\psi \in C^1([0,\infty))$ satisfy

$$\varphi'(u) > 0 \text{ (parabolicity)}, \quad \psi(u) > 0 \text{ for } u > 0, \quad \varphi(0) = \psi(0) = 0. \tag{6.8}$$

We now apply the results obtained in Chapter 4 in order to classify the 1D problem by means of intersection comparison with a family of TW solutions. The N-dimensional equation (6.7) with $N > 1$ does not admit, even in the radially symmetric case, TW solutions with bell-shaped form, which is a natural shape for blow-up solutions. Nevertheless, the 1D TWs can be used as solutions in a comparison argument to prove results on incomplete blow-up, thus generalizing the 1D results on incomplete blow-up described above. Furthermore, let the initial data u_0 be bounded and

$$u_0(x) \to 0 \quad \text{as } |x| \to \infty. \tag{6.9}$$

In order to motivate this assumption we recall that when $u_0(x) \to c > 0$ as $|x| \to \infty$, a simple comparison argument shows that complete blow-up must happen in finite time. We now recall the three hypotheses on the constitutive functions of the equation that are necessary for incomplete blow-up (Chapter 4):

$$\int^{\infty} \frac{\mathrm{d}s}{\psi(s)} < \infty \quad \text{(Osgood criterion)}, \tag{6.10}$$

$$F(u) = \frac{1}{u^2}\int_1^u \varphi'(s)\psi(s)\,\mathrm{d}s \quad \text{is uniformly bounded as } u \to \infty, \tag{6.11}$$

$$\int_1^{\infty} \frac{\mathrm{d}\varphi(s)}{s} < \infty. \tag{6.12}$$

(6.10)–(6.12) are valid for equation (6.1) precisely when $p > 1$ and $p + m \leq 2$. In particular, (6.12) excludes linear diffusion, probably the main reason why this phenomenon was unnoticed a long time ago.

Theorem 6.4 (Global continuation) *Let u be the proper solution to* (6.7) *under the assumptions* (6.8)–(6.12). *Then it can be continued in a nontrivial way for all times $t > 0$, i.e., $u(\cdot, t) \not\equiv \infty$ for all $t > 0$, even if u blows up at a time $T < \infty$. In fact, for every $t > 0$, the burnt zone $\mathcal{B}[u](t)$ is a bounded subset of $I\!R^N$ or the empty set.*

Proof. It is based on a 1D comparison argument, which relies essentially on the existence of certain singular TWs, ensured by the assumptions (6.10)–(6.12). We choose a one-dimensional, nonnegative, bounded, continuous, monotone decreasing function $U_0(x_1)$ such that $u_0(x) \leq U_0(x_1)$ in $I\!R^N$ and $U_0(\infty) = 0$. By comparison (which is true for proper solutions), we have

$$u(x,t) \leq U(x_1,t) \quad \text{in} \quad I\!R^N \times I\!R_+,$$

where U is the one-dimensional proper solution of the equation (6.7) for $t > 0$ with the initial data U_0. If we prove that the blow-up set of $U(x_1, t)$ is bounded above in $\mathbb{R}$ for every $t > 0$ (or empty), the conclusion of the theorem will follow by applying rotations in arbitrary directions. Therefore, we will assume that $N = 1$, and $u_0 = U_0$ is monotone decreasing and tends to 0 as $x \to \infty$ and will prove that, for every $t' > 0$, the proper solution is bounded in a rectangle of the form $\{(x,t) : x \geq R(t'), 0 < t < t'\}$. The rest of the proof is exactly the same as in Section 4.4, and such a one-sided comparison with singular TWs establishes the boundedness of $\mathcal{B}[u](t)$. □

It follows that the proper solution can be continued globally in time under a weaker directional version of (6.9), specifically when

$$u_0(x) \to 0 \quad \text{as} \quad x_1 \to +\infty$$

uniformly in $(x_2, ..., x_N)$; or any rotated version thereof. Of course, in such cases we can only assert a directional bound for the blow-up set.

6.4 On blow-up set in the limit case $p = 2 - m$

Consider in more detail the border-line case of incomplete blow-up

$$p + m = 2 \quad (m \in (0,1)). \tag{6.13}$$

Since $p > 1$ by Osgood's criterion, we have the fast diffusion case $m < 1$. In these circumstances blow-up actually occurs. More precisely, if $m \geq \frac{N-1}{N}$, then $m + p = 2$ implies that $p \leq p_F = m + \frac{2}{N}$, so any nontrivial solution blows up. On the other hand, when

$$m \in \left(\tfrac{N-2}{N}, \tfrac{N-1}{N}\right)$$

and $p + m = 2$, we are above the critical Fujita exponent and, accordingly, the blow-up of large solutions is incomplete, while small solutions never blow up. We will derive an optimal estimate of the burnt zone $\mathcal{B}[u](t)$ of an arbitrary compactly supported solution, not necessarily radially symmetric.

Theorem 6.5 *Let* (6.13) *hold. Let the initial function* u_0 *be bounded and*

$$u_0(x) = o(|x|^{-1/(1-m)}) \quad \text{as} \quad |x| \to \infty.$$

Then blow-up is incomplete and the burnt zone of the solution of (6.1) *propagates at most linearly in time. We have*

$$\text{diam}\,(\mathcal{B}[u](t)) \leq 4\sqrt{m}(t + c) \quad \text{for} \ t > T. \tag{6.14}$$

Proof. In the critical case (6.13) equation (6.1) admits explicit plane TWs (same as in 1D)

$$U(x,t) = [C((x \cdot \mathbf{n}) - \lambda t + a)_+]^{-1/(1-m)},$$

where $\mathbf{n}$ in an arbitrary vector of the unit sphere $S_1 \subset \mathbb{R}^N$, and

$$C = \tfrac{1-m}{2m}\left(\lambda \pm \sqrt{\lambda^2 - 4m}\right).$$

This imposes the restriction $\lambda \geq 2\sqrt{m}$. The value

$$\lambda_m = 2\sqrt{m}$$

corresponds to the minimal speed of propagation of blow-up interfaces.

Set now $\lambda = 2\sqrt{m}$ and fix an arbitrary $\mathbf{n} \in S_1$. There exists $a \in I\!R^N$ such that

$$u_0(x) \le U(x,0) \quad \text{in} \quad I\!R^N.$$

Then, by comparison of $U(x,t)$ with the proper solution (this corresponds to comparison with the minimal solution), we have that $u \le U$ everywhere in $I\!R^N \times I\!R_+$, whence incomplete blow-up and the estimate

$$\mathcal{B}[u](t) \cap \{(x,t): \ \ (x \cdot \mathbf{n}) - 2\sqrt{m}t + a > 0\} = \emptyset.$$

Since $\mathbf{n} \in S_1$ is arbitrary, we arrive at (6.14). $\square$

Explicit parabolic solutions on a linear invariant subspace. The TWs are interesting explicit solutions, but they are singular already at $t = 0$. Let us exhibit a real example, i.e., an explicit solution that blows up after a time $T > 0$. This construction is similar to that in Chapter 5 for the 1D case. Namely, we have that the pressure $v = \frac{m}{1-m}\, u^{m-1} > 0$ satisfies the quadratic equation

$$v_t = \mathbf{A}(v) \equiv (1-m)v\Delta v - |\nabla v|^2 - m.$$

The subspace $W_2 = \mathrm{Span}\{1, |x|^2\}$ is invariant under the quasilinear operator $\mathbf{A}$, so that the equation restricted to W_2 with solutions

$$\tilde{v}(x,t) = D_0(t) + D_1(t)|x|^2 \in W_2$$

reduces to the dynamical system

$$\begin{cases} D_0' = 2N(1-m)D_0D_1 - m, \\ D_1' = -2\delta D_1^2, \end{cases}$$

where $\delta = mN - (N-2) > 0$. Integrating it yields the exact solutions

$$\tilde{v}(x,t) = At^\gamma - \frac{mt}{1-\gamma} + \frac{|x|^2}{2\delta t}, \quad \text{if } \gamma = \tfrac{N(1-m)}{\delta} \ne 1, \tag{6.15}$$

and

$$\tilde{v}(x,t) = At - mt\log t + \frac{|x|^2}{2t}, \quad \text{if } \gamma = 1, \ \text{ i.e., } m = \tfrac{N-1}{N}, \tag{6.16}$$

where $A > 0$ is arbitrary. If $\gamma \ge 1$ ($m \ge \frac{N-1}{N}$), $\tilde{v}(x,t)$ vanishes in finite time $t = T$ at $x = 0$, i.e., $\tilde{u}(x,t)$ blows up at T.

Let us describe further properties of these invariant solutions. It follows from (6.15) that, for $t \gg 1$, the diameter of the corresponding burnt zone satisfies

$$\mathrm{diam}\,(\mathcal{B}[\tilde{u}](t)) = c_0\, t(1+o(1)),$$

where

$$c_0 = \sqrt{\tfrac{8\delta}{1-\gamma}} > 4\sqrt{m}.$$

Similarly, (6.16) implies that, for $t \gg 1$,

$$\mathrm{diam}\,(\mathcal{B}[\tilde{u}](t)) = t\sqrt{8m\log t}\,(1+o(1)).$$

Both equalities contradict (6.14), and hence these explicit solutions $\tilde{u}$ are not proper minimal for $t \gg 1$. The limit, minimal solution is somewhere below.

Finally, for $m < \frac{N-1}{N}$, we have $\gamma > 1$ and the large solutions with $A < 0$ blow-up, while the small ones ($A > 0$) do not, as expected.

6.5 Complete blow-up up to critical Sobolev exponent

We now consider the question of continuation after blow-up of the solutions to the Cauchy problem for equation (6.1) for values of p larger than $2-m$. We will make throughout the restriction to the class of radially symmetric solutions $u = u(r,t)$, $r = |x|$, with nonnegative, bounded initial data $u_0(r)$.

The main novelty of the N-dimensional problem with respect to the 1D one is the possibility of singularities focused at the centre of symmetry. The simplest example of such a phenomenon is given by *singular stationary solution* that has the form

$$U_s(x) = \frac{c_s}{|x|^{2/(p-m)}}, \quad c_s = [\gamma(N-2-\gamma)]^{1/(p-m)}, \ \gamma = \frac{2m}{p-m}, \tag{6.17}$$

and is defined if $N \geq 3$ and

$$p > p_{st} = m\,\frac{N}{N-2},$$

another critical exponent. Observe that $U_s(r)$ is locally integrable, $U_s \in L^1_{\rm loc}(\mathbb{R}^N)$, and moreover $U_s^p \in L^1_{\rm loc}(\mathbb{R}^N)$. Clearly, this is an example of incomplete blow-up of a very particular kind since it stays at $x = 0$, thus forming a *single-point*, stationary blow-up set. This kind of blow-up is a new feature of the many-dimensional case. An important characteristic that, in some sense, rules out this example is the fact that it starts from an already singular initial situation, while our assumptions are concerned with evolution finite-time blow-up starting from initially bounded functions. However, the stability properties of U_s will be of great importance in the discussion that follows.

The phenomenon of focusing at the origin will imply the appearance of the *critical Sobolev exponent* defined as

$$p_S - m\,\frac{N+2}{N-2} \quad \text{for } N \geq 3.$$

Observe that $2 - m < p_S$ precisely if $m > m_c = \frac{(N-2)_+}{N}$, and then $p_S > p_F$. Also, p_S is strictly larger than p_{st}. For $N = 1, 2$ we define $p_S = \infty$. The appearance of p_S as a critical exponent is due to the change of behaviour of the family of stationary solutions of (6.1) precisely for $p = p_S$, as we will explain in Sections 6.6 and 6.9.

In the range $p \leq 2 - m < p_S$ the focusing at $x = 0$ affects only the technique of analysis, but not the conclusion of *incomplete* blow-up. Let us state the main result on complete blow-up.

Theorem 6.6 *Let $u(r,t)$ be a proper solution of* (6.1) *under the above assumptions on $u_0(r)$ having blow-up time $T < \infty$. Assume that $m > m_c$ and*

$$2 - m < p \leq p_S. \tag{6.18}$$

Then complete blow-up occurs, i.e., $u(\cdot,t) \equiv \infty$ for any $t > T$. The same conclu-

sion is also true for $p > p_S$ if the blow-up set at time T is not just the origin,

$$\mathcal{B}[u](T) \neq \{0\}.$$

This settles the dimensions $N = 1$ and 2. The bounds in (6.18) are sharp. The lower bound is optimal because of the incomplete blow-up just studied; the upper bound p_S is also sharp and related to the appearance of other incomplete blow-up types, the *peaking solutions*, for higher p values if $N \geq 3$; see Remarks. We recall that, for $p > p_F$, small solutions do not blow-up at all.

The proof strongly depends on whether the blow-up set is focused at the origin $x = 0$ or not. In the latter case the 1D techniques described in Chapter 4 based on TWs can be used once conveniently adapted. This is not the case when the blow-up set $\mathcal{B}[u](T)$ is the origin. Then we need to perform a more delicate analysis of the formation of the singularity. It uses the method of stationary states, i.e., intersection comparison with the one-dimensional set B_S of stationary solutions, which are easily constructed for all $N \geq 1$. Then the *envelope* of the set B_S near a blow-up point gives a precise lower bound of the limit profile $u(x, T)$. Finally, for some parameter range, complete blow-up for $t > T$ directly follows by means of a local nonexistence result for evolution equations of the type (6.1). In the other range, using the same envelope analysis, we first prove that the blow-up interface (the boundary of the blow-up set) moves after $t = T$, and this makes it possible to reduce our problem to complete blow-up for a 1D equation already studied in Chapter 4.

6.6 Complete blow-up of focused solutions in the subcritical case

Let us proceed with the proof of Theorem 6.6 in the novel case where the blow-up set satisfies $\mathcal{B}[u](T) = \{0\}$.

Subset of stationary solutions and the envelope

For a fixed $\varepsilon > 0$, let $U = U(r; \varepsilon)$ be the radial stationary solution

$$\frac{1}{r^{N-1}}(r^{N-1}(U^m)')' + U^p = 0 \quad \text{for } r > 0, \quad U(0) = \varepsilon, \;\; U'(0) = 0.$$

Setting $U^m = V$ yields the semilinear equation

$$\frac{1}{r^{N-1}}(r^{N-1}V')' + V^q = 0, \qquad \text{where } \; q = \tfrac{p}{m},$$

with well-known properties; we refer to D.D. Joseph and T.S. Lundgren (1973) [207]. By scaling we have

$$U(r; \varepsilon) = \varepsilon U_1(\varepsilon^{(p-m)/2} r), \qquad \text{where } \; U_1(r) \equiv U(r; 1). \tag{6.19}$$

We will need the following result.

Lemma 6.7 *Let* $1 < p < p_S$. *Then* $U(r; \varepsilon)$ *vanishes at the finite point*

$$r_0(\varepsilon) = r_0(1)\varepsilon^{(m-p)/2} < \infty. \tag{6.20}$$

For a given constant $c \geq 0$*, there holds*

$$|(U^m)'_r| \to \infty \quad \text{as } \varepsilon \to \infty \text{ on the level set } \{U = c\}. \tag{6.21}$$

For convenience, in this case we formally extend $U(r;\varepsilon)$ to $r > r_0(\varepsilon)$ as $U = 0$ obtaining a continuous subsolution in $I\!R^N$. Let us next introduce the *envelope* $\mathcal{L}(r)$ of the set $B_S = \{U(\cdot;\varepsilon), \varepsilon > 0\}$,

$$\mathcal{L}(r) = \sup_{\varepsilon>0} U(r;\varepsilon).$$

From (6.19), it is easy to calculate that

$$\text{if } p < m, \quad \text{then } \mathcal{L}(r) \equiv \infty; \tag{6.22}$$

$$\text{if } p = m, \quad \text{then } \mathcal{L}(r) = \infty \text{ for } 0 \leq r < r_*, \text{ where } U_1(r_*) = 0; \tag{6.23}$$

$$\text{if } p > m, \quad \text{then } \mathcal{L}(r) = c_* r^{-2/(p-m)} \text{ with } c_* = c_*(m,p,N) > 0. \tag{6.24}$$

Intersection comparison in radial geometry

In order to make the main argument clear, we will prove the comparison result first under a simplifying assumption on the solution, namely that the solution becomes *eventually monotone* in the domain where it is large enough, i.e.,

$$\text{for } t \approx T^-, \ u(r,t) \text{ strictly decreases with } r \geq 0 \text{ in } \{u \gg 1\}. \tag{6.25}$$

This is a natural situation in blow-up problems, which occurs of course if u_0 is continuous and monotone decreasing with r. On the other hand, (6.25) holds if $u_0(r)$ intersects the level $[(p-1)T]^{-1/(p-1)}$ exactly once and $u_0(r) \to 0$ as $r \to \infty$. This follows from an elementary application of the intersection comparison with the family of flat (independent of x) blow-up solutions; see references in Remarks.

Lemma 6.8 *Let* $1 < p < p_S$ *and* (6.25) *hold. Then, for every* $\varepsilon \gg 1$*, we have in the positivity domain* $\{u > 0\}$

$$u(r,t) > U(r;\varepsilon) \quad \text{for all } \ t \approx T^-. \tag{6.26}$$

As a straightforward consequence, we conclude that $u(r,t) > U(r;u(0,t))$ for all $t \approx T^-$ in $\{u > 0\}$, and also that $u(0,t)$ is not decreasing with t.

Proof. Let $\text{Int}(t,U)$ be the number of intersections of the functions $u(r,t)$ and $U(r;\varepsilon)$ in the interval $\{0 \leq r < r_0(\varepsilon)\}$. In view of the regularity of u_0 and (6.20), (6.21), we have that, for $\varepsilon \gg 1$, the number of intersections satisfies $\text{Int}(0,U) \leq 1$. Hence, by the Sturm Theorem,

$$\text{Int}(t,U) \leq 1 \quad \text{for all } \ t \in (0,T). \tag{6.27}$$

It follows from the blow-up hypothesis that there exists a $t = t_\varepsilon > 0$ such that $u(0,t_\varepsilon) = \varepsilon = U(0;\varepsilon)$. Then from (6.27) we conclude that necessarily $\text{Int}(t_\varepsilon,U) = 0$, whence (6.26) for all $t \in [t_\varepsilon,T)$ by the usual comparison. The strict inequality in (6.26) in the positivity domain follows by the strong MP. By the definition, the proper solution satisfies (6.26) for all $t \geq T$. □

Since $\varepsilon \gg 1$ in (6.26) is arbitrary, we also have the following result.

Corollary 6.9 *For $1 < p < p_S$, the proper solution satisfies the estimate*

$$u(r,t) \geq \mathcal{L}(r) \quad \textit{for all} \quad t \geq T, \tag{6.28}$$

which is valid for all small $r > 0$ with the strict inequality if $p > m$, and for all $r \geq 0$ if $p \leq m$.

Estimate (6.28) still holds without the assumption of eventual monotonicity. The argument is based on the same ideas but becomes more involved in the details. We recall that, anyway, this should be a non-generic case. Assume therefore that $\mathcal{B}[u](T) = \{0\}$ and that (6.25) does not hold. Then there exist $\{t_n\} \to T^-$ and a monotone sequence $\{r_n\} \to 0$ such that $u(r_n, t_n) \to \infty$ and $r = r_n$ are maxima of the profiles $u(r, t_n)$. Using a general monotonicity result of large solutions, which says that $u_t > 0$ in $\{u \gg 1\}$, one can see that the unique intersection $r = \zeta_\varepsilon(t)$ between $u(r,t)$ and $U(r;\varepsilon)$ with $\varepsilon \gg 1$ satisfies $\zeta_\varepsilon(t) \to 0$ as $t \to T^-$. Therefore, we finally obtain that (cf. (6.26))

$$\lim_{t \to T^-} \inf u(r,t) \geq U(r;\varepsilon) \quad \text{for small } r > 0 \text{ and all } \varepsilon \gg 1.$$

This is enough to apply the envelope analysis as above to derive estimates (6.28). The rest of the proof of complete blow-up then needs only slight modification.

First result on complete blow-up

It immediately follows from (6.28) together with (6.22)–(6.24) that blow-up is *global* if $p < m$, and it is at least *regional* if $p = m$ (more precisely, $\text{meas}_r(\mathcal{B}[u](T)) \geq r_* > 0$). In all cases, by the MP, we deduce from (6.28) that the proper solution u satisfies the estimate

$$u \geq v \quad \text{in } I\!\!R^N \times (T, \infty),$$

where $v \geq 0$ is the solution of the Cauchy problem

$$v_t = \Delta v^m + \mathcal{L}^p(r) \quad \text{in } I\!\!R^N \times (T, \infty) \tag{6.29}$$

with $v(r,T) = \mathcal{L}(r)$ for small $r > 0$. From this we have complete blow-up in the range $1 < p \leq p_{st}$.

Proposition 6.10 *Let $m > 1$ and let u be eventually monotone. If $1 < p \leq p_{st}$, then complete blow-up occurs.*

Proof. It will be enough to prove that $v \equiv \infty$ for all $t > T$. We use a local nonexistence argument. It follows from (6.22)–(6.24) that $p \leq p_{st}$ implies that

$$\mathcal{L}^p(r) \notin L^1_{\text{loc}}(I\!\!R^N). \tag{6.30}$$

Obviously, this violates local solvability of equation (6.29) in L^1, which is a natural setting for such divergent quasilinear equations. We can prove that by comparison with explicit solutions. Assume for contradiction that there exists a bounded weak solution $v(r,t) \geq 0$ with bounded weak derivatives. Integrating equation (6.29) over the unit ball $\{r < 1\}$, we have that, for any small fixed $\delta > 0$, there holds

$$v(r, T+\delta) \notin L^1_{\text{loc}}(I\!\!R^N).$$

Therefore, by comparison we deduce that

$$v(r,t) \geq v_M(r,t) \quad \text{in } I\!R^N \times [T+2\delta,\infty), \tag{6.31}$$

where v_M is the ZKB solution of the PME

$$(v_M)_t = \Delta(v_M)^m \quad \text{in } I\!R^N \times (T+\delta,\infty) \tag{6.32}$$

with $v_M(x,T+\delta) = M\delta(x)$ in $I\!R^N$, where $\delta(x)$ is Dirac's delta function and the constant $M > 0$ can be arbitrarily large. v_M has the form

$$v_M(x,t) = [t-(T+\delta)]^{-kN}\theta(\xi), \quad \xi = x/[t-(T+\delta)]^k, \tag{6.33}$$

where

$$k = \tfrac{1}{N(m-1)+2} \quad \text{and} \quad \theta(\xi) = c_0(a^2 - |\xi|^2)_+^{1/(m-1)}.$$

The constant $a = a(M) > 0$ is calculated as

$$a = c_1 M^{(m-1)k} \to \infty \quad \text{as } M \to \infty$$

(here and later on c_0, c_1, ... denote different constants depending on the parameters). Hence, passing to the limit in (6.33) as $M \to \infty$ yields

$$v_M(r,T+2\delta) \to \infty \quad \text{as } M \to \infty \tag{6.34}$$

uniformly on compact subsets in r. From (6.31) we conclude that $v(r,T+2\delta) \equiv \infty$ for any small $\delta > 0$, and the result follows. □

We continue by considering the parameter range $p_{st} < p < p_S$. Eventual monotonicity in r is still assumed. It follows from Corollary 6.9 that, in this case, we have the strict inequality

$$u(r,T) > \mathcal{L}(r) = c_* r^{-2/(p-m)} \quad \text{for small } r > 0. \tag{6.35}$$

An important step is the following property of the blow-up interface.

Proposition 6.11 *Let* $p_{st} < p < p_S$. *If blow-up is not complete at* $t = T$, *then the blow-up interface is strictly increasing at* $t = T^+$.

Proof. (i) Let $\tilde{u} = \tilde{u}(r,t;\varepsilon)$ be the solution of (6.1) in $I\!R^N \times (0,T_\varepsilon)$ with initial function $U(r;\varepsilon)$. In this range of p, $\tilde{u}$ blows up in finite time T_ε. Indeed, since the derivative $z = \tilde{u}_t$ satisfies a linear parabolic equation in the positivity domain $\{\tilde{u} > 0\}$,

$$z_t = \Delta(m\tilde{u}^{m-1}z) + p\tilde{u}^{p-1}z,$$

after the necessary approximation, we conclude, by the strong MP, that $\tilde{u}_t > 0$ in $\{\tilde{u} > 0\}$. Therefore, assuming for contradiction that $\tilde{u}$ is global, by a standard Lyapunov argument we deduce that either $\tilde{u}(r,t)$ stabilizes as $t \to \infty$ to a stationary solution $U(r;\bar{\varepsilon})$ with $\bar{\varepsilon} > \varepsilon$ or to a singular stationary solution $U_s(r)$ defined for all $r > 0$ with $U_s(0) = \infty$. The former case is impossible since $U(r;\bar{\varepsilon})$ then intersects $U(r;\varepsilon)$. In the latter case using the monotonicity with time, $U_s(r) > \tilde{u}(r,0;\varepsilon) \equiv U(r;\varepsilon)$, one can easily see that after a small shifting by small $a \in I\!R^N$, $|a| \ll 1$, we still have the inequality $\tilde{u}(r,0;\varepsilon) \leq U_s(|x-a|)$ in $I\!R^N$. This implies the uniform boundedness of the solution,

$$\tilde{u} \leq \min\{U_s(|x|),\, U_s(|x-a|)\} \quad \text{for all } t > 0,$$

and we again arrive at the first case.

Thus we conclude that $\tilde{u} \to \infty$ as $t \to \infty$ uniformly on compact subsets of $I\!R^N$. One can see that such an unbounded solution cannot be global in time. This is easily proved by construction of blow-up subsolutions of the self-similar form; see Remarks. Hence, the blow-up time T_ε of $\tilde{u}(r,t;\varepsilon)$ is always finite.

(ii) In view of the scaling invariance of equation (6.1) and the initial function $U(r;\varepsilon)$, we have that

$$\tilde{u}(r,t;\varepsilon) = \varepsilon\tilde{u}(r\varepsilon^{(p-m)/2}, t\varepsilon^{p-1};1), \tag{6.36}$$

and, in particular, we deduce that

$$T_\varepsilon = T_1\varepsilon^{1-p} \to 0 \quad \text{as } \varepsilon \to \infty. \tag{6.37}$$

We now compare for $t > T$ the proper solution $u(x,t)$ and $\tilde{u}(x,t;\varepsilon)$. Fix an arbitrarily small $\delta > 0$ and the constant

$$\varepsilon = \big(\tfrac{\delta}{T_1}\big)^{-1/(p-1)},$$

so $T_\varepsilon = \delta$ and $\varepsilon(\delta) \to \infty$ as $\delta \to 0$. Since the inequality sign in (6.35) is strict, there exists $x_* = x_*(\delta) \in I\!R^N$, $x_* \neq 0$, such that

$$u(r,T) \geq U(|x - x_*|;\varepsilon) \quad \text{in } I\!R^N,$$

and therefore by comparison

$$u(r,t) \geq \tilde{u}(|x - x_*|, t - T;\varepsilon) \quad \text{in } I\!R^N \times (T, T+\delta).$$

By construction $\tilde{u}(|x - x_*|, t - T;\varepsilon)$ blows up at $t = T + \delta$ at the point $x = x_*$, and hence there holds

$$u(r_*,t) = \infty \quad \text{at } r_* = |x_*(\delta)| > 0 \text{ for all } t \geq T + \delta. \tag{6.38}$$

Since $\delta > 0$ is arbitrary, Proposition 6.11 follows. □

Proof of complete blow-up: subcritical Sobolev range

We now prove that a finite blow-up interface cannot exist. At this stage we can change from the stationary state analysis to the TW-analysis that allowed us to classify the 1D problem in Chapter 4. We consider a proper solution $u(r,t)$ of (6.1) satisfying the equation

$$u_t = (u^m)_{rr} + \tfrac{N-1}{r}\,(u^m)_r + u^p \quad \text{in } \{r > r_*\} \times \{t > T + \delta\} \tag{6.39}$$

and the singular "boundary condition" (6.38) at $r = r_*$. There are two steps: (i) construction of a 1D subsolution, and (ii) comparison.

(i) The simplest idea is to consider the TW solution $V(y,t) = f(y - \lambda t)$ with $\lambda \gg 1$ and $f(0) \gg 1$, $f'(0) = 0$, to the modified equation

$$v_t = (v^m)_{yy} + d_*(v^m)_y + v^p, \quad \text{where } d_* = \tfrac{N-1}{r_*} > 0. \tag{6.40}$$

This is the same equation

$$u_t = (u^m)_{yy} + u^p$$

that was studied in the 1D case in Chapter 4, plus an extra convection term to account for the new term in (6.39). Using 1D arguments, we easily show that the new term in (6.40) plays no role in the criterion for complete blow-up to (6.40), which is still

$$p + m > 2. \tag{6.41}$$

In fact, the ODE for f is

$$(f^m)'' + \lambda f' + d_*(f^m)' + f^p = 0\,. \tag{6.42}$$

By a phase-plane analysis similar to that in Section 4.2, we easily show that under condition (6.41), all TW profiles are always bounded, bell-shaped, and for large values of $f(0)$, these profiles $f(\xi)$ are monotone for $\xi > 0$ and are very steep on any given level set $f = c > 0$.

Unfortunately, one can see that $v(r,t)$ is a subsolution of equation (6.39) in any subset of the positivity domain where $v_r \le 0$, but it is not if $v_r > 0$. In order to avoid this difficulty and obtain a subsolution for both cases $v_r \le 0$ and $v_r > 0$, we introduce an extra term, which acts only for $v_r < 0$. Thus we look for a TW solution $V(y,t)$ of the equation

$$v_t = (v^m)_{yy} + F((v^m)_y) + v^p, \tag{6.43}$$

where

$$F(s) = \begin{cases} d_* s & \text{if } s \le 0, \\ 0 & \text{if } s > 0. \end{cases}$$

The operator F is continuous on the solutions that are smooth in the corresponding positivity domains. The corresponding TW f satisfies (6.42) in the domain $\{f' < 0\}$ and the equation with $d_* = 0$ in $\{f' > 0\}$. By the transformation $(f^m)' = -P$, the calculation reduces to integrating the ODE

$$\frac{\mathrm{d}P}{\mathrm{d}f} = \frac{1}{P}\left[\lambda P - m f^{p+m-1} + mF(P) f^{m-1}\right],$$

where

$$F(P) - \begin{cases} d_* P & \text{for } P \ge 0, \\ 0 & \text{for } P < 0. \end{cases}$$

One can check that under the criterion (6.41), all locally positive orbits $P = P(f)$ vanish at finite f (which implies complete blow-up for (6.40) in 1D), and are continued in a smooth way in $\{P < 0\}$. This means that, in the positivity domain, such TWs V have all derivatives entering (6.43) and they are classical solutions of (6.43) there. Hence, V is a classical subsolution of (6.39) in $\{V > 0\}$ as desired.

(ii) For a fixed small $\varepsilon > 0$, we compare a TW solution of (6.43), $V(r - (r_* - \varepsilon), t - (T + \delta))$ (a classical subsolution of (6.39)) with the solution $u(r,t)$ in the domain $\bar{Q} = \{V > 0\} \cap \{t \in (T + \delta, \infty)\}$. By (6.38) we have that $u(r, T + \delta) \ge V(r - (r_* - \varepsilon), 0)$. Moreover, $u \ge V = 0$ on both the lateral boundaries of $\bar{Q}$. Finally, we conclude by comparison that

$$u(r, T + 2\delta) \ge V(r - (r_* - \varepsilon), \delta) \quad \text{in } (r_*, \infty) \cap \{V(\delta) > 0\}. \tag{6.44}$$

Recall that, by construction, the wave speed λ of the wave V moving to the right is arbitrarily large. Therefore, using the same argument as in 1D, we conclude from

(6.44) with $\lambda = \infty$ that

$$u(r, T + 2\delta) = \infty \quad \text{for all} \quad r > r_*.$$

Using the fact that $\delta > 0$ is arbitrarily small, this completes the proof of Theorem 6.6 for $p < p_S$. $\square$

6.7 Complete blow-up in the critical Sobolev case

The behaviour of the stationary states is crucial in the study of focused blow-up. This behaviour undergoes a transformation at the critical exponent $p = p_S$. As a consequence, complete blow up will not be the sole blow-up possibility for $p > p_S$. Though the theorem is still valid for $p = p_S$, the different behaviour of the stationary states implies a new version and proof of the basic comparison result, Corollary 6.9. To be precise, in the critical case $p = p_S$ the stationary solutions are strictly positive and given explicitly

$$U(r; \varepsilon) = \left(\frac{\rho \sqrt{N(N-2)}}{\rho^2 + r^2} \right)^{\frac{N-2}{2m}}, \quad \rho = \sqrt{N(N-2)}\, \varepsilon^{-\frac{2m}{N-2}}.$$

This is a special case of exact integrability of the classical Emden-Fowler ODEs; see Remarks.

One can see that each $U(r; \varepsilon)$ intersects $U_s(r)$ exactly twice, and for any $\nu > 0$,

$$U(r; \varepsilon) \to 0 \quad \text{as } \varepsilon \to \infty \text{ uniformly on } [\nu, \infty). \tag{6.45}$$

It follows from (6.45) that the intersection hypothesis (6.21) is valid for all fixed positive level sets with $c > 0$. The envelope of the family of stationary solutions is still given by formula (6.24) with $c_* > c_s$. We then have the following result.

Lemma 6.12 *For any solution, estimate* (6.35) *is still valid for* $p = p_S$.

Proof. Since $U(r; \varepsilon) > 0$ for $r > 0$, (6.27) is not true. Nevertheless, using the property (6.45), we conclude that (6.27) is valid with $\text{Int}(U, t)$ being the number of intersections between the solutions $u(r, t)$ and $U(r; \varepsilon)$ in the domain $\{0 < r < r_0\}$ with $r_0 > 0$ small. Indeed, we have that, for all $t \approx T^-$, $u(r_0, t) \geq c > 0$, so, by (6.45), the difference $u(r, t) - U(r; \varepsilon)$ does not change sign at the lateral boundary $\{r = r_0,\ t \approx T^-\}$, and finally (6.27) holds for all $t \approx T^-$ provided that $\varepsilon \gg 1$. Therefore, (6.26) for $r \in (0, r_0)$ follows, whence the envelope estimates (6.28) and (6.35).

The proof of complete blow-up has no novelties now, since (as in the subcritical case) $\mathcal{L} > U_s$ and the technique above for $p < p_S$ applies directly. $\square$

6.8 Complete blow-up of unfocused solutions

Next, we consider the part of Theorem 6.6, where $m + p > 2$ and $\mathcal{B}[u](T) \neq \{0\}$ without any limitation from above on p. Then we have complete blow-up due to a purely 1D effect, and we are able to apply a variant of the TW analysis from Section 4.4 to get a clear-cut conclusion.

Blow-up on a sphere

First we consider the case where there exists $r_* > 0$ such that $\{|x| = r_*\} \subseteq \mathcal{B}[u](T)$ (blow-up on a sphere). This could happen for instance if $u_0(r)$ has a large single maximum at a point $r \approx r_*$ and a local minimum at the origin. As in Section 6.6, the proof of complete blow-up is based on comparison with 1D TWs that are subsolutions for the N-dimensional problem. We point out some slight necessary modifications of the proof. Fix a monotone sequence $\{t_n\} \to T^-$ and the corresponding sequence $\{r_n\} \to r_*$ such that $\alpha_n = u(r_n, t_n) \to \infty$. Recall that, due to the general monotonicity result of large solutions, we may assume that $u_t(r,t) > 0$ in $\{r \approx r_*,\, u \gg 1\}$, and therefore

$$u(r_n, t) > \alpha_n \quad \text{for } t \in (t_n, T).$$

Fix $n \gg 1$. We want to consider a TW $V(y,t) = f(y - \lambda t)$ with $\lambda = \lambda_n \to \infty$ as $n \to \infty$, and $f(0) = \alpha_n$, $f'(0) = 0$, of the 1D equation (6.40) with a different parameter $d_* = \frac{2(N-1)}{r_*} > 0$. The criterion of complete blow-up is again (6.41).

We use the same technique as in Section 6.6 in order to show complete blow-up of u. We choose $\lambda_n \gg 1$ such that

$$u(r, t_n) \geq V(r - (r_n - \varepsilon), 0) \quad \text{for } r \geq r_n.$$

Let us show that, if $t_n \approx T^-$, the TW is steep enough to guarantee this property that is necessary for the comparison. To prove this we need to perform an elementary analysis of the family of slightly perturbed 1D stationary solutions of (6.1). Namely, let $U_n(r)$ be the stationary solution satisfying

$$\frac{1}{r^{N-1}}\,(r^{N-1}(U^m)')' + U^p = 0 \quad \text{for } r > r_n, \quad U(r_n) = \alpha_n, \;\; U'(r_n) = 0.$$

Let $p > m$ (the case $p \leq m$ is similar and easier), and set $q = \frac{1}{2}(p-m) > 0$. After rescaling $U_n(r) = \alpha_n V_n(y)$, $y = (r - r_n)\alpha_n^q$, we obtain for the function $V = V_n$ the perturbed problem

$$(V^m)'' + V^p = -\frac{1}{\alpha_n^q}\,\frac{N-1}{r_n + \alpha_n^{-q} y}\,(V^m)' \quad \text{for } y > 0,$$

with $V(0) = 1$ and $V'(0) = 0$. In view of the continuous dependence of the ODE

$$(V^m)'' + V^p = 0 \quad \text{for } y > 0, \quad V(0) = 1, \;\; V'(0) = 0, \tag{6.46}$$

relative to a small (for $n \gg 1$) perturbation of the right-hand side, we conclude that

$$V_n(y) \to V_\infty(y), \quad V_n'(y) \to V_\infty'(y) \quad \text{as } \alpha_n \to \infty$$

uniformly on compact subsets from the positivity domain of V_∞, where V_∞ is given by the problem (6.46). Therefore,

$$U_n(r) \approx \alpha_n V_\infty((r - r_n)\alpha_n^q)$$

for $\alpha_n \gg 1$ in the positivity domain.

We now apply the same intersection comparison argument as in Section 6.6. Using obvious properties of the 1D stationary solution V_∞, we then deduce that the number of intersections $\mathrm{Int}(t, U_n)$ between $U_n(r)$ and $u(r,t)$ in the positivity domain of U_n satisfies $\mathrm{Int}(0, U_n) \le 2$. The Sturm Theorem implies that

$$\mathrm{Int}(t, U_n) \le 2 \quad \text{for all } t \in (0, T).$$

One can see that we can find a profile $\tilde{U}_n(r)$ with shifted parameter $\tilde{r}_n \in (\frac{1}{2}r_*, r_n]$ such that no intersections between $u(r, t_n)$ and $\tilde{U}_n(r)$ occur for $r \ge \tilde{r}_n$ in the positivity domain of $\tilde{U}_n$, i.e.,

$$u(r, t_n) \ge \tilde{U}_n(r) \quad \text{in} \quad \{r > r_n\} \cap \{u \gg 1\}.$$

We do not need to apply a result similar to Proposition 6.11. Indeed, the analysis above makes it possible to put, at time $t = t_n$, the TW $V(r - (r_n - \varepsilon), 0)$ (or with r_n replaced by $\tilde{r}_n$) below the profile $u(r, t_n)$ for $r \ge r_n$ to obtain

$$u(r, t_n) \ge V(r - (r_n - \varepsilon), 0) \quad \text{for} \quad r > r_n.$$

Since $V(r - (r_n - \varepsilon), t - t_n)$ is a "subsolution" of equation (6.39) in $(r_n, \infty) \times (t_n, \infty)$ (see the analysis in Section 6.6), we obtain, by the MP, that

$$u(r_n + y, t_n + \tau) \ge V(y + \varepsilon, \tau) \quad \text{in} \quad I\!\!R_+ \times I\!\!R_+.$$

Passing to the limit $n \to \infty$ in this inequality and using the fact that $\lambda_n \to \infty$, we obtain that, for $t > T$, the singular $\{u = \infty\}$-level propagates to the right with infinite speed, so

$$u(r, T + \delta) = \infty \quad \text{for all} \quad r > r_*$$

for arbitrarily small $\delta > 0$. Since $u(\cdot, T + \delta) \notin L^1_{\mathrm{loc}}(I\!\!R^N)$, this means that

$$u(r, T + 2\delta) \equiv \infty \quad \text{everywhere},$$

whence the complete blow-up.

Empty blow-up set

There is another possible case, $\mathcal{B}[u](T) = \emptyset$. Then there must exist a sequence $\{r_n\} \to \infty$ such that $u(r_n, t_n) \to \infty$, and the proof is the same and easier, with $d_*(n) \to 0$ as $n \to \infty$. Finally, we obtain that $u(r, T + \delta) \to \infty$ as $r \to \infty$. This means that there exists a finite r_* such that

$$\{|x| = r_*\} \subseteq \mathcal{B}[u](T + 2\delta),$$

and the rest of the proof is the same.

6.9 Complete blow-up in the supercritical case

Complete blow-up is not the only possibility in the supercritical case $p > p_S$ for $N \ge 3$. However, it is the common (stable and generic) occurrence, and, as we have just seen, we can give assumptions on the initial data, which ensure such a behaviour. In this section we exhibit two further sufficient conditions for complete blow-up. We recall that $m > m_c$.

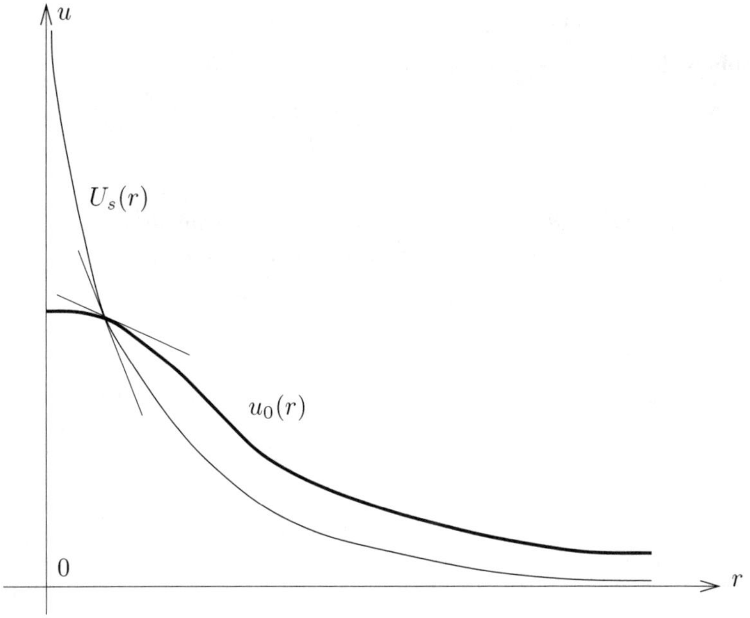

Figure 6.1 *If initial function* $u_0(r)$ *has a unique transversal intersection with the singular stationary solution* $U_s(r)$*, complete blow-up occurs.*

Theorem 6.13 (Complete blow-up) *Let* $p > p_S$ *and assume that* $u_0(r) \in C^1$ *is nonincreasing. Then blow-up is complete if* u_0 *intersects* $U_s(r)$ *exactly once and transversally, i.e.,* $u_0' > U_s'$ *at the intersection point, and furthermore,*

$$u_0(r) \geq (c_s + \kappa)\frac{1}{r^{2/(p-m)}} \quad \textit{for } r \gg 1 \textit{ with a constant } \kappa > 0. \tag{6.47}$$

This mutual location of $u_0(r)$ and $U_s(r)$ is shown in Figure 6.1. Observe that the conditions of the theorem are stable under small C^1-perturbations of the data. Furthermore, let us point out that the requirement of monotonicity of u_0 is too strong but convenient for the techniques at our disposal.

Theorem 6.14 (Complete blow-up) *Let* $p > p_S$ *and* $u_0(r)$ *be nonincreasing. Then complete blow-up occurs if the proper solution* $u(r,t)$ *is nondecreasing with time in* $I\!R^N \times (0,T)$.

In fact, in the proof we use the property of eventual monotonicity of the solution (6.25), instead of the more restrictive monotonicity assumption on $u_0(r)$ given in the statements. Before proceeding with the proof, let us temporarily consider some properties of the stationary solutions of (6.1) for $p > p_S$ (cf. Lemma 6.7). The present analysis needs still another critical exponent that we call p_u; see Remarks. For $N \geq 11$, it is finite and given by an explicit formula

$$p_u = m\left(1 + \frac{4}{N - 4 - 2\sqrt{N-1}}\right) \quad (N \geq 11). \tag{6.48}$$

Notice that $p_S < p_u$. For $N \le 10$ we will put $p_u = \infty$.

Lemma 6.15 *For $p > p_S$ the functions $U(r;\varepsilon)$ are positive and as $\varepsilon \to \infty$,*

$$\frac{U(r;\varepsilon)}{U_s(r)} \to 1 \quad \textit{and} \quad \frac{U'(r;\varepsilon)}{U_s'(r)} \to 1 \tag{6.49}$$

uniformly on $[\nu, \infty)$, $\nu > 0$. Moreover, for $p_S < p < p_u$, all solutions $U(r;\varepsilon)$ intersect each other and also intersect the singular one infinitely many times, while for $p \ge p_u$ the functions $U(\cdot;\varepsilon)$ are strictly monotone increasing in ε and

$$U(r;\varepsilon) \to U_s(r) \quad \textit{as } \varepsilon \to 0$$

from below uniformly for $r \in [\nu, \infty)$.

This analysis can be found in [207]. In all cases there exists the envelope and it is again given by (6.24). Observe that $c_* > c_s$ if $p \in [p_S, p_u)$ and $c_* = c_s$ if $p \ge p_u$, so

$$\mathcal{L}(r) = c_* \, \frac{1}{r^{2/(p-m)}} > U_s(r) = c_s \, \frac{1}{r^{2/(p-m)}} \quad \text{if } p \in (p_{st}, p_u), \tag{6.50}$$

$$\mathcal{L}(r) \equiv U_s(r) = c_s \, \frac{1}{r^{2/(p-m)}} \quad \text{if } p \ge p_u. \tag{6.51}$$

Proof of the first theorem on complete blow-up

Let us prove Theorem 6.13. Under the given assumptions on u_0, by virtue of (6.49), there holds

$$U(r;\varepsilon) \quad \text{intersects } u_0(r) \text{ exactly at one point for all } \varepsilon \gg 1. \tag{6.52}$$

In particular, this implies that $u_t(0,t) \ge 0$ for $t \approx T^-$. Therefore, by the same intersection comparison as in Section 6.6, inequality (6.35) is valid. In order to prove Proposition 6.11 we apply a similar idea. Since in our case $U(r;\varepsilon) > 0$ for all $r \ge 0$, we need some modifications.

Oscillatory case. Our construction is easier in the case $p \in (p_S, p_u)$ due to the properties of the stationary solutions given in Lemma 6.15. We introduce the function $\tilde{u} = \tilde{u}(r,t;1)$ as the solution of equation (6.1) for $0 \le r < l$, $t > 0$, with the boundary condition $\tilde{u}(l,t;1) = U(l;1)$ for $t > 0$, and perturbed initial data

$$\tilde{u}_0 = \max\{U(r;1),\, U(r;1+\nu)\},$$

ν being a small positive constant. Here $r = l$ is the unique tangency point of the function $U(r;1)$ and the envelope $\mathcal{L}(r)$. Then we have the same properties as in Section 6.6. Namely, $\tilde{u}_t \ge 0$ and $\tilde{u}$ blows up in finite time $T_1 > 0$. Indeed, using a similar idea, we have that, if $\tilde{u}$ is global, then either it stabilizes as $t \to \infty$ to a bounded stationary solution of the given form (which is impossible since any $U(r;\varepsilon)$ with $\varepsilon > 1$ intersects $U(r;1)$ in $(0,l)$) or to a stationary solution U singular at $r = 0$ (if any). Now the last conclusion cannot be true if this singular solution intersects $\tilde{u}_0(r)$ in $(0,l)$. The case

$$U_s > \tilde{u}_0 \quad \text{in} \quad [0,l]$$

(observe that then stabilization to U_s is impossible by regularity) is easily reduced by the shifting in x to the stabilization of $\tilde{u}$ to a bounded stationary solution; see the proof of Proposition 6.11. Finally, if

$$U_s > \tilde{u}_0 \quad \text{in} \quad [0, l) \quad \text{and} \quad U_s(l) = \tilde{u}_0(l),$$

then we arrive at the same contradiction replacing l by $\tilde{l}$ with $l - \tilde{l} > 0$ small.

Therefore, in view of the strict monotonicity with time and assuming that $\tilde{u}$ is global, we obtain that $\tilde{u}(r, t; 1) \to \infty$ as $t \to \infty$ uniformly on any compact subset from $[0, l)$. It is clear that, in this case, $\tilde{u}$ cannot be global and must blow up in finite time. Hence, we obtain the family of solutions (6.36) with blow-up times satisfying (6.37). Observe that, due to our construction, the envelope of the set $\{\tilde{u}_0(r; \varepsilon)\}$ with respect to the parameter ε is still $\mathcal{L}(r)$ for small $r > 0$, and the analysis based on the strict inequality (6.35) can be used in the same way.

Thus, in this case, Proposition 6.11 is proved by comparison with solutions $\tilde{u}$ shifted in $x_* \in I\!R^N$ as in Section 6.6. The final result follows in exactly the same way.

Monotone case. Let $p \geq p_u$ for $N \geq 11$. Then (6.51) and (6.52) hold, and to apply the above argument we need to derive a more precise lower bound of the final-time profile $u(r, T^-)$. First of all, assuming that $u \not\equiv \infty$ exists for $t \in (T, T + 4\delta)$, $\delta > 0$ small, one can check by the intersection comparison argument that under condition (6.52) the blow-up solution satisfies

$$u(r,t) > \mathcal{L}(r) \equiv U_s(r) \quad \text{for } r \in (0, r_0),\ r_0 > 0 \ \text{small},\ t \geq T. \tag{6.53}$$

It then follows from Theorem 6.6 (the last part) that $\mathcal{B}[u](t) = \{0\}$ for $t \geq T$. The function $v = u^m$ satisfies the equation

$$\tfrac{1}{m} v^{(1-m)/m} v_t = \Delta v + v^q \quad \text{in } I\!R^N \times (T, \infty),\ q = \tfrac{p}{m} > 1, \tag{6.54}$$

and the singular stationary solution becomes $V_s = U_s^m$ solving

$$\Delta V_s + V_s^q = 0.$$

Set $v = V_s + w$. We show that a linear instability analysis of equation (6.54) is enough to provide a necessary lower bound of the evolution singularity.

Linear instability analysis. In view of (6.53), we may assume that $w(r,t) > 0$ in $Q_\delta = (0, r_0) \times (T, T + 4\delta)$, and then we obtain the following parabolic inequality in Q_δ:

$$\begin{aligned} \tfrac{1}{m} (V_s + w)^{(1-m)/m} w_t &= \Delta V_s + \Delta w + (V_s + w)^q \\ &\equiv \Delta w + (V_s + w)^q - V_s^q \geq \Delta w + q V_s^{q-1} w. \end{aligned}$$

By the MP, we have that $w \geq 2c > 0$ in $Q_\delta^1 = (0, r_0) \times (T + \delta, T + 4\delta)$, so that w is not less than the solution of the problem

$$\tfrac{1}{m} (V_s + w)^{(1-m)/m} w_t = \Delta w + q V_s^{q-1} w \quad \text{in } Q_\delta^1,$$

with $w = c$ on the parabolic boundary ∂Q_δ^1. One can see that $w_t \geq 0$, and therefore w is not less than the solution of

$$\tfrac{1}{m} V_s^{(1-m)/m} w_t = \Delta w + q V_s^{q-1} w \quad \text{in } Q_\delta^1,\ w = c \text{ on } \partial Q_\delta^1. \tag{6.55}$$

We now study a minimal evolution singularity generated by the linear equation (6.55). Taking into account that

$$qV_s^{q-1} = \frac{A}{r^2}, \quad A = \tfrac{2q}{q-1}\left(N - 2 - \tfrac{2}{q-1}\right) > 0,$$

$$\tfrac{1}{m} V_s^{\frac{1-m}{m}} = Br^\gamma, \quad \gamma = \tfrac{2(m-1)}{m(q-1)}, \quad B > 0,$$

we deduce that, for $t > \tau = T + \delta$, the equation (6.55) admits self-similar solutions of the form

$$w_*(r,t) = (t-\tau)^\alpha g(\eta), \quad \eta = r/(t-\tau)^\beta, \; \beta = \tfrac{p-m}{2(p-1)}, \tag{6.56}$$

where $\alpha > 0$ is a parameter and the function g solves the linear ODE

$$g'' + (N-1)\frac{1}{\eta}g' + A\frac{1}{\eta^2}g + (\beta g'\eta - \alpha g)B\eta^\gamma = 0.$$

Consider first the case $p > p_u$. Then this equation admits two types of singular solutions with minimal and maximal singularities,

$$g_{1,2}(\eta) = \eta^{\mu_{1.2}}(1 + o(1)) \quad \text{as } \eta \to 0, \tag{6.57}$$

where $\mu_2 < \mu_1 < 0$ are the roots of the quadratic equation

$$\mu^2 + (N-2)\mu + A = 0.$$

The discriminant $D = (N-2)^2 - 4A > 0$ for $p > p_u$, and

$$\mu_2 < -\tfrac{N-2}{2} < \mu_1 < -\tfrac{2}{q-1}. \tag{6.58}$$

We now prove that the profile g_1 describes the minimal evolution singularity for (6.55). Fix $\alpha > 0$ and consider the function

$$g_\nu(\eta) = g_1(\eta) - \nu g_2(\eta) \quad \text{for} \quad \nu > 0 \quad \text{small}.$$

Then g_ν vanishes at $\eta = \eta_\nu \to 0$ as $\nu \to 0$, has a maximum at $\bar{\eta}_\nu \to 0$, and $g_\nu(\eta) > 0$ on the interval $(\eta_\nu, \eta_0]$ with a fixed small $\eta_0 > 0$.

It follows from (6.56) that $w_*(r,t)$ with the function g_ν satisfies $w_* \le c$ on the parabolic boundary of

$$\Omega_{\nu\delta} = (T+\delta, T+2\delta) \times \{\eta_\nu < \eta < \eta_0\}$$

provided that $\delta > 0$ is small. Therefore, $w \ge w_*$ in $\Omega_{\nu\delta}$. Passing to the limit $\nu \to 0$ and using the fact that $g_\nu(\eta) \to g_1(\eta)$ as $\nu \to 0$ uniformly on compact subsets $[\varepsilon, \frac{1}{\varepsilon}]$, we then conclude that $w \ge w_*$ with $g = g_1$ in $\Omega_{0\delta} = (T+\delta, T+2\delta) \times \{0 < \eta < \eta_0\}$. Setting here $t = T + 2\delta$ yields

$$w(r, T+2\delta) \ge \delta^\alpha g_1(r\delta^{-\beta}) = C_\delta r^{\mu_1}(1 + o(1)) \quad \text{as } r \to 0, \tag{6.59}$$

where $C_\delta > 0$ is a constant. Coming back to the original variable $u(r,t)$, (6.59) implies that

$$u(r, T+2\delta) \ge \tilde{C}_\delta r^\rho (1 + o(1)) \quad \text{as } r \to 0, \tag{6.60}$$

where $\rho = \frac{\mu_1}{m} < -\frac{2}{p-m}$ by (6.58). The rest of the proof is the same as in the case $p \in (p_S, p_u)$. Namely, fix $\lambda \approx 1^+$ and $r_\lambda > 0$ such that $\lambda U(r_\lambda; 1) > U_s(r_\lambda)$. Let

$\tilde{u}(r,t;1)$ be the solution to (6.1) in $(0,r_\lambda)\times I\!R_+$ with the initial data $\lambda U(r;1)$ and the boundary condition

$$\tilde{u}(r_\lambda,t;1)=\lambda U(r_\lambda;1) \quad \text{for} \quad t\geq 0.$$

Then in exactly the same way we prove that $\tilde{u}$ blows up in finite time T_1, and hence (6.37) is true for the corresponding family (6.36). Observe that, due to (6.60), the initial data for $\tilde{u}(r,t;\varepsilon)$ are below $u(r,T+2\delta)$ for small $r>0$. Finally by comparison, using the shifting $\tilde{u}$ in x, we deduce that

$$u(r,T+3\delta)=\infty \quad \text{for all small} \quad r>0,$$

whence complete blow-up as in Section 6.6.

If $p=p_u$, then

$$\mu_1=\mu_2=-\tfrac{1}{2}(N-2)<-\tfrac{2}{q-1}.$$

The minimal singularity $g_1(\eta)$ is again given by (6.57) and the maximal one is

$$g_2(\eta)=-\eta^{\mu_1}\log\eta(1+o(1)).$$

The rest of the proof is the same. □

Proof of Theorem 6.14

In view of the parabolic regularity for (6.1), $u(r,t)$ is nondecreasing with $t>0$ means that

$$\Delta u^m+u^p\geq 0 \quad \text{in} \ \{u>0\}. \tag{6.61}$$

Moreover, if $u_t\not\equiv 0$ (i.e., u is not a nontrivial stationary solution in $I\!R^N$ existing for $p\geq p_S$), then, by the strong MP, the sign in (6.61) is strict: $u_t>0$ in $\{u>0\}$. In view of this, we easily deduce that, for $t\geq T$, there holds $u(r,t)>\mathcal{L}(r)$ for small $r>0$. The rest of the proof is the same as in Theorem 6.13. □

6.10 Complete and incomplete blow-up for the equation with the p-Laplacian operator

The technique developed in previous sections can be applied in other circumstances. As an illustrative example, we now turn our attention to the Cauchy problem for the p-Laplacian equation with source

$$\boxed{u_t=\operatorname{div}(|Du|^\sigma Du)+u^p, \quad \sigma>-\tfrac{2}{N+1}, \quad p>1,} \tag{6.62}$$

and bounded initial data $u_0\geq 0$.

Theorem 6.16 *Let $u(r,t)$ be the proper minimal solutions of* (6.62). *Then:*
(i) Blow-up is complete if

$$\tfrac{1}{1+\sigma}<p\leq p_S=\tfrac{N(1+\sigma)+2+\sigma}{[N-(2+\sigma)]_+}.$$

(ii) Incomplete blow-up occurs if

$$1<p\leq\tfrac{1}{1+\sigma}. \tag{6.63}$$

Condition (6.63) coincides with the 1D result already discussed in Chapter 5. It is not difficult to translate Theorems 6.13 and 6.14 into the parameter ranges corresponding to equation (6.62). The proof of complete blow-up for (6.62) in the case $\sigma > 0$ and

$$1 < p \le p_{st} = \tfrac{N(1+\sigma)}{[N-(2+\sigma)]_+} \tag{6.64}$$

is exactly the same as the proof of Proposition 6.10. Consider a more delicate case of single point blow-up that occurs for $p > 1 + \sigma$ (see Remarks). It follows from the scaling invariance of the stationary equation

$$\frac{1}{r^{N-1}}\,(r^{N-1}|U'|^{\sigma}U')' + U^p = 0$$

with conditions $U(0) = \varepsilon$, $U'(0) = 0$, that the envelope of the family $\{U(r;\varepsilon)\}$ of stationary solutions takes the form

$$\mathcal{L}(r) \equiv \sup_{\varepsilon>0} U(r;\varepsilon) = \frac{c_*}{r^{(2+\sigma)/[p-(1+\sigma)]}} \quad \text{with } c_* > 0.$$

Thus (6.28) holds. Using the same comparison as above, we obtain that (6.64) implies (6.30) (local non-solvability) and hence complete blow-up. As the last step, we have to check the property (6.34) of the source type solution $v = v_M(x,t)$ of the equation

$$v_t = \mathrm{div}(|Dv|^{\sigma} Dv) \quad \text{in } I\!\!R^N \times (T+\delta, \infty),$$

with the initial data $v(x, T+\delta) = M\delta(x)$. The solution is again given by (6.33) with

$$k = \tfrac{1}{\sigma(N+1)+2}, \quad \theta(\xi) = c_0(a^{\alpha} - |\xi|^{\alpha})_+^{\frac{1+\sigma}{\sigma}},$$

where

$$\alpha = \tfrac{2+\sigma}{1+\sigma}.$$

The constant $a = a(M)$ then satisfies

$$a = c_2 M^{\sigma k} \to \infty \quad \text{as } M \to \infty.$$

Therefore, the solution $v_M(x,t)$ satisfies (6.34), thus concluding the proof of complete blow-up in the case (6.64).

The rest of the proof in the case $p_{st} < p \le p_S$ (or $\sigma \in (-1, 0)$) is the same as above. The problem of incomplete blow-up under the condition (6.63) again reduces to the 1D one.

The limit case of incomplete blow-up

Consider equation (6.62) with $\sigma \in (-1, 0)$ and the maximal exponent for incomplete blow-up

$$p = \tfrac{1}{1+\sigma} > 1.$$

The equation admits 1D TW solutions of the form

$$U(x,t) = [\mu_0 B((x \cdot \mathbf{n}) - \lambda t + a)_+]^{\frac{1+\sigma}{\sigma}} \quad \text{with } \mathbf{n} \in S^1,$$

where $\mu_0 = \frac{|\sigma|}{1+\sigma}$ and the constant $B > 0$ satisfies the algebraic equation

$$B^{2+\sigma} - \lambda B + 1 = 0.$$

We then calculate the minimal speed λ_m of propagation of TWs

$$\lambda_m = (2+\sigma)(1+\sigma)^{-\frac{1+\sigma}{2+\sigma}}.$$

Therefore, as in Section 6.4, by comparison we have the following estimate on the blow-up set of any proper solution $u(x,t)$ with bounded and compactly supported initial data:

$$\operatorname{diam} \mathcal{B}[u](t) \le 2\lambda_m (t+c) \quad \text{for } t > T.$$

6.11 Extinction problems in $\mathbb{R}^N$ and the criteria of complete and incomplete singularities

As in Chapter 4, we extend our techniques to another kind of singularities called finite-time extinction or quenching. We describe this effect in $\mathbb{R}^N$ by using nonlinear heat equations with singular absorption terms

$$\boxed{u_t = \Delta u^m - u^p \quad \text{in } S = \mathbb{R}^N \times \mathbb{R}_+, \quad m > 0, \; p < 0,} \tag{6.65}$$

and the equation with the p-Laplacian operator

$$\boxed{u_t = \operatorname{div}(|Du|^\sigma Du) - u^p, \quad \sigma > -1, \; p < 0.} \tag{6.66}$$

We take strictly positive, bounded initial data $u_0 = u_0(r)$ having an inverse bell-shaped form. As we already know, in this case the singularity in the equation may appear because of the extinction effect, and we assume that there exists the finite extinction time $T = T(u_0) < \infty$ such that $u > 0$ in $\mathbb{R}^N \times (0,T)$ and

$$\inf_{x \in \mathbb{R}^N} u(x,t) \to 0 \quad \text{as } t \to T^-.$$

A standard concept of a global proper (maximal) solution can be introduced by means of suitable approximations. The semigroup for the extinction problem is based on just the same ideas as in Section 6.2, but now the m.i.c. is replaced by the monotone decreasing convergence and $u = 0$ replaces $u = \infty$ as the singular level. The proper solution is now *maximal* instead of minimal.

We again arrive at the question of *complete* ($u \equiv 0$ for all $t > T$) or *incomplete extinction* ($u \not\equiv 0$ for $t > T$). As usual, we assume that the extinction set satisfies

$$\mathcal{E}[u_0] \ne \emptyset \quad \text{and} \quad \mathcal{E}[u_0] \ne \mathbb{R},$$

which is easily ensured by standard conditions on the data. Exactly as in 1D, an obvious transformation $u = \frac{1}{v}$ reduces the extinction problem to a blow-up one for a different non-divergent nonlinear parabolic equation, and the methods of the blow-up analysis can be applied. We study extinction in less detail, and first we focus our attention on equation (6.66) for a change.

Theorem 6.17 (Complete extinction) *Under the given assumptions on $u_0(r)$, the* *proper solutions* $u(r,t)$ *of* (6.66) *exhibit complete extinction if and only if*

$$p \le -1.$$

Proof. The analysis of complete extinction is similar to the proof of Theorem 6.6. As above, for simplicity, we may assume in view of the MP that

$$u_r > 0 \quad \text{for } r > 0,$$

so $r = 0$ is the extinction point and $u(0,T^-) = 0$. The stationary solutions $U(r;\varepsilon)$ satisfying

$$\frac{1}{r^{N-1}}(r^{N-1}|U'|^{\sigma}U')' - U^p = 0 \quad \text{for } r > 0, \quad U(0) = \varepsilon,\ U'(0) = 0,$$

can be easily estimated from above,

$$U(r) \le \bar{U}(r) \equiv \varepsilon + c_0 \varepsilon^{\frac{p}{1+\sigma}} r^{\frac{2+\sigma}{1+\sigma}}.$$

Therefore, the envelope of this family $\mathcal{L}(r)$ satisfies

$$\mathcal{L}(r) \le \bar{\mathcal{L}}(r) = \inf_{\varepsilon>0} \bar{U}(r;\varepsilon) = c_1 r^{\frac{2+\sigma}{1+\sigma-p}}.$$

More precisely, using a scaling invariance of the stationary equation, we have

$$\mathcal{L}(r) = c_* r^{\frac{2+\sigma}{1+\sigma-p}}, \tag{6.67}$$

where $c_* < c_1$. By the intersection comparison argument, we then obtain that the proper solution satisfies (cf. (6.35))

$$u(r,t) < \mathcal{L}(r) \quad \text{for small } r > 0 \ \text{ for all } \ t \ge T.$$

Therefore, by the same nonexistence argument as in Section 6.6, we deduce that no nontrivial continuation exists if (cf. (6.30))

$$(\mathcal{L}(r))^p \notin L^1_{\rm loc}(I\!R^N), \tag{6.68}$$

which, in view of (6.67), yields complete extinction for

$$p \le p_{st} = \tfrac{N(1+\sigma)}{(2+\sigma-N)_+}.$$

Indeed, assume for contradiction that the extinction set $\mathcal{E}[u](t)$ (defined similarly to the blow-up set) is bounded for small $t - T > 0$. Then we integrate equation (6.66) with $u = u_n$ over a ball B_R with

$$R \gg \operatorname{diam} \mathcal{E}[u](t)$$

and use the fact that, by the assumption, the sequence $\{u_n(r,t)\}$ is uniformly bounded away from zero for $r \approx R$ and hence is regular enough there. Finally, passing to the limit $n \to \infty$ and using the inequality $u < \mathcal{L}$ and (6.68), we obtain that

$$\frac{\mathrm{d}}{\mathrm{d}t} \int_{B_R} u(x,t)\,\mathrm{d}x \sim -\infty$$

simply meaning that

$$\int_{B_R} u(x,t)\,\mathrm{d}x = 0 \quad \text{for } t > T$$

contradicting the assumption on boundedness of $\mathcal{E}[u](t)$. The proof of complete extinction for $p_{st} < p \leq -1$ is the same as in Section 6.6. Incomplete extinction for $p < 1$ is proved by local comparison with 1D singular TW solutions as in Section 4.6. □

For equation (6.65) we have the following criterion proved similarly.

Theorem 6.18 (Complete extinction) *Under the given hypotheses on* $u_0(r)$, *the proper solution* $u(r,t)$ *of* (6.65) *exhibits complete extinction if and only if*

$$p \leq -m.$$

Remarks and comments on the literature

Main results are presented in [167].

§ **6.1.** Equation (6.1) is a popular mathematical model for different nonlinear reaction-diffusion phenomena, cf. the books [346], [43], [306], [317], [170]. If (6.1) is a nonlinear heat propagation model in a reactive medium, then u is a temperature.

The critical Fujita exponent $p_F = m + \frac{2}{N}$ for the quasilinear equations with $m > 1$ was proved to exist in [145]. These blow-up results are well known; see lists of related references in surveys [244] and [97] and in books [306, Chapter 4] and [267, Part II]. The blow-up set (6.2) corresponds to the idea of *burnt zone* in the theory of flame propagation, cf. [71].

§ **6.2.** The idea of applying the semigroup framework to nonlinear PDE problems is rather old and it has been extensively used in problems of nonlinear diffusion after the now classical works of Ph. Bénilan, H. Brezis and M.G. Crandall in the 1970s, cf. e.g. [44], [64] and [92].

For a semilinear heat equation, a concept of extension of a semigroup was used in [36], [37], where the main results essentially rely on the particular properties of the linear parabolic operator $\partial/\partial t - \Delta$ and the semigroups $e^{\Delta t}$, and therefore do not apply to quasilinear equations under consideration.

The present construction is general and applies to a number of similar situations like the Dirichlet and Neumann problems with zero lateral data. We can also change the equation and consider e.g. the p-Laplacian one with a source, which we already dealt with in the previous chapters. Ordinary differential equations are not out of the question as long as the monotonicity assumptions are satisfied. More general equations will be studied in Chapter 7.

In Definition 6.2 we introduce the term *proper* to denote such solutions. In [165] the authors used the name *viscosity solution*, a rather natural term for this case where regularization via truncation plays a "viscosity" role. But it could lead to misunderstanding with the current usage of that term, cf. [93], [94] and references therein.

In the application we have in mind that there is a certain compactness of the orbits whenever they are bounded, so that bounded proper solutions are easily proved to be weak solutions in the usual integration-by-parts sense. On the other hand, solutions can become identically infinite from a time on. This is precisely what we call complete blow-up. The cases of incomplete blow-up consist of proper solutions that become infinite in an (expanding) domain of positive measure after the blow-up time. Finally, the peaking solutions have an *"inessential"* blow-up ($u(t) \in L^\infty$ for all $t > 0$, $t \neq T$) since it only happens at precisely one moment of time and point, see similarity constructions in [167, Section 11]. In fact, these solutions do not exhibit any blow-up when we pass from the classical to the weak framework, since they are continuous functions $[0, \infty) \mapsto L^p(I\!R^N)$ for some $p \in [1, \infty)$ and global weak solutions of the equation.

§ **6.3.** The results and techniques are similar to those in 1D in Chapter 4.

§ **6.4.** Similar nonunique TWs were constructed in [126]. The analysis is a straightforward extension of 1D results from Chapter 5.

§ **6.5.** It seems that the first main contribution to the theory of complete singularities for parabolic equations is the work of P. Baras and L. Cohen (1985) [36], [37], who studied the problem of complete blow-up for semilinear heat equations

$$\boxed{u_t = \Delta u + \psi(u)}$$

and proved that, for $\psi(u) \sim u^p$, in the subcritical Sobolev range

$$1 < p < p_S = \tfrac{N+2}{N-2},$$

complete blow-up occurs, thus establishing the validity of a conjecture of H. Brezis in that range. Further results were obtained in [236]; see also the book [43] and the references therein. Peaking similarity solutions blowing up at $t = T$ only, for $p > p_S$, are described in [167], where further references are given.

§ **6.6.** A simple structure of envelopes in (6.22)–(6.24) follows from the scaling invariance of the stationary ODE. Envelope estimates can be easily derived, see [306, pp. 421-427]. The method of stationary states (intersection comparison of a blow-up solution with a continuous family of stationary profiles and the corresponding envelope analysis) was introduced in [127]; see different applications and extensions in [148], [152] (parabolic systems), [153] (p-Laplacian), [154] (a monotonicity result), [161], [166] (weakly quasilinear equations).

The intersection comparison proof of the eventual monotonicity of large solutions can be found in [166, Section 2]. For a similar monotonicity analysis in an extinction problem, see [162, Section 10]. Lemma 6.8 and Corollary 6.9 are typical for the method of stationary states; cf. [306, Chapter 7]. A general result of monotonicity in time of large solutions was proved in [150], [154] (1D equations) and in [149] (radial N-dimensional in the Sobolev subcritical range). The explicit ZKB-solution (6.33) was constructed in [347] for $N = 1, 3$ and in [39] for any $N \geq 1$. In Proposition 6.11 blow-up interfaces are not decreasing due to the monotonicity result mentioned above: if $p < p_S$, then [149]

$$u_t > 0 \quad \text{everywhere in the set} \quad \{u \gg 1\}.$$

In the proof we use the same comparison idea as in [306, p. 362]. Blow-up subsolutions were extensively used in [306, p. 215] in calculating critical Fujita exponents for quasilinear equations.

§ **6.7.** ODEs of R. Emden–R.H. Fowler type were introduced about a century ago (1907) [105], (1914) [119], and subsequently studied in detail in a number of monographs and textbooks on ODEs. These are famous exact solutions invariant under conformal and projective transformations; see [201] and [251].

Though for the subcritical range $1 < p < p_S$ $(m > 1)$, there exists a proof of complete blow-up of general non-radial solutions [326], the critical case $p = p_S$ in the non-radial geometry still remains an open problem.

§ **6.8.** Much of the above theory can be easily extended to the equation with general nonlinearities (6.7). The first part of the proof of the corresponding Theorem 6.6 is done along the same lines. To prove a result that is similar to Proposition 6.10 we need the corresponding envelope analysis in the case of rather general functions $\varphi(u)$ and $\psi(u)$ as it is done in the method of stationary states in Chapter 7 in [306]. Then, by the standard intersection comparison argument, we arrive at a lower estimate via the envelope similar to (6.28). Therefore for blow-up to be complete we need to check that (cf. (6.30))

$$\psi(\mathcal{L}(r)) \notin L^1_{\rm loc}(\mathbb{R}^N)$$

(we assume here that $\varphi'(u) \not\to 0$ as $u \to \infty$). The final step, which completes the proof of this general version of Proposition 6.10, consists of verifying whether the equation without the source term

$$v_t = \Delta\varphi(v) \quad \text{in } \mathbb{R}^N \times (T + \delta, \infty),$$

admits a source-type solution v_M satisfying the initial condition

$$v_M(x, T + \delta) = M\,\delta(x) \quad \text{in } \mathbb{R}^N$$

(cf. the problem (6.32)), and $v_M(x, T + 2\delta) \to \infty$ as $M \to \infty$ uniformly on compact subsets. This is a standard problem in the theory of filtration equations; see references in the survey [213].

In order to prove a stronger result similar to Theorem 6.6, according to our framework we need to check the following properties.

(i) To show that the stationary solutions $U = U(r; \varepsilon)$,

$$\Delta\varphi(U) + \psi(U) = 0 \quad \text{for } r > 0, \quad U(0) = \varepsilon, \;\; U'(0) = 0,$$

vanish at finite points $r = r_0(\varepsilon) > 0$ and satisfy the intersection hypothesis (6.21) naturally translated to the case of general φ. This is a standard problem of solvability of a semilinear elliptic equation in a bounded domain with Dirichlet boundary conditions, and it can be studied by S.I. Pohozaev's Identity [290]. Then, defining as above blow-up solutions $\tilde{u}(r, t; \varepsilon)$ that are monotone with time, we expect a natural result that $T_\varepsilon \to 0$ as $\varepsilon \to \infty$.

(ii) Finally, we prove strict monotonicity of the blow-up interface for $t = T^+$ (Proposition 6.11), and the problem of complete blow-up reduces to a similar problem for the 1D equation

$$v_t = (\varphi(v))_{yy} + F((\varphi(v))_y) + \psi(v),$$

which is solved by the technique from Section 4.4. The analysis of incomplete blow-up for (6.7) is exactly one-dimensional and a local comparison argument applies.

§ **6.9.** The exponent p_u is the critical exponent for the *uniqueness* of the singular stationary solution as a proper stationary solution: for $p \geq p_u$, U_s is proper, but for $p \in (p_{st}, p_u)$, it is not; see [167, Section 10]. p_u also plays a role in the study of peaking solutions with incomplete blow-up; see [167, Section 11].

§ **6.10.** Single point blow-up for (6.62) with $p > 1 + \sigma$ was proved in [153]; for $p = 1 + \sigma$ blow-up is regional; see [128] and a general localization theorem in [131]; for $p < 1 + \sigma$ blow-up is global [128].

§ **6.11.** As for precedents, it was known that, for (6.65) in the semilinear case $m = 1$, solutions may develop a nontrivial dead-core; see [288] and also references in [219] and [213]. The quasilinear case $m > 1$ is treated in [219], where a weak global solution is constructed under the assumption $p > -m$. Complete extinction in the critical case $p = -m$ (even for $m = 1$) has been an open problem for some years and has been solved in 1D in [165]. These results are explained in previous Chapter 4. Another class of quasilinear parabolic equations with singular absorption such as

$$u_t = \frac{u_{xx}}{1 + (u_x)^2} - \psi(u)$$

play an important role in the description of mean curvature flow. In particular, the case

$$\psi(u) = \tfrac{1}{u}$$

describes after a suitable surface parameterization the evolution of cylindrically symmetric hypersurfaces moving by mean curvature in $I\!R^3$, [110], [318], [5]; incomplete extinction is then proved by geometric methods; see [5]. It is curious that the TW analysis shows that extinction is *always* incomplete regardless the singularity term $\psi(u)$. One can take, say,

$$\psi(u) = e^{\frac{1}{u}}$$

and nevertheless there exists a nontrivial continuation with finite interfaces beyond single point extinction. Of course, equations of the singularity propagation depend on $\psi(u)$; see general examples in Section 7.11 of Chapter 7. This departs strongly from the results for (6.65) or (6.66) where the diffusion term is not strong enough to balance any type of singular absorption.

CHAPTER 7

Geometric Theory of Nonlinear Singular Parabolic Equations. Maximal Solutions

This chapter is a core of the geometric Sturmian theory of singular parabolic PDEs. Here we present all main geometric concepts that are crucial for the existence, nonexistence, uniqueness, differential properties and free-boundary propagation for nonlinear second-order parabolic equations with singularities. A detailed theory is developed for one-dimensional equations but some of the results are extended to equations in $\mathbb{R}^N$, where more detailed classification is available in the radial geometry.

Here we deal with the most general class of singular parabolic equations including those studied in detail in previous chapters with operators of the PME and the p-Laplacian equation and other types of quasilinear degenerate diffusion ones with extra lower-order terms from reaction-absorption-convection. We also consider some models from mean curvature flows and study classes of fully nonlinear equations from filtration and detonation theory.

We again note that the Sturmian theory is based on geometric ideas of intersection comparison with complete sets B of particular solutions, and actually does not deal with PDEs at all. Therefore, we do not hesitate to consider wide classes of fully nonlinear equations in $\mathbb{R}^N$ for which the MP holds if their 1D restrictions admit plane TW solutions or sub and supersolutions that are sufficient to prove existence-nonexistence results. In general, intersection techniques establish certain *a priori* estimates of proper solutions $u(x,t)$, which often can be constructed by regular approximations $\{u_n(x,t)\}$. If a monotone approximation is available and $u_n \downarrow u$, u is known to be a unique maximal solution. In the present context, dealing with general classes of equations with singular coefficients, we do not use and do not specify concepts of *viscosity solutions*, which are known to be effective for various parabolic and elliptic problems, in mean curvature flows and level set propagation. A list of related references is available in Remarks.

The analysis uses two concepts of the ordered geometric evolution explained in Section 1.4 without specifying parabolic PDEs. Therefore, the 1D geometric techniques and typical regularity results are similar for any reasonable classes of solutions if the first Sturm Theorem can be applied.

In the next three chapters, we will consider the following three classes of solutions of singular parabolic equations:

(i) nonnegative maximal solutions of the Cauchy problems (this chapter),

(ii) non-maximal solutions of one-phase and two-phase FBPs (Chapter 8), and

(iii) proper solutions of changing sign (Chapter 9).

In the present chapter we give a more detailed study of the most well-known class (i) of the maximal solutions, which includes weak solutions of the PME, the p-Laplacian equations and their generalizations. In the next two chapters we show that the geometric ideas can be applied to other solution classes, (ii) and (iii).

7.1 Introduction: Main steps and concepts of the geometric theory

We consider bounded solutions $u(x,t) \geq 0$ of a 1D parabolic PDE of the general form

$$u_t = \mathbf{F}(u) \equiv F(u, u_x, u_{xx}) \quad \text{in } S = I\!R \times I\!R_+, \tag{7.1}$$

where $F(p,q,r)$ is a given function being sufficiently smooth for $p > 0$ and satisfying the *parabolicity condition*

$$F_r(p,q,r) > 0 \quad \text{for } p > 0 \text{ and } q, r \in I\!R. \tag{7.2}$$

We pose for (7.1) the Cauchy problem with continuous initial data

$$u(x,0) = u_0(x) \geq 0 \quad \text{in } I\!R. \tag{7.3}$$

Let us specify the main hypothesis on the equations and initial data, which make it possible to study the existence and regularity questions in such a generality.

We suppose that singularities in the problem can occur at the only zero level $\{u = 0\}$ and that the solution $u = u(x,t) \geq 0$ is assumed to be sufficiently smooth for $u > 0$, i.e., away from the singularity. In the most of the cases, the initial function u_0 is assumed to be singular in the sense that there exists

$$\min_x u_0(x) = 0. \tag{7.4}$$

In the study of the regularity properties of singular interfaces, usually we assume that $u_0(x)$ is monotone increasing. Then a solution $u(x,t) \geq 0$ is monotone increasing with x for $t > 0$ by the MP, the singular propagation of finite interfaces starts at the initial moment $t = 0$, the interface $x = s(t)$ is unique, and we arrive at a (singular) free-boundary problem.

For suitable classes of positive (non-singular) initial data u_0 uniformly bounded away from $u = 0$, the Cauchy problem is supposed to admit a unique sufficiently smooth solution local in time, and known results from the parabolic theory such as uniqueness, interior regularity, comparison and the MP are assumed to apply to such classical solutions. For non-singular initial data, we then obtain another important problem of the occurrence of singularities in finite time $t = T >$) (not initially, at $t = 0$, as for data (7.4)). Thus, on any subset $\{0 < \delta \leq u \leq \frac{1}{\delta}\}$ with small $\delta > 0$, the equation is assumed to be sufficiently regular where the classical parabolic theory applies.

For convenience, we describe the main steps of the geometric theory to be developed in this chapter.

1. Set B of proper TWs and related geometric notions (Section 7.2). We first describe necessary hypotheses we need to impose on a set $B = \{V\}$ of particular solutions of (7.1). For autonomous equations (7.1), this is the set of the travelling wave (TW) solutions

$$V(x,t) = f_\lambda(\xi), \quad \xi = x - \lambda t - a, \tag{7.5}$$

where $\lambda \in I\!R$ is the constant speed of the TW and $a \in I\!R$ is the parameter of translation. Then continuous functions f_λ satisfy a nonlinear second-order ODE with the single parameter $\lambda \in I\!R$,

$$F(f, f', f'') + \lambda f' = 0 \quad \text{in } \{f > 0\}. \tag{7.6}$$

We assume that it can be studied by standard ODEs methods. Firstly, we choose proper TW profiles $f_\lambda(\xi)$ for λ from the existence subset denoted by $\Delta \subseteq I\!R$ such that their ranges satisfy

$$0 \in \overline{\mathrm{Im}}\, f_\lambda \quad \text{for any } \lambda \in \Delta.$$

These profiles $f_\lambda(\xi)$ are heteroclinic connections of the singular points at $\{f = 0\}$ with regular ones in $\{f > 0\}$.

Secondly, we study *singular* TWs with finite propagation on the singular level for which

$$\{\xi \in I\!R: \ f_\lambda(\xi) = 0\} \neq \emptyset \quad \text{for } \lambda \in \Delta_0 \subseteq \Delta.$$

We detect continuous branches of the *proper* TW solutions f_λ, $\lambda \in \Delta_0$, which can be constructed as the limit of suitable sequences of non-singular solutions of the ODE (7.6). We show that proper branches are strictly monotone decreasing with λ. Indeed, this is a comparison property driven by the MP for uniformly parabolic equations.

Let B be a two-parameter set of all singular proper TW solutions

$$B = \{V = f_\lambda(x - \lambda t - a), \ \lambda \in \Delta_0, \ a \in I\!R\}$$

with the straight line interfaces $s_\lambda(t) = \lambda t + a$. As in Chapter 2, we assume that B is complete, continuous and monotone. Then the set $\{f_\lambda(\xi), \ \lambda \in \Delta_0\}$, where all functions f_λ have their interfaces at the origin $\xi = 0$, is called the *proper TW-bundle* or the *B-bundle*.

Next, using the TW-bundle, we introduce the basic notions:

(i) the pressure variable,

(ii) the interface operators $\mathbf{M}(u)$ and $\mathbf{N}(u)$ and the interface slopes S of the first or second order,

(iii) the TW-diagram $\lambda = A_0^{-1}(S) \in \Delta_0$, which is the Rankine–Hugoniot condition of the singular propagation for the parabolic PDEs,

(iv) the gradient function (G-function) characterizing optimal Bernstein-type estimates in functional B-classes, and so on.

2. Proper maximal solution. Existence if $\Delta \neq \emptyset$ (Section 7.3). A unique proper maximal solution $u(x,t)$ in $S_T = I\!R \times (0,T)$ is constructed as the limit of a monotone sequence of smooth non-singular solutions $\{u_n\}$ of the regularized equation with regularized initial data. The regularized Cauchy problem is assumed to be well-posed in the sense that the standard theory for smooth nonlinear uniformly parabolic equations applies. The existence of a nontrivial limit $u = \lim u_n \not\equiv 0$ in the case $\Delta \neq \emptyset$ (which means *incomplete singularity* for (7.1) at $u = 0$) follows by comparison with the TW solutions f_λ, $\lambda \in \Delta$. If $\Delta_0 \neq \emptyset$, the proper solutions u exhibit finite propagation on the singular level $\{u = 0\}$. Existence results can be applied to singular parabolic equations in $I\!R^N$ such as

$$u_t = F(u, |\nabla u|, \Delta u) \quad \text{in } I\!R^N \times I\!R_+. \tag{7.7}$$

3. Sturm Theorem and intersection comparison with B, nonexistence if $\Delta = \emptyset$, further geometric notions (Sections 7.5 and 7.6). The main geometric tool of the analysis is the intersection comparison of a singular proper solution $u(x,t)$ with the complete set B of the TW solutions. In this intersection comparison we use the first Sturm Theorem, which can be applied to such singular equations in view of the monotone approximations via regular problems.

As a first result, we show that, in the case $\Delta = \emptyset$ for any u_0, the proper solutions can be trivial, $u = \lim u_n \equiv 0$ for arbitrarily small $t > 0$. This means that, at $\{u = 0\}$, there occurs a *complete* singularity. This nonexistence analysis is extended to general parabolic equations (7.7). In the existence case $\Delta \neq \emptyset$, we show that proper complete sets B define the property of B-concavity preserved in time, the corresponding sign-invariants and the B-number.

4. Optimal Bernstein estimates in B-classes as transversality conditions (Sections 7.7–7.9). A complete set B of the TW solutions makes it possible to introduce two functional classes $B^+_{\rm loc}(c)$ and $B^-_{\rm loc}(c)$ where $c \in \Delta_0$ is a parameter. The local classes $B^\pm_{\rm loc}(c)$ described by spatial shapes of the TWs $f_{\pm c}(x)$, are defined according to a single intersection (from above or below) near the interface of the solution profile $u(x,t)$ with TWs $f_{\pm c}(x - a)$ for $a \approx 0^+$ or $a \approx 0^-$. The intersection comparison then implies that

$$u_0 \in B^\pm_{\rm loc}(c) \implies u(t) \in B^\pm_{\rm loc}(c) \ \text{ for small } t > 0. \tag{7.8}$$

Under extra assumptions, this establishes the *instantaneous smoothing effect* for arbitrary initial data: for any $t_0 \in (0, T)$

$$\text{there exists a } \ c(t_0) \in \Delta_0 \ \text{ such that } \ u(x, t_0) \in B^\pm_{\rm loc}(\mp c). \tag{7.9}$$

These classes are shown to provide us with optimal *gradient estimates* on proper solutions near the singular level and a bound on the interfaces. Global B-classes $B^\pm(c)$ describe the behaviour in x of the proper solutions everywhere in $I\!R$. Some transversality and concavity results apply to radial solutions of the N-dimensional equation (7.7).

5. Interface operators and interface equation (Section 7.10). The above estimates in B-classes extended up to the singular interface

$$s(t) = \inf\{x \in I\!R : \ u(x,t) > 0\} \quad \text{for } t > 0$$

give bounds that are sufficient to define the *interface operators* introduced beforehand by the TW set B. As a consequence, we obtain uniform bounds of the right-hand and left-hand derivatives $D^\pm s(t)$ of the interface and derive either the *first-order interface equation* such as

$$D^+ s(t) = \mathbf{N}_1(u(\cdot, t)) \quad \text{for } t > 0, \tag{7.10}$$

or the *second-order interface equation*, which is a system of two equations

$$\begin{cases} \mathbf{M}_1(u(\cdot,t)) = A_0 = \text{const.}, \\ D^+ s(t) = \mathbf{N}_2(u(\cdot,t)). \end{cases} \tag{7.11}$$

Here the interface operators $\mathbf{N}_1$ and $\mathbf{M}_1$ are of first order, while $\mathbf{N}_2$ is of second order. Equation (7.10) and system (7.11) are the *Rankine–Hugoniot conditions* of finite singular propagation for arbitrary solutions $u(x,t)$.

6. Applications (Section 7.11). Optimal regularity estimates and the interface equations are established in the maximal generality provided that detailed properties of the whole set of TW profiles $B = \{f_\lambda\}$ satisfying the ODE (7.6) are available. We consider several applications of the method to quasilinear and fully nonlinear singular parabolic equations.

7.2 Set B of singular travelling waves and related geometric notions: pressure, slopes, interface operators, TW-diagram

Proper and improper TWs in one dimension

The geometric analysis uses the sets $B = \{V\}$ of reasonably simple particular solutions of the PDE exhibiting free-boundary propagation on the singular level. In the study of autonomous equations (7.1), we take the simplest two-dimensional set of the TW solutions (7.5) with two arbitrary parameters $\lambda, a \in I\!R$. We then obtain a nonlinear ODE for the continuous function $f = f_\lambda$ with the condition of the singularity connection,

$$F(f, f', f'') + \lambda f' = 0 \quad \text{in} \quad \{f > 0\}, \quad 0 \in \overline{\text{Im}\, f}. \tag{7.12}$$

Setting $f' = P$ in this autonomous ODE yields the first-order equation

$$F\left(f, P, P\frac{\mathrm{d}P}{\mathrm{d}f}\right) = -\lambda P.$$

By the parabolicity condition, it can be resolved relative to the derivative,

$$\frac{\mathrm{d}P}{\mathrm{d}f} = \frac{1}{P} F^{-1}(f, P, -\lambda P) \quad \text{for} \quad f > 0, \tag{7.13}$$

where F^{-1} denotes the inverse relative to the third variable r in $F(p,q,r)$. The problem (7.12) consists of determining a *singular orbit* $P = P(f)$ (a heteroclinic connection of zero) of the first-order ODE (7.13) defined for all small $f > 0$. Denote

$$\Delta = \{\lambda \in I\!R : \quad \text{there exists a singular orbit of (7.13)}\}. \tag{7.14}$$

In what follows the condition $\Delta \neq \emptyset$ is of principle importance; otherwise, if $\Delta = \emptyset$, the PDE (7.1) may not admit any nontrivial singular solutions.

Consider next the problem of finite propagation for the TW solutions when equation (7.12) admits a continuous solution $f(\xi)$ satisfying

$$f(0) = 0. \tag{7.15}$$

Then, for convenience, we extend it beyond singularity by setting $f(\xi) \equiv 0$ for

$\xi > 0$ or $\xi < 0$ as the context dictates. Denote

$$\Delta_0 = \{\lambda \in \Delta : \ (7.12), (7.15) \ \text{ has a continuous solution}\}. \tag{7.16}$$

Thus, for any TW speed $\lambda \in \Delta_0$, there exists a local in ξ continuous solution $f = f_\lambda(\xi) \not\equiv 0$ in a small neighbourhood of the origin $\xi = 0$. These continuous TWs $V(x,t) = f_\lambda(x - \lambda t)$ with straight line interfaces $s_\lambda(t) = \lambda t$ explain possible connections of the singular level $\{u = 0\}$ with the space of the positive regular orbits. We assume that, for any $\lambda \in \Delta_0 \subseteq I\!R$, there exists a local in ξ solution $f_\lambda(\xi)$ of (7.12), (7.15) that is continuous on the maximal interval $[0, \xi_\lambda)$ or $(\xi_\lambda, 0]$ and is strictly positive at interior points. If ξ_λ is finite, then either $f_\lambda(\xi) \to 0$ or $f_\lambda(\xi) \to \infty$ as $\xi \to \xi_\lambda$.

In order to study evolution properties of the left-hand interface of solutions $u(x,t)$, we consider strictly monotone increasing TW profiles $f_\lambda(\xi)$. The set B of such TW profiles $\{f_\lambda(\xi),\ \lambda \in \Delta_0\}$ describes the TW-propagation on the singular level.

Proper TW profiles. We now show how to choose special *proper* ODE profiles f_λ satisfying a *monotonicity* property by the MP. This is directly related to the construction of proper maximal solutions of the PDE in Section 7.3. For ODEs such a construction is easier. For definiteness, we consider the case of the monotone increasing profiles $f_\lambda(\xi)$, i.e., $P(f) = f' > 0$ at least for small $f > 0$ ($\xi > 0$).

Definition 7.1 A singular TW $f_\lambda(\xi)$ satisfying the ODE problem (7.12) is said to be *proper* (or *maximal*) if on a sufficiently small interval $\xi \in [0, c]$ it can be constructed as the limit of a sequence $\{f_\lambda^n\}$ of strictly positive (non-singular) smooth solutions of the ODE (7.12).

Without loss of generality we suppose that f_λ^n satisfy conditions

$$f_\lambda^n(0) = \tfrac{2}{n}, \quad (f_\lambda^n)'(0) = 0, \tag{7.17}$$

as shown on Figure 7.1.

For a fixed $\lambda \in \Delta_0$, the singular ODE problem may admit many solutions in general. It is not difficult to distinguish the proper profiles. Under certain natural restrictions on F, one can prove that continuous branches of proper maximal profiles $\{f_\lambda\}$ are always *minimal* and *decreasing* in λ. These results follow from the standard phase-plane analysis of the ODE (7.12).

Proposition 7.1 *Fix a $\lambda \in \Delta_0$. Let the set of all singular solutions of* (7.12), (7.15) *be a discrete ordered subset $\{f_1, f_2, ..., f_k\}$ and f_1 be the minimal one, i.e., $f_1(\xi) < f_j(\xi)$ for small $\xi > 0$ for all $j = 2, ..., k$. Then f_1 is the proper profile and f_j for $j \geq 2$ are not.*

Proof. The singular solutions $f_1, ..., f_k$ are given by continuous orbits $\{P_1, ..., P_k\}$ on the $\{f, P\}$-plane, $P_1(f) \leq P_j(f)$, $P_1 \not\equiv P_j(f)$ for small $f > 0$, which are extended up to $f = 0$. Let $\{P^n(f)\}$ be the corresponding monotone increasing sequence of strictly positive orbits defined for $f \geq \frac{2}{n}$ and satisfying (cf. (7.17))

$$P^n(\tfrac{2}{n}) = 0. \tag{7.18}$$

Since $P_1(f)$ is the minimal orbit, it follows from the phase-plane that the sequence

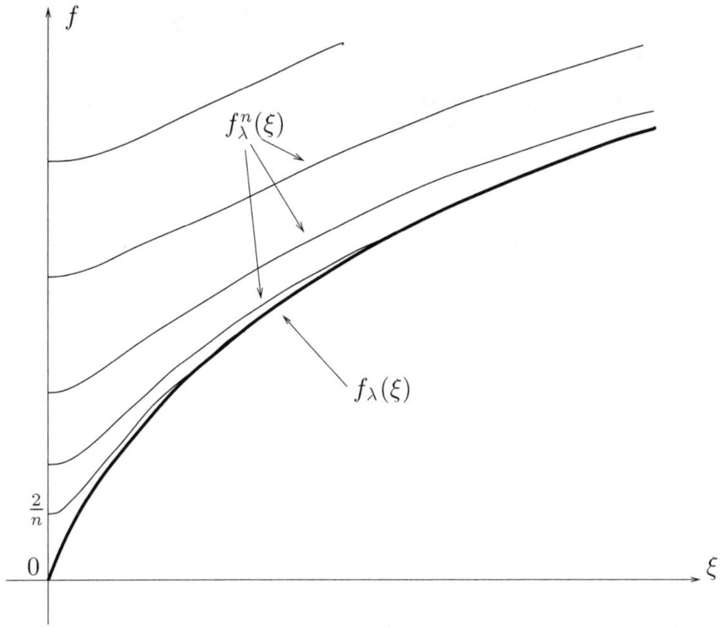

Figure 7.1 *Proper singular TW $f_\lambda(\xi)$ with $f_\lambda(0) = 0$ can be approximated by strictly positive regular TWs $\{f_\lambda^n(\xi)\}$ satisfying $f_\lambda^n(0) = \frac{2}{n}$ and $(f_\lambda^n)'(0) = 0$.*

$\{P^n\}$ cannot converge to P_j for any $j \geq 2$, and converges to the minimal orbit P_1 by comparison of the ODE orbits. □

Let $P_{\min}(f) = \lim P_n(f) > 0$ be the minimal positive orbit of the ODE (7.13) defined for all small $f > 0$. Then the proper TW profiles $f(\xi)$ are given by the ODE,

$$\frac{\mathrm{d}f}{\mathrm{d}\xi} = P_{\min}(f) > 0 \quad \text{for } \xi > 0. \tag{7.19}$$

Note that, in the case $\Delta_0 = \emptyset$ but $\Delta \neq \emptyset$, it gives strictly positive solutions with infinite interfaces at $\xi = -\infty$.

Choosing now all λ-branches of solutions of (7.12), (7.15), we next characterize the following λ-monotonicity of the proper TW profiles.

Proposition 7.2 *Each proper TW profile $f_\lambda(\xi)$ for $\xi > 0$ is strictly decreasing with $\lambda \in \Delta_0$.*

Proof. Fix arbitrary $\lambda_1, \lambda_2 \in \Delta_0,\ \lambda_1 < \lambda_2$, and let f_1 and f_2 be the corresponding singular TW profiles. Let $\{f_1^n\}$ and $\{f_2^n\}$ be sequences of approximating non-singular solutions satisfying (7.17). Then, by the parabolicity condition on F, we have

$$\frac{\mathrm{d}P}{\mathrm{d}f} = \frac{1}{P} F^{-1}(f, P, -\lambda_1 P) > \frac{1}{P} F^{-1}(f, P, -\lambda_2 P), \tag{7.20}$$

so that, by the standard comparison for the ODEs with smooth coefficients, we have that $P_1^n(f) > P_2^n(f)$ for $f \geq \frac{2}{n}$. Passing to the limit, by the strict inequality in (7.20), we obtain that minimal orbits satisfy $P_1(f) > P_2(f)$ and on integration we obtain $f_1(\xi) > f_2(\xi)$ for small $\xi > 0$. $\square$

Improper profiles. *Improper* branches of TW profiles that are increasing with λ generate TW solutions of the PDE that do not satisfy the comparison principle. Indeed, assume that there exist $\lambda_1 < \lambda_2$ such that $f_1(\xi) \le f_2(\xi)$ so that, at $t = 0$, the TWs $V_i(x,t) = f_i(x - \lambda t)$ satisfy

$$V_1(x,0) \equiv f_1(x) \le V_2(x,0) \equiv f_2(x). \tag{7.21}$$

Since the singular interfaces of $V_1(x,t)$ and $V_2(x,t)$ are $s_1(t) = \lambda_1 t < s_2(t) = \lambda_2 t$ for $t > 0$, the solutions V_1, V_2 are not ordered for $t > 0$ as they were in (7.21) at $t = 0$. Indeed, they intersect each other for any small $t > 0$. In Section 6.2 we showed that, in general, the proper solutions of the PDE must satisfy the usual comparison, which is a PDE selection principle of proper solutions obviously related to the above ODEs principle.

According to these results, we denote by $B = \{f_\lambda,\ \lambda \in \Delta_0\}$ the set of proper TW profiles. For any $\lambda \in \Delta_0$, the profile $f_\lambda(\xi)$ is unique (the minimal one). By $\bar{B} = \{f_\lambda,\ \lambda \in \bar{\Delta}_0\}$ we denote the set of all *improper* TW profiles, where $\bar{\Delta}_0 \subseteq \Delta_0$.

Steep and flat approximations in the nonexistence case. Fix a $\lambda \notin \Delta \neq I\!R$. Then the minimal orbit $P_{\min}(f)$ does not exist and for any $\lambda \in I\!R \setminus \Delta$, the equation (7.12) does not admit a connection of the singular set $\{f = 0\}$. In terms of approximation (7.17), we distinguish two types of nonexistence of a proper TW profile motivated by applications in Section 7.11.

(i) "Steep limit" of regular approximation (Figure 7.2 (a)): $P^n(f)$ is defined for $f > \frac{2}{n}$ and for any fixed $f > 0$, $P^n(f) \to \infty$ as $n \to \infty$, so that $f^n(\xi)$ for $\xi > 0$ becomes arbitrarily steep as $n \to \infty$.

(ii) "Flat limit" of regular approximation (Figure 7.2 (b)): $P^n(f)$ is defined at least locally for $f > 0$ and $f^n(\xi) \to 0$ as $n \to \infty$.

There are simple relations between existence and two types of nonexistence for different values of parameter $\lambda \in I\!R$. In particular, we have:

Proposition 7.3 *Let $\lambda_1 \notin \Delta$ correspond to the steep limit of approximation. Then $(-\infty, \lambda_1) \cap \Delta = \emptyset$ and any $\lambda < \lambda_1$ also corresponds to the steep limit.*

Proof. Arguing by contradiction, assume that there exists a profile $f_\lambda \in B$ with a $\lambda < \lambda_1$. Then there exists a constant $a \ge 0$ such that the steep approximation satisfies $f^n_{\lambda_1}(x) \ge f_\lambda(x - a)$, so that $f^n_{\lambda_1}(x - \lambda_1 t) \ge f_\lambda(x - \lambda t - a)$ for $t > 0$ by comparison. This contradicts the assumption $\lambda < \lambda_1$. $\square$

Let us consider a typical example describing properties of proper and improper TWs. We dealt with similar equations in previous chapters.

Example 7.1: proper and improper TW branches. Consider a parabolic equation with a separate Hamilton-Jacobi operator

$$u_t = \varphi(u, u_x, u_{xx}) + H(u_x), \tag{7.22}$$

where φ and H are smooth functions, $\varphi_r(p,q,r) > 0$ for $p > 0$ and $\varphi(p,q,0) \equiv$

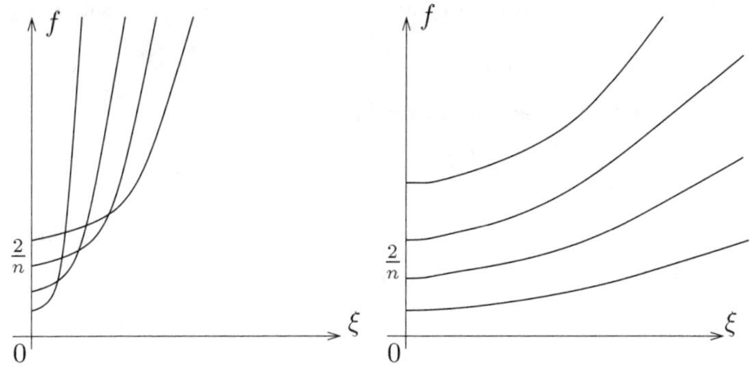

Figure 7.2 *Approximating TW profiles $f^n(\xi)$ with $f^n(0) = \frac{2}{n} \to 0$ as $n \to \infty$ in the nonexistence case: (a) steep limit, and (b) flat limit.*

0. Then the TW equation

$$\varphi(f, f', f'') + H(f') + \lambda f' = 0 \tag{7.23}$$

has linear solutions

$$f_\lambda(\xi) = (S\xi)_+. \tag{7.24}$$

Here $S = S(\lambda) > 0$ is the spatial slope of the solution on the interface satisfying the algebraic equation

$$\frac{H(S)}{S} - \lambda. \tag{7.25}$$

Due to the specific structure of the second-order PDE (7.22), this coincides with the Rankine–Hugoniot condition of the weak discontinuity propagation at $\{u = 0\}$ for the first-order Hamilton-Jacobi equation

$$u_t = H(u_x).$$

In particular, in the case of the PME, where u denotes the pressure variable and $H(S) = S^2$, we have the quasilinear equation

$$u_t = (m-1)uu_{xx} + (u_x)^2 \quad (m > 1), \tag{7.26}$$

and obtain a unique positive solution $f_\lambda(\xi) = ((-\lambda)\xi)_+$ with $S = -\lambda > 0$ for any $\lambda \in (-\infty, 0) = \Delta_0$. This branch is monotone decreasing with λ so that these are the proper TW solutions. The same conclusion is true for the general equation (7.22) with $H(u_x) = (u_x)^2$ provided that (7.23) admits linear solutions (7.24) only, or these are the only minimal ones. It is easy to see that any $\lambda \geq 0$ corresponds to the nonexistence with the flat limit of regular approximation.

In Chapter 5 we studied propagation of blow-up interfaces on the singular level $\{v = \infty\}$ for the fast diffusion equation with a superlinear reaction term

$$\boxed{v_t = (v^m)_{xx} + v^{2-m}, \quad m \in (0, 1),} \tag{7.27}$$

or the singular zero propagation for the pressure $u = \frac{m}{1-m}\, v^{m-1}$ satisfying

$$u_t = (1-m)uu_{xx} - (u_x)^2 - m, \quad \text{where } H(S) = -S^2 - m. \tag{7.28}$$

For $\lambda > 0$, the algebraic equation for the interface slope S

$$-\frac{H(S)}{S} \equiv S + \frac{m}{S} = \lambda$$

admits two continuous monotone branches of solutions. The decreasing one for $S \in (0, \sqrt{m}]$ is proper with the TW unique solution

$$f_\lambda(\xi) = \tfrac{1}{2}\,(\lambda - \sqrt{\lambda^2 - 4m})\xi \quad \text{for } \lambda \in \Delta_0 = [2\sqrt{m}, \infty). \tag{7.29}$$

The set of improper TW profiles $\bar{B}$ consists of functions

$$f_\lambda(\xi) = \tfrac{1}{2}\,(\lambda + \sqrt{\lambda^2 - 4m})\xi \quad \text{for } \lambda \in \bar{\Delta}_0 = (2\sqrt{m}, \infty). \tag{7.30}$$

No proper or improper TW profile exists for $\lambda < 2\sqrt{m}$, where we have the steep limit of approximation and $P^n(f) \to +\infty$ as $n \to \infty$. We showed that the C^2-discontinuity of the extended proper branch at $\lambda = 2\sqrt{m}$ (as on Figure 5.1) implies that the blow-up interfaces $x = s(t)$ are given by $C^{1,1}$, not C^2 functions, and $s''(t)$ becomes discontinuous when the interface reaches the minimal speed

$$D^+ s(t) = 2\sqrt{m}.$$

For more general functions H, the equation (7.23) can admit several continuous decreasing branches of proper TW profiles depending on the character of non-monotonicity of $-\frac{H(S)}{S}$; see further examples below.

Plane TWs for equations in $I\!R^N$

Consider a general parabolic equation

$$u_t = \mathbf{F}(u) \equiv F(u, \nabla u, D^2 u) \quad \text{in } I\!R^N \times I\!R_+, \tag{7.31}$$

where $D^2 u = \|u_{x_i x_j}\|$ is the symmetric Hessian matrix. We assume similar regularity hypotheses so that $F(p, q, r)$ is sufficiently smooth in $I\!R_+ \times I\!R^N \times M(N)$, where $M(N)$ is the space of symmetric real $N \times N$ matrices. The parabolicity condition is

$$F(p, q, r + s) \ge F(p, q, r) \quad \text{for any } s \in M(N), \; s \ge 0, \tag{7.32}$$

and as usual we assume the strict uniform parabolicity on compact subsets bounded away from the singularity level $\{p = 0\}$.

Consider a plane TW propagating in the x_1-direction

$$V(x, t) = f(\xi), \quad \xi = x_1 - \lambda t. \tag{7.33}$$

Then $\nabla V = (f', 0, ..., 0)$ and the only non-zero element of $D^2 V$ is the first one $V_{x_1 x_1} = f''$. Hence, f solves a second-order ODE written in the form

$$F(f, f', 0, ..., 0, f'', 0, ..., 0) + \lambda f' = 0, \quad \xi \in I\!R. \tag{7.34}$$

We then assume that it can be classified by the above 1D analysis.

Pressure, interface operators, slopes and TW-diagram

We return to the 1D problem. In general, the interface equation for propagation of the TWs on the singularity level follows from the ODE (7.12),

$$\lambda = -\lim_{f \to 0+} \frac{F(f, f', f'')}{f'}. \tag{7.35}$$

This formally gives the interface velocity in terms of the asymptotic behaviour of the TW profiles $\{f_\lambda,\ \lambda \in \Delta_0\}$ near the interface at $\xi = 0$. For general solutions of the PDE, such limits existing imposes strong assumptions on the regularity in x of proper solutions $u(x,t)$, which are difficult to prove. Actually, this requires regularity assumptions that are still unjustified for several well-known quasilinear models of reaction-diffusion-absorption type to say nothing about singular fully nonlinear equations under consideration.

Instead, we present a classification of the TW interface equations using the *algebraic* expansion of the TW profiles near the interface. Later on, this classification will be applied to proper solutions of the parabolic PDEs.

Following typical nonlinear models, we assume that there exists a smooth strictly monotone increasing function

$$Q : [0, \infty) \to [0, \infty), \quad Q(0) = 0, \quad Q'(s) > 0 \quad \text{for } s > 0,$$

such that $Q(f_\lambda(\xi))$ has a behaviour that is asymptotically linear in ξ near the singularity,

$$g(\xi) \equiv Q(f_\lambda(\xi)) = A_0(\lambda)\xi + \rho(\xi, \lambda) \quad \text{for } \xi > 0, \tag{7.36}$$

where $A_0(\lambda) \neq 0$. Here, for any $\lambda \in \Delta_0$, $\rho(\xi,\lambda) = o(\xi)$ and $\rho'_\xi(\xi,\lambda) = o(1)$ as $\xi \to 0^+$. Then, by Proposition 7.2, $A_0(\lambda)$ is monotone decreasing with λ. For linear solutions (7.24), $Q(f) \equiv f$ is the identity. We assume that, on different TW-branches, Q is independent of λ, though in some cases it has to be defined separately on each continuous branch of proper profiles from B. Thus there exists a finite limit

$$\mathbf{M}_1(f) \equiv \lim_{\xi \to 0+} \tfrac{1}{\xi}\, Q(f_\lambda(\xi)) = A_0 > 0. \tag{7.37}$$

Using the terminology from the PME theory, we call $v = Q(u)$ the *pressure variable*. By L'Hospital's rule, we have that on TW profiles,

$$\mathbf{M}_1(f) = (Q(f))'(0)$$

is a first-order differential operator. We distinguish two cases.

(I) First-order TW interfaces: *$A_0(\lambda)$ depends on $\lambda \in \Delta_0$, is strictly monotone decreasing and the inverse A_0^{-1} exists.* Then we introduce the *first-order interface operator*

$$\lambda = \mathbf{N}_1(f) \equiv A_0^{-1}\left(\lim_{\xi \to 0} \tfrac{1}{\xi}\, Q(f(\xi))\right) \tag{7.38}$$

or, in the equivalent differential form,

$$\mathbf{N}_1(f) = A_0^{-1}((Q(f))'(0)).$$

Denoting $S = \mathbf{M}_1(f)$, which is the *first-order slope* of the TW, the graph

$$\lambda = A_0^{-1}(S) \in \Delta_0 \quad (S > 0) \tag{7.39}$$

is called the (proper) *TW-diagram* on the $\{S, \lambda\}$-plane. In this case, it represents a standard first-order Rankine–Hugoniot condition of the singular propagation for nonlinear parabolic PDEs, which is similar to that in the theory of conservation laws. The TW-diagram is strictly monotone decreasing in its domain of definition and can be composed of several continuous monotone decreasing branches of proper TW solutions.

Next, we modify Example 7.1 to illustrate some other discontinuity properties of proper TW solutions of equations with analytic nonlinearities.

Example 7.2: discontinuous TW-diagram. Consider (7.22) with the operator

$$H(u_x) = -\frac{(u_x)^2 + 2}{(u_x)^2 + 20}.$$

Assume that piecewise linear solutions (7.24) are still the minimal ones. Then the algebraic equation for the first-order slope (7.25) takes the form

$$\lambda = \frac{S^2 + 2}{S(S^2 + 20)} \equiv F(S) \quad \text{for } S > 0. \tag{7.40}$$

This graph $\lambda = F(S)$ is approximately shown on Figure 7.3. The function $F(S) > 0$ is strictly monotone on the intervals $(0, 2)$, $(2, \sqrt{10})$, and $(\sqrt{10}, \infty)$, attains its local minimum at $S = 2$ and a local maximum at $S = \sqrt{10}$ with the values $F(2) = \frac{1}{8}$, $F(\sqrt{10}) = 2/5\sqrt{10}$, and $F(4) = \frac{1}{8}$. The proper TW-diagram consists of two minimal, monotone decreasing branches given by

$$\lambda = A_0^{-1}(S) \equiv F(S) \quad \text{for } S \in (0, 2] \text{ and } S \in (4, \infty).$$

Hence, $\Delta_0 = I\!R_+$. On this TW-diagram the interface speed λ changes continuously, while the corresponding interface slope S is discontinuous and has a jump $[S] = 2$, when λ crosses $\lambda_0 = \frac{1}{8}$.

(II) Second-order TW interfaces: $A_0 > 0$ *does not depend on* λ *on a fixed proper TW-branch.* Then $\rho(\cdot, \lambda)$ is monotone decreasing with λ. We set for convenience $\rho_1(\xi, \lambda) = \frac{1}{\xi}\rho(\xi, \lambda)$ and by $\rho_1^{-1}(\xi, s)$ denote the inverse relative to the variable λ, i.e., $\rho_1(\xi, \rho_1^{-1}(\xi, s)) \equiv s$. We then define the *second-order interface operator*

$$\mathbf{N}_2(f) \equiv \lim_{\xi \to 0+} \rho_1^{-1}\left(\xi, \tfrac{1}{\xi}\, Q(f_\lambda(\xi)) - A_0\right) = \lambda \tag{7.41}$$

that contains two indeterminacies in the limit.

If the higher-order term $\rho(\xi, \lambda)$ in (7.36) is a power-like function, as happens in several applications in Section 7.11, we proceed as follows (similar computations can be performed in the general case). We assume that, for a continuous function $R : I\!R \to I\!R$, $R(0) = 0$, there exists the limit

$$\mathbf{M}_2(f) \equiv \lim_{\xi \to 0+} \tfrac{1}{\xi} \left[R\left(\tfrac{1}{\xi}\, Q(f_\lambda(\xi)) - A_0\right)\right] = B_0(\lambda), \tag{7.42}$$

where $B_0(\lambda)$ is continuous monotone decreasing and the inverse B_0^{-1} is well-

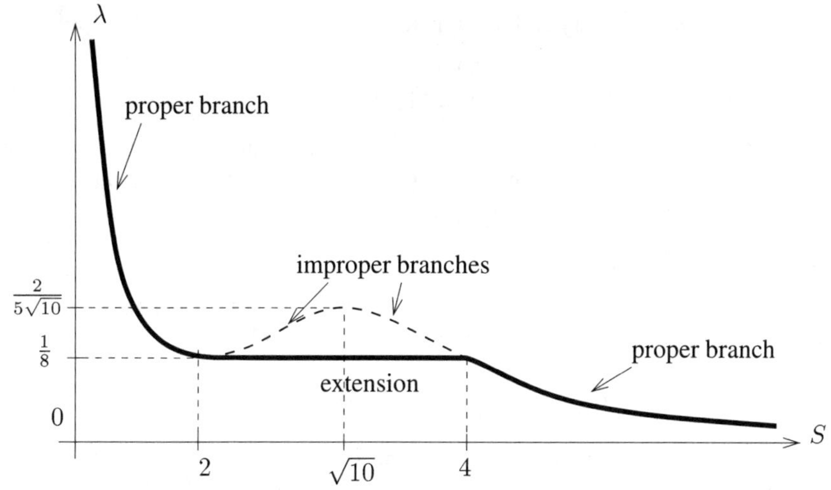

Figure 7.3 *Rankine–Hugoniot diagram* (7.40) *consisting of two proper and two improper monotone branches.*

defined. In many applications R is independent of λ. Then the second-order interface operator takes the form

$$\lambda = \mathbf{N}_2(f) \equiv B_0^{-1}\left(\lim_{\xi \to 0^+} \tfrac{1}{\xi}\left[R\left(\tfrac{1}{\xi}\, Q(f(\xi)) - A_0\right)\right]\right) \tag{7.43}$$

with two indeterminacies as $\xi \to 0^+$. Since $\xi = \frac{1}{A_0} Q(f_\lambda(\xi))(1+o(1))$, it can be written down in the algebraic form

$$\mathbf{N}_2(f) \equiv B_0^{-1}\left(\lim_{\xi \to 0^+} \frac{A_0}{Q(f_\lambda(\xi))}\left[R\left(\tfrac{1}{\xi}\, Q(f_\lambda(\xi)) - A_0\right)\right]\right).$$

In order to prescribe the corresponding differential form of these operators, we differentiate expansion (7.36) in ξ to obtain

$$(Q(f_\lambda))' - A_0 = \rho'(\xi, \lambda)$$

and assume that there exists a continuous monotone function $\tilde{R} : I\!R \to I\!R$, $\tilde{R}(0) = 0$, performing the second "straightening" near the interface, i.e.,

$$\tilde{R}(\rho'(\xi, \lambda)) = B_0(\lambda)\xi + o(\xi) \quad \text{as} \quad \xi \to 0^+.$$

Differentiating once more yields the differential interface operator

$$\mathbf{N}_2(f) = B_0^{-1}([\tilde{R}((Q(f_\lambda))' - A_0)]')(0) \quad ('= \mathrm{d}/\mathrm{d}\xi \equiv \mathrm{d}/\mathrm{d}x). \tag{7.44}$$

In this case the proper TW-diagram expresses the *second-order* Rankine–Hugoniot condition of the singular propagation and is given by a strictly monotone decreasing graph

$$\lambda = B_0^{-1}(S) \in \Delta_0 \quad (S > 0), \tag{7.45}$$

where $S = \mathbf{M}_2(f)$ is thus the corresponding *second-order slope* of the TW solution. Actually, in this case the TW interface equation is a system of two equations consisting of the stationary and the dynamical one,

$$\begin{cases} \mathbf{M}_1(f) = A_0, \\ \lambda = \mathbf{N}_2(f). \end{cases} \tag{7.46}$$

Our goal will be to prove that these interface operators are well-defined not only on the TW solutions but also on general proper solutions of the parabolic PDE.

Gradient function

Given a proper TW profile $f_\lambda(\xi)$ with a $\lambda \in \Delta_0$, for a fixed small $\varepsilon > 0$, we denote by $G(\lambda, \varepsilon)$ the *gradient function* (the G-function)

$$G(\lambda, \varepsilon) = \frac{\mathrm{d}}{\mathrm{d}\xi} Q(f_\lambda(\xi)) \,|_{Q(f_\lambda(\xi))=\varepsilon}, \tag{7.47}$$

where the derivative on the right-hand side is calculated at the first (minimal) intersection between $Q(f_\lambda(\xi))$ and the ε-level $\{Q(f_\lambda) = \varepsilon\}$. The G-function (7.47) is necessary to detect the algebraic structures of optimal gradient bounds on proper solutions.

We assume that $G(\lambda, \varepsilon)$ depends continuously upon the parameters $\lambda \in \Delta_0$ and $\varepsilon > 0$. We need the following monotonicity property of the G-function.

Lemma 7.4 *Given a fixed small $\varepsilon > 0$, $G(\lambda, \varepsilon)$ is strictly monotone decreasing with $\lambda \in \Delta_0$, where G is defined.*

Proof. Introducing in (7.13) the pressure variable $g = Q(f) > 0$, we obtain the ODE

$$\psi(g) P \frac{\mathrm{d}P}{\mathrm{d}g} = F^{-1}(Q^{-1}(g), P, -\lambda P), \quad \text{where } \psi(g) = Q'(Q^{-1}(g)). \tag{7.48}$$

Consider TW-orbits in the first quadrant of the $\{g, P\}$-plane. Fix arbitrary $\lambda_2 < \lambda_1$ from Δ_0. Let $P_1(g)$ and $P_2(g)$ be the corresponding minimal orbits. We have to prove the strict inequality

$$P_1(g) < P_2(g) \quad \text{for } g > 0. \tag{7.49}$$

By approximation, using the strict monotonicity of $F^{-1}(p, q, r)$ in $r > 0$, we have that

$$P_2^n(g) > P_1^n(g) \quad \text{for } g \approx (\tfrac{2}{n})^+.$$

Assume that $P_1^n(g)$ intersects $P_2^n(g)$ for the first time at a finite $g = \bar{g}$ so that $P_1^n(\bar{g}) = P_2^n(\bar{g}) = \bar{P} > 0$ and $P_1^n(g) < P_2^n(g)$ for $g \in (\frac{2}{n}, \bar{g})$. Subtracting equations (7.48) for the functions P_2^n and P_1^n, we obtain that $-\lambda_2 \bar{P} > -\lambda_1 \bar{P}$ at $g = \bar{g}$, so, by parabolicity,

$$\psi(\bar{g}) \bar{P} \left(\frac{\mathrm{d}P_2^n}{\mathrm{d}g} - \frac{\mathrm{d}P_1^n}{\mathrm{d}g} \right)$$
$$= F^{-1}(Q^{-1}(\bar{g}), \bar{P}, -\lambda_2 \bar{P}) - F^{-1}(Q^{-1}(\bar{g}), \bar{P}, -\lambda_1 \bar{P}) > 0.$$

This gives a contradiction since the left-hand side is nonpositive. Thus

$$P_2^n(g) > P_1^n(g) \quad \text{for } g > \tfrac{2}{n}.$$

Passing to the limit $n \to \infty$, we first obtain the inequality $\leq$ in (7.49). The strict inequality follows by the strict monotonicity of F^{-1}. $\square$

7.3 On construction of proper maximal solutions

Limit semigroups and maximal solutions

As in Section 6.2, by proper solutions of singular parabolic PDEs that do not satisfy equations in the classical sense in the singularity subset we mean those that can be constructed by regular approximations of both data and the equations. Monotone decreasing approximations lead to maximal solutions. Some particular examples of such a construction have been discussed in detail in previous chapters.

For convenience we review the main steps of the construction for a general parabolic equation (7.31) with the same regularity and singularity assumptions and properties of F specified in previous sections. The initial function u_0 is assumed to be continuous and compactly supported.

(i) Regular monotone approximation of the equation. We assume that there exists a monotone decreasing sequence $\{F_n\}$ of smooth non-singular functions satisfying

$$F_n(p, q, r) \to F(p, q, r) \quad \text{as } n \to \infty \tag{7.50}$$

uniformly on compact subsets from $I\!R_+ \times I\!R^N \times M(N)$, and that the equations

$$u_t = F_n(u, \nabla u, D^2 u) \quad \text{in } S \tag{7.51}$$

are uniformly parabolic and satisfy necessary properties for the unique global classical solvability with given smooth non-singular initial data. The corresponding semigroups $\mathcal{S}_n(t)$ for the Cauchy problem (or any standard Dirichlet initial-boundary value problem with non-singular boundary conditions) are assumed to be order-preserving and solutions $u(x,t)$ to satisfy the comparison principle.

(ii) Approximation of initial data. We fix a monotone decreasing sequence $\{u_{0n}\}$ of smooth non-singular approximations of the data satisfying

$$u_{0n} \geq \tfrac{2}{n} \quad \text{and } u_{0n} \to u_0 \quad \text{as } n \to \infty \text{ uniformly in } I\!R^N. \tag{7.52}$$

(iii) Monotone approximation: existence. We assume that, for any n, the Cauchy problem for (7.51) with initial data u_{0n} is well-posed and has a unique global classical positive solution $u_n(x,t) \geq \frac{1}{n}$ thus defining the approximating semigroup

$$u_n(x,t) = \mathcal{S}_n(t) u_{0n}(x) \quad \text{in } S. \tag{7.53}$$

Both approximations (7.50) and (7.52) are monotone decreasing with n, so, by comparison, there exists a finite nonnegative limit

$$u(x,t) \equiv \mathcal{T}(t) u_0(x) = \lim_{n \to \infty} \mathcal{S}_n(t) u_{0n}(x), \tag{7.54}$$

which is the *proper solution* of the Cauchy problem and $\mathcal{T}(t)$ is the corresponding *limit semigroup*.

(iv) Uniqueness. The proper solution does not depend on monotone approximations. Indeed, consider another suitable monotone approximation $\{\tilde{F}_k, \tilde{u}_{0k}\}$, which leads to a sequence $\{\tilde{u}_k \geq \frac{1}{k}\}$ of global smooth positive solutions. Denote $S_T = S \cap \{t < T\}$ with a bounded $T > 0$. By the monotone construction, for any $k \gg 1$, there exist $n_2(k) > n_1(k) \gg 1$ such that

$$F_{n_2} \leq \tilde{F}_k \leq F_{n_1}$$

in the corresponding bounded ranges of regularized solutions in S_T relative to arguments u, ∇u and $D^2 u$, and $u_{0n_2} \leq \tilde{u}_{0k} \leq u_{0n_1}$. By comparison, this implies that

$$u_{n_2(k)} \leq \tilde{u}_k \leq u_{n_1(k)} \quad \text{in } S_T.$$

Hence, as $k \to \infty$, $\{\tilde{u}_k\}$ converges to the same proper solution (7.54).

(v) Comparison principle for proper solutions. Obviously, the limit semigroup $\mathcal{T}(t)$ is order-preserving in the sense that

$$\mathcal{T}(t)u_0 \geq \mathcal{T}(t)\tilde{u}_0 \quad \text{if } u_0 \geq \tilde{u}_0, \tag{7.55}$$

a property, which the limit semigroup inherits from the regularizing order-preserving ones. In the proof one can use suitable ordered approximations of both proper solutions.

(vi) Proper solutions are maximal. This means that $u(x,t)$ is the maximal one among any other singular solutions understood in any "weak" or generalized sense. This follows from the construction. Assume that there exists another continuous singular "solution" $\tilde{u}(x,t)$ of (7.31). Since on any subset $S_n = S \cap \{u \geq \frac{1}{n}\}$ equation (7.31) is smooth and uniformly parabolic, using the allowed comparison of non-singular solutions of initial-boundary value problems with ordered initial data $u_{0n} \geq \tilde{u}_0$ and obviously ordered boundary data at the set $\{u = \frac{1}{n}\}$, where $u_n \geq \frac{1}{n}$, and ordered operators $F_n \geq F$, we have that

$$\tilde{u} \leq u_n \quad \text{in } S_n.$$

Passing to the limit $n \to \infty$ yields

$$\tilde{u} \leq u \quad \text{in } S.$$

The concept of proper maximal solutions is *very* weak. It assumes no regularity and any involved *a priori* bounds on the monotone sequence $\{u_n\}$ in order to pass to the limit in (7.54), which always exists and is finite. For constructing suitable non-singular regular approximations of the equation and solutions one should rely on the advanced theory of fully nonlinear uniformly parabolic equations. A list of references devoted to their classical or viscosity solutions is given in Remarks. Note that according to this construction, maximal solutions can be discontinuous in t as future examples show. If this happens at $t = 0^+$, they cannot satisfy the initial condition in any sense. For instance, this is precisely the case of complete singularities to be studied in Section 7.5.

For a number of singular operators like in the PME with absorption, the function $F(p,q,r)$ itself is not singular in $\{p > 0\}$, so that we can take approximations

$$F_n(p,q,r) \equiv F(p,q,r) \quad \text{in } \{p \geq \tfrac{2}{n}\} \times I\!\!R^N \times M(N). \tag{7.56}$$

Furthermore, in the domain $\{\frac{1}{n} \leq p \leq \frac{2}{n}\}$, a necessary regularization is performed preventing singularity formations for the approximating solutions $\{u_n\}$.

Condition (7.56) simplifies the geometric constructions since the TWs $V(x,t) = f_\lambda(\xi)$ become exact solutions of regularized equations in $\{f_\lambda \geq \frac{2}{n}\}$.

Notice that, in the present geometric analysis, we cannot involve or assume any kind of interaction between $\{u = 0\}$ and other singularities. Such interaction can create new types of singularity patterns that cannot be described by TWs with a straight line characteristic propagation.

In the 1D case, for simplicity, we often consider monotone increasing (or decreasing) initial data $u_0(x)$. Therefore, the derivative $z = (u_n)_x$ satisfies a uniformly parabolic equation with regular coefficients

$$z_t = F_r z_{xx} + F_q z_x + F_p z, \tag{7.57}$$

to which the strong MP applies. Then, choosing a strictly increasing approximation $u_{0n}(x)$, we have $(u_n)_x \geq 0$. Hence, $u(x,t)$ is increasing with x for all $t > 0$. Moreover, assuming that the strong MP applies in the positivity domain, we have the strict inequality $u_x > 0$ in the positivity set $\{u > 0\}$.

7.4 Existence: incomplete singularities in $I\!R$ and $I\!R^N$

As the first step of the geometric theory, we focus on existence of a nontrivial solution and derive first bounds on u. By definition (7.54), there exist two possibilities: either

(i) **Incomplete singularity**: the proper solution is nontrivial and $u(x,t) \not\equiv 0$ for small $t > 0$, or

(ii) **Complete singularity (nonexistence)**: $u(x,t) \equiv 0$ for arbitrarily small $t > 0$.

Notice that, in the latter case, the solution does not satisfy the initial condition and $u(x,0^+) \equiv 0$ for any singular initial function u_0. Hence, as we have seen, the limit semigroup $\mathcal{T}(t)$ is discontinuous at $t = 0$.

Incomplete singularity and existence in 1D

We begin with the 1D equation and first state some typical assumptions on the initial data, to ensure applications of the intersection comparison with TWs.

Namely, we assume that the initial data $u_0(x)$ have less growth as $x \to \infty$ than typical TW profiles $f_\lambda(x)$ for any $\lambda \in \Delta$. Assuming that the approximating semigroups are well defined on functions with such a growth condition, we have by comparison that new intersections between $u(x,t)$ and $f_\lambda(x - \lambda t)$ cannot occur for $x \gg 1$. Furthermore, without loss of generality we assume that $u_0(x)$ is strictly monotone increasing and

$$u_0(x) = 0 \quad \text{for} \;\; x \leq 0 \quad \text{and} \quad u_0(x) > 0 \quad \text{for}\; x > 0. \tag{7.58}$$

Since the existence analysis is essentially local in x, i.e., is performed in an arbitrarily small neighbourhood of the singular interface, this is not a restrictive assumption. In order to control finite propagation of the unique left-hand interface $x = s(t)$, by Δ_0 we then mean the subset of those speeds λ for which there exist singular proper TW profiles $f_\lambda(\xi)$ increasing with ξ.

Theorem 7.5 (Incomplete singularity) *Let* $\Delta_0 \neq \emptyset$. *Then under the given hypotheses on* u_0, *the Cauchy problem* (7.1), (7.3) *has a nontrivial proper solution* $u(x,t) \not\equiv 0$ *for any small* $t > 0$ *with finite propagation.*

Proof. First of all, by monotonicity of the approximating sequence $\{u_n\}$, $u(x,t) \le u_1(x,t)$ so that u is bounded above. Let us construct a lower bound. Fix a $\lambda \in \Delta_0$.

In the simple case where $u_0(x) \ge f_\lambda(x)$ in $I\!\!R$, the result is straightforward since, by approximation and comparison, in view of the monotonicity of $\{F_n\}$, we have $u(x,t) \ge f_\lambda(x - \lambda t)$ everywhere. Hence, $u(x,t) \not\equiv 0$ for $t > 0$.

Assume now that $u_0(x) > 0$ for $x > 0$ intersects $f_\lambda(x - a)$ with some $a \gg 1$ at a point $x = x_1 \gg 1$. Then we need a slight modification to the application of the strong MP that now can be applied locally, in a small neighbourhood of singularity points.

We may assume that $u_0(x_1)$ is bounded away from zero, say, $u_0(x_1) \ge 2$. We have that $f_\lambda(\xi)$ is monotone increasing with $\xi > 0$ and for simplicity we assume that the intersection is unique. For convenience, we set $\xi = x - \lambda t$ so that $f_\lambda(\xi - a)$ becomes a stationary solution of the equation for $v(\xi,t) \equiv u(\xi + \lambda t, t)$,

$$v_t = F(v, v_\xi, v_{\xi\xi}) + \lambda v_\xi. \tag{7.59}$$

Assume now that given u_0, the singularity is complete so that $v(\xi,t) \equiv 0$ for arbitrarily small $t > 0$. This means that the approximating solutions satisfy

$$v_n(\xi,t) \to 0 \quad \text{as } n \to \infty \text{ for any } \ t > 0. \tag{7.60}$$

We will show that this is impossible by the strong MP.

We compare two solutions of the regularized equation (7.59), $v_n(\xi,t)$ with $n \gg 1$ and $f_\lambda(\xi - a)$, in the domain $\{v \ge \frac{2}{n}\}$. Without loss of generality, we can use the assumption (7.56). There exists $\xi_1 \approx x_1$ such that

$$v_{0n}(\xi_1) = f_\lambda(\xi_1 - a), \quad v_{0n}(\xi) > f_\lambda(\xi - a) \quad \text{for all } \ \xi < \xi_1.$$

Fix an arbitrary small constant $\mu > 0$ assuming that $\frac{2}{n} \ll \mu < 1$ for $n \gg 1$. Choose now ξ_2 such that $f_\lambda(\xi_2 - a) = \mu$; see Figure 7.4 (a). Since equation (7.59) is uniformly parabolic for solutions v_n in any domain where $v_n \ge \mu$, we have that, by the strong MP for small $t > 0$,

$$v_n(\xi,t) > f_\lambda(\xi - a) \quad \text{on} \quad (\xi_2, \xi_3), \quad \xi_3 = \tfrac{1}{2}(\xi_2 + \xi_1), \tag{7.61}$$

provided that the same inequality is true for $\xi \le \xi_2$. In other words, in the present situation, the strong MP implies that inequality (7.61) can be violated at some small $t > 0$ provided that it has been violated at a smaller $t_n \ll 1$ in a neighbourhood $\xi \in (a, \xi_2)$ of the TW interface, i.e., close to the singular level. Choosing the minimal value of such a moment denoted again by t_n, from the initial configuration on Figure 7.4 (a) we necessarily arrive at a "tangent" configuration as shown on Figure 7.4 (b). Notice that this occurs at some small $t_n \to 0^+$ as $n \to \infty$ (note that $\mu \gg \frac{2}{n}$ is arbitrary and we can choose $\mu = \mu_n \to 0^+$). Obviously, this configuration is impossible by the local comparison of solutions involved (the strong MP for uniformly parabolic equations does not accept such tangency points between solutions). In other words, we have that, if a single nontrivial singular solution V exists, the local comparison and the strong MP ensure that, for the approximations

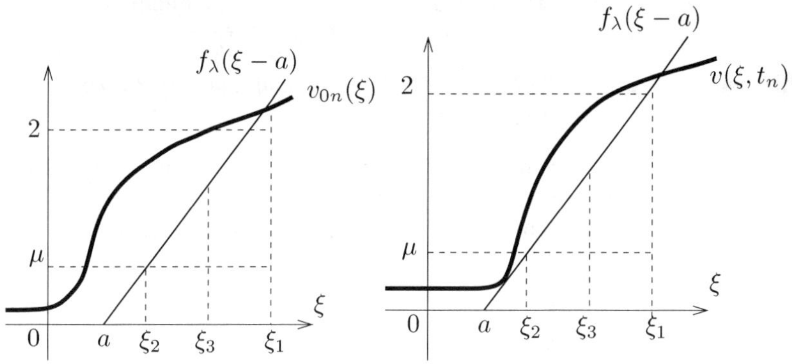

(a) initial configuration (b) impossible evolution

Figure 7.4 *Illustration of the proof of Theorem 7.5: (a) at $t = 0$, the mutual location of initial data $v_{0n}(\xi)$ and the singular TW $V = f_\lambda(\xi - a)$, and (b) at $t = t_n \ll 1$, we then have contradiction with the strong MP since $v(\xi, t)$ cannot touch the TW in the domain of uniform parabolicity.*

of arbitrary solutions, the singular limit (7.60) is not possible. Otherwise, it is possible if another singularity phenomenon occurs for some $u > 0$ (nonexistent by the present assumptions on the uniqueness of the singular level $\{u = 0\}$).

Finite propagation follows by the standard local comparison with the TW solutions. □

Thus, under the assumption $\Delta_0 \neq \emptyset$, the singular propagation is finite and moreover, the singular interface moves continuously. This is proved in exactly the same way by local comparison with existing continuous singular TWs. Indeed, any discontinuous "jumping" of the interface and hence the discontinuity of the proper solution would contradict local strong MP as explained on Figure 7.4. This means that the solution must change continuously in a neighbourhood of singularity points. Hence, the proper solution $u(x, t) \not\equiv 0$ is a continuous function, and, by the assumed interior regularity results for uniformly parabolic equations, u is sufficiently smooth on any subset where the solution is uniformly bounded away from zero. This yields the following conclusion.

Corollary 7.6 *Under the above assumptions, if $\Delta_0 \neq \emptyset$, then the limit semigroup $\mathcal{T}(t)$ of maximal solutions is continuous at $t = 0$, i.e.,*

$$\mathcal{T}(0^+)u_0 = u_0.$$

Consider now a special case $\Delta_0 = \emptyset$ but $\Delta \neq \emptyset$. Then, in the proof of existence, we compare u with the strictly positive TW $f_\lambda(\xi - a)$. One can check that this implies no novelties in the local MP analysis illustrated on Figure 7.4 (note that there we can deal with strictly positive approximations f_λ^n). Alternatively, we can use comparison with approximating sequence of non-maximal TWs $\{f_\lambda^k(\xi)\}$ having finite interfaces to be introduced in the next section. Summing up, we obtain the following optimal existence result in the 1D geometry.

Theorem 7.7 *Let $\Delta \neq \emptyset$. Then under given assumptions, the Cauchy problem* (7.1), (7.3) *has a proper solution $u(x,t) \not\equiv 0$ for small $t > 0$.*

Note that, in this case, we cannot guarantee that the singular interface is finite. Nevertheless, the same standard comparison from above with strictly positive TWs shows that even in the special case $\Delta \neq \emptyset$, $\Delta_0 = \emptyset$, the limit semigroup $\mathcal{T}(t)$ is continuous at $t = 0$.

Existence for equations in $\mathbb{R}^N$

We now consider singular parabolic PDEs in $\mathbb{R}^N$. An easy way to extend existence results for nontrivial maximal solutions $u(\cdot,t) \not\equiv 0$ of equations like (7.7) in $\mathbb{R}^N$ is based on comparison with radial TW subsolutions satisfying the 1D equations.

Analysis in radial geometry: comparison with TW subsolutions. Let $u = u(r,t) \geq 0$, $r = |x| \geq 0$, be a maximal radial solution of equation (7.7)

$$u_t = F(u, u_r, \Delta_r u) \quad \text{in } S = \mathbb{R}_+ \times \mathbb{R}_+, \tag{7.62}$$

where $\Delta_r u = u_{rr} + \frac{N-1}{r} u_r$, with a compactly supported continuous monotone decreasing initial function $u_0(r)$. Then, by the MP,

$$u_r \leq 0 \quad \text{in } S \cap \{u > 0\}. \tag{7.63}$$

Without loss of generality, let $u_0(r) > 0$ on $[0,2]$. It follows from (7.63) that

$$\Delta_r u \geq u_{rr} + (N-1)u_r \quad \text{in } \{r \geq 1\} \cap \{u > 0\}. \tag{7.64}$$

Consider monotone decreasing TW solutions

$$V(r,t) = f_\lambda(\xi), \quad \xi = r - \lambda t,$$

of the parabolic equation

$$v_t = F(v, v_r, v_{rr} + (N-1)v_r) \quad \text{in } \{r > 1\} \times \mathbb{R}_+, \tag{7.65}$$

where f_λ solves the ODE

$$F(f, f', f'' + (N-1)f') + \lambda f' = 0. \tag{7.66}$$

The parabolicity condition and (7.64) imply that, in $S \cap \{r > 1\}$, V is a subsolution of equation (7.62). Comparison with this TW is performed exactly as in the 1D case and we obtain a similar existence result for radial solutions. Recall that we use the usual comparison only, no intersection analysis is involved, and the comparison is essentially local near the singular interface, where the TW is well defined and is a subsolution.

Existence in $\mathbb{R}^N$. We next easily extend the existence result to arbitrary nonradial solutions.

Theorem 7.8 (Incomplete singularity) *Let* $\Delta_0 \neq \emptyset$ *for the ODE* (7.66). *Then the Cauchy problem for* (7.7) *with compactly supported initial data* $u_0(x) \not\equiv 0$ *admits a maximal solution* $u(x,t) \not\equiv 0$ *for small* $t > 0$.

Proof. As we have checked, the hypotheses of the theorem guarantee the existence of nontrivial proper radial solutions. In order to treat general data $u_0(x)$, we choose a small radial initial function $\tilde{u}_0(r) \le u_0(x)$ in $I\!R^N$ satisfying the necessary properties, and hence $u(x,t) \ge \tilde{u}(r,t) \not\equiv 0$ for sufficiently small $t > 0$ by comparison of maximal solutions. □

A similar comparison applies to general equations (7.31). For instance, assume that it admits a parabolic radial "sub-equation"

$$v_t = \underline{F}(v, v_r, \Delta_r v) \tag{7.67}$$

in the sense that, for sufficiently smooth non-singular radial decreasing functions $v = v(r)$, there holds

$$F(v, \nabla v, D^2 v) \ge \underline{F}(v, v_r, \Delta_r v). \tag{7.68}$$

Assuming that equation (7.67) satisfies necessary assumptions for proper radial solutions, we next introduce the parabolic equation

$$v_t = \underline{F}(v, v_r, v_{rr} + (N-1)v_r) \quad \text{in } \{r > 1\} \times I\!R_+ \tag{7.69}$$

for the TWs that are radial subsolutions of both the radial equation (7.67) and the general equation (7.31), and we obtain the following result.

Theorem 7.9 *Let* $\Delta_0 \neq \emptyset$ *for the ODE* (7.66) *with* $F = \underline{F}$ *as in* (7.69) *satisfying* (7.68). *Then the Cauchy problem for* (7.31) *with compactly supported* $u_0 \not\equiv 0$ *has a maximal solution* $u(x,t) \not\equiv 0$ *for small* $t > 0$.

7.5 Complete singularities in $I\!R$ and $I\!R^N$. Infinite propagation and pathological equations

In the nonexistence analysis in 1D (i.e., the discontinuity of the limit semigroup at some $t = T > 0$) we will need to use intersection comparison techniques. Therefore, let us recall a basic principle of applying the Sturmian argument to proper solutions of general nonlinear parabolic PDEs. Given a proper solution $u(x,t)$ and a set B of proper TW profiles, we use the first Sturm Theorem and count the number of intersection between them. Then, for any $V \in B$,

$$\text{Int}(t, u, V) \;\; \text{does not increase with time.} \tag{7.70}$$

For singular proper maximal solutions, this is proved by approximation. Namely, given two smooth positive solutions $u_1, u_2 \ge \delta > 0$ of equation (7.1) with smooth right-hand side, the difference $w = u_1 - u_2$ satisfies a linear parabolic equation. The number of zeros of the difference $w(\cdot, t)$ is then assumed to satisfy (7.70). By monotone approximations, this is true for maximal solutions $u(x,t) = \lim u_n(x,t)$ since, by construction, it holds for any pair of sufficiently smooth regularized solutions $u_{1n}(x,t)$ and $u_{2n}(x,t)$. Notice also that, since, in any domain $\{u_1, u_2 \ge \delta > 0\}$, the strong MP for uniformly parabolic equations applies to the equation for the difference $w = u_1 - u_2$, for $t > 0$, intersections are points and form continuous intersection curves in the $\{x, t\}$-plane. New intersection curves cannot appear for any $t_0 > 0$, so that each one is originated at the initial moment $t = 0$.

Complete singularity (nonexistence) in 1D

Complete initial singularity. We begin with a 1D analysis. In the case $\Delta = \emptyset$ for the ODE (7.12) (plus an extra condition) the TWs cannot connect the singular level with regular points. Our goal is then to prove that the singularity at $\{u = 0\}$ is *complete* in the PDE sense.

As usual in comparison analysis with the TWs, we consider the Cauchy problem with monotone increasing initial data $u_0(x)$ satisfying (7.58) and exhibiting slower growth as $x \to \infty$ than the steep TW profiles $f_\lambda(x)$ with $\lambda \gg 1$. This guarantees that new intersections between solutions cannot appear for $x \gg 1$.

Theorem 7.10 (Complete singularity) *Let*

$$\Delta = \emptyset$$

for the corresponding ODE (7.6) *and any* $\lambda \in I\!R$ *be in the steep limit of regular approximation. Then the Cauchy problem* (7.1), (7.3) *has the maximal solution* $u(x, t) \equiv 0$ *for arbitrarily small* $t > 0$.

In particular, this means that the maximal solutions are entirely singular and cannot satisfy the initial condition in the usual sense. One may refer to this case as to the *nonexistence* of a nontrivial local solution of the Cauchy problem.

Proof. Fix an arbitrary large $\lambda \gg 1$. Under the above assumptions, the approximating TW sequence $\{f_\lambda^n(\xi)\}$ of profiles satisfying (7.17) does not have a finite limit as $n \to \infty$. For $n \gg 1$ and any fixed small $f > 0$, we have $P^n(f) \to +\infty$ as $n \to \infty$, as Figure 7.2 (a) shows. This means that the TW profiles become arbitrarily steep on any finite level for $n \gg 1$. Therefore, given any initial data satisfying (7.58), choosing the approximation $u_{0n}(x) \equiv \frac{2}{n}$ for $x \le 0$, we obtain that, for any $n \gg 1$, $u_{0n}(x) \le f_\lambda^n(x + 1)$ in an arbitrarily large neighbourhood of $x = 0$, as shown on Figure 7.5 (a). Then $u_n(x, t) \le f_n(x - \lambda t + 1)$ for $t > 0$ there by comparison; Figure 7.5 (b). Since $\lambda \gg 1$ is arbitrary, this implies that, for arbitrarily small $t > 0$, any positive level of $u(x, t)$ propagates with infinite speed and hence $u(x, t) \equiv 0$ for $t > 0$. □

Complete evolution singularity. We next consider a finite-time evolution formation of complete singularity that is not posed initially at $t = 0$. Such evolution singularity phenomena are important for understanding blow-up and extinction singularities for general PDEs.

We temporarily change usual properties of the initial function $u_0(x)$ (cf. (7.58)) and assume that

$$u_0(x) \ge 1 \quad \text{in } I\!R \tag{7.71}$$

and that u_0 has the inverse bell-shaped form. We suppose that the solution $u(x, t)$ is non-singular and sufficiently smooth in the interval $t \in (0, T)$ and also has a

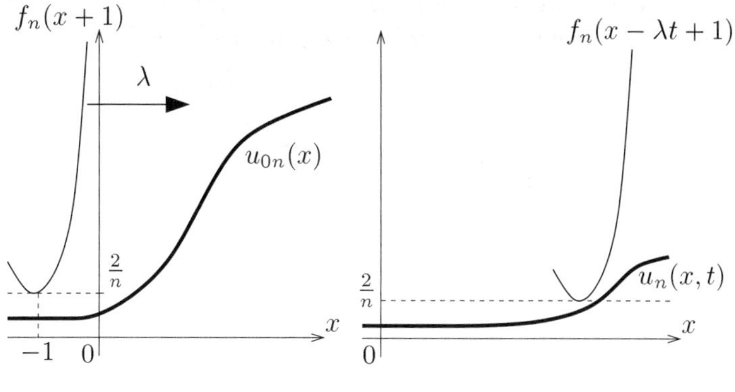

(a) initial location (b) evolution

Figure 7.5 *(a) Illustration of the proof of Theorem* 7.10*:* $t = 0$*, (b) for* $t > 0$*, by comparison, the steep TW* $f_n(x - \lambda t + 1)$ *moving to the right with speed* $\lambda \gg 1$ *sweeps away any solution below.*

bell-shaped form by the MP for the derivative u_x. Here T is the first moment of time when the solution touches the zero-level at a finite point, i.e.,

$$u(0,T) = 0 \quad \text{and} \quad u(x,t) > 0 \;\text{ for any }\; t \in [0,T). \tag{7.72}$$

We then arrive at a similar discontinuity (nonexistence) result where we need to apply the intersection comparison approach.

Proposition 7.11 (Complete singularity) *Let* (7.71) *and* (7.72) *hold. Let* $\Delta = \emptyset$ *for the ODE* (7.6) *and any* $\lambda \in I\!R$ *be in the steep limit of regular approximation. Then the proper maximal continuation of the solution for* $t > T$ *is trivial:* $u(x,t) \equiv 0$.

Proof. It is based on a slightly more refined intersection comparison with the family of steep TW solutions $\{f_\lambda^n\}$ and is essentially the same as in the TWs construction with $|\lambda| \gg 1$ in Section 4.4 (Theorem 4.7), where this eventual B-concavity property is applied to quasilinear heat equations

$$u_t = (\varphi(u))_{xx} + \psi(u). \tag{7.73}$$

□

Nonexistence in $I\!R^N$

The proof of nonexistence in 1D is based on the usual comparison with "steep" sequences of approximating TWs $\{f_\lambda^n\}$ and no intersection comparison techniques have been used. Therefore, the results apply to any N-dimensional parabolic equation satisfying the MP and the comparison principle.

Theorem 7.12 (Complete singularity) *Let* $u(x,t)$ *be the maximal solution of the Cauchy problem for* (7.31) *with bounded continuous compactly supported* u_0*. Let* *the subset* Δ *of the singular plane TWs* (7.33) *satisfying the ODE* (7.34) *be empty and any* $\lambda \in I\!R$ *be in the steep limit of approximation. Then* $u(x,t) \equiv 0$ *for any* $t > 0$.

It follows from the 1D proof that the one-directional nonexistence hypothesis (formulated for the x_1-direction only) is sufficient to guarantee the trivial maximal solution $u \equiv 0$ provided that the initial interface of u_0 is bounded in this direction.

Infinite propagation and pathological PDEs

We now discuss a special case of parabolic PDEs satisfying

$$\Delta_0 = \emptyset, \quad \text{but} \quad \Delta \neq \emptyset. \tag{7.74}$$

This means that proper singular TW solutions exist (so that the proper solution $u(x,t)$ is nontrivial by Theorem 7.7) but all of them exhibit infinite propagation: for any $\lambda \in \Delta$, $f_\lambda(\xi) > 0$ and $f_\lambda(\xi) \to 0^+$ as $\xi \to -\infty$. Assuming that $(\mathrm{d}f_\lambda/\mathrm{d}\xi)(\xi) \to 0$ as $\xi \to -\infty$, we have that positive minimal orbits satisfy

$$P_{\min}(0) = 0. \tag{7.75}$$

1. Infinite propagation. In general, this implies infinite singular propagation for the parabolic PDE. We prove the following result.

Proposition 7.13 *Let* $\Delta_0 = \emptyset$ *and* $(-\infty, -b) \subset \Delta$ *for some constant* $b \geq 0$. *Then any maximal solution of* (7.1) *with monotone* $u_0 \not\equiv 0$ *is strictly positive:* $u(x,t) > 0$ *for* $t > 0$.

Proof. The result is based on a suitable approximation of the minimal orbit $P_{\min}(f)$ from *above* and uses a natural monotonicity for first-order ODEs. It follows from the general structure of the phase-plane for equation (7.13) that besides a positive approximation of $P = P_{\min}(f)$ by $\{P^n(f)\}$ from below, such minimal orbit admits an approximation from above by orbits generating non-maximal TW profiles with finite interfaces.

We thus consider a sequence $\{\lambda_k\} \subset \Delta$, $\lambda_k \to -\infty$, such that, for any $k \gg 1$, each positive profile $f^k \equiv f_{\lambda_k}$ can be approached from *below* by a sequence $\{f_j^k(\xi) \geq 0\}$ of functions vanishing at finite points denoted by $\xi_{kj} \to -\infty$ as $j \to \infty$. On Figure 7.1, this would correspond to approximation of f_λ from below (not from above as shown). Then, by (7.75), we obtain

$$f_j^k(\xi + \xi_{kj}) \to 0 \quad \text{as } j \to \infty, \tag{7.76}$$

uniformly on compact subsets. Let us prove infinite propagation. Given initial data $u_0(x)$ satisfying (7.58), i.e., with $s(0) = 0$, we choose $k \gg 1$ such that, by (7.76) for any $n \gg 1$,

$$u_{0n}(x) \geq f_j^k(x + \xi_{kj} - 1).$$

Therefore, passing to the limit $n \to \infty$, by the usual comparison, the maximal solution u satisfies, for any small $t > 0$,

$$u(x,t) \geq f_j^k(x - \lambda_k t + \xi_{kj} - 1),$$

so that the interface is such that $s(t) \le \lambda_k t + 1$. Passing now to the limit $k \to \infty$, since $\lambda_k \to -\infty$, we deduce that $s(t) = -\infty$ for arbitrarily small $t > 0$. $\square$

The classical example from the 1960s [209], [287] of a degenerate parabolic equation with infinite propagation is the filtration equation

$$u_t = (\varphi(u))_{xx}, \tag{7.77}$$

where $\varphi(u) > 0$ for $u > 0$ satisfies $\varphi(0) = 0$ but

$$\int_0 \frac{\varphi'(s)}{s}\, \mathrm{d}s = \infty.$$

The positivity result is true for any N-dimensional equation (7.31) provided that the ODE (7.34) for the plain TWs satisfies the hypotheses of Proposition 7.13.

2. Pathological class of PDEs. Let us now introduce the so-called "pathological" class of PDEs in the case (7.74). For blow-up solutions of the quasilinear heat equations (7.73), the pathological phenomenon was detected in Section 4.3. Now we detect a general class of such pathological PDEs.

The corresponding pathological behaviour of solutions takes place if a finite-time singularity can occur in the problem for an initially strictly positive solution with $u_0(x) > 0$. The simplest type of such singularities is exhibited by solutions $u = V(t)$ that are flat in x and satisfy the ODE

$$V' = F(V, 0, 0) \quad \text{for } t > 0, \quad V(0) = c > 0. \tag{7.78}$$

Assume that, for any small $c > 0$, the unique solution $V(t)$ becomes singular in finite time $T_c < \infty$: $V(t) > 0$ on $(0, T_c)$ and $V(T_c^-) = 0$. This is guaranteed by Osgood's criterion: $F(V, 0, 0) < 0$ for $V > 0$ and

$$\int_0 \frac{\mathrm{d}s}{F(s, 0, 0)} < \infty.$$

On the other hand, consider a singular behaviour for nontrivial spatially inhomogeneous initial data $u_0(x) > 0$ satisfying

$$u_0(x) \ge f_\lambda(x - a) \quad \text{for all } x \in I\!R \tag{7.79}$$

for some TW speed $\lambda \in \Delta$ and $a \gg 1$ (we assume that $f_\lambda > 0$ in $I\!R$). Then

$$u(x,t) \ge f_\lambda(x - \lambda t - a) > 0 \quad \text{in } S \tag{7.80}$$

by comparison of positive sufficiently smooth solutions.

This is what we call the *pathological* property of the PDEs satisfying (7.74): *while all flat solutions* (7.78) *become singular in finite time, the spatially inhomogeneous* (*e.g. inverse bell-shaped*) *solutions with initial data satisfying* (7.79) *never create a singularity.*

The proof of the existence of pathological quasilinear heat equations (7.73) with blow-up needed a special construction in Section 4.3. The result is true for equation (7.31) in $I\!R^N$ if the ODE (7.34) exhibits the pathological behaviour of plain TWs.

7.6 Further geometric notions: B-concavity, sign-invariants, B-number

We return to the general 1D parabolic equation (7.1) in the case of the existence of continuous proper solutions, and will develop the corresponding geometric theory, where intersection comparison techniques play a crucial role. In doing so, we will use a number of geometric notions developed in Section 7.2 via TWs and will use others introduced in the previous chapters for particular classes of PDEs.

Completeness of proper B-bundles

In our analysis, the set B of TWs is assumed to be complete in the sense of existence and uniqueness of tangent solutions (see below). As it was shown in Chapters 2 and 3 for a number of quasilinear PDEs, such a set generates invariant properties of B-concavity/convexity for general solutions in appropriate functional classes. Then the B-concavity of $u(x,t_0)$ means that it is concave with respect to the functions $\{V(x,t_0)\}$. We add to the set B the flat solution $V_*(t)$ satisfying the ODE (7.78). It is invariant under translation $t \mapsto T_0 + t$ and can be formally treated as a special limit of the TW-profiles $f_\lambda(\xi - a)$ with $\lambda = \infty$. Denote such an extended TW-set by

$$B_* = B \cup \{V_*\}.$$

Consider the question of *completeness*. Fix an increasing with x solution $u(x,t) \geq 0$ in S with $u_x \geq 0$. Recall that the completeness of B_* means that, for any $P_0 = (x_0, t_0) \in \bar{S}$ such that $u(x_0,t_0) > 0$, there exists a unique tangent solution $V \in B_*$ to u at P_0, such that

$$V(x_0,t_0) = u(x_0,t_0) = \mu, \quad V_x(x_0,t_0) = u_x(x_0,t_0) = \nu, \tag{7.81}$$

where $\mu > 0$, $\nu \geq 0$ can be arbitrary, and $V(x,t)$ is defined for all $t \in [0, t_0]$.

This "tangential" system (7.81) is convenient to state in terms of the pressure variable $v = Q(u)$ on the corresponding proper branch of the TW-diagram. The completeness condition can then be expressed in terms of the gradient function.

Proposition 7.14 *Let the G-function $G(\lambda, \varepsilon)$ given in (7.47) be continuous and for any fixed $\varepsilon > 0$,*

$$\mathrm{Im}\,\{G(\lambda,\varepsilon),\ \lambda \in \Delta_0\} = (0,\infty).$$

Then the set B_* is complete.

Proof. The unique solvability of the tangential system (7.81) for $v = Q(u)$ reduces to the existence and uniqueness of $\lambda \in \Delta_0$ such that a unique TW-profile $f_\lambda(x)$ satisfies, for some $x_0 > 0$,

$$Q(f_\lambda)(x_0) = Q(\mu), \quad (Q(f_\lambda))'(x_0) \equiv G(\lambda, Q(\mu)) = Q'(\mu)\nu. \tag{7.82}$$

Completeness is a straightforward consequence of the continuity and the monotonicity property of G established in Proposition 7.2. If $\nu \neq 0$, then $f_\lambda \in B$. For $\nu = 0$, the unique tangent solution is given by V_*. $\square$

We say that the TW set B is complete if (7.82) admits a unique solution for any $\mu > 0$ and $\nu > 0$. We then identify the proper set $B = \{V\}$ with the proper

B-bundle (the *TW-bundle*) $B = \{f_\lambda(\xi),\ \lambda \in \Delta_0\}$ of the proper TW profiles with their interfaces fixed at the origin $\xi = 0$. The completeness of B implies that, for any $\xi_0 > 0$ and $\mu > 0$, there exists a unique $f_\lambda(\xi) \in B$ such that

$$Q(f_\lambda(\xi_0)) = Q(\mu), \tag{7.83}$$

i.e., through any point (ξ_0, μ) on the $\{\xi, f\}$-plane there passes a unique curve $f_\lambda(\xi)$ from the TW-bundle. In addition, the complete set $\{f_\lambda\}$ is ordered (strictly monotone decreasing with λ) for small $\xi > 0$ and is continuous relative to λ.

B-concavity

Without loss of generality we assume that $u(x, t_0) \in C^1$ in the positivity domain for any $t_0 \geq 0$. We recall the property introduced in Section 2.6 for a class of filtration type equations.

Definition 7.2 The solution profile $u(x, t_0)$ is said to be B-concave (resp. B-convex) if, for any $x_0 \in \{x :\ u(x, t_0) > 0\}$, there holds

$$u(x, t_0) \leq V(x, t_0) \quad (\text{ resp. } u(x, t_0) \geq V(x, t_0)) \quad \text{in } I\!R, \tag{7.84}$$

where $V(x, t_0)$ is the tangent profile at the point (x_0, t_0).

Assume that the initial function $u_0(x)$ is chosen so that new intersections between the proper solution and the functions $V \in B$ cannot appear for $x \gg 1$. This is necessary for application of intersection comparison with TWs. Then the property of B-concavity (convexity) is preserved in time.

Proposition 7.15 *Let B be complete. Then*

$$u_0(\cdot) \text{ is } B\text{-concave} \implies u(\cdot, t) \text{ is } B\text{-concave for } t > 0. \tag{7.85}$$

Proof. It is established in Section 2.6 that given a set of particular solutions B, the B-concavity (convexity) is preserved in time under assumptions of the completeness, continuity and monotonicity of B relative to the parameters. The last two properties are obviously true for the set of TW solutions. The proof is completed as usual. Namely, any B-concave (or B-convex) solution $u(x, t)$ satisfies

$$\text{Int}(t, V) \leq 2 \quad \text{for } t > 0 \text{ and for any } V \in B. \tag{7.86}$$

Therefore, given a B-concave (convex) initial data u_0, we conclude that, if $u(x, t_0)$ with a $t_0 \in (0, T)$ is not B-concave (convex) in the sense of (7.84), then this evidently contradicts (7.86). □

Sign-invariants

It follows from (7.84) that, for solutions $u(x, t) \in C^2$ in $\{u > 0\}$ for $t \geq 0$, the B-concavity is equivalent to the differential inequality

$$u_{xx} \leq V_{xx} \quad \text{in } S \cap \{u > 0\}. \tag{7.87}$$

The B-convexity gives the opposite inequality sign. Calculating the derivative V_{xx} from the corresponding ODE for the TW profiles with the tangency conditions

$u = V, u_x = V_x$,

$$F(u, u_x, V_{xx}) + \lambda u_x = 0, \tag{7.88}$$

where $\lambda = \lambda(u, u_x)$ is uniquely determined from the "tangential" system (7.81), we obtain $V_{xx} = F^{-1}(u, u_x, -\lambda u_x)$, and so (7.87) implies

$$u_{xx} \le F^{-1}(u, u_x, -\lambda(u, u_x)u_x) \quad \text{in } S \cap \{u > 0\}. \tag{7.89}$$

This means that the operator

$$H_B(u) \equiv u_{xx} - F^{-1}(u, u_x, -\lambda(u, u_x)u_x) \le 0 \tag{7.90}$$

preserves the negative sign on proper parabolic orbits. It also preserves the positive sign due to the opposite property of the B-convexity. Therefore, H_B is a *sign-invariant* of the parabolic PDE (7.1) generated by the complete set B of its TW solutions. One can see that, by the construction of the sign-invariant,

$$H_B(V) \equiv 0 \quad \text{for any } V \in B \tag{7.91}$$

in the positivity domain. B-concavity/convexity properties exhibit the following connection with the MP.

Proposition 7.16 *The function $J(x,t) = H_B(u(x,t))$ satisfies a quasilinear homogeneous parabolic equation*

$$J_t = \mathbf{A}(J) \equiv aJ_{xx} + bJ_x + cJ \quad \textit{in } S \cap \{u > 0\}, \tag{7.92}$$

where the coefficients $a > 0$, b and c depend on u and J and are sufficiently smooth.

Proof. Differentiating $J = H_B(u) \equiv \Phi(u, u_x, u_{xx})$ yields

$$J_t = \Phi_p u_t + \Phi_q u_{xt} + \Phi_r u_{xxt}.$$

Calculating the derivatives u_t, u_{xt} and u_{xxt} from $u_t = F(u, u_x, u_{xx})$ and using a standard linearization procedure, we conclude that J solves a parabolic equation

$$J_t = \mathbf{A}(J) + W, \tag{7.93}$$

where $W = W(u, u_x)$ is the corresponding continuous non-homogeneous term independent of J. Using (7.91), we obtain that, for any $V \in B$,

$$W(V, V_x) \equiv 0.$$

Let us show that $W \equiv 0$ by completeness of B by a formal application of the MP. Indeed, assuming that $W(\mu, \nu) > 0$ for some $\mu > 0, \nu > 0$, we choose a TW $V(x,t)$ such that $V(x_0, 0) = \mu$, $V_x(x_0, 0) = \nu$ and, by the construction, $J(\cdot, 0) \equiv 0$. Then $J(\cdot, t) \equiv 0$ for $t > 0$ by (7.91) and this contradicts equation (7.93) since, by regularity, $J \equiv H_B(V) > 0$ for all $t > 0$ small and $x \approx x_0$. □

By the strong MP, any solution $J(x,t) \not\equiv 0$ (i.e., $u \not\equiv V \in B$ by (7.91)) of the quasilinear parabolic equation (7.92) has isolated zeros in the positivity domain for any $t > 0$. Therefore, the results on the B-concavity/convexity follow from the MP, and $J(x, 0) \ge 0$ (≤ 0) implies that $J(x,t) \ge 0$ (≤ 0) for $t > 0$. The analysis near the singular interfaces needs some extra properties obtained by the intersection comparison with the singular proper solutions $V \in B$; see further comments below.

Example 7.3 Consider a general filtration equation (cf. Section 2.2) in the pressure representation

$$u_t = F(u)u_{xx} + (u_x)^2, \quad F(u) > 0 \;\; \text{for} \;\; u > 0. \tag{7.94}$$

Under some natural assumptions on F, a complete proper TW-bundle is given by the liner profiles

$$f_\lambda(\xi) = (-\lambda)\xi, \quad \text{where } \lambda \in (-\infty, 0) = \Delta_0,$$

and hence $Q(u) \equiv u$. The tangential system (7.81) takes the form $(-\lambda)\xi = u$, $-\lambda = u_x$, so that $\lambda(u, u_x) = -u_x$. Equation (7.88) reads

$$F(u)V_{xx} + (u_x)^2 - u_x u_x = 0,$$

and obviously we obtain $V_{xx} = 0$. Hence, the sign-invariant is

$$H_B(u) = u_{xx}.$$

A complete set B of solutions that are piecewise linear in x generates the standard notions of concavity/convexity and both ones are invariant under the parabolic flow. This is the result of Section 2.2.

B-number

Consider more general proper solutions that are not B-concave or B-convex so that $H_B(u(x,t))$ changes sign in the positivity domain. Using sign-invariant H_B, we introduce a new geometric characteristic of proper solutions $u(x,t)$. We assume that $u(x, t_0) \in C^2$ in the positivity domain for any $t_0 \geq 0$.

Definition 7.3 For the given complete set B, the B-number $Z_B(t_0)$ of the proper profile $u(x, t_0)$ is the number of sign changes of $H_B(u(x, t_0))$ in the positivity domain.

At any point $P_0 = (x_0, t_0) \in \{u > 0\}$, $t_0 > 0$, where $H_B(u(x_0, t_0)) > 0$ (resp. < 0), the tangent solution $V \in B$ satisfies the strict inequality in (7.87)

$$u_{xx}(x_0, t_0) < V_{xx}(x_0, t_0) \quad (> V_{xx}(x_0, t_0)). \tag{7.95}$$

This is followed by the strong MP applied to the linear parabolic equation (7.92) for the sign-invariant $J = H_B(u)$. On the other hand, if $H_B(u(x_0, t_0)) = 0$ and $H_B(u(x, t_0))$ changes sign at $x = x_0$, this point is an *inflection* (intersection) point between the profile $u(x, t_0)$ and the tangent solution $V(x, t_0)$. Therefore, $Z_B(t)$ is equal to the total number of the inflection points that the profile $u(x, t_0)$ has with all tangent TW profiles $V(x, t_0)$ at any $(x_0, t_0) \in \{u > 0\}$.

We next formulate the main property of the B-number.

Theorem 7.17 *The finite B-number $Z_B(u(\cdot, t))$ of a proper solution $u(x,t)$ does not increase.*

Proof. It is not difficult to see that the appearance of a new zero of $H_B(u)$ at a point P_0 would mean that a new intersection between $u(x,t)$ and the tangent solution $V(x, t)$ at P_0 occurs for $t > t_0$ contradicting Sturm's Theorem. The proof follows the lines of the detailed analysis in the proof of Theorem 2.2 in Section 2.2. The same is true at the singular interfaces and the proof is similar to that in Lemma 2.10. □

For convenience, we state the last result separately.

Lemma 7.18 *If $Z_B(0) < \infty$, no new sign changes of the sign-invariant $H_B(u(\cdot, t))$ are generated at singular interfaces.*

The B-number gives a useful intersection bound.

Theorem 7.19 *Let B be complete and $Z_B(0) < \infty$. Then, for any $t > 0$,*

$$\mathrm{Int}(t, V) \leq Z_B(t) + 2 \quad \textit{for all } V \in B. \tag{7.96}$$

Proof. This is a typical conclusion from Chapters 2 and 3. We argue by contradiction. Assume that there exists a TW solution $V = f(\xi - a, \lambda) \in B$ such that $\mathrm{Int}(t_0, V) \geq Z_B(t_0) + 3$. Let $x_1 < x_2 < x_3$ be any three neighbouring points of intersections. Using the continuous evolution of V upon the parameters a and λ, we conclude that there exists at least one inflection point $\tilde{x} \in (x_1, x_3)$ between the profile $u(x, t_0)$ and the corresponding tangent profile $\tilde{V}(x, t_0)$. Therefore, there exist more than $Z_B(t_0)$ inflection points on the profile $u(x, t_0)$ that are sign changes of $H_B(u(x, t_0))$. This contradicts the assumption. □

Eventual B-concavity

As a new phenomenon, we now study the *eventual* B-concavity of proper solutions, which in the pure concavity form was established in Section 2.6 for filtration equations.

We assume that $u_0(x)$ is compactly supported and has a bell-shaped form with a single maximum. This bell-shaped form is preserved in time (follows from the MP for the linear parabolic equation (7.57) for the derivative $z = u_x$). Then $u(x, t)$ has two interfaces. Since u_x now takes both the positive, $u_x > 0$, and the negative sign, $u_x < 0$, we introduce the tangent TW solutions defined by (7.81) with $\nu > 0$. For simplicity we impose the condition

$$F(p, -q, r) \equiv F(p, q, r), \tag{7.97}$$

so that the PDE (7.1) is invariant under the reflection transformation $x \mapsto -x$. This is true for general quasilinear heat operators (or filtration ones)

$$F(u, u_x, u_{xx}) = (\varphi(u))_{xx} + \psi(u) \equiv \varphi'(u)u_{xx} + \varphi''(u)(u_x)^2 + \psi(u). \tag{7.98}$$

Tangent solutions with $\mu > 0$ and $\nu < 0$ are given by the reflection $f_\lambda(\xi) \mapsto f_{-\lambda}(-\xi)$. We also impose the condition

$$F(p, 0, 0) \equiv 0. \tag{7.99}$$

The flat solution given by the ODE (7.78) is a constant one, $V_*(t) \equiv \mu$ for any $t \geq 0$. One can consider a more general case where $F(p, 0, 0) \leq 0$, so that $\psi(p) \leq 0$ in (7.98), and then $V_*(t)$ is decreasing.

In order to follow the lines of the proof of the eventual concavity for the filtration equation in Section 2.3, we assume that $u(x,t)$ exists in $I\!R \times I\!R_+$ and

$$\sup_{x \in I\!R} u(x,t) \to 0 \quad \text{as } t \to \infty.$$

Furthermore, we assume that the set B with $\Delta_0 = I\!R_+$ is complete and the profiles $f_\lambda(\xi)$ are strictly monotone and are defined for all $\xi > 0$.

Proposition 7.20 *Assume that the initial B-number is finite, $Z_B(0) < \infty$. Then there exists a finite $t_* \geq 0$ such that $u(\cdot,t)$ is B-concave and $Z_B(t) = 0$ for any $t \geq t_*$.*

Proof. As in Section 2.3, the proof is based on a compactness argument establishing that all inflection curves, on which $H_B(u(x,t)) = 0$, disappear in finite time. We also should take into account that the interfaces of $u(x,t)$ for $t \gg 1$ propagate slower than any travelling wave. This follows by comparison with the remote TWs $f_{\pm\lambda}(x \pm \lambda t \pm a)$, $a \gg 1$. Indeed, $u_0(x) \leq f_{\pm\lambda}(x \pm a)$ in $I\!R$ provided that $a \gg 1$ whence the necessary bounds on the interfaces. □

Strong Maximum Principle for interfaces

Proposition 7.21 *Let $B = \{f_\lambda,\ \lambda \in \Delta_0\}$ be complete and sufficiently smooth in λ. Assume that, for any $\lambda \in \Delta_0$,*

$$\frac{\partial f_\lambda}{\partial \lambda}(\xi) = o\left(\frac{\partial f_\lambda}{\partial \xi}(\xi)\right) \quad as \;\; \xi \to 0^+. \tag{7.100}$$

Given a TW solution $V(x,t) = f_c(x - ct - a) \in B$, assume that, for some $t_0 > 0$, the interfaces of $u(x,t_0)$ and $V(x,t_0)$ coincide: $s(t_0) = s_c(t_0) = ct_0 + a$, and the B-number is finite, $Z_B(t_0) < \infty$. If $s(t) \not\equiv s_c(t)$ on $[0,t_0]$, then

$$s(t) \not\equiv s_c(t) \quad \text{on any small interval } (t_0, t_0 + \varepsilon) \text{ with } \varepsilon > 0. \tag{7.101}$$

Proof. Since $Z_B(t_0) < \infty$, assuming that $u(\cdot,t_0)$ is, say, B-convex near the interface, we consider tangent TW profiles. Let $s(t_0) = 0$. Namely, for any small $x_0 > 0$, we consider the tangential system

$$u(x_0,t_0) = f_\lambda(\xi_0), \quad u_x(x_0,t_0) = f'_\lambda(\xi_0), \tag{7.102}$$

where $\xi_0 > 0$ and $\lambda < c$ are uniquely determined and depend on $x_0 \approx 0^+$. Such tangent solutions can be written down in the form $V(x,t) = f_\lambda(x - \lambda(t-t_0) - (x_0 - \xi_0))$, and hence the tangent TW profiles at the points (x_0,t_0) are given by $V(x,t_0) = f_\lambda(x - (x_0 - \xi_0))$. By the B-convexity of $u(x,t_0)$ near the interface, we have that $u(x,t_0) \geq f_c(x)$ for small $x \geq 0$ and $u(x,t_0) \geq f_\lambda(x - (x_0 - \xi_0))$ for any $x_0 \approx 0^+$. By the standard local comparison of the solutions $u(x,t)$ and any tangent solution $V(x,t)$, we conclude that these inequalities are true for small $x \approx 0$ and $t - t_0 > 0$. Therefore, the interface $s_\lambda(t) = \lambda(t-t_0) + (x_0 - \xi_0)$ of the tangent solution V overtakes the interface $s_c(t) = c(t-t_0)$ of the fixed TW after the time

$$\Delta t(x_0) = \frac{x_0 - \xi_0}{c - \lambda}.$$

We now pass to the limit $x_0 \to 0^+$. Then, by continuity, $\lambda \to c^-$ and $\xi_0 \to 0$. Using L'Hospital's rule, we conclude that

$$\lim_{x_0 \to 0+} \Delta t(x_0) = \lim_{x_0 \to 0+} \left(1 - \frac{\mathrm{d}\xi_0}{\mathrm{d}x_0}\right)\left(-\frac{\mathrm{d}\lambda}{\mathrm{d}x_0}\right)^{-1}. \tag{7.103}$$

Differentiating the first equation in (7.102) with respect to x_0 yields

$$\frac{\partial u}{\partial x}(x_0, t_0) = \frac{\partial f_\lambda}{\partial \xi}\frac{\mathrm{d}\xi_0}{\mathrm{d}x_0} + \frac{\partial f_\lambda}{\partial \lambda}\frac{\mathrm{d}\lambda}{\mathrm{d}x_0}.$$

From the second equation we have

$$\frac{\mathrm{d}\xi_0}{\mathrm{d}x_0} = 1 - \frac{\mathrm{d}\lambda}{\mathrm{d}x_0}\left(\frac{\partial f_\lambda}{\partial \lambda}\right)\left(\frac{\partial f_\lambda}{\partial \xi}\right)^{-1}.$$

Substituting this equality into (7.103), we obtain

$$\lim_{x_0 \to 0+} \Delta t(x_0) = - \lim_{x_0 \to 0+} \left(\frac{\partial f_\lambda}{\partial \lambda}\right)\left(\frac{\partial f_\lambda}{\partial \xi}\right)^{-1}(\xi_0) = 0$$

by assumption (7.100) and the smoothness of $f_\lambda(\xi)$ in λ. Finally, $s(t) < s_c(t) = c(t - t_0)$ for any small $t - t_0 > 0$. $\square$

7.7 Regularity in B-classes by transversality: gradient estimates, instantaneous smoothing, Lipschitz interfaces, optimal moduli of continuity

We now begin the regularity analysis of proper solutions by the geometric approach. We consider the 1D parabolic equation (7.1) in the case $\Delta_0 \neq \emptyset$, where continuous proper solutions $u(x,t)$ exist. In what follows we assume that the proper B-bundle is as smooth as necessary for the intersection comparison reasons.

B-classes, transversality and gradient estimates

Given the set $B = \{f_\lambda, \ \lambda \in \Delta_0\}$, we introduce two functional classes $B^\pm(c)$, where $c \in \Delta_0$ is a parameter. We first consider *local* B-classes that describe some differential properties of the proper solutions near the singular interface, in a small ν-neighbourhood of $x = s(t)$, $\nu > 0$. The analysis is local in space and without loss of generality we assume that, for all $\lambda \in \Delta_0$, the proper profiles $f_\lambda(\xi)$ are monotone increasing with $\xi \geq 0$. Fix a $t_0 \geq 0$ and, as usual, assume that the solution profile $u(x, t_0)$ is continuous and monotone with its interface at $x = s(t_0) = 0$.

By $\mathrm{Int}_\nu(t_0, V)$ we denote the number of intersections between $u(x, t_0)$ and $V(x, t_0)$ in the right-hand ν-neighbourhood of the interface at $x = 0$. The definition of B-classes is local in space and applies to rather arbitrary solutions, e.g., to bounded, compactly supported and bell-shaped profiles $u(x, t_0)$ with the left-hand interface satisfying $s(t_0) = 0$.

Definition 7.4 Let $c \in \Delta_0$. (i) We say that $u(x, t_0)$ belongs to the *local class*

$B^+_{\rm loc}(c)$ if there exists a small $\nu > 0$ such that the TW solutions $V(x,t;c,a) \equiv f_c(x - c(t - t_0) - a) \in B$ satisfy

$$\mathrm{Int}_\nu(t_0, V) = 1 \quad \text{for all } a \approx 0^+ \text{ and } \mathrm{Int}_\nu(t_0, V) = 0 \ \text{ if } a = 0. \tag{7.104}$$

(ii) $u(x,t_0)$ belongs to the local class $B^-_{\rm loc}(c)$ if

$$\mathrm{Int}_\nu(t_0, V) = 1 \quad \text{for all } a \approx 0^- \text{ and } \mathrm{Int}_\nu(t_0, V) = 0 \ \text{ if } a = 0. \tag{7.105}$$

It follows from the definition that, in the local class $B^+_{\rm loc}(c)$ at $t = t_0$, the TW profiles $\{f_c(x-a)\}$ with $a \approx 0^+$ are steep enough to intersect the solution profile $u(x,t_0)$ transversally and exactly once in a right ν-neighbourhood of the interface. On the contrary, in the second local class $B^-_{\rm loc}(c)$, the TW profiles are flat enough to intersect $u(x,t_0)$ once if they are shifted to the left with $a \approx 0^-$. In both cases, by continuity, no intersection exists in the ν-neighbourhood with the TW without shifting, i.e., $a = 0$. Thus the *character* of intersections of $f_c(x-a)$ and $u(x,t_0)$ are different for both B-classes. Namely, the intersections are from *below* in the first class (steep TWs) and are from *above* for the second one (flat TWs).

From the monotonicity of the G-function upon λ we easily derive the following gradient estimates in B-classes. Here $v = Q(u)$ is the pressure variable.

Proposition 7.22 *(i) If $u(x,t_0) \in B^+_{\rm loc}(c)$, then, for small $x - s(t_0) \ge 0$,*

$$\overline{D}_x v(x,t_0) \equiv \limsup_{x \to x_0} \frac{v(x,t_0) - v(x_0,t_0)}{x - x_0} \le G(c, v(x,t_0)). \tag{7.106}$$

(ii) If $u(x,t_0) \in B^-_{\rm loc}(c)$, then, for small $x - s(t_0) \ge 0$,

$$\underline{D}_x v(x,t_0) \equiv \liminf_{x \to x_0} \frac{v(x,t_0) - v(x_0,t_0)}{x - x_0} \ge G(c, v(x,t_0)). \tag{7.107}$$

If $v(x,t_0) \in C^1$ in $\{v > 0\}$ (by the assumptions, this is true by the interior regularity if $t_0 > 0$), then these estimates mean that, for $x \approx s(t_0)^+$,

$$(i) \ v_x \le G(c,v) \quad \text{and} \quad (ii) \ v_x \ge G(c,v). \tag{7.108}$$

Proof. (i) Let $t_0 = 0$. It follows from the definition and the monotonicity property of the G-function (Lemma 7.4) that steep profiles $Q(V(x,0)) \equiv Q(f_c(x-a))$ intersect $v_0(x) = Q(u_0(x))$ from below, i.e., for $a \approx 0^+$ at the intersection point, the differences $Q(V(x,0)) - v_0(x)$ change sign from minus to plus when x crosses the intersection point. This implies (7.106). (ii) Similarly, if $u_0 \in B^-_{\rm loc}(c)$, then, for $a \approx 0^-$ near the interface, the intersection is from above and (7.107) follows.
□

By continuity of the solution $u(x,t)$, we have that transversality is locally preserved in time. Indeed, since, by Sturm's Theorem, new intersections cannot occur for $x \approx 0$, the geometric configurations (7.104) and (7.105) remain valid for all small $t > 0$.

Proposition 7.23 *If $u_0 \in B^+_{\rm loc}(c)$ (resp. $B^-_{\rm loc}(c)$), then $u(x,t) \in B^+_{\rm loc}(c)$ (resp. $B^-_{\rm loc}(c)$) for small $t > 0$.*

Let the spatial profile $u(x,t_0)$ for $t_0 > 0$ be sufficiently smooth in the positivity domain. Then the inclusions $u(x,t_0) \in B^+_{\rm loc}(c_1)$ and $u(x,t_0) \in B^-_{\rm loc}(c_2)$,

$c_{1,2} \in \Delta_0$, imply that $v = v(x, t_0)$ satisfies the following gradient bound near the singular interface where $v_x > 0$:

$$\frac{v_x}{G(c_1, v)} \le 1 \quad \text{and} \quad \frac{v_x}{G(c_2, v)} \ge 1. \tag{7.109}$$

Both bounds give Bernstein-type upper and lower estimates on the first derivative near the singularity. These estimate are *optimal* since the corresponding equalities are true for the set of the TW solutions.

The above definition of the local classes is extended to the global one when we count all the intersections in $I\!R$ of the proper solution with the TW solutions.

Definition 7.5 Let $c \in \Delta_0$ and $t_0 \ge 0$. (i) $u(x, t_0)$ belongs to the *global class* $B^+(c)$ if

$$\text{Int}(t_0, V) \le 1 \quad \text{for all } \ a \in I\!R \ \text{ and } \ u(x, t_0) \le f_c(x) \ \text{ in } \ I\!R. \tag{7.110}$$

(ii) $u(x, t_0)$ belongs to the global class $B^-(c)$ if

$$\text{Int}(t_0, V) \le 1 \quad \text{for all } \ a \in I\!R \ \text{ and } \ u(x, t_0) \ge f_c(x) \ \text{ in } \ I\!R. \tag{7.111}$$

By the transversality of intersections, global B-classes control the character of intersections in the whole space and this gives gradient estimates in $I\!R$. Evidently, (7.110) and (7.111) impose certain restrictions on the behaviour of the solutions $u(x, t_0)$ as $x \to \infty$.

The intersection analysis in global B-classes gives the following result. As usual, we assume that, due to the natural properties of the proper solutions in the positivity domain, new intersections between $u(x, t)$ and the TW solutions $f_\lambda(x - \lambda t - a)$ cannot appear at $x = \infty$.

Proposition 7.24 *If* $u_0 \in B^\pm(c)$*, then* $u(\cdot, t) \in B^\pm(c)$ *for all* $t > 0$.

Instantaneous smoothing phenomenon in B-classes

Smoothing phenomena in parabolic problems, when solutions exhibit better regularity than initial data, are classical in the general theory. We now present a geometric interpretation of smoothing via intersection comparison with families of TWs. In the intersection comparison proof we use such families only, and so the result holds for arbitrary parabolic PDEs or other B-equations admitting such particular solutions.

Using the local interior regularity for positive solutions, we state the *instantaneous smoothing phenomenon* as follows. By $\xi_*(\lambda, \varepsilon) > 0$, we denote the minimal intersection point of the TW pressure profile $Q(f_\lambda(\xi))$ with the positive ε-level,

$$Q(f_\lambda(\xi_*)) = \varepsilon.$$

Theorem 7.25 *(i) Let* $(-\infty, b] \subseteq \Delta_0$ *and for any fixed small* $\varepsilon > 0$,

$$G(\lambda, \varepsilon) \to \infty \quad as \ \lambda \to -\infty. \tag{7.112}$$

Given any continuous u_0*, for any* $\delta > 0$*, there exists* $c = c(\delta) \in \Delta_0$ *such that* $u(\cdot, t) \in B^+_{\rm loc}(c)$ *for small* $t - \delta > 0$.

(ii) Let $[b, \infty) \subseteq \Delta_0$ *and for any fixed small* $\varepsilon > 0$,

$$G(\lambda, \varepsilon) \to 0 \quad as \ \lambda \to \infty, \quad and \tag{7.113}$$

$$\lim_{\lambda \to \infty} \frac{1}{\lambda} \xi_*(\lambda, \varepsilon) \to 0 \quad as \ \varepsilon \to 0. \tag{7.114}$$

Then, for any $\delta > 0$, *there exists a* $c(\delta) \in \Delta_0$ *such that* $u(\cdot, t) \in B^-_{\rm loc}(c)$ *for small* $t - \delta > 0$.

(iii) If $u_0 \in B^{\pm}_{\rm loc}(c)$, *then* $u(\cdot, t) \in B^{\pm}_{\rm loc}(c)$ *for all small* $t > 0$.

Proof. Let $s(\frac{\delta}{2}) = 0$. (i) We apply intersection comparison with a subset $\{f_\lambda\} \subset B$ where $\lambda \ll -1$. The profile $u(x, \frac{\delta}{2})$ is continuous and there exists a $\lambda \in \Delta_0$, $\lambda \le -2$, such that $u(x, \frac{\delta}{2}) \le f_\lambda(x + \frac{\delta}{2})$ in a neighbourhood of the origin, say, for $x \in (-1, 1)$. Since the solution $u(x, t)$ is uniformly smooth in the positivity domain where $u \ge \mu = u(\delta, \frac{\delta}{2}) > 0$, there exists a constant $c \ll 2\lambda$ such that the steep profiles $f_c(x - a)$ intersect $u(x, \frac{\delta}{2})$ exactly once, say, for any $a \in [\frac{\delta}{2}, \frac{1}{2}]$, as shown in Figure 7.6 (a). Denote such a subset of TW solutions $V(x, t) = f_c(x - c(t - \frac{\delta}{2}) - a)$ by $B_{c,\delta} \subset B$.

Since the speeds of these chosen TWs, f_λ and f_c, differ strongly, $c \ll -2|\lambda|$, it follows that after the *time of smoothing*

$$\Delta t = -\frac{\delta}{c - \lambda} \le \frac{\delta}{|\lambda|} \le \tfrac{1}{2}\, \delta,$$

the solution profile $u(x, \delta)$ is completely "covered" near the interface by the profiles $V(x, \delta) \in B_{c,\delta}$ in the sense that, for a small $\nu > 0$ and $t = \delta$,

$$\text{Int}_\nu(t, V) \le 1 \quad \text{for any} \ \ V \in B_{c,\delta}. \tag{7.115}$$

Moreover, for any $x_* \in (s(\delta), \frac{1}{2})$, there exists a unique $V_* \in B_{c,\delta}$ such that $V_*(x, \delta)$ intersects $u(x, \delta)$ exactly once at $x = x_*$ with

$$\text{Int}_\nu(\delta, V_*) = 1.$$

This is shown in Figure 7.6 (b). Since, by construction, the steep profile $V_*(x, \delta)$ intersects $u(x, \delta)$ at $x = x_*$ from below, i.e., the difference $V_*(x, \delta) - u(x, \delta)$ changes sign from minus to plus when x crosses x_*, we conclude that $u(x, \delta) \in B^+_{\rm loc}(c)$. By the same intersection comparison argument and continuity of the solutions, this remains true for $t - \delta > 0$ sufficiently small.

(ii) The proof is similar. We fix a small $\mu = u(\delta, \frac{\delta}{2}) > 0$ and we first find values of the TW speed $c \gg 1$ and the translational parameter $a = a_0 < 0$ of the TW profile $f_c(x - a_0)$ such that

$$(Q(f_c))'(a_0) \ll 1 \quad \text{at a point where} \ \ Q(f_c) = Q(\mu) = \varepsilon. \tag{7.116}$$

These equations mean that on the ε-level the profile $Q(f_c(x - a_0))$ is flat enough to intersect the pressure profile $v(x, \frac{\delta}{2})$ exactly once in a neighbourhood of $x = 0$, excluding the points $x \in [0, \delta]$ near the singular interface at $x = s(\frac{\delta}{2}) = 0$, where the regularity of $u(x, \frac{\delta}{2})$ is supposed to be unknown. In the positivity set $\{v \ge \varepsilon\}$ the monotone profile $v(x, \frac{\delta}{2})$ is sufficiently smooth. Hence, by the strong

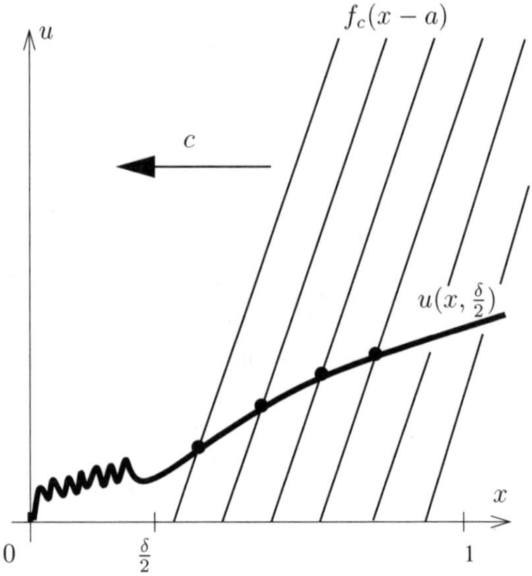

(a) u is not smooth near the interface

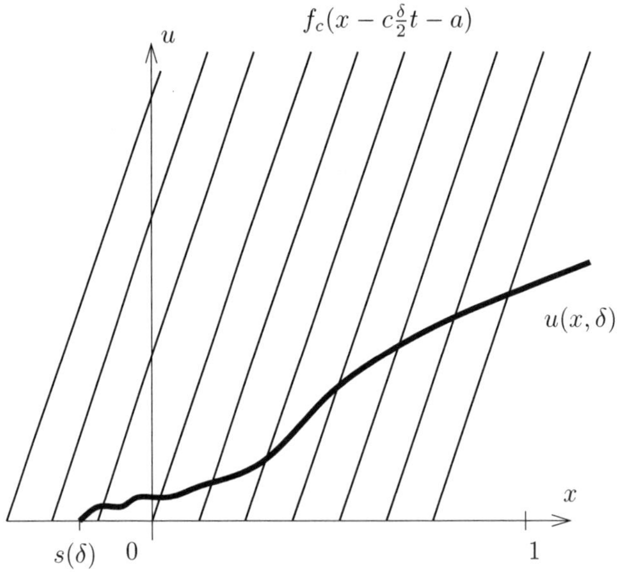

(b) u becomes smooth everywhere

Figure 7.6 *Illustration of the proof of Theorem* 7.25 *(instantaneous smoothing): (a)* $u(x,\frac{\delta}{2})$ *is not smooth near interface at* $x=0$*, and (b)* $u(x,\delta)\in B^+_{\rm loc}(c)$*, when fast steep TWs* $f_c(x-c\frac{\delta}{2}t-a)$ *cover the whole interface neighbourhood.*

MP applied to the linear parabolic equation for the derivative $z = v_x$, it is assumed to be strictly positive and bounded away from zero, say, $v_x \geq \nu_\delta > 0$.

It follows from the properties of the TW profiles that such constants $c \gg 1$ and a_0 exist, and moreover, for a fixed $c \gg 1$, one can chose $a_0 = -\xi_*(c, \varepsilon) < 0$. We then form the corresponding subset $B_{c,\delta} \subset B$ of the TW solutions $V(x,t) = f_c(x - c(t - \frac{\delta}{2}) - a)$ with $a \leq a_0$. With such a choice of the parameters, the time of smoothing is estimated as follows

$$\Delta t \sim \tfrac{1}{c}\, \xi_*(c, \varepsilon), \tag{7.117}$$

so that we need assumption (7.114) implying that the smoothing time can be arbitrarily small for $\varepsilon \ll 1$. This means that after the time $\Delta t \leq \frac{\delta}{2}$, in a right-hand neighbourhood of the interface $x = s(\delta)$ the profile $u(x, \delta)$ will be completely covered by the TW profiles $f_c(x - c\frac{\delta}{2} - a)$ with a single intersection from above. This means that $u(x, \delta) \in B^-_{\rm loc}(c)$. By the intersection comparison with the subset $B_{c,\delta}$ and by continuity of the solutions, we have the same result for small $t - \delta > 0$.

In the case (iii) the initial data u_0 are already smooth enough and the proof is straightforward. □

For convenience, we state the following general result on the gradient estimates near the interface and on the interface for singular proper solutions.

Theorem 7.26 *Let $(-\infty, -b] \cup [b, \infty) \subseteq \Delta_0$ for some $b \geq 0$ and (7.112)–(7.114) hold. Then, for any small $\delta > 0$, there exists a $c = c(\delta) \gg 1$ such that, for $t \geq \delta$, the following estimates hold near the interface:*

$$G(c, u) \leq (Q(u))_x \leq G(-c, u), \tag{7.118}$$

$$f_c(x - s(t)) \leq u(x,t) \leq f_{-c}(x - s(t)). \tag{7.119}$$

Proof (7.118) is a consequence of (7.109). The estimates on the interface (7.119) follow from the intersection comparison with the steep ($-c \ll -1$) and the flat ($c \gg 1$) TW solutions as in the proof of the smoothing phenomenon (cf. the definition of local B-classes). □

Lipschitz continuity of interfaces and level propagation

Fix a small $\varepsilon > 0$ and let $l_\varepsilon = \{s_\varepsilon(t),\ t > 0\}$ be the level set curve

$$u(s_\varepsilon(t), t) = \varepsilon \quad \text{for} \quad t > 0 \quad (s_0(t) \equiv s(t)). \tag{7.120}$$

Theorem 7.27 *Under the hypothesis of Theorem 7.26 (i) $s(t)$ is uniformly Lipschitz continuous on bounded intervals $[\delta, C]$ with small $\delta > 0$, and (ii) for some $c = c(\delta) > 0$,*

$$s'_\varepsilon(t) \equiv -\frac{F(u, u_x, u_{xx})}{u_x}\Big|_{u=\varepsilon} \in [-c, c] \quad on \quad [\delta, C]. \tag{7.121}$$

Proof. (i) By intersection comparison with steep and flat TWs, we have that $-c(t - t_0) \leq s(t) - s(t_0) \leq c(t - t_0)$ for $t \geq t_0$ and any $t_0 \in [\delta, C]$. This implies Lipschitz continuity of $s(t)$.

(ii) Given a $t_0 > 0$, we have $s'_\varepsilon(t_0) = \lambda$ where $\lambda = \lambda(t_0)$ is the TW speed of the corresponding tangent TW profile f_λ at (x_0, t_0) with $u(x_0, t_0) = \varepsilon$. Since $u(x, t_0) \in B^{\pm}_{\rm loc}(c)$ by Theorem 7.25, the slope of the tangent profile on the level set is always in between the slopes of the profiles with the speeds c and $-c$. The interface (the 0-level set) also satisfies this estimate, $D^+ s(t) \in [-c, c]$, if it exists. □

Optimal moduli of continuity in x and t

The results below are a straightforward consequence of the regularity estimates obtained by comparison of general solutions with a complete TW-bundle. Actually, these estimates imply that optimal moduli of continuity, in both independent variables x and t of $u(x, t)$ near singular interfaces, are the same as the modulus of continuity in ξ of the corresponding TW profile $f_\lambda(\xi)$ having the same slope. Hence, moduli in x and t must coincide.

1. Modulus of continuity in x. This is easier and follows from first-order Bernstein estimates. We will use some hypotheses, but one can obtain a similar result under weaker conditions.

Theorem 7.28 *Under the hypotheses of Theorem 7.26, for any small $t - \delta > 0$ there holds: (i) For any $x_1, x_2 \geq s(t)$ in a neighbourhood of the interface,*

$$|\Phi_{-c}(u(x_2, t)) - \Phi_{-c}(u(x_1, t)| \leq |x_2 - x_1|, \tag{7.122}$$

where the function Φ_{-c} is given by the convergent integral

$$\Phi_{-c}(z) = \int_0^z \frac{Q'(y)dy}{G(-c, Q(y))} \quad \text{for } z \geq 0.$$

(ii) In a small neighbourhood of the interface, for $x_1, x_2 \approx s(t)^+$, the pressure $Q(u(x, t))$ is Lipschitz continuous:

$$|Q(u(x_2, t)) - Q(u(x_1, t))| \leq C|x_2 - x_1|. \tag{7.123}$$

Proof. (i) Integrating estimate in (7.118) with $v = Q(u)$ yields (7.122).

(ii) By the definition of the pressure (7.36) and the G-function (7.47), we have that

$$G(-c, Q(y)) = A_0 + o(1) \quad \text{as } y \to 0,$$

where A_0 may depend on the parameter $\lambda = -c$. Substituting this expansion into the definition of Φ_{-c}, in a small neighbourhood of the interface, we obtain (7.123) with $C = 2A_0$. □

The lower bound in (7.118) gives a similar estimate from below if we replace $-c$ by $c \gg 1$.

2. Modulus of continuity in t. We first state a particular result explaining that the optimal modulus of continuity in t is prescribed by a majorizing TW profile and that the pressure is Lipschitz continuous in t.

Proposition 7.29 *Let $(-\infty, 1] \subseteq \Delta_0$. Fix a $t_1 \geq \delta$ and let $s(t_1) = 0$. Then, for*

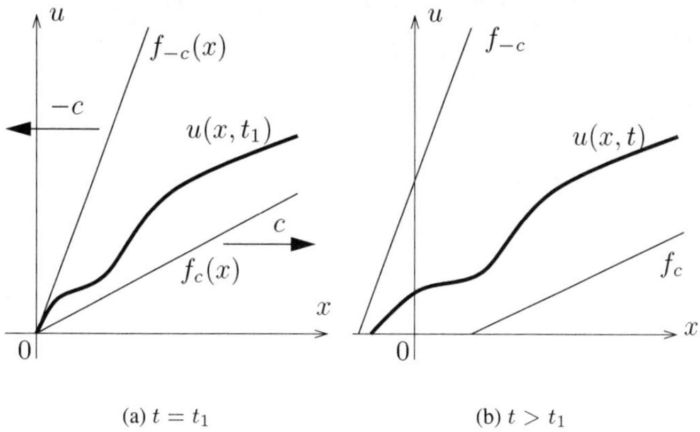

Figure 7.7 *Comparison establishing optimal moduli of continuity.*

$t_2 > t_1$, *there exists a constant* $c = c(\delta) \gg 1$ *such that*

$$|u(0,t_2) - u(0,t_1)| \le f_{-c}(c(t_2 - t_1)).$$

Proof. The estimate follows by comparison since $u(x,t) \le V(x,t) = f_{-c}(x + c(t - t_1))$ for $t > t_1$. □

3. Modulus of continuity in x and t. The analysis is similar and follows by comparison with the TW-bundle.

Theorem 7.30 *Let* $(-\infty, 1] \cup [1, \infty) \subseteq \Delta_0$. *Let* $t_1 \ge \delta$ *and* $s(t_1) = 0$. *Then, for any* x_1, x_2 *and* $t_2 \ge t_1$, *there exists a* $c = c(\delta) \gg 1$ *such that*

$$|u(x_2,t_2) - u(x_1,t_1)|$$
$$\le \max\{f_{-c}(x_2 + c(t_2 - t_1)) - f_c(x_1),\ f_{-c}(x_1) - f_c(x_2 - c(t_2 - t_1))\}.$$

Proof. It follows that $u(t_1, x) \in B^{\pm}(c)$ and hence

$$f_c(x) \le u(x,t_1) \le f_{-c}(x) \quad \text{in } I\!R,$$

as in Figure 7.7 (a). Therefore,

$$f_c(x - c(t - t_1)) \le u(x,t) \le f_{-c}(x + c(t - t_1)) \quad \text{for } x \in I\!R,\ t \ge \delta,$$

by comparison (Figure 7.7 (b)), and the estimate follows. □

Eventual smoothing and waiting time phenomena

The absence of the instantaneous smoothing can lead to singular effects like *waiting time* for the PME. This is possible if at least one class $B^+_{\rm loc}(-c)$ or $B^-_{\rm loc}(c)$ is not defined for $c \gg 1$ in view of the nonexistence of corresponding proper TW solutions or in view of their undesirable properties. In particular, for the PME, the class $B^-_{\rm loc}(c)$ is not available since singular TW profiles f_λ do not exist for any $\lambda \ge 0$, i.e., $\Delta_0 = (-\infty, 0)$.

We now describe the *eventual smoothing* phenomenon. The statement below is motivated by the waiting time effect for the PME and for convenience we choose the same range Δ_0 of the TWs.

Proposition 7.31 *Let $\Delta_0 = (-\infty, 0)$ and the proper set $B = \{f_\lambda\}$ be complete. Assume (7.112) holds. Then, for $t_0 > 0$, there holds:*

(i) $u(\cdot, t_0) \in B^+_{\rm loc}(c(t_0))$,

(ii) $\overline{D}^+ s(t_0) \equiv \limsup_{t \to t_0^+} \frac{s(t)-s(t_0)}{t-t_0} \le 0$, and

(iii) if u_0 is B-convex and $s(t_0) \ne s(0)$, then $u(\cdot, t_0) \in B^-(c(t_0))$.

The last property (iii) means that once the interface moves, the solution attains the lower gradient bound (the upper one is always available for any $t > 0$ by (i)).

Proof. (i) The result follows from Theorem 7.25 (i).

(ii) This is proved by local comparison. By completeness of B, $f_\lambda(\xi) \to 0$ as $\lambda \to 0^-$, so, for any $a \approx s(t_0)^+$, there exists a $\lambda_a \approx 0^-$ such that

$$u(x, t_0) \ge f(x - a, \lambda_a) \quad \text{for} \quad x \approx s(t_0)^+.$$

Hence, $u(x, t) \ge f(x - \lambda_a(t - t_0) - a, \lambda_a)$ by comparison, so that

$$s(t) - s(t_0) \le \lambda_a(t - t_0) + a - s(t_0),$$

whence the result by passing to the limit $a \to s(t_0)^+$.

(iii) Let $s(0) = 0$ and let $u_0(x)$ touch 0 at $x = 0$ so smoothly that $u_0 \notin B^-(c)$ for any $c < 0$ (otherwise, if such $c < 0$ exists, the result is true for any $t_0 > 0$). Assume $s(t_0) < 0$. Set $a_0 = \frac{1}{2}s(t_0) < 0$ and choose $c = \frac{s(t_0)}{2t_0} < 0$, $c \in \Delta_0$. The TW solution $V(x, t) = f_c(x - ct - a_0)$ with the interface at $a_0 = \frac{1}{2}s(t_0) < 0$ for $t = 0$ and at $s(t_0)$ for $t = t_0$, has a unique intersection with the B-convex initial function u_0, i.e.,

$$\text{Int}(0, V) = 1.$$

This follows from the property (7.86) of the B-convexity. Indeed, if $\text{Int}(0, V) \ge 2$, then using the mutual location of the interfaces of both solutions, by slightly varying the parameters c and a_0, one can find a new solution $V' \in B$ (e.g. $V'(x, 0) = f_c(x - \varepsilon)$, $\varepsilon > 0$ small) such that $\text{Int}(0, V') \ge 3$ contradicting (7.86). The unique intersection between $u(x, t)$ and $V(x, t)$ for $t \in (0, t_0)$ must disappear at $t = t_0$ where the interfaces coincide. Indeed, if the intersection stays uniformly away from the interface $x = s(t)$ for $t \approx t_0^-$, then, by shifting of $V(x, t)$ in x to the right-hand side, by comparison, we conclude that the interface $s(t)$ cannot reach $s(t_0)$ at $t = t_0$.

Thus, at $t = t_0$, we have $\text{Int}(t_0, V) = 0$, so that

$$u(x, t_0) \ge V(x, t_0) = f_c(x - s(t_0)).$$

Since the profile $u(x, t_0)$ is B-convex, from (7.86) we conclude that it belongs to $B^-(c)$ according to Definition 7.5, (ii). By Proposition 7.24, $u(\cdot, t) \in B^-(c)$ for $t > t_0$. □

We now present a simple condition where the above eventual smoothing effect becomes the instantaneous one since no waiting time is available. We again impose the convenient assumption of B-convexity, which is not restrictive and in the local B-classes can be dispensed with.

Proposition 7.32 *Let $\Delta_0 = (-\infty, 0)$ and $B = \{f_\lambda\}$ be proper and complete. Let u_0 be B-convex, $s(0) = 0$, and let $u_0(x)$ be the envelope of the family $\{f(x - a, \lambda)\}$ of proper TW profiles, so, for any $x_0 > 0$, $f(x - a(x_0), \lambda(x_0))$ is the tangent profile to $u_0(x)$ at $x = x_0$. Assume*

$$\limsup_{x_0 \to 0+} \frac{a(x_0)}{\lambda(x_0)} \geq 0. \tag{7.124}$$

Then $s(t) < 0$ and $u(\cdot, t) \in B^-(c(t))$ for any small $t > 0$.

Proof. Fix a small $t > 0$. By the assumption, there exists a sequence $\{x_{0k}\} \to 0^+$ such that

$$\frac{a(x_{0k})}{\lambda(x_{0k})} \geq -\tfrac{1}{2}t \quad \text{for} \quad k \gg 1.$$

By the B-convexity, $u_0(x) \leq f(a - a(x_{0k}), \lambda(x_{0k}))$, so that

$$s(t) \leq \lambda(x_{0k})t + a(x_{0k}) \equiv \lambda(x_{0k}) \left[t + \frac{a(x_{0k})}{\lambda(x_{0k})} \right].$$

Choosing $k \gg 1$, we deduce that $s(t) < 0$. Then Proposition 7.31 (iii) applies, to guarantee that $u(\cdot, t) \in B^-(c)$. □

7.8 Transversality and smoothing in the radial geometry in $I\!R^N$

First-order transversality estimates. We show how to extend the 1D transversality results to radial monotone solutions of equation (7.7) in $I\!R^N$, i.e., assuming that (7.63) holds. Consider the monotone decreasing TWs

$$V(r, t) = f_\lambda(\xi), \quad \xi = r - \lambda t$$

satisfying the 1D parabolic equation (notice that $\Delta_r v$ is replaced by v_{rr})

$$v_t = F(v, v_r, v_{rr}) \quad \text{in } I\!R \times I\!R_+, \tag{7.125}$$

where $f_\lambda \geq 0$ solve the corresponding ODE

$$F(f, f', f'') + \lambda f' = 0 \quad \text{with } \lambda \in \Delta_0. \tag{7.126}$$

These TWs are strict plain supersolutions of (7.7) for $r > 0$. Since $u_r < 0$ for $r > 0$ in the positivity domain, by the parabolicity of the equation, we have

$$V_t > F(V, V_r, \Delta_r V) \quad \text{in } \{V > 0\} \cap \{r > 0\}. \tag{7.127}$$

Let us show that the transversality first-order estimates can be derived by comparison with the family of steep TW supersolutions $B = \{V = f_\lambda\}$.

Proposition 7.33 *Let $\Delta_0 \neq \emptyset$ for the ODE (7.126), and for some $c \in \Delta_0$, $u_0(r) \in B^+(c)$. Assume that*

$$Q_1 \equiv \frac{\mathrm{d}}{\mathrm{d}r} \left[F\left(V, V_r, V_{rr} + \tfrac{N-1}{r} V_r\right) - F(V, V_r, V_{rr}) \right] \geq 0 \tag{7.128}$$

in $\{V > 0\} \cap \{r > 0\}$ for the profile $V = f_c(r)$ satisfying (7.126). Then $u(r, t) \in B^+(c)$ for $t > 0$.

Proof. As in the 1D case, we compare $u(r,t)$ with the family of translated TW supersolutions $\{f_c(\xi - a)\}$ and perform a typical construction based on the MP. We show that such a gradient estimate can be violated if a special *inflection* point with a TW is created with evolution.

By approximation, we exclude the analysis on the singularity level and always deal with regular positive approximations of the solutions involved. Let $t_0 > 0$ be the first moment, at which $u(\cdot, t) \notin B^+(c)$ for some t in an arbitrarily small right-hand neighbourhood of $t = t_0$. It then follows that, at $t = t_0$ there exists a TW profile $V(r,t) = f_c(\xi - a)$ such that $u(r, t_0)$ has an inflection with V at $r = r_0$, i.e., the difference $v(r,t_0) \equiv u(r,t_0) - V(r,t_0)$ satisfies

$$v = v_r = v_{rr} = 0 \quad \text{at } P_0 = (r_0, t_0), \tag{7.129}$$

and $v(r,t_0) < 0$ for $r < r_0$, $v(r,t_0) > 0$ for $r > r_0$. Since at some $t = t_0^+$, $r \approx r_0$, the gradient estimate

$$u_r \geq V_r|_{V=u} \quad (u_r, V_r < 0)$$

is going to fail for the first time relative to the TWs $f_c(\xi - a) \approx V$, we have that, at $t = t_0$, this is still true everywhere and

$$w(r,t) \equiv (u_r - V_r)(r, t_0) \leq 0 \quad \text{at intersections for } r \geq 0.$$

Consider the equation for the difference $v = u - V$ (hence $u = V + v$)

$$v_t = F(V + v, V_r + v_r, V_{rr} + v_{rr} + \tfrac{N-1}{r}(V_r + v_r)) - F(V, V_r, V_{rr}). \tag{7.130}$$

Differentiating it in r, performing a standard linearization in v and using (7.129), we obtain that, at P_0,

$$w_t = \mathcal{L}_1 w + Q_1,$$

where $\mathcal{L}_1$ is a linear second-order elliptic (ordinary differential) operator and Q_1 is given by (7.128). It follows by construction that, at P_0,

$$w = w_r = 0 \quad \text{and } w_{rr} \geq 0,$$

since $r = r_0$ is a global minimum of $w(r, t_0)$.

Assume first that the inequality sign in (7.128) is strict, $Q_1 > 0$. Hence,

$$w_t \geq Q_1 > 0 \quad \text{at } P_0,$$

so that, by regularity of (approximating) solutions, $w_t > 0$ in a neighbourhood of P_0 contradicting the choice of t_0. In the case $Q_1 \geq 0$, as in typical proofs of the MP for linear parabolic equations with smooth bounded coefficients (cf. A. Friedman's book [121]), we perform a standard change of variable $w = e^{-\mu t}\tilde{w}$ with $\mu \gg 1$, which will imply that such tangency points for $\tilde{w}$ cannot occur as $t \to t_0^-$. □

Similar to the 1D equation, once we have obtained first-order Bernstein estimates near singular interfaces, this gives the modulus of continuity of maximal solutions of the radial PDE via the modulus for solutions of the ODE (7.126).

Proposition 7.34 *Under the hypotheses of Proposition* 7.33, *for monotone data* $u_0(r)$, *the moduli of continuity in* $r > 0$ *and* $t > 0$ *of radial solutions of* (7.7) *are the same as that of solutions* $f_c(\xi)$ *of the ODE* (7.126).

For a local (close to interfaces) analysis in $B^-(c)$ classes, we check that $v = u - V$ satisfies $w \equiv v_r = u_r - V_r \leq 0$ at the intersection points, which gives a lower gradient bound

$$u_r \leq V_r|_{V=u} \quad (u_r, V_r < 0).$$

Therefore, by the same construction, at P_0 there holds

$$v = v_r = v_{rr} = 0, \quad v_{rrr} \equiv w_{rr} \leq 0,$$

since $v_r \leq 0$ for $r \approx r_0$. By the MP, we need to have that

$$Q_1 \leq 0 \quad \text{on } V = f_c. \tag{7.131}$$

Example 7.4 Consider a general radial filtration equation in the pressure form

$$u_t = \varphi(u)\left(u_{rr} + \tfrac{N-1}{r}\, u_r\right) + (u_r)^2 \equiv F(u, u_r, \Delta_r u), \tag{7.132}$$

where $\varphi(u) > 0$, $\varphi'(u) > 0$ for $u > 0$, $\varphi(0) = 0$, and $\varphi(u)$ is sufficiently smooth for $u > 0$. The TW supersolutions V satisfying the quadratic Hamilton-Jacobi equation

$$V_t = (V_r)^2$$

are piecewise linear,

$$V(r,t) = [\lambda(\lambda t - r)]_+ \quad \text{for any } \lambda \in I\!R_+ = \Delta_0. \tag{7.133}$$

Then $V_{rr} = V_{rrr} \equiv 0$ (in the limit of approximation) and calculations in (7.128) become simpler

$$Q_1|_V = \frac{\mathrm{d}}{\mathrm{d}r}\left[\varphi(V)\tfrac{N-1}{r}\, V_r\right] = (N-1)\frac{V_r}{r^2}[r\varphi'(V)V_r - \varphi(V)].$$

Since $V_r \equiv -\lambda$, we have

$$Q_1|_V = \frac{N-1}{r^2}\lambda[\lambda r\varphi'(V) + \varphi(V)] > 0 \quad \text{in } \{V > 0\}, \tag{7.134}$$

so that, by Proposition 7.33, the upper transversality bound

$$|u_r| \leq |V_r| \quad \text{at } \{V = u\}$$

is exactly the same as in 1D. It follows from inequality (7.134) that a local *lower* gradient Bernstein estimate for which (7.131) must be valid, cannot be proved by such a comparison with TW supersolutions and needs other families of TW super or subsolutions.

Instantaneous smoothing. For general radial parabolic PDEs, the instantaneous smoothing phenomenon also needs subsets of *subsolutions*. Namely, in order to study smoothing of the solution profile near the interface $r = s(t) \approx 2$, we take TWs f_λ satisfying (7.65) that are subsolutions in the domain $\{r \geq 1\}$. We then reconstruct the following result.

Proposition 7.35 *Let the hypotheses of Theorem* 7.25 *(i) hold, where* Δ_0 *is generated by singular TWs of* (7.65). *Assume that*

$$\tilde{Q}_1 \equiv \frac{\mathrm{d}}{\mathrm{d}r}\left[F\left(V, V_r, V_{rr} + \tfrac{N-1}{r}\, V_r\right)\right.$$

$$-F(V, V_r, V_{rr} + (N-1)V_r)] \geq 0 \tag{7.135}$$

in $\{V > 0\} \cap \{r > 1\}$ *for the profile* $V = f_c(r)$ *satisfying the ODE* (7.66). *Then, for any* $\delta > 0$, *there exists* $c(\delta) \in \Delta_0$ *such that* $u(\cdot, t) \in B^+_{\rm loc}(c)$ *for* $t > \delta$.

Proof. We need a slight modification to the 1D proof. Choosing a steep TW subsolution

$$V(r,t) = f_c(r - c(t - \tfrac{\delta}{2}) - a)$$

with $a \approx s(\frac{\delta}{2})^-$, as in the proof of Proposition 7.33, we establish that the first minimal intersection point $r = r_1(t, V)$ with $u(r,t)$ always stays transversal and no inflection can occur. Hence,

$$\mathrm{Int}(t, u, V) \leq 1 \quad \text{for } t > \tfrac{\delta}{2}.$$

Note that, by construction, $u \geq V$ in $\{r > r_1(t, V)\}$, where V is a subsolution. The rest of the geometric analysis is the same as in 1D. Since $c \gg 1$, at $t = \delta$, the set $B = \{f_c(r - c(t - \frac{\delta}{2}) - a),\ a \approx s(\frac{1}{2}\delta)\}$ completely covers the profile $u(r, \delta)$ near the interface, thus creating intersections of suitable shape and establishing the corresponding Bernstein bound from above. □

7.9 B-concavity in the radial geometry in $I\!R^N$

Let us now show that the B-concavity analysis applies to radial solutions of (7.7) in $I\!R^N$ under a certain extra condition on the operator. Here, by $B = \{V\}$ we denote a subset of TW supersolutions satisfying the 1D equation (7.125).

Proposition 7.36 *Let* B *be complete and*

$$Q_2 = \frac{d^2}{dr^2}\left[F\left(V, V_r, V_{rr} + \tfrac{N-1}{r}\,V_r\right) - F(V, V_r, V_{rr})\right] \leq 0 \tag{7.136}$$

for any TW profile $V \in B$ *given by* (7.126). *Then* (7.85) *holds, i.e., the* B-*concavity is evolutionary invariant.*

Proof. It is similar to that of the previous transversality analysis. The difference is that, at $t = t_0$, when the B-concavity is going to be violated for the first time as $t \to t_0^+$ at $r \approx r_0$, there exists a $V \in B$ such that $V(r, t_0)$ is tangent to $u(r, t_0)$ and the difference $v = u - V$ forms a higher-order *tangency* point relative to $v \equiv 0$, i.e.,

$$v = v_r = 0 \quad \text{at } P_0 = (r_0, t_0). \tag{7.137}$$

Moreover, the second derivative $w = v_{rr}$ satisfies the conditions of a local maximum at $r = r_0$

$$w = w_r = 0 \quad \text{and } w_{rr} \leq 0 \quad \text{at } P_0. \tag{7.138}$$

We recall that, by construction, we still have the comparison

$$V(r, t_0) \geq u(r, t_0) \quad \text{for } r \approx r_0,$$

so that $w_{rr} \leq 0$ at P_0. On the other hand, one can see that, if $w_r \neq 0$ at P_0, then $t = t_0$ is not the first moment of time, at which the B-concavity is going to be violated on any interval $(t_0, t_0 + \varepsilon)$.

We now deal with smooth approximations of the solutions. Differentiating equation (7.130) twice in r, we arrive at a linear parabolic equation

$$w_t = \mathcal{L}_2 w + Q_2.$$

If the inequality in (7.136) is strict, we directly proceed to obtain that, at P_0 by (7.137) and (7.138), there holds

$$w_t \le Q_2 < 0 \quad \text{at } P_0.$$

Hence, by construction, $w = (u - V)_{rr} < 0$ at any $P \approx P_0$ contradicting the above choice of t_0. If $Q_2 \le 0$, as in the proof of the MP, setting

$$w = e^{\mu t}\tilde{w}$$

with a constant $\mu \gg 1$, we establish that such a tangency cannot occur as $t \to t_0^-$. $\square$

Observe that, in the case $N = 1$, we have $Q_2 \equiv 0$ and the B-concavity is invariant, which is proved by intersection comparison.

Example 7.5 For the radial filtration equation, (7.132) taking the same piecewise linear TW supersolutions (7.133), we have that

$$Q_2 = (N-1)\frac{V_r}{r^3}[\varphi''(V)(V_r)^2 r^2 - 2\varphi'(V) r V_r + 2\varphi(V)].$$

Since $V_r \equiv -\lambda$,

$$Q_2 = -\lambda(N-1)\frac{1}{r^3}[\varphi''(V)\lambda^2 r^2 + 2\varphi'(V)\lambda r + 2\varphi(V)] < 0$$

for any $\lambda > 0$, provided that

$$\varphi''(u) \ge 0 \quad \text{for } u > 0.$$

In particular, this is true for the PME, where

$$\varphi(u) = (m-1)u.$$

Therefore, for any increasing convex nonnegative coefficients $\varphi(u)$, the B-concavity relative to piecewise TW supersolutions of (7.132) is preserved in time.

In other words, (7.136) is the necessary and sufficient condition for the following property: in the radial geometry in $\mathbb{R}^N$, the B-concavity relative to a complete subset $B = \{V\}$ of TW supersolutions satisfying (7.125), is invariant under the parabolic flow generated by the radial equation (7.7).

As we have seen, the proof uses the MP for a single parabolic differential inequality. In the general non-radial case of equation (7.7), a similar invariance B-concavity test reduces to an application of the MP to systems of N parabolic differential inequalities and the analysis becomes more complicated.

7.10 Interface operators and equations, uniqueness

We now return to the second half of the geometric regularity theory concerning the most delicate questions of the refined behaviour of proper solutions near interfaces and dynamic interface equations or systems. In what follows for the 1D equation (7.1), we use the interface operators defined in Section 7.2 for the TW solutions. For general proper solutions $u(x,t)$ with the interface at $x = s(t)$, we perform the shifting $x \mapsto s(t) + \xi$ to use the same formulae.

The case $\Delta_0 = I\!R$

Theorem 7.37 *Let the set* B *with* $\Delta_0 = I\!R$ *be proper and* (7.112)–(7.114) *hold. Then:*

(i) Interface operators are finite for the proper solution $u(\cdot,t)$ *for* $t > 0$. *If TW interfaces are of first order, the interface equation is*

$$D^+ s(t) = \mathbf{N}_1(u(\cdot,t)) \equiv A_0^{-1}((Q(u))_x(s(t),t)) \quad \textit{for } \; t > 0. \tag{7.139}$$

If TW interfaces are of second order, the interface equation is the following system for $t > 0$*:*

$$\begin{cases} \mathbf{M}_1(u(\cdot,t)) = A_0, \\ D^+ s(t) = \mathbf{N}_2(u(\cdot,t)). \end{cases} \tag{7.140}$$

(ii) The one-sided derivatives $D^{\pm} s(t)$ *are uniformly bounded on* $[\delta, C]$ *for arbitrarily small* $\delta > 0$ *and the interfaces are uniformly Lipschitz continuous on such bounded intervals.*

Proof. (i) We first prove that interface operators introduced in Section 7.2 by the TW-bundles are well-defined on the proper solutions. Fix a $t_0 > 0$ and consider the solution profile $u(x,t_0)$ with $s(t_0) = 0$ on the complete TW-bundle. Denote

$$\Lambda(t_0) = \{\lambda \in \Delta_0 : \;\; u(x,t_0) \;\; \text{intersects} \;\; f_\lambda(x)\} \neq \emptyset. \tag{7.141}$$

In view of the completeness of B, $\Lambda(t_0) = \emptyset$ would mean $u(x,t) \equiv f_\lambda(\xi)$ for some $\lambda \in \Delta_0$ by uniqueness of the proper solution. We exclude this trivial case from the consideration. Moreover, by the strong MP for uniformly parabolic equations, we have that $u(x,t_0) \not\equiv f_\lambda(x)$ on any compact subset in x from the positivity domain of both solutions.

The set $\Lambda(t_0)$ is open by continuity. It follows from Theorem 7.25 that

$$\Lambda(t_0) \;\; \text{is bounded.} \tag{7.142}$$

Given a small $X > 0$, we introduce two functions

$$\lambda_-(X) = \inf\{\lambda \in \Lambda(t_0) : \;\; u(x,t_0) \;\; \text{intersects} \;\; f_\lambda(x) \;\; \text{on} \; (0,X)\},$$

$$\lambda_+(X) = \sup\{\lambda \in \Lambda(t_0) : \;\; u(x,t_0) \;\; \text{intersects} \;\; f_\lambda(x) \;\; \text{on} \; (0,X)\}.$$

Then $\lambda_-(X) < \lambda_+(X)$, $\lambda_-(X)$ is increasing and $\lambda_+(X)$ is decreasing. It follows that

$$f(x,\lambda_+(X)) \le u(x,t_0) \le f(x,\lambda_-(X)) \quad \text{for} \;\; x \in (0,X). \tag{7.143}$$

By monotonicity, there exist finite limits

$$\lim_{X \to 0+} \lambda_\pm(X) = \lambda^\pm, \quad \text{where} \;\; \lambda^+ \ge \lambda^-. \tag{7.144}$$

If $\lambda^- < \lambda^+$, then, for any $\lambda \in (\lambda^-, \lambda^+)$, $u(x,t_0)$ intersects TW profiles $f_\lambda(x)$ infinitely many times, which contradicts the following easy observation.

Proposition 7.38 *For $t_0 > 0$, the proper solution profile $u(x, t_0)$ cannot have infinitely many intersections with two different TW profiles $f(x, \lambda_1)$ and $f(x, \lambda_2)$ with $\lambda_1 \neq \lambda_2$.*

Proof. We argue by contradiction. Consider the time-evolution of the intersections of the solution $u(x, t)$ and two TWs $V_1(x, t) = f(x - \lambda_1(t - t_0), \lambda_1)$ and $V_2(x, t) = f(x - \lambda_2(t - t_0), \lambda_2)$. By the Sturm Theorem, at $t = 0$ the continuous initial function $u_0(x)$ must have an infinite number of intersections with both $V_1(x, 0)$ and $V_2(x, 0)$, which now have the disjoint interfaces, $s_1(0) = -\lambda_1 t_0 \neq s_2(0) = -\lambda_2 t_0$. Therefore, the intersections with either $V_1(x, 0)$ or $V_2(x, 0)$ are concentrated in a domain of the uniform strict positivity of the corresponding solutions. Therefore, at $t = \frac{1}{2}t_0$ such an infinite number becomes finite due to Theorem 1.4 for uniformly parabolic equations. Hence, for $t > \frac{1}{2}t_0$, one of these numbers is finite contradicting the assumption. □

Notice that this analysis does not rely on the phenomenon of disappearance for any $t > 0$ of an infinite number of intersections between a fixed pair of solutions $u(x, t)$ and $V(x, t)$ with coinciding interfaces. This is known for uniformly parabolic equations and some degenerate ones like the PME (Chapter 1). For singular fully nonlinear equations (7.1), this generates a linearized parabolic equation with a strong moving degeneracy of the coinciding interfaces and such a result on finite number of intersections is not straightforward at all.

We continue the proof of (i). Thus

$$\lambda^- = \lambda^+ = \lambda_0(t_0), \tag{7.145}$$

as is shown in Figure 7.8. Consider the complete, continuous and ordered TW-bundle

$$B = \{f_\lambda(\xi),\ \lambda \in \Delta_0\}.$$

It follows from (7.141)–(7.143) that the first interface operator $\mathbf{M}_1$ given by (7.37) is well-defined for the profile $u(x, t_0)$ and

$$\mathbf{M}_1(u(\cdot, t_0)) = A_0. \tag{7.146}$$

In the case of first order TW interface equations this means that there exists (cf. (7.38))

$$\mathbf{N}_1(u(\cdot, t_0)) = \lambda_0. \tag{7.147}$$

Indeed, if this limit is not defined, $u(x, t_0)$ would have infinitely many intersections with an uncountable subset of TWs from the proper bundle, which is impossible by Proposition 7.38.

Then we obtain that $D^+ s(t_0) = \lambda_0$ by the standard comparison for small $t - t_0 > 0$ with the TW solutions $f_{\lambda_0 \pm \varepsilon}(x - (\lambda_0 \pm \varepsilon)(t - t_0) - s(t_0))$ from the TW-bundle, where we let $\varepsilon \to 0^+$.

In the case of second-order TW interface equations, in addition, we obtain from (7.143) as $X \to 0^+$ that

$$\mathbf{N}_2(u(\cdot, t_0)) = \lambda_0 \in I\!R, \tag{7.148}$$

where $\lambda_0 = D^+ s(t_0)$ by the usual comparison.

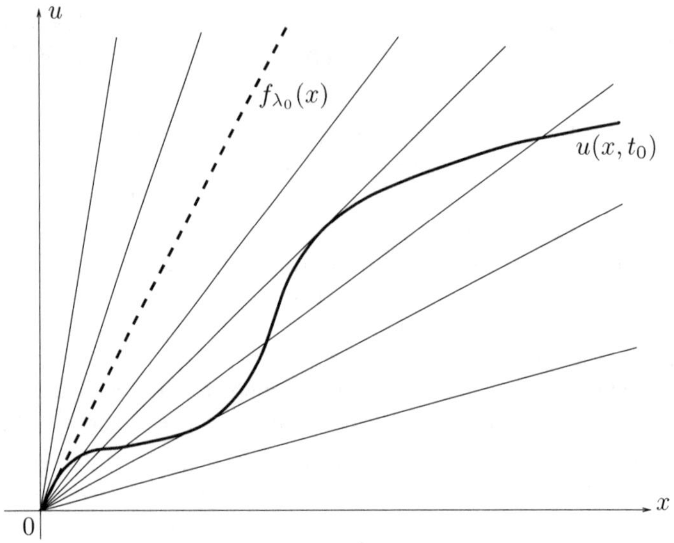

Figure 7.8 *Analysis in the B-bundle: there exists a unique TW* $f_{\lambda_0}(x)$ *(the dotted line) "tangent" to* $u(x,t_0)$ *at the interface* $x = 0$*. This guarantees the existence of suitable interface slopes and operators.*

(ii) We have already proved that $D^+s(t)$ exists and is finite for $t > 0$. Moreover, we have shown that on $[\delta, C]$

$$|D^+s(t)| \le c, \tag{7.149}$$

where c depends on small $\delta > 0$. Consider the left-hand derivative

$$D^-s(t_0) = \lim_{t \to t_0^-} P(t), \quad P(t) = \frac{s(t) - s(t_0)}{t - t_0}.$$

Assuming that there exist two partial limits $P(t_k^{1,2}) \to \lambda_{1,2}$ with $\lambda_1 < \lambda_2$ along two different sequences $\{t_k^{1,2}\} \to t_0^-$ as $k \to \infty$, we have that

$$s(t_k^{1,2}) - s(t_0) = (\lambda_{1,2} + o(1))(t_k^{1,2} - t_0),$$

so that, for $k \gg 1$, the interface $s(t)$ oscillates infinitely many times around the interface $s(t_0) + \lambda(t - t_0)$ of the TWs

$$V(x,t) = f_\lambda(x - \lambda(t - t_0) - s(t_0))$$

for arbitrary $\lambda \in (\lambda_1, \lambda_2)$. This means that $u(x,t)$ has an infinite number of intersections with a subset of these TW solutions V, which is impossible by Proposition 7.38. Therefore, $D^-s(t_0) \in \bar{I\!R}$ exists.

Finally, it is easy to see from (7.149) that $D^-s(t_0)$ is finite and satisfies the same bound. This implies the uniform Lipschitz continuity of the interface on $[\delta, C]$ (cf. Theorem 7.27) and completes the proof. □

Intersection techniques also guarantee some extra regularity of interfaces. Let $S(t) = \mathbf{M}_i(u(\cdot,t))$ with $i = 1$ or 2 be the interface slope (see Section 7.2) corresponding to the velocity $D^+s(t)$. For instance, the following regularity from the right is straightforward.

Proposition 7.39 *Under the hypothesis of Theorem* 7.37*,*

$$S(t) \to S_0 = S(t_0) \quad as \quad t \to t_0^+ > 0.$$

Proof. If there exists a sequence $\{t_k\} \to t_0^+$ such that $S(t_k) \to S_1 \neq S_0$, e.g., $S_1 > S_0$, new intersections would occur at the interface with an infinite number of TWs $f_\lambda(x - \lambda(t - t_0) - a)$ for $\lambda - \lambda_0 > 0$ ($\lambda_0 = D^+s(t_0)$) and $a - s(t_0) > 0$ small. □

Let us state more regularity results proved by geometric techniques.

Theorem 7.40 *Under the hypotheses of Theorem* 7.37*, for* $t_0 > 0$*,*

$$\lim_{t \to t_0^-} D^+s(t) = \lim_{t \to t_0^-} D^-s(t) = D^-s(t_0), \tag{7.150}$$

$$\lim_{t \to t_0^+} D^+s(t) = \lim_{t \to t_0^+} D^-s(t) = D^+s(t_0). \tag{7.151}$$

Proof. It is based on a compactness argument similar to that applied for the eventual concavity in Section 2.3. See details in [160]. □

It is interesting to identify what kind of optimal regularity of the interfaces can be obtained in principle by purely geometric techniques only by intersection comparison with TWs, i.e., without using more advanced methods based on the particular structure of the parabolic PDE under consideration.

Interfaces in the case $\Delta_0 \neq I\!R$

We now discuss more special properties of interfaces occurring in the case

$$\Delta_0 \neq I\!R.$$

The first conclusion is motivated by the results in Chapter 5, where the blow-up equation (7.28) was considered. Then the proper TW subset B is given by (7.29) with $\Delta_0 \neq I\!R$ and the improper set $\bar{B}$, (7.30), is non-empty. Observe that this implied (Section 5.6) that the blow-up interface of a proper solution is not a C^2-function.

Proposition 7.41 *Let* $\Delta = \Delta_0 = [b, \infty)$ *for some* $b \in I\!R$*, and any* $\lambda \approx b^-$ *is in the steep limit of regular approximations. Then, for any* $t_0 \geq 0$*,*

$$\underline{D}^+s(t) \equiv \liminf_{t \to t_0^+} \frac{s(t) - s(t_0)}{t - t_0} \geq b. \tag{7.152}$$

Proof. Let $\{u_n \geq \frac{1}{n}\}$ be a monotone sequence of the regularized solutions and $u = \lim u_n$. Fix an arbitrarily small $\varepsilon > 0$ and consider the monotone increasing sequence of the regular TW profiles $\{f^n(\xi, b - \varepsilon)\}$ satisfying the corresponding ODE with the conditions (7.17). Since $b - \varepsilon \notin \Delta$ by assumption, we have that $P_n(f) \to \infty$ as $n \to \infty$ for any $f > 0$. The profile $f^n(\xi, b - \varepsilon)$ becomes arbitrarily steep on any fixed level for $n \gg 1$. Therefore, for any $n \gg 1$, there exists a small $\delta = \delta(\varepsilon) > 0$ such that

$$u_{0n}(x) \le f^n(x+\delta, b-\varepsilon) \quad \text{for small } x > 0.$$

Hence,

$$u_n(x,t) \le f^n(x-(b-\varepsilon)t+\delta, b-\varepsilon) \quad \text{for small } t > 0 \text{ and } x \approx 0$$

by comparison. Passing to the limit $n \to \infty$ yields $u(x,t) \le H_\infty(x-(b-\varepsilon)t+\delta)$, where H_∞ is the infinite-step Heaviside function

$$H_\infty(z) = \begin{cases} \infty & \text{for } z > 0, \\ 0 & \text{for } z \le 0. \end{cases}$$

Therefore $s(t) \ge (b-\varepsilon)t - \delta$, and we obtain (7.152) by passing to the limit $\varepsilon \to 0^+$ (then $\delta(\varepsilon) \to 0^+$). Furthermore, we obtain that

$$s(t) \ge bt. \tag{7.153}$$

□

As a typical property of proper solutions in the case $\Delta_0 \ne I\!R$, we give the following result on the uniform interface propagation generated by the initial improper TW.

Proposition 7.42 *Let $\Delta = \Delta_0 = [b, \infty)$, any $\lambda \approx b^-$ be in the steep limit and $\bar{\Delta}_0 \ne \emptyset$. Fix an arbitrary $c \in \bar{\Delta}_0 \subseteq \Delta_0$ and the corresponding improper TW profile $\bar{f}_c \in \bar{B}$. Let $u(x,t)$ be the proper solution with the initial data $u_0(x) = \bar{f}_c(x)$. Then*

$$s(t) \equiv bt. \tag{7.154}$$

Proof. By the strict monotonicity in λ of $B = \{f_\lambda, \lambda \in \Delta_0\}$, we have that $\bar{f}_c(x) \ge f_b(x) \ge f_c(x)$. Therefore, $u(x,t) \ge f_b(x-bt)$ and hence $s(t) \le bt$. Combining with (7.153) yields (7.154). □

Thus, if the TW-diagram consists of a continuous proper branch and an improper one, we observe interfaces moving with constant speed for all $t > 0$, representing moving singular internal layers. Such layers can occur in finite time and then the interface loses C^2-regularity at the moment of straightening.

The case of the nonexistence of the proper TW of flat type motivated by the PME is simpler.

Proposition 7.43 *Let $\Delta = \Delta_0 = (-\infty, b)$, $b \in I\!R$, and $\lambda = b$ be in the flat limit of regular approximations. Then*

$$\underline{D}^+ s(t) \le b. \tag{7.155}$$

Proof. We compare $u_n(x,t)$ with arbitrarily small solutions $f_b^n(x - bt - a_n)$ and pass to the limit $n \to \infty$. □

On interface velocity estimates in $I\!R^N$

Consider a maximal solution $u(x,t) \ge 0$ of equation (7.7). Assume that its singular interface in $I\!R^N$, $\Gamma(t) = \partial \text{supp}\, u(\cdot,t)$ given by the equation $t = \Sigma(x)$, is C^2-smooth at $t = t_0$ in a neighbourhood of x_0, $t_0 = \Sigma(x_0)$. We perform a straightforward construction of estimates on its normal velocity $v^\perp$ by comparison with radial solutions satisfying

$$v_t = F(v, v_r, \Delta_r v).$$

An upper estimate. We take a ball $B_{r_0}(x_1)$, $x_1 \notin \operatorname{supp} u(\cdot, t)$ such that

$$\bar{B}_{r_0}(x_1) \cap \operatorname{supp} u(\cdot, t_0) = \{x_0\},$$

i.e., the ball touches the singular surface at x_0 from outside the support. Setting $r = |x - x_1|$ (then $r_0 = |x_0 - x_1|$), we fix a monotone increasing radial supersolution of equation (7.7)

$$V(r,t) = f(r - \bar{\lambda}(t - t_0) - r_0), \quad f(0) = 0, \quad f'(\xi) > 0 \ \text{ for } \xi > 0,$$

with some $\bar{\lambda} \in \Delta_0$, $\bar{\lambda} \ll -1$, to be obtained from the 1D parabolic equation

$$V_t = F(V, V_r, V_{rr} + \tfrac{N-1}{r_0} V_r) \quad \text{in } \{r > r_0\} \times \{t > t_0\}.$$

Proposition 7.44 *Assume that, given a* $\bar{\lambda} \ll -1$, $u(x, t_0) \le f(r - r_0)$ *for all* $x \approx x_0$ *in the positivity domain. Then the normal interface velocity* $v_+^\perp(x_0, t_0)$, *understood in the sense of* $\limsup_{t \to t_0^+}$, *satisfies*

$$v_+^\perp(x_0, t_0) \le -\bar{\lambda}. \tag{7.156}$$

Proof. By comparison, $u(x,t) \le f(r - r_0(t))$, where $r_0(t) = \bar{\lambda}(t - t_0) + r_0$, for any $t - t_0 > 0$ small and $x \approx x_0$. Therefore,

$$\operatorname{supp} u(\cdot, t) \cap B_{r_0(t)}(x_1) = \emptyset,$$

which implies (7.156). □

A lower estimate. Consider a radial decreasing TW subsolution of (7.7)

$$V(r,t) = f(r - \underline{\lambda}(t - t_0) - r_0), \quad f(0) = 0, \quad f'(\xi) < 0 \ \text{ for } \xi < 0,$$

where $r = |x - x_1|$ and $r_0 = |x_0 - x_1|$. Let $x_1 \in \operatorname{supp} u(\cdot, t_0)$ be such that $B_{r_0}(x_1)$ touches the singular surface at x_0 from inside. Then $V(r, t_0)$ stays below $u(x, t_0)$ for all $x \approx x_0$. This gives a lower estimate on the normal velocity

$$v_-^\perp(x_0, t_0) \ge \underline{\lambda}, \tag{7.157}$$

which is understood in the $\liminf_{t \to t_0^+}$ sense.

Uniqueness for FBPs for maximal solutions

By construction, the proper maximal solution is unique for a general class of singular equations. Once we have established in Theorem 7.37 a precise statement of the corresponding FBPs and some typical regularity properties of interfaces, one needs an extra uniqueness result for such FBPs for maximal solutions. Recall that the regularity class of interfaces $\{x = s(t)\}$ here is as follows:

(i) $s(t)$ is continuous on $[0, C]$ for any $C > 0$, and

(ii) $D^\pm s(t)$ are uniformly bounded on $[\delta, C]$ for any $\delta > 0$ and $s(t)$ is uniformly Lipschitz continuous.

Theorem 7.45 *Let B with $\Delta_0 = I\!R$ be proper and* (7.112)–(7.114) *hold. Assume that the TW-diagrams* (7.39) *or* (7.45) *are sufficiently smooth and strictly monotone decreasing for all slopes S. Then the corresponding FBP* (7.139) *or* (7.140) *for equation* (7.1) *with initial data u_0 admits a unique solution.*

Proof. Such a solution exists and it is the proper maximal one satisfying regularity properties stated in Theorem 7.37. With the above interface regularity class, the proof of uniqueness for this FBP coincides with that of Proposition 5.17 in Section 5.6. It suffices to replace the interface slope $S = v_x$ either by $S = \mathbf{M}_1(u)$ (for first-order interfaces) or by $S = \mathbf{M}_2(u)$ (for second-order ones), and instead of s' we write $D^+ s$. $\square$

7.11 Applications to various nonlinear models with extinction and blow-up singularities in $I\!R$ and $I\!R^N$

In this section we apply the main results of the geometric intersection comparison analysis to a number of nonlinear singular equations whose right-hand sides are composed of different nonlinear operators of diffusion and reaction-absorption-convection types.

Let us comment on some specific properties of the nonlinear models involved. Firstly, in several cases the construction of regularized sequences of positive non-singular solutions is done by replacing the strong absorption term, say $-u^p$, or similar ones, by a smooth uniformly Lipschitz continuous function

$$-\psi_n(u) \equiv -u^p \quad \text{for } u \geq \tfrac{2}{n}, \quad \psi_n(u) = 0 \quad \text{for } u \leq \tfrac{1}{n},$$

where $\{-\psi_n(u)\}$ is strictly monotone decreasing with n. By the MP, regular solutions $\{u_n\}$ do not reach the singular level, and standard known results from the theory of quasilinear and fully nonlinear equations guarantee the unique global classical solvability of the regularized Cauchy problems. By monotonicity, the proper solution $u = \lim u_n$, trivial (entirely singular) or nontrivial, is defined, and next we describe its regularity properties using the G-theory. For instance, in general, we have that interfaces are Lipschitz continuous, derivatives $D^{\pm} s(t)$ exist and uniformly bounded on time intervals $[\delta, T - \delta]$ bounded away from the extinction time T and from the waiting time T_w (if any).

Secondly, in equations, where singularities can occur on subsets like $\{u_x = 0\}$ or $\{u_{xx} = 0\}$ or others, we have to assume that such singularity formation sets either coincide with the fixed one $\{u = 0\}$ (e.g., u_x or u_{xx} can vanish at $u = 0$ only), or are disjoint so that we never observe evolution interactions between different types of singularities and singularity subsets. Such complicated singularity phenomena cannot be described by TWs and possibly need families of other similarity or group-invariant solutions. For instance, in order to fix the unique singularity formation level $\{u = 0\}$, we consider special classes of initial data for which such singularity interaction is impossible. This can be controlled by the MP applied to parabolic equations satisfied by the derivatives u_x, u_{xx} or other differential expressions of solutions.

As usual, we present a detailed description of 1D models, and from time to time will comment on existence/nonexistence and regularity results for the corresponding N-dimension equations.

Quasilinear heat equations with absorption

We begin with the 1D equation of non-stationary filtration of non-Newtonian fluids with absorption

$$\boxed{u_t = [|(u^m)_x|^\sigma (u^m)_x]_x - u^p, \quad m > 0, \ \sigma > -1,} \tag{7.158}$$

where monotone increasing smooth initial data $u_0(x)$ satisfy necessary hypotheses. By $T = T(u_0) > 0$ we then denote finite or infinite extinction time. In general, the behaviour of solutions as $t \to T^-$ represents an interesting, difficult and challenging problem of the asymptotic theory, which cannot be treated by using TWs only and hence fall out of the scope of the present geometric theory. We refer to [170, Chapter 5] for results and the literature. Therefore, describing typical regularity properties of proper solutions $u(x,t)$, we always take $t < T$.

If $\sigma > 0$, we assume that $u_x \neq 0$ in an open right-hand neighbourhood of the interface so that $\{u = 0\}$ is the only possible singular level. The TW solutions $V(x,t) = f(x - \lambda t)$ are determined from the second-order ODE

$$[|(f^m)'|^\sigma (f^m)']' - f^p + \lambda f' = 0. \tag{7.159}$$

Setting $P = \mathrm{d}f^m/\mathrm{d}\xi > 0$ for $f > 0$ yields the first-order ODE

$$(1+\sigma)P^{1+\sigma}\frac{\mathrm{d}P}{\mathrm{d}f} = G(f) - \lambda P, \tag{7.160}$$

where

$$G(f) = mf^{p+m-1}.$$

Consider first the problem of the existence/nonexistence of a nontrivial proper solution of the Cauchy problem for (7.158). These results are extended to its N-dimensional counterpart

$$\boxed{u_t = \nabla \cdot (|\nabla u^m|^\sigma \nabla u^m) - u^p,}$$

with compactly supported initial data; cf. Theorem 7.12 (nonexistence), and Theorem 7.8 (existence).

Theorem 7.46 *There holds:*
(i) If $p \le -m$, then $\Delta = \emptyset$ and $u(x,t) \equiv 0$ for $t > 0$.
(ii) If $p > -m$, then $\Delta \neq \emptyset$ and $u(x,t) \not\equiv 0$ for small $t > 0$.

In the case (ii) the proper solutions can have finite interfaces (if $\Delta_0 \neq \emptyset$) or can be strictly positive (if $\Delta_0 = \emptyset$). See the classification below. The theorem follows from the criterion of the existence of the minimal orbit $P_{\min}$.

Lemma 7.47 *The convergence of the integral*

$$\int_0^1 G(s)\,\mathrm{d}s < \infty \Longleftrightarrow p > -m \tag{7.161}$$

is the necessary and sufficient condition for the existence of the minimal orbit $P_{\min} = \lim P_n < \infty$.

Proof. (i) *Nonexistence.* Since $G(f) \geq 0$ we have

$$(1+\sigma)P^{\sigma}\frac{\mathrm{d}P}{\mathrm{d}f} \geq -\lambda.$$

Integrating this inequality over $(0, \delta)$, we conclude that $P(0)$ exists and is finite. Therefore in the domain $\{P \leq c\}$

$$(1+\sigma)P^{\sigma}\frac{\mathrm{d}P}{\mathrm{d}f} = \frac{G(f)}{P} - \lambda \geq \frac{G(f)}{c} - \lambda.$$

Thus, if the integral diverges,

$$\int_0^1 G(s)\,\mathrm{d}s = \infty, \tag{7.162}$$

for any $\lambda \in I\!R$, the minimal orbit does not exist and $P_n(f) \to \infty$ as $n \to \infty$ for any small $f > 0$ (the steep limit).

(ii) *The existence of* $P_{\min}$. Obviously, under convergence (7.161) $P_{\min}$ exists for $\lambda = 0$. Let us prove that it then also exists for any $\lambda > 0$. Define the function

$$Y(f) = A + \frac{1}{1+\sigma}\frac{1}{A^{1+\sigma}}\int_0^f G(s)\,\mathrm{d}s, \quad A > 0. \tag{7.163}$$

Then

$$\frac{\mathrm{d}P}{\mathrm{d}f}\Big|_{P=Y(f)} = \frac{G(f)}{(1+\sigma)Y^{1+\sigma}(f)} - \frac{\lambda}{(1+\sigma)Y^{\sigma}(f)} < \frac{G(f)}{(1+\sigma)A^{1+\sigma}} \equiv \frac{\mathrm{d}Y}{\mathrm{d}f}.$$

This means that the orbits intersecting the curve $P = Y(f)$ stay in the domain $\{P > Y(f),\ f > 0\}$ and are well-defined for all $f > 0$ small. Then the minimal orbit $P = P_{\min}(f)$ is the separatrix that is constructed as the limit $P_{\min} = \lim P_n$. □

Consider the existence case $p > -m$ and let us perform a more detailed analysis of the TW set B with finite propagation $\Delta_0 \neq \emptyset$. We begin with the most interesting case of the second-order interface equation where the proper solutions exhibit a special behaviour near the interfaces.

1. Second-order interface equation, $\Delta_0 = I\!R$**.** We impose the conditions

$$-m < p < p_c = \tfrac{2+\sigma}{1+\sigma} - m, \quad p < m(1+\sigma), \tag{7.164}$$

where p_c is a critical exponent. We will show that it separates the parameter ranges with the interfaces of second and first order. The first inequality says that the first term $G(f)$ on the right-hand side of (7.160) is the leading one for small $f > 0$ and, as $f \to 0$, the minimal orbit has the form

$$P_{\min}(f) = f^{\gamma}[c_0 - \lambda c_1 f^{\gamma+1-p-m}(1+o(1))], \quad \gamma = \tfrac{p+m}{2+\sigma}, \tag{7.165}$$

where $c, \tilde{c}, A_0, c_0, c_1, c_2, \ldots$ denote different positive constants depending on the parameters m, σ and p, and

$$c_0 = \left[\tfrac{m(2+\sigma)}{(1+\sigma)(p+m)}\right]^{\frac{1}{2+\sigma}}.$$

Integrating the equation

$$\frac{\mathrm{d}f^m}{\mathrm{d}\xi} = P_{\min}(f)$$

and taking into account the second inequality in (7.164), we obtain that, for any $\lambda \in I\!R$, there exists a unique TW profile with the following behaviour near the interface as $\xi \to 0^+$:

$$f^\alpha(\xi) = A_0\xi - \lambda c_2 \xi^\beta (1 + o(1)), \tag{7.166}$$

where

$$\alpha = \tfrac{m(1+\sigma)-p}{2+\sigma} > 0, \;\; \beta = \tfrac{(2+\sigma)(1-p)}{m(1+\sigma)-p} > 1, \;\; A_0 = \tfrac{m(1+\sigma)-p}{m(2+\sigma)}\, c_0.$$

Proposition 7.48 *If* (7.164) *holds, then* $\Delta_0 = I\!R$ *and the TW-bundle* B *is complete, sufficiently smooth and strictly monotone decreasing with* λ*.*

It follows from (7.166) that the pressure variable is $v = Q(u) = u^\alpha$ and the following two interface operators are well defined on the complete TW-bundle: for any $f \in B$ there exist the first-order operator

$$\mathbf{M}_1(f) \equiv \lim_{\xi \to 0+} \frac{1}{\xi} f^\alpha(\xi) \equiv (f^\alpha)'(0) = A_0, \tag{7.167}$$

and the second-order one

$$\mathbf{N}_2(f) \equiv \frac{1}{c_2}\left\{ \lim_{\xi \to 0+} \frac{1}{\xi}[A_0\xi - f^\alpha(\xi)]^{\frac{1}{\beta}} \right\}^\beta = \lambda > 0. \tag{7.168}$$

The corresponding differential form, which applies to the TW profiles, is obtained from (7.166). Differentiating once yields

$$(f^\alpha)' = A_0 - \lambda \tilde{c} \xi^{\beta-1} + \dots .$$

Straightening once more, we have

$$[(f^\alpha)' - A_0]^{1/(\beta-1)} = [(-\lambda)\tilde{c}]^{1/(\beta-1)}\xi + \dots \quad \text{for } \lambda < 0,$$

and differentiating again, we obtain the second-order interface operator

$$\mathbf{N}_2(f) = \lambda = -\tilde{c}\{[((f^\alpha)' - A_0)^{1/(\beta-1)}]'\}^{\beta-1}.$$

Similarly, we derive the operator for $\lambda > 0$. The interface of the TWs propagates via the dynamic equation

$$s'(t) = \mathbf{N}_2(V(s(t), t)).$$

Next, let us determine the G-function. It follows from (7.166) that, for small $\varepsilon > 0$,

$$G(\lambda, \varepsilon) \equiv \frac{\mathrm{d}}{\mathrm{d}\xi} Q(f)\,|_{Q(f)=\varepsilon} = A_0 - \lambda c_3 \varepsilon^{\beta-1}(1 + o(1)). \tag{7.169}$$

We now perform the regularity analysis in the B-classes. One can conclude from (7.169) and equation (7.160) that (7.112) and (7.113) hold. In order to check (7.114) (which is necessary for the instantaneous smoothing in $B^-_{\rm loc}$) we observe that given small $\varepsilon > 0$ we have for $\lambda \gg 1$ that $P_{\min}(f) \approx P_0(f)^-$ for $f \gg 1$, with the zero cline

$$P_0(f) = \tfrac{m}{\lambda}\, f^{m+p-1}.$$

Therefore

$$\frac{{\rm d}f}{{\rm d}\xi} \approx \tfrac{m}{\lambda}\, f^p \quad \text{for } \xi \gg 1.$$

Integrating, we conclude that $f^\alpha(\xi) \approx h(\frac{\xi}{\lambda})$ for $\xi \gg 1$, where h is a smooth monotone function satisfying $h(0) = 0$. Solving the algebraic equation for the intersection point $\xi_*(\lambda, \varepsilon)$: $h(\frac{\xi_*}{\lambda}) \approx \varepsilon$, we have $\frac{\xi_*}{\lambda} \approx h^{-1}(\varepsilon) \to 0$ as $\varepsilon \to 0$.

We summarize the results on the regularity properties of solutions to the PDE (7.158) as follows.

Theorem 7.49 *In the range* (7.164), *on* $[\delta, T-\delta]$ *with any small* $\delta > 0$ *there holds:*

(i) The gradient Bernstein estimate near the interface has the form

$$|(u^\alpha)_x - A_0| \le c(\delta) u^\mu, \quad \textit{where} \quad \mu = \alpha(\beta - 1) > 0. \tag{7.170}$$

(ii) The interface equation is the system

$$\begin{cases} \mathbf{M}_1(u(\cdot,t)) = A_0, \\ D^+ s(t) = \mathbf{N}_2(u(\cdot,t)). \end{cases} \tag{7.171}$$

In particular, (7.170) implies that $u^\alpha(x,t)$ is uniformly Lipschitz continuous in x, which gives the optimal exponent $\nu = \min\{1, \frac{1}{\alpha}\}$ of the Hölder continuity in x of the solutions $u(x,t)$. As we have proved in Section 7.7, the modulus of continuity in t is the same as in x.

2. First-order interface equation for $p = p_c$**,** $\Delta_0 = I\!\!R$**.** In the critical case

$$p = p_c \quad (m(1+\sigma) > 1) \tag{7.172}$$

for any $\lambda \in \Delta = I\!\!R$, the ODE (7.160) admits the exact solution

$$P_{\min}(f) = A_0(\lambda) f^{1/(1+\sigma)}, \tag{7.173}$$

where $A_0(\lambda) > 0$ is the unique positive root of the algebraic equation

$$A_0^{2+\sigma} + \lambda A_0 - m = 0. \tag{7.174}$$

The function $A_0(\lambda)$ is well-defined, strictly monotone and continuous for all $\lambda \in I\!\!R$. The inverse A_0^{-1} exists and the TW-diagram consists of a single analytic monotone decreasing branch

$$\lambda = A_0^{-1}(S) \equiv -S^{1+\sigma} + \frac{m}{S} \quad \text{for } S > 0.$$

Integrating equation

$$(f^m)' = A_0(\lambda) f^{1/(1+\sigma)}$$

yields the proper TW-bundle

$$f^\alpha(\xi) = c A_0(\lambda) \xi, \quad \alpha = \tfrac{m(1+\sigma)-1}{1+\sigma} > 0, \;\; c = \tfrac{\alpha}{m}. \tag{7.175}$$

Thus the pressure variable stays the same, $v = u^\alpha$, but due to (7.175), the interface is of first order. We introduce the first-order interface operator

$$\mathbf{M}_1(f) = \lim_{\xi \to 0+} \frac{1}{\xi} f^\alpha(\xi) \equiv (f^\alpha)'(0) = cA_0(\lambda) \tag{7.176}$$

and set

$$\mathbf{N}_1(f) \equiv A_0^{-1}(\tfrac{1}{c}\mathbf{M}_1(f)). \tag{7.177}$$

The G-function does not depend on the level $\{f^\alpha = \varepsilon\}$ and

$$G(\lambda, \varepsilon) \equiv \frac{\mathrm{d}f^\alpha}{\mathrm{d}\xi} = cA_0(\lambda). \tag{7.178}$$

Using the completeness, continuity and monotonicity of B, we next introduce the local B-classes. Conditions (7.112) and (7.113) are obviously valid. Consider (7.114). Computing the intersection point ξ_* with the ε-level, we have that, for solutions (7.175), there holds

$$\frac{\xi_*(\lambda, \varepsilon)}{\lambda} \sim \frac{\varepsilon}{\lambda A_0(\lambda)} \quad \text{for } \lambda \gg 1.$$

Since, by (7.174), $A_0(\lambda) \approx \frac{m}{\lambda}$ for $\lambda \gg 1$, we have that $\lambda A_0(\lambda) \to m$ as $\lambda \to \infty$ so that (7.114) is true.

Theorem 7.50 *In the critical case* (7.172) *on intervals* $[\delta, T - \delta]$, *we have:*
(i) The gradient estimate near the interface is

$$0 < c_2(\delta) \le (u^\alpha)_x \le c_1(\delta).$$

(ii) The interface equation is $D^+ s(t) = \mathbf{N}_1(u(\cdot, t))$.

3. First-order Interfaces for $p \in (p_c, 1)$, $\Lambda_0 = I\!R$**.** Consider the supercritical case

$$p_c < p < 1 \quad (m(1+\sigma) > 1). \tag{7.179}$$

The minimal orbit is

$$P_{\min} = \begin{cases} [(-\lambda)f]^{1/(1+\sigma)}(1 + o(1)) & \text{for } \lambda < 0, \\ cf^{(p+m)/(2+\sigma)} & \text{for } \lambda = 0, \\ \lambda^{-1} m f^{p+m-1}(1 + o(1)) & \text{for } \lambda > 0. \end{cases}$$

The TW-bundle is composed of the profiles (here A_0 is the constant in (7.166))

$$f^\alpha = (\alpha m)^{-1}(-\lambda)^{\frac{1}{1+\sigma}}\xi(1 + o(1)), \ \lambda < 0, \ \alpha = \tfrac{m(1+\sigma)-1}{1+\sigma}; \tag{7.180}$$

$$f^\alpha(\xi) = A_0\xi, \ \lambda = 0, \ \alpha = \tfrac{m(1+\sigma)-p}{2+\sigma};$$

$$f^{1-p}(\xi) = \tfrac{1}{\lambda}(1 - p)\xi(1 + o(1)), \ \lambda > 0.$$

Hence, $\Delta_0 = I\!R$, the pressure is $Q(u) = u^\alpha$ for $\lambda < 0$, $Q(u) = u^\gamma$ for $\lambda = 0$ and $Q(u) = u^{1-p}$ for $\lambda > 0$. Calculating the G-function and checking the necessary conditions for the instantaneous smoothing in the B-classes, we obtain

Theorem 7.51 *In the parameter range* (7.179) *on* $[\delta, T - \delta]$ *there holds:*
(i) The gradient bound near the interface is

$$(u^\alpha)_x \le c_1(\delta) \quad \textit{and} \quad (u^{1-p})_x \ge c_2(\delta).$$

(ii) The interface equation is $D^+s(t) = \mathbf{N}_1(u(\cdot,t))$, *where*

$$\mathbf{N}_1(u(\cdot,t)) = \begin{cases} -[\alpha m(u^\alpha)_x(s(t),t)]^{1+\sigma} < 0 & \text{if } (u^\alpha)_x > 0, \\ 0 & \text{if } (u^\gamma)_x = 0, \\ (1-p)[(u^{1-p})_x(s(t),t)]^{-1} > 0 & \text{if } (u^{1-p})_x > 0. \end{cases}$$

4. First-order interfaces for $p \geq 1$, $\Delta_0 = (-\infty, 0]$. In the case

$$1 \leq p < (1+\sigma)m \quad (m(1+\sigma) > 1), \tag{7.181}$$

for any $\lambda < 0$, the first term on the right-hand side of (7.160) is negligible and we obtain the same solution as in (7.180). If $\lambda = 0$, we arrive at the profile (7.166). The instantaneous smoothing effect in the local B^+-class is no longer available and the waiting time phenomenon is possible.

Theorem 7.52 *In the parameter range* (7.181) *on* $[T_w + \delta, T - \delta]$, *where* $T_w \in [0,T)$ *is the waiting time, there holds:*

(i) The gradient bound near the interface is

$$c_2(\delta) \leq (u^\alpha)_x \leq c_1(\delta).$$

(ii) The interface equation is

$$D^+s(t) = \mathbf{N}_1(u(\cdot,t)) \equiv -[\alpha m(u^\alpha)_x(s(t),t)]^{1+\sigma} < 0.$$

5. First-order interfaces for $p \geq m(1+\sigma)$, $\Delta_0 = (-\infty, 0)$. Here a complete TW-bundle is composed of functions (7.180), and the previous theorem holds.

6. Quasilinear diffusion-absorption-convection equation. As a new extension, we add an extra convection term to the quasilinear equation

$$\boxed{u_t = [|(u^m)_x|^\sigma (u^m)_x]_x - u^p - (u^\nu)_x.} \tag{7.182}$$

Then the first-order ODE takes the form

$$(1+\sigma)P^{1+\sigma}\frac{\mathrm{d}P}{\mathrm{d}f} = G(f) - \lambda P + \nu f^{\nu-1}P, \tag{7.183}$$

where

$$G(f) = mf^{p+m-1}.$$

The analysis of (7.183) is the same. In particular, one can check that a similar second-order interface with $\Delta_0 = I\!R$ occurs if the convection term is not that strong and, in addition to (7.164), there holds

$$\nu > \nu_c = \tfrac{(1+\sigma)(p+m)}{2+\sigma}.$$

In the critical case (7.172) and $\nu = \nu_c$, the exact solution (7.173), where $A_0(\lambda) > 0$ satisfies

$$A_0^{2+\sigma} + (\lambda + \nu_c)A_0 - m = 0,$$

generates a complete TW-bundle with the first-order interface equation. The rest of the cases are straightforward.

On the other hand, if $\nu < \nu_c$, $\nu < 1$ $(\nu < m(1+\sigma))$, the last term in (7.183) is the leading one, and we arrive at the second-order interface equation governed by the convection term.

7. A fully nonlinear generalization. Consider now the fully nonlinear equation

$$\boxed{|u_t|^{\gamma-1}u_t = [|(u^m)_x|^\sigma (u^m)_x]_x - u^p, \quad m > 0, \ \sigma > -1, \ \gamma > 0.} \tag{7.184}$$

The ODE for the TW profiles $V = f(\xi)$ takes the form

$$[|(f^m)'|^\sigma (f^m)']' - u^p + B_0|f'|^{\gamma-1}f' = 0, \quad B_0(\lambda) = |\lambda|^{\gamma-1}\lambda.$$

Setting $P = (f^m)'$ and using that $P > 0$ and $\frac{\mathrm{d}P}{\mathrm{d}f} > 0$ near the interface, we obtain the equation

$$(1+\sigma)P^{1+\sigma}\frac{\mathrm{d}P}{\mathrm{d}f} = G(f) - B_0 m^{1-\gamma} f^{(1-m)(\gamma-1)} P^\gamma, \tag{7.185}$$

where

$$G(f) = m f^{p+m-1}.$$

Lemma 7.53 *There holds*

$$\Delta \neq \emptyset \Longleftrightarrow \int_0^1 G(s)\,\mathrm{d}s < \infty \Longleftrightarrow p > -m. \tag{7.186}$$

It follows from Theorem 7.12 that, if $p \leq -m$, then, for any compactly supported initial data, the equation

$$\boxed{|u_t|^{\gamma-1}u_t = \nabla\cdot(|\nabla u^m|^\sigma \nabla u^m) - u^p \quad \text{in } I\!R^N \times I\!R_+}$$

admits the trivial maximal solution $u(x,t) \equiv 0$ only.

Let us perform necessary computations for the second-order interfaces in the subcritical range:

$$-m < p < p_c = \gamma\,\tfrac{2+\sigma-m(1+\sigma)}{2+\sigma-\gamma}, \quad p < m(1+\sigma) \quad (\gamma < 2+\sigma). \tag{7.187}$$

Integrating (7.185) two times, we obtain a complete proper TW-bundle:

$$f^\alpha(\xi) = A_0\xi - B_0(\lambda)c_2\xi^\beta(1+o(1)), \ \alpha = \tfrac{m(1+\sigma)-p}{2+\sigma}, \tag{7.188}$$

$$\beta = \tfrac{(2+\sigma)(1-p)+(\gamma-1)[p+m-(2+\sigma)(m-1)]}{m(1+\sigma)-p},$$

where $\alpha > 0$ and $\beta > 1$. Next, using operators (7.167) and $\tilde{\mathbf{N}}_2$ given by (7.168), we define the main second-order interface operator as follows

$$\mathbf{N}_2(f) = B_0^{-1}(\tilde{\mathbf{N}}_2(f)). \tag{7.189}$$

Differentiating (7.188), we obtain the G-function as $\varepsilon \to 0$

$$G(\lambda,\varepsilon) = A_0 - B_0(\lambda)c_3\varepsilon^{\beta-1}(1+o(1)).$$

Checking the conditions of the smoothing in the B-classes, finally we arrive at Theorem 7.49. The exponent in the Bernstein estimate is $\mu = \alpha(\beta-1) > 0$. In the critical range $p = p_c$, where equation (7.185) admits exact solutions, and in the supercritical one $p > p_c$ the interfaces are of first-order and the analysis is similar.

Blow-up interfaces for quasilinear equation with source

We now apply the results to solutions of the quasilinear reaction-diffusion equation

$$\boxed{v_t = [|(v^m)_x|^\sigma (v^m)_x]_x + v^p, \quad m > 0, \ \sigma > -1, \ p > 1.} \tag{7.190}$$

The transformation $u = \frac{1}{v}$ reduces the blow-up level $\{v = +\infty\}$ to the zero level $\{u = 0\}$ for the following quasilinear equation with strong absorption and non-divergent diffusion operator:

$$u_t = -u^2[|(u^{-m})_x|^\sigma (u^{-m})_x]_x - u^q, \quad q = 2 - p < 1. \tag{7.191}$$

The corresponding ODE for the TW solutions $V = f(\xi)$ is

$$-f^2[(-(f^{-m})')^\sigma (f^{-m})']' - f^{2-p} + \lambda f' = 0.$$

Setting $P = -(f^{-m})' = mf^{-m-1}f' > 0$ yields the first-order ODE

$$(1+\sigma)P^{1+\sigma}\frac{\mathrm{d}P}{\mathrm{d}f} = G(f) - \lambda f^{-2}P, \quad G(f) = mf^{-p-m-1}. \tag{7.192}$$

Since $\int_0 G(s)ds$ diverges, unlike the absorption case, we always have that

$$\Delta \cap (-\infty, 0] = \emptyset. \tag{7.193}$$

Furthermore, we conclude that the blow-up interfaces are always of first order. By Theorem 7.12, the equation in $I\!R^N$

$$u_t = -u^2\nabla \cdot (|\nabla u^{-m}|^\sigma \nabla u^{-m}) - u^{2-p}, \quad p > p_c = \tfrac{2+\sigma}{1+\sigma} - m,$$

with compactly supported u_0 admits trivial solutions only, $u \equiv 0$ for $t > 0$.

1. The critical case $p = p_c$**.** For the critical exponent, there exists the exact minimal orbit

$$P_{\min} = A_0 f^{-1/(1+\sigma)},$$

where $A_0(\lambda) > 0$ solves the algebraic equation

$$A_0^{2+\sigma} - \lambda A_0 + m = 0.$$

The proper profiles correspond to the minimal branch so that

$$\Delta = \Delta_0 = [\lambda_0, \infty), \quad \lambda_0 = (2+\sigma)\left(\tfrac{m}{1+\sigma}\right)^{\frac{1+\sigma}{2+\sigma}},$$

and the slope $S = A_0(\lambda)$ is also given by the minimal branch. This means that we always have $D^+ s(t) \geq \lambda_0$, by Proposition 7.41. The TW-bundle is composed of the profiles

$$f^\alpha(\xi) = cA_0(\lambda)\xi, \ \lambda \in \Delta_0; \quad \alpha = \tfrac{1-m(1+\sigma)}{1+\sigma} > 0, \ c = \tfrac{\alpha}{m}.$$

The TW-diagram consists of a monotone decreasing branch

$$\lambda = S^{1+\sigma} + \frac{m}{S}, \quad 0 < S \leq S_0 = \left(\tfrac{m}{1+\sigma}\right)^{\frac{1}{2+\sigma}}.$$

This gives the corresponding first-order interface operator, the G-function and we arrive at a first-order interface equation. The waiting time phenomenon is not possible (since $\{\lambda \gg 1\} \subset \Delta_0$) and the interface is moving for all $t \in (0,T)$.

2. Subcritical range $1 < p < p_c$**.** In the subcritical case, the minimal orbit

$$P_{\min} = \tfrac{m}{\lambda}\, f^{-p-m+1}(1+o(1)), \quad \lambda \in \Delta_0 = I\!R_+,$$

generates the TW-bundle

$$f^{p-1}(\xi) = \tfrac{1}{\lambda}\,(p-1)\,\xi(1+o(1)),$$

the corresponding interface operator and the G-function. The interface is moving for arbitrarily small $t > 0$.

The dual PME with absorption

Consider the dual PME with absorption

$$\boxed{u_t = |u_{xx}|^{m-1}u_{xx} - u^p, \quad m > 0.} \tag{7.194}$$

For $m > 1$, this equation degenerates on the set $\{u_{xx} = 0\}$. Therefore, we consider the class of solutions with no spatial inflections for small $u > 0$. On the other hand, one can introduce related "regularized" equations such as

$$u_t = \left[u^2 + (u_{xx})^2\right]^{\frac{m-1}{2}} u_{xx} - u^p$$

(or any similar one) with the only singular level at $u = 0$. For (7.194), the interaction between the singular sets $\{u = 0\}$ and $\{u_{xx} = 0\}$ (focusing or collision-like patterns) needs self-similar solutions creating distinct blow-up singularities.

Consider the TW solutions of (7.194). Setting $f' = P$ in the ODE for the TW profiles

$$|f''|^{m-1}f'' - f^p + \lambda f' = 0,$$

we obtain

$$|PP_f'|^{m-1}PP_f' = f^p - \lambda P.$$

The proper solutions satisfy $f'' > 0$ near the interface, so that, for small $f > 0$ the first-order ODE takes the form

$$P\frac{\mathrm{d}P}{\mathrm{d}f} = (f^p - \lambda P)^{\frac{1}{m}}. \tag{7.195}$$

Lemma 7.54 $\Delta \neq \emptyset$ *if and only if* $p > -m$.

The proof is the same as above. By Theorem 7.12, for the N-dimensional equation

$$\boxed{u_t = |\Delta u|^{m-1}\Delta u - u^p \quad \text{in } I\!R^N \times I\!R_+}$$

with $p \leq -m$, the only maximal singular solution is $u(x,t) \equiv 0$. For $p > -m$ there exist nontrivial maximal solutions; see Theorem 7.8.

Let $N = 1$ and $p > -m$, hence $\Delta \neq \emptyset$.

1. Second-order interfaces, $p < p_c$**.** Consider first the subcritical range

$$-m < p < p_c = \tfrac{m}{2m-1} > 0, \quad p < m, \;\; m > \tfrac{1}{2}. \tag{7.196}$$

The minimal orbit is given by

$$P_{\min} = f^\gamma[c_0 - \lambda c_1 f^{\gamma-p}(1+o(1))], \quad \gamma = \tfrac{p+m}{2m}, \;\; c_0 = \sqrt{\tfrac{1}{\gamma}}.$$

The TW-bundle with $\Delta_0 = I\!R$ is given by (7.166) where

$$\alpha = \tfrac{m-p}{2m}, \quad \beta = \tfrac{2m(1-p)}{m-p} > 1, \quad A_0 = \tfrac{m-p}{\sqrt{2m(p+m)}}.$$

The pressure is $v = u^\alpha$ and we introduce the interface operators (7.167), (7.168) and the G-function (7.169). Finally, Theorem 7.49 is true with the exponent μ in the Bernstein-type estimate replaced by

$$\mu = \alpha(\beta - 1) = \tfrac{m-(2m-1)p}{2m} > 0.$$

2. The critical exponent $p = p_c$**.** Then the ODE (7.195) admits the exact minimal orbit

$$P_{\min}(f) = A_0(\lambda) f^p,$$

where $A_0(\lambda) > 0$ solves the algebraic equation

$$p^m A_0^{2m} + \lambda A_0 - 1 = 0.$$

The TW-bundle is given by

$$f^\alpha(\xi) = cA_0(\lambda)\xi, \quad \lambda \in \Delta_0 = I\!R, \quad \alpha = c = \tfrac{m-1}{2m-1} > 0.$$

Introducing the first-order operator (7.176), (7.177) and the G-function (7.178), we obtain Theorem 7.50.

3. Supercritical range $p \in (p_c, 1)$**.** In this range the TW-bundle is

$$f^\alpha(\xi) = c(-\lambda)^{1/(2m-1)}\xi(1+o(1)), \; \lambda < 0, \; \alpha = \tfrac{m-1}{2m-1};$$
$$f^\alpha(\xi) = A_0\xi, \; \lambda = 0, \; \alpha = \tfrac{m-p}{2m};$$
$$f^{1-p}(\xi) = \tfrac{1}{\lambda}\,(1-p)\,\xi(1+o(1)), \; \lambda > 0.$$

We then determine the corresponding pressures, the G-functions and the first-order interface operators. Finally, we arrive at the result similar to Theorem 7.51.

4. Non-extinction range $p \geq 1$**.** The analysis is the same. We have that $\Delta_0 = (-\infty, 0]$ for $p < m$, $\Delta_0 = (-\infty, 0)$ for $p \geq m$, and Theorem 7.52 applies.

5. A more fully nonlinear generalization. Consider a generalization of equation (7.194) of the form

$$\boxed{|u_t|^{\gamma-1}u_t = |u_{xx}|^{m-1}u_{xx} - u^p, \quad \gamma > 0.} \tag{7.197}$$

We then obtain the ODE for $P = f' > 0$ and $f > 0$ small,

$$P\frac{\mathrm{d}P}{\mathrm{d}f} = (f^p - B_0 P^\gamma)^{\frac{1}{m}}, \quad \text{where } B_0(\lambda) = |\lambda|^{\gamma-1}\lambda.$$

The computations are similar. For instance, $\Delta \neq \emptyset$ if $p > -m$. The second-order interface equation takes place if

$$p < p_c = \tfrac{\gamma m}{2m-\gamma}, \quad m > \tfrac{\gamma}{2}.$$

If $p = p_c > 0$, then the explicit complete TW-bundle is given by

$$f^\alpha(\xi) = cA_0(\lambda)\xi, \quad \lambda \in I\!R, \quad \alpha = \tfrac{\gamma - p}{\gamma} > 0,$$

where $A_0(\lambda) > 0$ solves the algebraic equation

$$(\tfrac{p}{\gamma})^m A_0^{2m} + |\lambda|^{\gamma-1}\lambda A_0^\gamma - 1 = 0.$$

This TW-bundle determines the first-order interface operators, the G-function and the interface equation. The analysis for $p > p_c$ is straightforward.

Blow-up for the dual PME with source

Consider the blow-up propagation for the dual PME with source

$$\boxed{v_t = |v_{xx}|^{m-1}v_{xx} + v^p, \quad m \in (0,1), \ \ p > 1.} \tag{7.198}$$

The function $u = \frac{1}{v}$ solves the fully nonlinear equation with strong absorption

$$u_t = -u^2|(\tfrac{1}{u})_{xx}|^{m-1}(\tfrac{1}{u})_{xx} - u^q, \quad q = 2 - p < 1. \tag{7.199}$$

Setting $P = f^{-2}f' > 0$ in the ODE for the TW solutions $V = f(\xi)$, one obtains

$$P\frac{\mathrm{d}P}{\mathrm{d}f} = -\left(\lambda P - f^{-p}\right)^{\frac{1}{m}}.$$

It follows that $\Delta = \emptyset$ if

$$p > p_c = \tfrac{m}{1-2m} \quad (m \in (0, \tfrac{1}{2})),$$

and hence for the equation in $I\!R^N$

$$u_t = -u^2|\Delta \tfrac{1}{u}|^{m-1}\Delta \tfrac{1}{u} - u^{2-p}$$

the only compactly supported solutions are $u \equiv 0$.

In the critical case $p = p_c > 1$ for $m \in (\frac{1}{3}, \frac{1}{2})$, we obtain the exact minimal orbit $P_{\min} = A_0 f^{-p}$, where $A_0(\lambda) > 0$ is the minimal root of the algebraic equation

$$p^m A_0^{2m} - \lambda A_0 + 1 = 0.$$

Since $2m < 1$, one can see that there exists a unique continuous monotone decreasing branch of $A_0(\lambda)$ for $\lambda > 0$ that gives proper maximal TWs. By Theorem 7.8, this guarantees the existence of nontrivial maximal solutions of the parabolic PDE in $I\!R^N$. The TW-bundle is

$$Q(f) \equiv \tfrac{1}{p-1} f^{p-1} = A_0(\lambda)\xi \quad \text{for } \lambda > 0$$

with the first-order slope $S = A_0(\lambda)$. The proper TW-diagram $\lambda = A_0^{-1}(S)$ consists of the unique branch

$$\lambda = p^m S^{2m-1} + \frac{1}{S} \quad \text{for } S > 0.$$

In the subcritical case $p \in (1, p_c)$, $m \in (\frac{1}{3}, \frac{1}{2})$, we have the proper TW-bundle for $\lambda > 0$,

$$f^\alpha(\xi) = c\lambda^{1/(2m-1)}\xi(1 + o(1)) \quad \text{with } \alpha = \tfrac{3m-1}{1-2m}.$$

The interface equation is of first order and the interfaces are moving for $t > 0$.

Similar properties are true for more general fully nonlinear equations such as

$$\boxed{|v_t|^{\gamma-1}v_t = |v_{xx}|^{m-1}v_{xx} + v^p, \quad \gamma > 0.}$$

General quasilinear heat equation with absorption

We consider the general quasilinear parabolic equation with arbitrary nonlinearities in both the diffusion and the absorption operators

$$\boxed{u_t = (\varphi(u))_{xx} - \psi(u)} \tag{7.200}$$

and the corresponding N-dimensional equation

$$\boxed{u_t = \Delta\varphi(u) - \psi(u),} \tag{7.201}$$

where $\varphi(u)$ and $\psi(u)$ are given sufficiently smooth functions satisfying the natural conditions including $\varphi'(u) > 0$ (parabolicity), $\psi(u) > 0$ for $u > 0$, $\varphi(0) = 0$. The plain TWs for equations like (7.201) with arbitrary nonlinearities were studied in detail in Chapters 2 and 3.

Let us perform a full classification of singular propagation in 1D. We begin with the properties of the TW profiles. Setting $P = d\varphi(f)/d\xi > 0$ yields

$$P\frac{\mathrm{d}P}{\mathrm{d}f} = G(f) - \lambda P, \quad \text{where} \quad G(f) = \varphi'(f)\psi(f). \tag{7.202}$$

Lemma 7.55 $\Delta \neq \emptyset \iff \int_0^1 G(s)\,\mathrm{d}s < \infty$.

By Theorem 7.12, in the case

$$\int_0^1 \varphi'(s)\psi(s)\,\mathrm{d}s = \infty$$

for any compactly supported initial data, the equation (7.201) has the trivial maximal solution $u \equiv 0$. If this integral converges, Theorem 7.8 establishes the existence of nontrivial proper solutions in any space dimension $N \geq 1$.

1. Subcritical range: second-order interface equation in 1D. Let us derive a criterion for this the most interesting case.

Lemma 7.56 *Assume that*

$$G'(s) \to \infty \quad as \;\; s \to 0. \tag{7.203}$$

Then $\Delta = I\!R$ and as $f \to 0$,

$$P_{\min}(f) = Y(f) - \frac{\lambda}{Y(f)}\int_0^f Y(\eta)\,\mathrm{d}\eta + \dots, \tag{7.204}$$

where $Y(f) = \sqrt{2\int_0^f G(s)\,\mathrm{d}s}$.

In this case, one can see that the first term on the right-hand side of (7.202) is the leading one. Integrating the equation

$$\varphi'(f)f'_\xi = P_{\min}(f)$$

yields that $\Delta_0 = I\!R$ provided that

$$\int_0^1 \frac{\varphi'(s)}{Y(s)}\, ds < \infty.$$

Let $f_0(\xi)$ be the TW profile with $\lambda = 0$ given by the pressure equality

$$Q(f_0) \equiv \int_0^{f_0} \frac{\varphi'(s)}{Y(s)}\, ds = \xi,$$

i.e., $f_0(\xi) = Q^{-1}(\xi)$. Using a standard iteration procedure, under necessary hypotheses, we conclude that, for any $\lambda \in I\!R$, there holds

$$Q(f) = \xi - \lambda R(\xi)(1 + o(1)), \quad \text{where} \tag{7.205}$$

$$R(\xi) = \int_0^\xi \left(\frac{1}{Y^2(f_0(\eta))} \int_0^{f_0(\eta)} Y(\zeta)\, d\zeta \right) d\eta,$$

which gives the complete TW-bundle. Thus $v = Q(u)$ is the pressure and as $\varepsilon \to 0^+$, the G-function is given by

$$G(\lambda, \varepsilon) = 1 - \lambda R'(\varepsilon)(1 + o(1)). \tag{7.206}$$

Therefore, the optimal gradient Bernstein bound on the proper solutions near the singular interface takes the form

$$|(Q(u))_x - 1| \le cR'(Q(u)). \tag{7.207}$$

Using this bound, we define the first-order operator

$$\mathbf{M}_1(f) = (Q(f))_x(0) \equiv 1. \tag{7.208}$$

The expansion (7.205) makes it possible to define the second-order operator as follows

$$\mathbf{N}_2(f) \equiv \lim_{\xi \to 0^+} \frac{1}{R(\xi)} [\xi - Q(f(\xi))] = \lambda. \tag{7.209}$$

After checking the conditions of the instantaneous smoothing in B-classes, we arrive at the interface system

$$\begin{cases} \mathbf{M}_1(u(\cdot,t)) \equiv (Q(u))_x(s(t),t) = 1, \\ D^+ s(t) = \mathbf{N}_2(u(\cdot,t)) \equiv \lim_{\xi \to 0^+} \frac{1}{R(\xi)} [\xi - Q(u(s(t)+\xi, t))]. \end{cases} \tag{7.210}$$

2. The critical case. In the case of general nonlinearities in the equation, we define the critical case as follows: equation (7.202) admits an exact solution of the form $P_{\min}(f) = A_0(\lambda) G(f)$. Obviously, substituting into the equation, we conclude that the critical case occurs if the coefficients φ and ψ satisfy the condition

$$G'(s) = (\varphi'\psi)'(s) \equiv a_0 > 0. \tag{7.211}$$

Then $A_0(\lambda) > 0$ for any $\lambda \in I\!R = \Delta$ is uniquely determined from the quadratic equation

$$a_0 A_0^2 + \lambda A_0 - 1 = 0.$$

The TW-diagram is presented by the monotone analytic branch

$$\lambda = A_0^{-1}(S) = -a_0 S + \frac{1}{S} \quad \text{for} \quad S > 0.$$

In the case of finite propagation we obtain the TW-bundle

$$Q(f(\xi)) = A_0(\lambda)\xi, \quad \lambda \in I\!R = \Delta_0, \quad \text{where} \;\; Q(f) = \int_0^f \frac{\mathrm{d}s}{\psi(s)}.$$

Hence, the Bernstein estimate is

$$c_2(\delta) \le (Q(u))_x \le c_1(\delta)$$

and the first-order interface equation takes the form

$$D^+ s(t) = A_0^{-1}((Q(u))_x(s(t), t)).$$

3. The supercritical range. We assume that, unlike (7.203) or (7.211), there holds

$$G'(s) \to 0 \quad \text{as} \;\; s \to 0. \tag{7.212}$$

Then the minimal proper orbit is given by

$$P_{\min} = (-\lambda) f(1 + o(1)), \;\; \lambda < 0; \quad P_{\min} = \tfrac{1}{\lambda} G(f)(1 + o(1)), \;\; \lambda > 0.$$

This gives the TW-bundle

$$Q_1(f) \equiv \int_0^f \frac{\varphi'(s)}{s} \, \mathrm{d}s = (-\lambda)\xi(1 + o(1)), \;\; \lambda < 0,$$

$$Q_0(f) \equiv \int_0^f \frac{\varphi'(s)}{Y(s)} \, \mathrm{d}s = \xi, \;\; \lambda = 0,$$

$$Q_2(f) \equiv \int_0^f \frac{\mathrm{d}s}{\psi(s)} = \lambda^{-1}\xi(1 + o(1)), \;\; \lambda > 0.$$

Therefore, $\Delta_0 = I\!R$ provided that all pressure functions are well defined, and $\Delta_0 = (-\infty, 0]$ or $\Delta_0 = (-\infty, 0)$ if the corresponding integrals diverge. We conclude that after a waiting time (if any) on $[T_w + \delta, T - \delta]$, $T_w \in [0, T)$, the gradient bound has the form

$$(Q_1(u))_x \le c_1(\delta), \quad (Q_2(u))_x \ge c_2(\delta),$$

where the pressure $Q_{1,2}$ depends on the slope (or, which is the same, on the interface direction). In the first-order interface equation

$$D^+ s(t) = \mathbf{N}_1(u(\cdot, t))$$

the operator $\mathbf{N}_1$ is determined by the above TW-bundle.

4. Generalizations. The analysis applies to doubly nonlinear equations such as

$$u_t = [\Phi((\varphi(u))_x)]_x - \psi(u), \quad \Phi' \ge 0,$$

or to fully nonlinear equations

$$Z(u_t) = [|(\varphi(u))_x|^\sigma (\varphi(u))_x]_x - \psi(u), \quad \sigma > -1, \;\; Z' \ge 0.$$

Different types of interface equations and optimal regularity bounds can be established for the equations with extra convection terms

$$u_t = (\varphi(u))_{xx} - (\rho(u))_x - \psi(u).$$

Nonexistence-existence results, as well as Bernstein and B-concavity estimates in the radial geometry, are true for the corresponding N-dimensional extensions of these models.

Applications to equations from mean curvature flows

We consider the equation

$$u_t = \frac{u_{xx}}{1+(u_x)^2} - \psi(u) \quad \text{in } S, \tag{7.213}$$

where $\psi(u) > 0$ for $u > 0$ is a given decreasing function satisfying $\psi(u) \to \infty$ and $\psi'(u) \to -\infty$ as $u \to 0^+$. In the case

$$\psi(u) = \tfrac{N-2}{u}, \quad N \geq 3, \tag{7.214}$$

equation (7.213) describes the evolution of rotational symmetric hypersurfaces moving by mean curvature in $I\!R^N$. Due to the strong absorption singularity $-\psi(u)$ in the equation, given uniformly positive initial data $u_0(x) \geq \delta > 0$, the parabolic evolution generates a finite-time singularity when u vanishes at $t = T$ meaning blow-up of the curvature. For $t > T$ we arrive at continuous (non-classical) maximal solutions with singular interfaces. By the geometric theory, we can treat the N-dimensional equation

$$u_t = \frac{\Delta u}{1+|\nabla u|^2} - \psi(u) \quad \text{in } S = I\!R^N \times I\!R_+. \tag{7.215}$$

It turns out that, for any singular absorption functions $\psi(u)$, there exists a nontrivial continuation of maximal solutions beyond singularity.

Proposition 7.57 *For any compactly supported* $u_0 \not\equiv 0$, (7.215) *has a nontrivial maximal solution* $u(x,t) \not\equiv 0$ *for any small* $t > 0$.

In order to classify the singular interfaces, we consider the 1D TWs

$$V(x,t) = f(\xi) \implies \frac{f''}{1+(f')^2} - \psi(f) + \lambda f' = 0,$$

where $P = f'$ satisfies

$$P\frac{\mathrm{d}P}{\mathrm{d}f} = (1+P^2)(\psi(f) - \lambda P).$$

1. First-order interfaces. If

$$\int_0 \psi(s)\,\mathrm{d}s = \infty, \tag{7.216}$$

which is true for the mean curvature case (7.214), then $\Delta_0 = I\!R_+$. We distinguish the following three cases depending on the limit (finite or infinite)

$$\kappa = -\lim_{s\to\infty} \tfrac{\psi'(s)}{\psi^2(s)} \geq 0.$$

(i) $\kappa > 0$ is finite. This is true for (7.214) with $\kappa = \frac{1}{N-2}$, and in this case

$$P_{\min}(f) = \tfrac{1+\kappa}{\lambda}\, \psi(f) + \dots .$$

(ii) If $\kappa = 0$, then

$$P_{\min}(f) = \tfrac{1}{\lambda}\, \psi(f) + \dots .$$

(iii) If $\kappa = \infty$, then, assuming furthermore that

$$\lim_{s \to 0} \frac{\psi''(s)\psi(s)}{(\psi'^2)(s)} = 2,$$

we obtain the minimal orbit

$$P_{\min}(f) = -\tfrac{1}{\lambda}\, \tfrac{\psi'(f)}{\psi(f)} + \dots .$$

For example, if

$$\psi(f) = \frac{1}{f}\, |\log f|^{\alpha} \quad \text{for } 1 < f \ll 1,$$

where $\alpha \ge -1$, we have (i) if $\alpha = 0$, (ii) if $\alpha > 0$ and (iii) if $\alpha \in [-1, 0)$. In these three cases, the gradient function and the proper TW-bundle are given by

$$Q(f) \equiv \int_0^f \frac{\mathrm{d}s}{P_{\min}(s)} = \tfrac{1}{\lambda}\, \xi + \dots .$$

The interface equation is of first order with the single dynamic equation

$$D^+ s(t) = \frac{1}{[Q(u)]_x|_{x=s(t)}} \quad \text{for } t > 0.$$

2. Second-order interfaces. If the integral in (7.216) converges, the proper TW-bundle is similar to that in (7.205) with

$$Q(f) \equiv \int_0^f \frac{\mathrm{d}z}{(2\int_0^z \psi(s)\,\mathrm{d}s)^{1/2}} = \xi - \lambda R(\xi) + \dots ,$$

where $\lambda \in \Delta_0 = I\!R$,

$$R(\xi) = \tfrac{1}{4}\, \frac{\int_0^f (2\int_0^z \psi(s)\,\mathrm{d}s)^{1/2}\,\mathrm{d}z}{\int_0^f \psi(s)\,\mathrm{d}s},$$

and $f = Q^{-1}(\xi)$. This leads to a second-order interface system similar to that in (7.210).

On a generalization with discontinuous limit semigroup

Nonexistence and discontinuity of the limit semigroup (not available for the model (7.215)) can be revealed by the following generalized equation:

$$\boxed{u_t = \varphi(|\nabla u|)\Delta u - \psi(u) \quad \text{in } S = I\!R^N \times I\!R_+,} \tag{7.217}$$

where ψ is the same as above and $\varphi(q) > 0$ for $q > 0$ is a continuous function. The TW ODE takes the form

$$\varphi'(f')f'' - \psi(f) + \lambda f' = 0$$

and setting $P = f'$ yields

$$\varphi(P)P\tfrac{\mathrm{d}P}{\mathrm{d}f} = \psi(f) - \lambda P. \tag{7.218}$$

Theorem 7.58 *Let* (7.216) *hold and*

$$\int^{\infty} \varphi(z)\,\mathrm{d}z = \infty. \tag{7.219}$$

Then the limit semigroup of (7.217) *is trivial: for any compactly supported* u_0, $\mathcal{T}(t)u_0 \equiv 0$ *for* $t > 0$

Proof. We need to check that, for the ODE (7.218), $\Delta = \emptyset$ and any $\lambda \in I\!R$ is in the steep limit. It follows that, for any fixed λ, $\varphi(P)\mathrm{d}P/\mathrm{d}f \geq -\lambda$. Integrating over $(0, f)$, we obtain that

$$\Phi(P(f)) - \Phi(P(0)) \geq -\lambda f,$$

where $\Phi(s) = \int_0^s \varphi(z)\,\mathrm{d}z$. Since $\Phi(\infty) = \infty$ by (7.219), we have that $P(0)$ is finite, i.e., the minimal orbit is bounded: $P_{\min}(f) \leq c$ for $f \approx 0$. Then

$$\varphi(P)P\tfrac{\mathrm{d}P}{\mathrm{d}f} = \psi(s) + ... \quad \text{for small } f > 0.$$

Integrating near $f = 0$ yields a contradiction in view of the divergence in (7.216). □

Fully nonlinear equation from detonation theory

Finally, we consider a class of fully nonlinear parabolic equations

$$\boxed{u_t - g(uu_{xx}) - (u_x)^2 - \psi(u),} \tag{7.220}$$

where $g'(u) > 0$ and $\psi(u) > 0$ for $u > 0$. This equation with special coefficients

$$g(s) = 2\log\left(\frac{e^s - 1}{s}\right) \quad \text{and} \quad \psi(u) = -2\log u \quad \text{for } u \in (0, 1), \tag{7.221}$$

was derived by J.D. Buckmaster (1989) [70], as a model describing the evolution of perturbations of Zel'dovich-von Neumann-Doering's (ZDN) square wave occurring during a detonation in a duct.

We consider (7.220) assuming arbitrary nonlinearities g and ψ. Similar to the ZDN-model, we impose the condition of singular absorption

$$\psi(f) \to +\infty \quad \text{as } f \to 0.$$

The ODE for the minimal orbit $P = f'$ takes the form

$$P\tfrac{\mathrm{d}P}{\mathrm{d}f} = \tfrac{1}{f}\, g^{-1}(\psi(f) - \lambda P + P^2). \tag{7.222}$$

We then conclude that

$$\Delta \neq \emptyset \quad \text{if} \quad \int_0^1 G(s)\,\mathrm{d}s < \infty, \tag{7.223}$$

where $G(s) = \frac{1}{s}\, g^{-1}(\psi(s))$.

For the coefficients of the ZDN-wave equation

$$g(s) = 2(s - \log s + ...) \quad \text{as } s \to +\infty, \tag{7.224}$$

so that the integral diverges. Therefore, the proper maximal solutions cannot be continued in a nontrivial way beyond singularity, i.e., for any singular initial data, $u(x,t) \equiv 0$ for arbitrarily small $t > 0$. This means that the limit semigroup is discontinuous at $t = 0$. Then the corresponding multi-dimensional equations

$$u_t = g(u\Delta u) - |\nabla u|^2 - \psi(u) \quad \text{in } I\!R^N \times I\!R_+,$$

with any compactly supported u_0, also admit trivial solutions only, $u(\cdot,t) \equiv 0$ for $t > 0$.

Let us describe the finite propagation in 1D, $\Delta_0 \neq \emptyset$, under the hypothesis

$$\int_0^1 \frac{\mathrm{d}s}{Y(s)} < \infty, \quad \text{where } Y(s) = \sqrt{\int_0^s G(\eta)\,\mathrm{d}\eta}.$$

Let us describe the most striking second-order interfaces. This means that on the right-hand side of (7.222) the function $\psi(f)$ is the leading term. Comparing it with other terms, we derive the following sufficient condition of the second-order interface propagation:

$$\left(\frac{g^{-1}(\psi(s))}{\psi(s)\psi'(s)}\right)' \to 0 \quad \text{as } s \to 0. \tag{7.225}$$

Then $f' = P_{\min} = \sqrt{2}Y(f)(1 + o(1))$. Hence, the pressure function is

$$v = Q(u) = \tfrac{1}{\sqrt{2}} \int_0^u \tfrac{\mathrm{d}s}{Y(s)}.$$

Consider, for instance, equation (7.220) with the function g given in (7.221), and so (7.224) holds. The condition (7.225) reads

$$\left(\tfrac{1}{\psi'(s)}\right)' \to 0 \quad \text{as } s \to 0.$$

The ODE for small $f > 0$ takes the asymptotic form

$$P\tfrac{\mathrm{d}P}{\mathrm{d}f} = \tfrac{1}{2f}\left[\psi(f) - \lambda P + ...\right] \quad \text{for } \lambda \neq 0. \tag{7.226}$$

We thus arrive at a perturbed two-term ODE similar to (7.202), which is studied in the same way. In particular, we obtain a similar minimal orbit

$$P_{\min} = X(f) - \frac{\lambda}{2X(f)} \int_0^f \frac{X(s)}{s}\,\mathrm{d}s + ..., \quad X(f) = \sqrt{\int_0^f \frac{\psi(s)}{s}\,\mathrm{d}s}.$$

The pressure variable is $v = Q(u) = \int_0^u \frac{\mathrm{d}s}{X(s)}$. Integrating equation $f' = P_{\min}(f)$, we obtain a complete TW-bundle

$$Q(f) = \xi - \tfrac{1}{2}\lambda R(\xi) + ...,$$

where

$$R(\xi) = \int_0^\xi \left(\frac{1}{X^2(f_0(\eta))} \int_0^{f_0(\eta)} \frac{X(s)}{s}\,\mathrm{d}s\right)\mathrm{d}\eta, \tag{7.227}$$

and $f_0(\xi) = Q^{-1}(\xi)$. If $\lambda = 0$, the quadratic term P^2 in (7.222) has to be taken into account.

The analysis of the TW-bundle with function (7.227) is standard. Under necessary conditions, we arrive at the G-function, derive the optimal gradient estimate similar to (7.207) and the interface operators similar to (7.206)–(7.209). Finally, we obtain the interface system.

Remarks and comments on the literature

Main concepts and results of the geometric theory were introduced in [139].

§ **7.1.** Some aspects of the geometric analysis have been separately discussed and applied to particular quasilinear parabolic equations in a series of papers, [127]–[134], [137], [148]–[154], [159]–[168]. These results are summarized in the books [306, Chapter 4] and [170, Chapters 2, 9, 10]. In particular, we use some ideas and results on the TW-analysis and on intersection comparison with the families of TW solutions presented in [126], [137], [159], [160], [164]–[168].

It is important that similar geometric concepts apply in the case of infinite propagation where the left-hand interface is always at infinity, $s(t) \equiv -\infty$. This means that $\Delta \neq \emptyset$ (the TW-bundle is not empty) but $\Delta_0 = \emptyset$. Then, if a singularity does not occur at finite points (but at $x = -\infty$), the unknown existence-regularity properties and interface propagation are replaced by similar questions on the evolution of another "infinite interface" parameter denoted by $s(t)$ again. It describes the motion of the general evolution orbit of the PDE in the complete set of particular orbits B. Then the equation for $s(t)$ governing the motion of an orbit $u(\cdot, t)$ in B (replacing the natural interface equation in the case of infinite propagation) and existence-regularity questions in the B-classes are to be studied by a similar intersection comparison with the solutions from B. In this sense, there is not much difference between the regularity theories for the FBPs and for standard problems where interfaces are always at infinity.

The geometric analysis does not rely necessarily on the subsets of TWs, and the geometric theory can be developed on the basis of any suitable complete set B of particular solutions to the nonlinear PDEs. For some special nonlinear parabolic equations, other sets B of group-invariant (self-similar) solutions or those on linear subspaces invariant under nonlinear operators can be used. Such examples were presented in [164], [137], see also special new families of exact solutions in [134], [136] and [156]. In these cases particular solutions from B are given by reasonably simple nonlinear ODEs so that their properties can be studied by standard ODEs methods. Introducing an appropriate set $B = \{V\}$ of proper particular solutions we always take into account that the general proper solutions u under consideration belong to the corresponding functional class where the singular behaviour can be described by the solutions from B.

The geometric approach exhibits certain aspects of a kind of a "nonlinear superposition principle" for the second-order parabolic PDEs, where some properties (like existence, differential properties, concavity, etc.) of suitable sets $B = \{V\}$ of particular solutions are translated to the *general* solutions.

§ **7.2.** A detailed TW-analysis for quasilinear reaction-diffusion-convection equations

$$u_t = (a(u))_{xx} + (b(u))_x + c(u), \tag{7.228}$$

is performed in [175] and [176], where necessary and sufficient conditions of $\Delta_0 \neq \emptyset$ for (7.228) are derived and an extended survey on TW solutions for quasilinear heat equations is available.

§ **7.3, 7.4.** Approximation techniques are well known for quasilinear equations. In the case of divergent differential operators this gives weak solutions defined via integration by parts. See the survey paper [213] and the books [99], [122], [306], [170]. In most singular cases (say, for blow-up solutions in the L^∞-sense), a weak formulation beyond blow-up time is often not available. Such an approximation in the blow-up case leads to a unique proper minimal solution. We refer to the extended semigroup theory of monotone order-preserving semigroups; see Section 6.2 and [167]. In this case regularized truncated problems are non-singular (no blow-up) and can be solved by known methods. General results for fully nonlinear parabolic equations are presented in [298], [248], [233], [100], [49], [253], [345], [94].

For classes of second-order PDEs, the concept of continuous viscosity solutions is developed in [93], [73], [345] and [94] (see also references therein on recent development of L^p-viscosity solutions). The main mathematical tool to study the generalized motion of interfaces in $I\!R^N$ is level-set approximation, where the viscosity theory is quite natural; see [282], [110] (for the mean curvature flows), [87], [41], [109] (applications to semilinear reaction-diffusion equations). For a general review see the book [319]. Viscosity theory for the PME and related quasilinear parabolic equations is developed in [79], where advantages and difficulties of the viscosity approach are discussed.

The proper solutions are maximal viscosity solutions. For general essentially singular equations, the viscosity definition of a unique solution must include some rather delicate properties of solutions and singular interface propagation, which are usually unknown *a priori* and are the main feature of the singular problem. In many cases, viscosity definitions (e.g. S.N. Kruzhkov's entropy solutions for the first-order conservation laws [232], [317]) express the fact that a good solution is the one that can be constructed by regular approximation, and others should be excluded from consideration. For second-order parabolic PDEs with arbitrary nonlinearities, we prefer to use the concept of approximation without specifying a viscosity version of such definition, though this can be done for a class of equations, if we have already obtained sufficiently detailed information on the behaviour of proper maximal solutions near singular interfaces. In a general case, where equations can generate singularities not only at the fixed level $\{u = 0\}$ but at a singularity subset in the tangential space $\{u, \nabla u\}$ (we consider some such examples), i.e., we do not know *a priori* where and which type of singularities are going to occur, we prefer not to deal with the equation itself, which is not well posed at a unknown singularity subset.

§ **7.5.** Let us mention some first papers on blow-up solutions for semilinear and quasilinear heat equations where intersection comparison ideas played a key role.

These are [127], [129], [123], [130], [150], [76]; see also more recent papers [16], [188], [132], [133], [162], [165]–[168], [306, Sections 4, 7], [170, Chapters 9, 10] and references therein.

A result similar to Proposition 7.13 was first derived for the filtration equations (7.77), [209], [287]; see references in the survey [213]. For the quasilinear equation (7.228) such a general conclusion based on the TW-analysis was performed in [176, Theorem 2].

§ **7.6.** A notion of sign-invariants was introduced in [134], where the corresponding backward problem (given a sign-invariant, determine the corresponding set B of particular solutions) was solved for some classes of quasilinear parabolic equations and new exact solutions were found.

The idea of intersection comparison with families of simple particular solutions with envelopes is effective in the study of essentially nonstationary combustion processes. The first results on intersection comparison with a complete set of stationary solutions (the method of stationary states [148]) of the quasilinear heat equations are presented in [127], [130]; see also [306, Chapter 7]. The general monotonicity property of large solutions in [150], [154] can be treated as the eventual B-concavity result. Other examples are presented in [137].

§ **7.7–7.10.** As a comment on Theorem 7.26, we note, that it seems that higher-order regularity estimates on the second derivative of the pressure v_{xx} and on v_t cannot be proved by the intersection comparison with TWs. On the other hand, such higher order regularity results are proved by the geometric comparison with a set B provided that it is complete in the tangential space $I\!\!R^3$. See an example of such intersection comparison with parabolic solutions in the proof of Proposition 5.11.

Optimal modulus of continuity in x of the solutions depends on their generic behaviour near the singular interfaces. It was proved by D.G. Aronson (1969) [29] by a modification to the Bernstein method that, in the case of nonnegative compactly supported solutions of the PME $u_t = (u^m)_{xx}$, $m > 1$, the exponent of Hölder continuity of solutions in x is $\nu = \min\{1, \mu\}$, $\mu = \frac{1}{m-1}$, and so

$$|u(x_1, t) - u(x_2, t)| \le C|x_1 - x_2|^\nu.$$

It is directly related to the actual regularity of the TWs $V(x,t) = (A_0\xi)_+^\mu$, $\xi = x - \lambda t$, where $A_0 = -\frac{\lambda(m-1)}{m}$ and $\lambda \in \Delta_0 = (-\infty, 0)$. Near the interface, for $x_1 = \lambda t$ and $x_2 \approx (\lambda t)^+$, the behaviour corresponds to the TW exponent $\mu = \frac{1}{m-1}$. For the PME the exponents of Hölder continuity in x and t are known to coincide [98] (see also earlier key papers on this subject [231], [174]) and are given by the actual regularity of the TW solutions.

§ **7.11.** In the examples considered in this section, equations can admit other types of singularities or degeneracy beyond the level $\{u = 0\}$ depending on the diffusion-like terms. We then consider a class of solutions where those stay away from the main singular zero-level. Indeed, TWs with the straight line characteristic propagation cannot explain the interaction between different types of singularities. Such new singular patterns can be studied by using other sets $\tilde{B}$ of exact solutions (like self-similar ones) that contain the desired focusing-type collision of singularities. These are typical questions of blow-up; see e.g. [16], [127]–[133], [148]–[154], [162], [167], [168] and books [306, Chapter 4], [170, Chapters 9, 10].

Equation (7.158) with $\sigma = 0$,

$$\boxed{u_t = (u^m)_{xx} - u^p,} \tag{7.229}$$

has been studied in [159] and [160], where the second-order interface equation for $p \in (-m, 2-m)$ was derived. First gradient bound of a typical form

$$|(u^\alpha)_x| \le C \quad \text{with the exponent } \alpha = \tfrac{1}{2}(m-p) \;\text{ for }\; 0 < p < 2-m,$$

was proved by A.S. Kalashnikov (1974) [212] by the classical Bernstein technique. As (7.170) shows, this estimate is a first approximation of the optimal transversality estimate via a complete TW-bundle. A formal analysis of second-order interface equations and the behaviour of solutions near their interfaces for the equation (7.229) with $p < 2 - m$ was performed in [215]. This analysis was subsequently justified for TW solutions in [187].

In [315] a second-order interface equation and further regularity results were established for the divergent PME with convection

$$\boxed{u_t = (u^m)_{xx} - (u^\nu)_x, \quad m > 1, \;\; \nu \in (0,1),}$$

where the method of Lagrangian coordinates applied. See also the last section in [160]. TW solutions for quasilinear heat equations (7.200) were studied in a number of papers; see [126], [165], [175] and references in [176].

Mean curvature equations (7.213), (7.214) with viscosity solutions generate different types of singularities and were studied in [110], [318], [5] and [19]; see extra references to § 7.3 given above.

Equation (7.220), (7.221) is known to create finite-time singularity, when initially positive solutions touch the singular level $\{u = 0\}$. A mathematical asymptotic theory of such singular extinction behaviour is presented in [141] where a list of further references is available. It was proved that the formation of such a finite-time singularity as $t \to T^-$ occurs in accordance with the Hamilton-Jacobi equation

$$u_t = -(u_x)^2 + 2\log u,$$

and the fully nonlinear dissipativity term $g(uu_{xx})$ plays a role of an exponentially small, singular perturbation of this first-order equation. Observe that this original Buckmaster's equation always admits complete singularity, so that the unique maximal continuation for $t > T$ is $u(x,t) \equiv 0$, which corresponds to a high instability of the square detonation wave; see [141, Section 9]. We thus show that certain slight perturbations of the nonlinear coefficients of this detonation model can lead to a continuous limit semigroup where singular interfaces exhibit finite propagation on $\{u = 0\}$.

CHAPTER 8

Geometric Theory of Generalized Free-Boundary Problems. Non-Maximal Solutions

In this chapter, using intersection techniques, we describe regularity properties of *non-maximal* solutions of nonlinear singular parabolic equations. In general, such non-maximal solutions satisfy free-boundary problems (FBPs). The whole class of these FBPs includes the classical Stefan or Florin problems, as well as other generalized ones with various dynamic interface equations.

As in the previous chapter, a correct choice of proper subsets B of TWs detects key existence and regularity properties of general solutions. We show that each proper subset B of TWs generates a FBP that is correctly posed in the corresponding regularity B-class and its solutions obey comparison and intersection principles.

8.1 Introduction: One-phase free-boundary Stefan and Florin problems

Our strategy is to demonstrate that the geometric analysis of the free-boundary propagation based on intersection comparison with complete sets B of TWs can be applied to classes of FBPs. We again consider the general nonlinear singular parabolic equation

$$u_t = \mathbf{F}(u) \equiv F(u, u_x, u_{xx}) \tag{8.1}$$

with the same assumptions on F and initial data $u_0 \geq 0$ as in Section 7.1.

As we have shown, the maximal solutions (Chapter 7) satisfying special free-boundary conditions at the singularity level $\{u = 0\}$ are uniquely determined by any monotone regular approximations. Then the free boundary equations need not be involved in the construction and can be identified *a posteriori*.

On the other hand, one can specify other free-boundary conditions. The important question is as follows: which free-boundary conditions are sufficiently well posed for a given parabolic PDE (8.1)?

For instance, let us prescribe at the interface $x = s(t)$ on the singular level $\{u = 0\}$ a dynamic condition of the form (similar to some of conditions for maximal solutions in Chapter 7):

$$D^+ s(t) = \mathbf{G}(u(s(t), t)) \quad \text{for } t > 0, \tag{8.2}$$

where $\mathbf{G}(u) = G(u, u_x, u_{xx})$ is a fixed second-order interface operator depending on a suitable interface slope. Here, as usual, the initial function $u_0(x)$ is assumed to have a finite left-hand interface at $x = 0$ and $u_0(x) > 0$, $u_0'(x) > 0$ for $x > 0$, and so a continuous free boundary at $s(t)$ (if any) satisfies $s(0) = 0$.

Formally, we then arrive at a FBP and hence at a number of typical questions on existence-nonexistence, uniqueness and regularity of solutions, i.e., the same three questions **(Q1)–(Q3)** from the Introduction studied in Chapter 7 for maximal solutions. But the first question, which does not occur at all in the theory of maximal solutions, is to determine a correct statement of the free-boundary conditions in (8.2) for which the FBP makes sense. More precisely, this is to determine which and how many free-boundary conditions (e.g. described by operator $\mathbf{G}$) should be prescribed at the interface $x = s(t)$ to obtain a *well-posed* FBP having sufficiently regular global solutions. In other words, firstly, one needs to answer the following question. Given a singular operator $\mathbf{F}$ in the parabolic PDE (8.1), describe the whole class of free-boundary conditions (e.g. interface operators $\mathbf{G}$ in (8.2)) such that the corresponding FBPs are well-defined and obey typical parabolic properties such as the MP, comparison and Sturmian properties, at least, locally in time.

We begin with two classical examples of FBPs and next discuss possible ways of extending the free-boundary conditions.

Example 8.1: the one-phase Stefan problem. These kind of problems were formulated and studied by J. Stefan and others in the nineteenth century (references are given in Remarks). For convenience, stressing the attention to the free-boundary conditions, we pose the problem on the unbounded interval $\{x > s(t)\}$. Typically, the Stefan problem is posed on a bounded interval with a standard Neumann or Dirichlet boundary condition at the finite right-hand end point.

Thus we consider nonnegative solutions of the heat equation (HE)

$$u_t = u_{xx} \quad \text{in } S = \{x > s(t),\ t > 0\} \quad \text{with } u(s(t), t) \equiv 0 \tag{8.3}$$

and the following dynamic boundary condition given by the first-order interface operator $\mathbf{G}(u) = ku_x$ with a constant $k \neq 0$:

$$s'(t) = ku_x(s(t), t) \quad \text{for } t > 0. \tag{8.4}$$

There is a great amount of literature devoted to existence, nonexistence, uniqueness and regularity properties for the Stefan FBPs; see Remarks. Here, we will mainly discuss some special geometric aspects of the free-boundary conditions that are important for future applications of the intersection techniques.

Now $k = -1$ corresponds to the expanding wave, where

$$s'(t) = -u_x \equiv H(u_x) < 0.$$

Then the corresponding Rankine–Hugoniot diagram describing the dependence of the interface velocity s' on the interface slope $S = u_x$ takes the form

$$s' = -S \equiv H(S),$$

where $H(S)$ is a strictly monotone decreasing function of the first-order slope $S = u_x$. As we have seen by using maximal TWs (Proposition 7.2), this is a necessary condition for a singular problem to be properly posed at the singularity level. One can see that such a monotonicity means that this free-boundary condition does not violate the MP, and the comparison principle on interfaces, as well as the Sturm

Theorem (new intersections of solutions cannot occur on free boundaries) hold. For any $k < 0$, the Stefan problem is well-posed and has a unique global classical solution.

On the other hand, if $k = 1$ (or $k > 0$), the diagram

$$s' = u_x \equiv H(u_x) > 0$$

exhibits an incorrect monotonicity dependence of the interface velocity upon the interface slope $S = u_x$, where $H'(S) \equiv 1$. Hence, this Stefan problem loses both the comparison and Sturmian properties of solutions and new intersections can occur on interfaces. Such a *supercooled* Stefan problem is not well-posed and special singularities of $s(t)$ can occur in finite time; see references in Remarks.

Note that the first-order Stefan boundary condition (8.4) can be equivalently formulated by using a second-order operator. Namely, we substitute

$$u_x = \tfrac{1}{k}\, s'$$

into the identity

$$u_x s' + u_t = 0$$

obtained by differentiating $u(s(t), t) = 0$. Then using the HE yields

$$(s')^2 = -k u_{xx}.$$

We always take into account boundary operators of the lowest order.

Example 8.2: the Florin problem. This was formulated by V.A. Florin (1951) [118] where the free-boundary condition for the heat equation (8.3) takes the form

$$u_x(s(t), t) = 1 \quad \text{for} \ \ t > 0. \tag{8.5}$$

Differentiating $u(s(t), t) = 0$ yields

$$s' = -\frac{u_t}{u_x} \equiv -u_t$$

and hence, by the heat equation and (8.5), we have the second-order dynamic condition on the interface

$$s' = -u_{xx} \equiv H(u_{xx}). \tag{8.6}$$

This is (8.2) with the interface operator

$$\mathbf{G}(u) = -u_{xx}.$$

In (8.6) $S = u_{xx}$ plays a role of the generalized interface slope. The corresponding diagram $\lambda = H(S) \equiv -S$ is strictly monotone decreasing with respect to the slope S and $H'(S) = -1$, and hence the interface equation satisfies the conditions of comparison. It is not difficult to show that Sturm's Theorem is also true for this well-posed Florin problem.

One can obtain a free-boundary condition of higher order. Differentiating $u_x(s, t) = 1$ yields

$$u_{xx} s' + u_{xt} = 0,$$

where

$$u_{xt} = u_{xxx}$$

by the heat equation. This gives the third-order condition

$$s' = -\frac{u_{xxx}}{u_{xx}},$$

or, which is the same,

$$(s')^2 = u_{xxx}.$$

On sufficiently smooth solutions both are equivalent to the second-order one (8.6)

Dynamic boundary conditions and geometric *a priori* bounds. The well-posed Stefan problem with $k = -1$ is known to have a unique smooth global solution. It is important that, dealing with general singular parabolic PDEs, we may assume that proper solutions can be constructed by regular approximations. Let us discuss such a scheme for the analysis for FBPs.

The FBPs belong to the class of parabolic problems with *nonlinear dynamic boundary conditions*. Indeed, assuming that $u(x,t)$ is strictly monotone increasing with x (as we know, this property is controlled by the MP), we introduce the variable

$$x = X(u,t), \quad \text{where} \quad X_u > 0.$$

This is the famous *von Mises transformation* in the *Prandtl boundary layer theory*. Then the HE yields the following quasilinear parabolic equation:

$$X_t = \frac{X_{uu}}{(X_u)^2} \quad \text{in } S = \mathbb{R}_+ \times \mathbb{R}_+, \tag{8.7}$$

with initial data $X_0(u)$. The Stefan free-boundary condition reduces to the dynamic boundary condition at the fixed point $u = 0$

$$X_t(0,t) = \frac{k}{X_u(0,t)} \quad \text{for } t > 0. \tag{8.8}$$

This is a second-order boundary condition, i.e, of the same order as the HE since (8.7) yields

$$X_{uu} = kX_u \quad \text{at } u = 0 \text{ for } t > 0. \tag{8.9}$$

Together with standard Dirichlet and Neumann initial-boundary value problems, local existence and uniqueness theory for such dynamic boundary conditions are known in the general parabolic theory; see references in Remarks. Smooth solutions satisfy the usual comparison if it is true on the parabolic boundary, and in this sense the boundary conditions are well-posed. One can see by the standard MP analysis that the comparison on the boundary at $u = 0$ is only true for negative $k = -1$. Then new intersections cannot occur on the boundary, and Sturm's Theorem applies. For the positive $k = 1$ neither the parabolic monotonicity nor the comparison on the boundary are valid and finite-time singularities can occur there.

Dealing with singular FBPs, it is important that non-maximal solutions can be constructed by regular approximations using more standard initial-boundary value problems. Unlike the maximal solutions, this is not a straightforward procedure for other FBPs. For instance, for the Stefan problem with $k = -1$, consider the following approximation of the solution $X(u,t)$ by regular solutions $\{X^{(n)}(u,t)\}$

of parabolic problems with the boundary condition with a "frozen" coefficient:

$$X_u^{(n)} = -X_{uu}^{(n-1)} \quad \text{at } u = 0 \ \text{ for } t > 0.$$

Though such recurrent relations are not perfect in some parabolic problems, we consider it as a formal example of approximation. For each n we arrive at the Neumann problem for the quasilinear equation (8.7) for $X^{(n)}$. If a positivity *a priori* estimate of the gradient on the boundary

$$X_u^{(n)}(0,t) \geq \delta > 0$$

is not available, we introduce the regularized equation for $\{X^{(n)}\}$

$$X_t = \frac{X_{uu}}{\frac{1}{n^2} + (X_u)^2}.$$

Assume that this gives a sequence $\{X^{(n)}\}$ of global solutions. It follows that it is a compact subset if the derivatives X_{uu} and X_u of the limit solutions obtained along a subsequence $X = \lim_{k\to\infty} X^{(n_k)}$ do not blow-up on the boundary $u = 0$. On the other hand, in the original problem such blow-up is forbidden by transversality techniques based on intersection comparison with complete TW-bundles, which we are going to perform next. In the limit, we then obtain a sufficiently smooth proper solution $X(u,t)$.

Notice that we cannot guarantee the monotonicity of the sequence $\{X^{(n)}\}$ in general. It seems that, in the most of the cases, such a monotonicity can be easily achieved for maximal solutions only. Nevertheless, it follows from the MP for uniformly parabolic equations that, under some hypothesis, such a convergence relative to the regularizing parameter can be monotone with, possibly, different monotonicity directions on different subsets in the $\{x,t\}$-space.

Notice that, as in the case of maximal solutions, we do not need estimates to guarantee that $\{X^{(n)}\}$ is a compact subset. Indeed, we need boundedness of the sequence $\{X^{(n)}(0,t)\}$ at the boundary "singular" point only. Then the convergence $X^{(n_k)}(0,t) \to X(0,t)$ will imply the convergence on compact subsets by the classical parabolic theory. The uniqueness then follows from the MP applied to equation (8.7), which is uniformly parabolic and non-singular if X_u is uniformly bounded away from zero.

8.2 Classification of free-boundary problems for the heat equation

We begin with the heat equation and perform a first application of the geometric analysis by using a complete TW-bundles. The construction of the TW-bundle is easy. The TW solutions

$$V(x,t) = f(\xi), \quad \xi = x - \lambda t,$$

satisfy the linear ODE

$$f'' + \lambda f' = 0, \quad f(0) = 0, \tag{8.10}$$

and hence

$$f_\lambda(\xi) = \begin{cases} C(e^{-\lambda\xi} - 1) & \text{for } \lambda \neq 0, \\ C\xi & \text{for } \lambda = 0, \end{cases} \tag{8.11}$$

where $C \neq 0$ is arbitrary constant.

Let us discuss the bundle by using the geometric notions from Section 7.2. First of all, we have obtained a three-parameter subset of TWs

$$B_3 = \{f_\lambda(\xi - a),\ \lambda, C, a \in I\!R\},$$

where a denotes the translational parameter. The TW interface $\bar{s}(t) \equiv \lambda t$ is governed by the dynamic equation

$$\lambda = \mathbf{N}_1(f) \equiv -\frac{1}{C}\, f'(0) \equiv -\frac{1}{C}\, V_x(s(t), t). \tag{8.12}$$

In order to describe the left-hand interface, we consider TW profiles $f_\lambda(\xi) > 0$ for $\xi > 0$, and assume that $C \neq 0$ is such that

$$\lambda C < 0. \tag{8.13}$$

Since B_3 is three-parametric, tangent solutions are not unique and the corresponding B-bundle is overdetermined. Moreover, one can see from (8.11) and (8.12) that the standard order-preserving (comparison) property can be violated on the interfaces.

As usual for such cases, we have to choose a two-dimensional proper subset $B_2 \subset B_3$ and determine the corresponding *geometric evolution* via characteristic functions $V \in B_2$ obeying the first Sturm Theorem.

This can be done by choosing a suitable function $C = C(\lambda)$. Then the constant one $C(\lambda) \equiv 1$ corresponds to the proper (undercooled) Stefan problem with the first-order interface equation following from (8.12),

$$s'(t) \equiv \lambda = -V_x(s(t), t).$$

The function $C(\lambda) = -\frac{1}{\lambda}$ gives the Florin problem with the free-boundary condition $V_x = 1$, which is equivalent to the second-order dynamic interface equation

$$s'(t) \equiv \lambda = -V_{xx}.$$

We now assume that $C = C(\lambda) : I\!R \to I\!R$ is arbitrary, so that

$$B_2(C) = \{f_\lambda(\xi),\ \ C = C(\lambda), \lambda, a \in I\!R\}$$

is a two-dimensional family. We first check *completeness* of B_2 in the hodograph plane by solving the tangential system:

$$V(x,t) = \mu > 0, \quad V_x(x,t) = \nu > 0$$

or, which is the same,

$$C(e^{-\lambda\xi} - 1) = \mu, \quad -\lambda C e^{-\lambda\xi} = \nu.$$

Then one obtains a single equation for λ,

$$R(\lambda) \equiv -\frac{\nu}{\lambda} - C(\lambda) = \mu. \tag{8.14}$$

We next verify under what assumptions on $C(\lambda)$, the set B_2 is monotone decreasing with λ. Similar to proper maximal solutions, this is necessary for the parabolic monotonicity at the interfaces when new intersections cannot appear on interfaces. Observe that local monotonicity at the interfaces implies global monotonicity of TW profiles, which is easily proved by intersection comparison (and also follows from the corresponding ODE).

Proposition 8.1 *If the profiles $f_\lambda(\xi)$ of the TW-bundle B_2 are properly ordered, i.e., are monotone decreasing in λ for small $\xi > 0$, they are ordered for all $\xi > 0$.*

Proof. Let us present an elementary PDE proof. Assume for contradiction that there exist $\lambda_1 < \lambda_2$ such that $f_{\lambda_1}(\xi)$ intersects $f_{\lambda_2}(\xi)$ at some $\xi = \xi_*$. Then there exists a shift $a > 0$ such that the solutions

$$V_1(x,t) = f_{\lambda_1}(x - \lambda_1 t - a) \quad \text{and} \quad V_2(x,t) = f_{\lambda_2}(x - \lambda_2 t)$$

create two new intersection at some $t = t_* > 0$ in the positivity domain of both solutions, which is impossible by the usual comparison for the HE. Obviously, this contradicts uniqueness for the corresponding ODE. □

It follows from (8.11) that

$$f(\xi) = -\lambda C\xi + \tfrac{1}{2}\lambda^2 C\xi^2 + O(\xi^3) \quad \text{for small } \xi > 0, \tag{8.15}$$

and hence

$$\frac{\mathrm{d}}{\mathrm{d}\lambda} f(\xi) = -\frac{\mathrm{d}}{\mathrm{d}\lambda}(\lambda C)\xi + \tfrac{1}{2}\frac{\mathrm{d}}{\mathrm{d}\lambda}(\lambda^2 C)\xi^2 + O(\xi^3).$$

Therefore, the strict monotonicity,

$$\frac{\mathrm{d}}{\mathrm{d}\lambda} f(\xi) < 0 \quad \text{for } \xi > 0 \tag{8.16}$$

is valid provided that, for all $\lambda \in I\!R$,

$$(\lambda C(\lambda))' \geq 0 \quad \text{and if} \quad (\lambda C(\lambda))' = 0, \quad \text{then} \quad (\lambda^2 C(\lambda))' = \lambda C < 0. \tag{8.17}$$

Both completeness and monotonicity conditions are true for

(i) $C(\lambda) \equiv 1$, the proper Stefan problem, and

(ii) $C(\lambda) = -\frac{1}{\lambda}$, the Florin problem.

Consider more examples.

Example 8.3: a generalized Stefan problem. Let

$$C(\lambda) = 1 + \lambda^2 \quad \text{for } \lambda < 0.$$

The monotonicity (8.17) holds and the tangency equation (8.14) takes the form

$$R(\lambda) \equiv -\nu\frac{1}{\lambda} - 1 - \lambda^2 = \mu, \tag{8.18}$$

where the function $R(\lambda)$ must be strictly monotone, i.e.,

$$R'(\lambda) = \nu\frac{1}{\lambda^2} - 2\lambda > 0 \quad \text{for all } \lambda < 0.$$

Therefore, (8.18) has a unique solution $\lambda < 0$ for all hodograph coordinates $\{\mu > 0, \nu > 0\}$.

We obtain a proper (complete and monotone) TW-bundle $B_2(C)$ with the interface propagation governed by the first-order interface equation

$$H^{-1}(\lambda) \equiv -\lambda(1+\lambda^2) = f'(0) \equiv V_x \quad \text{for } \lambda = s'(t) < 0$$

that uniquely defines the Rankine–Hugoniot condition $\lambda = H(V_x)$ with monotone decreasing H.

Finally, we conclude that this proper TW-bundle generates a FBP for the HE with the interface equation

$$D^+ s(t) = H(u_x) \quad \text{at } x = s(t), \ t > 0. \tag{8.19}$$

By the intersection comparison of $u(x,t)$ with the proper TW-bundle, similar to the analysis in Sections 7.7 and 7.10, we have that proper solutions $u(x,t)$ exhibit all the regularity properties prescribed by TWs such as Bernstein bounds near the interface and instantaneous smoothing in B-classes. Moreover, the interface operators in (8.19) are well defined and the interfaces are Lipschitz continuous on $[\delta, C]$, and so on.

Example 8.4: a Stefan-Florin problem. Let

$$C(\lambda) = \begin{cases} -\frac{1}{\lambda} & \text{for } |\lambda| < 1, \\ -\frac{1}{\lambda} + \varepsilon(\lambda) & \text{for } |\lambda| \geq 1, \end{cases}$$

where ε is a smooth even function satisfying $\varepsilon(1) = \varepsilon'(1) = 0$ and $\varepsilon(\infty) = 0$. Here $\varepsilon(\lambda)$ stands for a perturbation of Florin's function $C(\lambda) = -\frac{1}{\lambda}$.

The monotonicity condition (8.17) is valid for $|\lambda| < 1$ as for the Florin problem. For $|\lambda| > 1$, it holds provided that ε satisfies

$$(\lambda\varepsilon)' > 0 \quad \text{for } |\lambda| \geq 1. \tag{8.20}$$

Consider the tangency equation (8.14) with

$$C(\lambda) = -\frac{1}{\lambda} + \varepsilon,$$

where this equation takes the form

$$R(\lambda) \equiv \frac{1-\nu}{\lambda} - \varepsilon(\lambda) = \mu.$$

We restrict our analysis to a class of solutions with uniformly bounded gradient satisfying

$$\nu = u_x \leq \gamma < 1.$$

Then

$$R'(\lambda) = -(1-\nu)\frac{1}{\lambda^2} - \varepsilon'(\lambda) \leq -(1-\gamma)\frac{1}{\lambda^2} - \varepsilon'(\lambda) < 0$$

provided that

$$\varepsilon'(\lambda) > -\frac{1-\gamma}{\lambda^2} \quad \text{for } |\lambda| \geq 1. \tag{8.21}$$

Under hypotheses (8.20), (8.21) on $\varepsilon(\lambda)$, the TW-bundle $B_2(C)$ is proper and we arrive at a well-posed Stefan-Florin problem for the HE. It has first-order Stefan interface condition

$$H^{-1}(\lambda) \equiv \lambda\varepsilon(\lambda) - 1 = -V_x \quad \text{for } |\lambda| > 1$$

and the Florin second-order one

$$\lambda = -V_{xx} \quad \text{if } |\lambda| < 1.$$

The Rankine–Hugoniot condition changes the order at the critical speeds $\lambda = \pm 1$. It follows by the G-theory that this FBP is well-posed for the HE and general proper solutions $u(x,t)$ satisfy *a priori* regularity induced by the proper TW-bundle $B_2(C)$.

In both examples we do not present the corresponding regularity estimates that are straightforward for the HE. We will do this for the PME to be studied next.

8.3 Classification of free-boundary problems for the quadratic porous medium equation

We now study proper FBPs for the quadratic PME ($m = 2$)

$$\boxed{u_t = (u^2)_{xx}}. \tag{8.22}$$

Travelling waves. Setting $V = f(x - \lambda t)$, we obtain the ODE

$$(f^2)'' + \lambda f' = 0$$

and integrating yields

$$f' = -\frac{\lambda}{2} + \frac{C}{2f}, \quad \text{where } C > 0 \text{ is arbitrary.} \tag{8.23}$$

Note that the constant $C = 0$ corresponds to the maximal proper TWs studied in detail in the previous chapter.

Equation (8.23) prescribes a two-parameter TW set B with the following behaviour near the interface

$$f(\xi) = \sqrt{C\xi}\left(1 - \lambda\sqrt{\tfrac{\xi}{2C}} + \ldots\right). \tag{8.24}$$

In order to derive the interface operators, we introduce the pressure $v = V^2$, which is asymptotically linear close to the interface,

$$f^2(\zeta) = C\xi - \lambda\sqrt{C\xi^3} + \ldots \quad \text{as } \xi \to 0.$$

This gives two differential interface operators of the first and second order

$$\mathbf{M}_1(V) \equiv (V^2)_x = C, \quad \lambda^2 = \mathbf{N}_2^2(V) \equiv \frac{4}{9}\frac{1}{(V^2)_x}\{[(V^2)_x - C]^2\}_x. \tag{8.25}$$

The algebraic form of the second-order operator $\mathbf{N}_2$ is derived by means of two indeterminacies as $\xi \to 0^+$ as follows:

$$\lambda = \mathbf{N}_2(f) \equiv -\frac{1}{\sqrt{C}} \lim_{\xi \to 0^+} \frac{1}{\sqrt{\xi}}\left(\frac{f^2(\xi)}{\xi} - C\right), \tag{8.26}$$

where

$$C = (f^2)'|_{f=0}.$$

Proper TW-bundles. As usual, we fix a proper TW-bundle $B_2 = B_2(C)$ by choosing smooth functions $C = C(\lambda) : I\!R \to I\!R_+$. It follows from (8.24) that such a bundle is monotone decreasing with λ provided that

$$C(\lambda) > 0 \quad \text{and} \quad C'(\lambda) \le 0. \tag{8.27}$$

The completeness follows from (8.23), giving the following tangency equation at $f = \mu > 0,\ f' = \nu > 0$:

$$\nu = -\frac{\lambda}{2} + \frac{C(\lambda)}{2\mu} \equiv R(\lambda),$$

where $R : I\!R \to I\!R_+$ is strictly monotone,

$$R'(\lambda) = -\tfrac{1}{2} + \frac{C'(\lambda)}{2\mu} < 0.$$

It follows that there exists the unique tangent TW solution f_λ with a function $\lambda = \lambda(\mu, \nu)$.

On proper solutions. We need to extend the regularity properties of the complete TW-bundle $B_2(C)$ to solutions of the PME with such free-boundary conditions. The quadratic pressure variable $v = u^2$ satisfies

$$v_t = 2\sqrt{v}\, v_{xx}.$$

Setting $x = X(v,t)$ and assuming the strict monotonicity of $v(x,t)$ in x, we arrive at the quasilinear parabolic equation

$$X_t = 2\sqrt{v}\, \frac{X_{vv}}{(X_v)^2} \quad \text{in } S = I\!R_+ \times I\!R_+, \tag{8.28}$$

and the dynamic boundary condition at $v = 0$

$$X_v = \frac{1}{C(X_t)} \quad \text{or} \quad (X_t)^2 = \mathbf{N}_2^2(X) \equiv \frac{4}{9}\left\{\left[\frac{1}{X_v} - C(X_t)\right]^2\right\}_v. \tag{8.29}$$

Assuming that $X_v(0,t) \neq 0$, we have that (8.28) degenerates at the boundary $v = 0$ in the independent spatial variable v. The degeneracy of the order $O(\sqrt{v})$ is known to keep necessary properties of solutions such as the strong MP and the Oleinik-Hopf Boundary Point Lemma (see the last part of Remarks to Chapter 1 devoted to multiple zeros of linear and quasilinear parabolic equations). Note that, at the same time, the equation is non-singular at the boundary in terms of the solution X. If necessary, we perform a usual regularization in the denominator on the right-hand side of (8.28). We assume that a solution can be constructed by a suitable approximation in terms of regular parabolic problems (e.g. with frozen coefficients).

Classification of proper FBPs

Thus a complete proper TW-bundle $B_2(C)$ with functions $C(\lambda)$ satisfying (8.27) generates an FBP that is well posed relative to TWs. In order to prescribe the propagation of the interface, one can either use the first-order interface operator

$$(u^2)_x = C(D^+ s), \tag{8.30}$$

or put the dynamic condition with the second-order operator

$$D^+ s = \mathbf{N}_2(u), \tag{8.31}$$

where $\mathbf{N}_2$ is defined according to the algebraic form (8.26). By intersection arguments from Section 7.10, these interface operators are well defined on general proper solutions. If the inverse function C^{-1} exists, (8.30) is the first-order interface equation

$$D^+s = C^{-1}((u^2)_x).$$

If $C(\lambda) \equiv \text{const.}$ on some interval, the interface speed is governed by the second-order interface equation (8.31).

Example 8.5: Florin problem. It follows from (8.30) that choosing

$$C(\lambda) \equiv 1 \quad \text{for} \quad \lambda \in \Delta_0 = I\!R$$

gives a correctly posed Florin problem for the PME with the interface equation

$$(u^2)_x = 1.$$

Example 8.6: proper Stefan problem. Choosing the monotone decreasing function

$$C(\lambda) = -\lambda > 0 \quad \text{for} \quad \lambda \in \Delta_0 = (-\infty, 0),$$

we obtain the proper TW-bundle $B_2(-\lambda)$ generating the Stefan problem for the PME with the interface equation

$$D^+s = -(u^2)_x.$$

Example 8.7: supercooled, improper Stefan problem. The TW-bundle $B_2(\lambda)$ with

$$C(\lambda) = \lambda > 0 \quad \text{for} \quad \lambda \in \Delta_0 = I\!R_+$$

is not the proper one since C does not satisfy (8.27). This means that solutions of the Stefan FBP with

$$D^+s = (u^2)_x \tag{8.32}$$

do not obey the MP on the interface and finite-time singularities of $s(t)$ can occur. This is an example of the supercooled Stefan problem for the PME. Note that the formation of finite-time singularities at the interfaces are not described by TW-bundles and special approximate similarity blow-up structures admitted by the PME are necessary. This remains an open problem.

Singular supercooled-type phenomena can occur for any FBP generated by TW-bundles $B_2(C)$ if $C(\lambda)$ violates the monotonicity assumption (8.27) for $\lambda \gg 1$.

Example 8.8: transition to maximal solutions. As a link to maximal solutions, let us fix a monotone decreasing function $C(\lambda) \geq 0$ satisfying

$$C(\lambda) > 0 \quad \text{for } \lambda < 1 \quad \text{and } C(\lambda) \equiv 0 \quad \text{for } \lambda \geq 1. \tag{8.33}$$

Then if the interface propagates with the speed $D^+s < 1$, it is governed by the FBP with the interface equation (8.30) or (8.31). Once $D^+s > 1$, the solution becomes the proper maximal one with the already known interface equation that is the classical Darcy law. Actually, in this case we do not need to specify any interface equation and the solution is determined by any monotone regular approximation.

On typical regularity properties. By the G-theory, a proper B-bundle $B_2(C)$

generates the usual regularity results for the FBP. Consider solutions $u(x,t)$ of the PME with the interface conditions (8.30) or (8.31). Regularity estimates are derived similarly to those for the maximal solutions in Section 7.7. For instance, let us reconstruct the corresponding Bernstein bound. The G-function of the TW-bundle takes the form

$$G(\lambda,\varepsilon) \equiv (f^2)'_\xi|_{f^2=\varepsilon} = C(\lambda) - \tfrac{3}{2}\,\lambda\sqrt{\varepsilon} + ... \quad \text{as } \varepsilon \to 0.$$

Hence, the Bernstein bound for $t \geq \delta > 0$ in the corresponding local B-class has the form (we put $\varepsilon = u^2$)

$$|(u^2)_x - C(D^+ s)| \leq c\,u, \quad \text{where } c = c(u_0,\delta) > 0 \text{ is a constant.}$$

The phenomenon of the instantaneous smoothing has the same geometric nature with the same proof as for the maximal solutions. For bounded initial data u_0, we can guarantee that all regularity estimates are valid on intervals $t \in [\delta, C]$ with any $\delta > 0$.

8.4 On general one-phase free-boundary problems

The above examples show an approach to a proper formulation of generalized FBPs for nonlinear parabolic equations (8.1). The strategy of such a geometric approach is the same as for the maximal solutions in Chapter 7. We present brief comments on these techniques and avoid repetition of the corresponding propositions and theorems.

(1) We first construct a full set B of singular TW solutions depending on two parameters λ and $C \in I\!R$.

(2) Next, we fix a proper (complete and monotone) subset $B_2 \subset B$ by choosing a function $C = C(\lambda)$ for which $B_2(C)$ is complete and is monotone decreasing with λ. Such a proper B-bundle defines regularity properties of TW solutions and TW interface equations.

(3) The intersection comparison with the B-bundle maps the properties of the ODE bundle onto a class of general solutions of the PDE accomplished with these free-boundary conditions. The regularity bounds and the interface behaviour are *a priori* properties of solutions to such FBPs.

(4) It then follows that proper solutions satisfy both the comparison principle and the Sturmian intersection argument and we use proper complete sets of TWs B in comparison and intersection comparison analysis establishing the corresponding PDE–ODEs duality.

Unlike the maximal solutions, where monotone approximations can be performed in rather general situations, approximations of singular non-maximal solutions lead to some non-standard initial-boundary value problems for regularized second-order uniformly parabolic equations. For some FBPs such a regularization implies a parabolic equation with dynamic boundary conditions. A general problem of approximation of non-maximal solutions is individual for each kind of quasilinear and fully nonlinear equations. In this case, the intersection geometric approach establishes main *a priori* estimates of such solutions as for the maximal solutions in Sections 7.7 and 7.10 (since the G-theory deals with proper functional B-bundles regardless the equations and FBPs under consideration). Translation of several results on extended limit semigroups of maximal solutions from the previous chapter to generalized FBPs is straightforward. We then obtain a number of *a priori* regularity results induced by proper TW-bundles as in the case on maximal solutions.

Under these assumptions, for the 1D equations (8.1) admitting a complete proper subset B of particular solutions, we can adopt the concept of *B-solutions* $u(x,t)$, i.e., those satisfying intersection comparison properties relative to any characteristic function $V(x,t) \in B$. We then finally arrive at general results similar to those in Theorem 7.37 establishing the following properties:

(i) The TW interface operators are well defined on proper solutions of the FBP (these provide us with optimal gradient or higher-order estimates near singular interfaces in the corresponding functional B-classes; cf. Section 7.7), and

(ii) $s(t)$ is continuous on $[0, C]$, $D^+ s(t)$ is uniformly bounded on intervals $[\delta, C]$ for any $\delta > 0$ and $s(t)$ is Lipschitz continuous there.

Furthermore, similar to Theorem 7.45 we deduce the uniqueness of such B-solution:

(iii) In the above regularity class, a *maximal* B-solution, defined in a natural way, is unique. The proof uses the usual comparison only.

On free-boundary problems in $I\!R^N$. In the class of proper maximal solutions, several results from 1D have been extended to the corresponding N-dimensional equations. This can also be done for non-maximal solutions of the FBPs. In particular, a nonexistence (semigroup discontinuity) result is true for the most general parabolic equation

$$u_t = F(u, \nabla u, D^2 u) \quad \text{in } I\!R^N \times I\!R_+, \tag{8.34}$$

with typical regularity and monotonicity properties of F that we have used in the previous chapter. In particular, the following result is straightforward (establishing that, if the maximal solution is entirely singular, the same is true for all non-maximal ones).

Theorem 8.2 *Assume that a one-dimensional restriction of* (8.34) *satisfies the assumptions of Theorem* 7.12. *Then any FBP for* (8.34) *assuming propagation on the singular level* $\{u = 0\}$ *admits the trivial solution only:* $u(x,t) \equiv 0$ *for* $t > 0$.

The existence Theorem 7.8 also admits a natural extension to the FBPs by comparison with arbitrarily small nontrivial radial TW subsolutions. A number of other transversality results (Propositions 7.33 and 7.35) and the concavity ones (Proposition 7.36) are true *a priori* for radial solutions $u = u(r,t)$. The construction of solutions of non-standard FBPs, which are equivalent to nonlinear parabolic equations with second-order dynamic boundary conditions, become much more complicated in $I\!R^N$; see Remarks. In particular, estimates of normal velocities $v_\pm^\pm$ of singular interfaces are obtained by comparison with radial super and subsolutions as in Proposition 7.44. By the geometric analysis, typical regularity results are justified without specifying PDEs and interface conditions. In these cases we observe a one-to-one map of the results for maximal solutions and for proper FBPs, thus establishing *a priori* regularity estimates on general classes of maximal and non-maximal singular solutions of nonlinear parabolic PDEs.

Consider a couple of further examples of FBPs exhibiting some new properties of singular interface propagation that are not available for the maximal solutions.

8.5 Higher-order free-boundary problems for the porous medium equation with absorption

First we describe a class of FBPs that are proper for the PME with strong absorption

$$\boxed{u_t = (u^m)_{xx} - u^p, \quad m > 0, \ p > -m.} \tag{8.35}$$

As we have shown, if $p \le -m$, then $\Delta = \emptyset$ for maximal solutions. Therefore, for any singular initial data u_0, the maximal solution is $u(x,t) \equiv 0$ for every $t > 0$ and the same is true for any non-maximal solution. Hence, FBPs do not make sense for $p \le -m$ in any space dimension for the equation

$$\boxed{u_t = \Delta u^m - u^p.}$$

In the existence case $p > -m$ we follow the lines of the TW study in Section 7.11. The ODE for $V(x,t) = f(\xi)$ takes the form

$$(f^m)'' - f^p + \lambda f' = 0, \tag{8.36}$$

and setting $P = (f^m)'$ yields the first-order equation

$$P\frac{\mathrm{d}P}{\mathrm{d}f} = m f^{p+m-1} - \lambda P. \tag{8.37}$$

We begin with a new class of FBPs.

Third-order interface equation and corresponding Bernstein estimates. Let

$$\tfrac{1}{2}(1-2m) < p < 1-m.$$

Consider the orbit $P(f)$ satisfying $P(0) = C > 0$. Then for small $f > 0$,

$$P(f) = C + \frac{m}{(p+m)C} f^{p+m} - \lambda f + \dots .$$

Hence, for any $\lambda \in I\!R$, there exist a family of TWs with the following behaviour near the interface:

$$f^m(\xi) = C\xi + a_1 C^{\frac{p}{m}} \xi^{\frac{p+2m}{m}} - \lambda a_2 C^{\frac{1}{m}} \xi^{\frac{m+1}{m}} + \dots, \tag{8.38}$$

where $C > 0$ is arbitrary and $a_1 = \frac{m^2}{(p+m)(p+2m)}$, $a_2 = \frac{m}{m+1}$. One can see that $C = 0$ corresponds to maximal TWs in Section 7.11.

As usual for FBPs for the PME, the pressure variable is $Q(f) = f^m$. The TW speed λ appears in the third expansion term, so that we resolve three indeterminacies by means of the following three interface operators:

$$\mathbf{M}_1(f) = \lim_{\xi \to 0} \frac{1}{\xi} f^m(\xi) \equiv (f^m)'(0),$$

$$\mathbf{N}_2(f) = \frac{1}{a_1} \lim_{\xi \to 0} \frac{1}{\xi^{\frac{p+m}{m}}} \left[\frac{1}{\xi} f^m(\xi) - C \right],$$

$$\mathbf{N}_3(f) = -\frac{1}{a_2 C^{\frac{1}{m}}} \lim_{\xi \to 0} \xi^{\frac{p+m-1}{m}} \left[\frac{\frac{1}{\xi} f^m(\xi) - C}{\xi^{\frac{p+m}{m}}} - a_1 C^{\frac{p}{m}} \right], \tag{8.39}$$

where, in the algebraic sense, $\mathbf{N}_3$ is a *third-order* interface operator. Hence, the TW interface $s(t) \equiv \lambda t$ satisfies a system of three equations, two stationary and one dynamic,

$$\begin{cases} \mathbf{M}_1(V) = C, \\ \mathbf{N}_2(V) = C^{\frac{p}{m}}, \\ s' \equiv \lambda = \mathbf{N}_3(V). \end{cases} \tag{8.40}$$

One can obtain differential forms for both operators $\mathbf{N}_2$ and $\mathbf{N}_3$, which are of second and third order respectively (though they cannot be used for the solutions since they assume too much regularity).

We now choose a smooth monotone decreasing function

$$C = C(\lambda) : I\!R \to I\!R_+,$$

and define a complete, monotone, proper subset $B_2(C)$.

Example 8.9 A simple choice is $C \equiv 1$ leading to the proper Florin free-boundary condition

$$(u^m)_x = 1. \tag{8.41}$$

The geometric analysis in the TW-bundle $B_2(1)$ shows that the dynamic interface equation is of third order

$$D^+ s = \mathbf{N}_3(u).$$

Example 8.10 If $C(\lambda) = -\lambda$ for $\lambda \in \Lambda_0 = (-\infty, 0)$, we arrive at the proper Stefan problem with the interface equation

$$D^+ s = -(u^m)_x.$$

Optimal, three-term Bernstein estimate. In general, a proper $B_2(C)$ determines an FBP for equation (8.35), which is well-posed for any of the three free-boundary conditions

$$\mathbf{M}_1(u) = C(D^+ s), \quad \mathbf{N}_2(u) = (C(D^+ s))^{\frac{p}{m}}, \quad \text{or } D^+ s = \mathbf{N}_3(u). \tag{8.42}$$

Recall that the problem is well-posed in the sense that, for bounded monotone initial data u_0, the solution stays in a local $B_2(C)$-class, has the same Bernstein estimate as the corresponding TWs and exhibits the same moduli of continuity in x and t as the TW profiles $\{f_\lambda\} \subset B_2(C)$ in ξ, etc. We also obtain instantaneous smoothing phenomena, the strong MP for interfaces, as well as other conclusions of the G-theory, Section 7.7.

For example, let us derive the corresponding Bernstein estimate near the interfaces. The G-function is given by

$$G(\lambda, \varepsilon) = (f^m)'_\xi|_{f^m = \varepsilon}$$

$$= C(D^+ s) + a_3 C^{-1}(D^+ s)\varepsilon^{\frac{p+m}{m}} - \lambda a_4 \varepsilon^{\frac{1}{m}} + \dots, \tag{8.43}$$

so that on $[\delta, T-\delta]$, $\delta > 0$ ($T \le \infty$ is the extinction time) we obtain the estimate

$$\left| (u^m)_x - C(D^+ s) - \frac{a_3}{C(D^+ s)} u^{p+m} \right| \le c\, u, \tag{8.44}$$

where c is a constant depending on the initial data and δ. Recall a simple proof of this non-standard gradient estimate: if this is not true (i.e., such a constant c does not exist), then changing the parameter λ in the proper TW-bundle, we deduce that the profile $u(x,t)$ has an infinite number of intersections with an uncountable subset of TWs $f_\lambda \in B$. This is impossible for any continuous solution (Proposition 7.38).

The first-order differential operator on the left-hand side of (8.44) consists of three different terms and therefore it describes the optimal regularity of solutions. As a consequence, we have that near interfaces, proper solutions are Hölder continuous with exponent $\frac{1}{m}$, in both x and t. Indeed, the TW-bundle contains many other refined regularity properties of general solutions.

Fourth and higher-order interface operators. Let

$$\tfrac{1}{3} - m < p < \tfrac{1}{2} - m.$$

Then, in the expansion of the orbit $P(f)$ with $P(0) = C > 0$, we have to keep four terms as $f \to 0$,

$$P(f) = C + \tfrac{m}{(p+m)C}\, f^{p+m} - \tfrac{m^2}{2(p+m)^2 C^3}\, f^{2(p+m)} - \lambda f + \dots. \tag{8.45}$$

This gives TWs with a parameter $C > 0$

$$f^m(\xi) = C\xi + a_1 C^{\frac{p}{m}} \xi^{\frac{p+2m}{m}} - a_3 C^{\frac{2p-m}{m}} \xi^{\frac{2p+3m}{m}} - \lambda a_2 C^{\frac{1}{m}} \xi^{\frac{m+1}{m}} + \dots. \tag{8.46}$$

In order to specify the speed λ by algebraic manipulations, we now have to use four interface operators, and, finally, λ is expressed by an algebraic fourth-order one. The rest of the analysis stays the same as above. The left-hand side of the optimal Bernstein bound similar to (8.44) then contains four terms.

Higher $(k+2)$th order interface operators occur if

$$\tfrac{1}{k+1} - m < p < \tfrac{1}{k} - m,$$

where expansions such as (8.45) and (8.46) have $k+2$ terms on the right-hand side in order to display the speed λ in the last one. The first-order Bernstein bound such as (8.44) contains $k+2$ different terms, etc.

One can see that, in the differential sense, the dynamic interface equations always have, formally, a second-order representation

$$u(s(t),t) \equiv 0 \implies s'(t) = \lim_{u \to 0} \frac{(u^m)_{xx} - u^p}{u_x}.$$

It turns out that this formula assumes too much regularity of solutions and is difficult to use in applications. We have shown instead that the geometric analysis in complete TW-bundles $B_2(C)$ dealing with algebraic indeterminacies generates the precise and more refined higher-order interface operators and multi-term Bernstein estimates near singularities.

Second-order interfaces. This is an easier case. One can see from (8.36) and (8.38) that, for all

$$p \geq 1 - m,$$

the interface equation is of second order. In particular, in the critical case $p = 1 - m$ the TW-bundle has a two-term expansion

$$f^m(\xi) = C\xi + a_1 C^{\frac{1}{m}} \left(\frac{m}{C} - \lambda\right) \xi^{\frac{m+1}{m}} + \ldots,$$

where $C = C(\lambda)$ is a smooth monotone decreasing function. This expansion defines two interface operators, one stationary and one dynamic. The intersection comparison with the proper bundle $B = B_2(C)$ establishes that such a generalized FBP is well-posed and possesses a number of regularity properties.

8.6 Higher-order free-boundary problems for the dual porous medium equation with singular absorption

Consider now the following fully nonlinear model from Section 7.11:

$$\boxed{u_t = |u_{xx}|^{m-1} u_{xx} - u^p, \quad m > 0, \quad -m < p < 0.} \tag{8.47}$$

We have proved that, for $p \leq -m$, maximal solutions are trivial, $u \equiv 0$, for any initial data u_0 vanishing on an arbitrarily small interval. Hence, the same is true for any FBP posed on the singularity level in any dimension $N \geq 1$ for the equation

$$\boxed{u_t = |\Delta u|^{m-1} \Delta u - u^p.}$$

For the TW $V(x,t) = f(\xi)$ we obtain the ODE

$$|f''|^{m-1} f'' - f^p + \lambda f = 0.$$

Then $f'' > 0$ near the interface, and setting $f' = P$, we obtain the first-order ODE

$$P \frac{\mathrm{d}P}{\mathrm{d}f} = (f^p - \lambda P)^{\frac{1}{m}}.$$

If $p > -\frac{m}{1+m}$, the orbit $P(f)$ with $P(0) = C > 0$ has the following three-term expansion as $f \to 0$:

$$f' = P(f) = C + a_1 \tfrac{1}{C} f^{\frac{p+m}{m}} - a_2 \lambda f^{\frac{p(1-m)+m}{m}} + \ldots.$$

Integrating yields TW profiles depending on two parameters $\lambda \in I\!R$ and $C > 0$:

$$f(\xi) = C\xi + a_3 C^{\frac{p}{m}} \xi^{\frac{p+2m}{m}} - a_4 \lambda C^{\frac{p(1-m)+m}{m}} f^{\frac{p(1-m)+2m}{m}} + \ldots. \tag{8.48}$$

Choosing a monotone decreasing function $C(\lambda)$ yields a proper TW-bundle $B_2(C)$ that generates a well-posed FBP with the dynamic interface equation of third order. The Florin problem with $C \equiv 1$ is of this type. The Bernstein gradient estimate follows from (8.48) and, similar to (8.44), contains three terms. The interface system is derived from (8.48) exactly as above. The rest of the regularity results are similar to the PME with absorption. For

$$p \in (-m, -\tfrac{m}{1+m}),$$

the expansion such as (8.48) can include more than three terms and we obtain higher-order interface operators, equations and optimal Bernstein bounds.

In a similar way, classes of well-posed FBPs are constructed for other nonlinear parabolic equations studied in Section 7.11 studied there from the point of view of maximal solutions.

8.7 On generalized two-phase free-boundary problems

In a typical setting, a *two-phase* FBP assumes that the solution $u(x,t)$ changes sign once at the interface $x = s(t)$, where $u(s(t),t) = 0$, and

$$u(x,t) > 0 \quad \text{for } x > s(t) \;\text{ and }\; u(x,t) < 0 \quad \text{for } x < s(t). \tag{8.49}$$

The corresponding dynamic interface equation depends on both the right- and left-hand slopes of the solution on the interface.

Two-phase FBPs for the heat equation

Consider a solution $u(x,t)$ of the HE satisfying (8.49) and impose the following two-phase Stefan free-boundary condition at the interface:

$$s' = -[u_x] \equiv -(D_x^+ u - D_x^- u) \quad \text{at } x = s(t) \;\text{ for } t > 0. \tag{8.50}$$

Smooth bounded initial data u_0 are assumed to change sign once at a finite $x = s(0)$. Condition (8.50) is properly posed in the sense that it does not violate the usual comparison of interfaces. Indeed, given two classical solutions satisfying, at some $t_0 > 0$,

$$u_2(x,t_0) \geq u_1(x,t_0) \quad \text{in } I\!R$$

and $s_1(t_0) = s_2(t_0)$, we have that $s_2'(t_0) \leq s_1'(t_0)$ so that the partial order cannot fail at the interface (and at any $x \neq s(t)$, where the standard comparison and the MP for the heat equation takes place). Moreover, the strong MP for the heat equation implies that $s_2(t) < s_1(t)$ for $t > t_0$ provided that $u_2(x,t_0) \not\equiv u_1(x,t_0)$.

Proper generalizations of Stefan FBPs. Prescribing the interface evolution by a general equation of the form

$$s' = G(D_x^+ u, D_x^- u) \quad \text{at } x = s(t),\; t > 0, \tag{8.51}$$

the parabolicity condition on a smooth function $G : I\!R^2 \to I\!R$ is

$$G(d^+, d^-) \quad \text{is decreasing with } d^+ \text{ and is increasing with } d^-. \tag{8.52}$$

Otherwise, the free-boundary condition (8.51) would violate the usual comparison of solutions on the interfaces.

We now need to describe the TW-bundle for the two-phase FBP that takes the form

$$f_\lambda(\xi) = \begin{cases} C_\pm(e^{-\lambda\xi} - 1) & \text{for } \lambda \neq 0, \\ C_\pm \xi & \text{for } \lambda = 0, \end{cases} \tag{8.53}$$

where we take C_+ for $\xi > 0$ and C_- for $\xi < 0$. Therefore, in view of (8.49), we impose the condition

$$C_\pm \lambda < 0 \quad \text{for any } \lambda \in \Delta_0,\; \lambda \neq 0. \tag{8.54}$$

The following geometric property of the bundle is straightforward.

Proposition 8.3 *The TW-bundle* (8.53) *is complete on the interface in the sense that, for any given first-order slopes* $d^\pm = D_x^\pm u$ *on the interface, there exists a unique tangent TW* f_λ *satisfying the free-boundary condition* (8.51).

Proof. We have that the speed $\lambda = G(d^+, d^-) \neq 0$ is uniquely determined by the slopes and hence (8.53) gives

$$D^\pm f_\lambda \equiv -\lambda C_\pm = d^\pm \implies C_\pm = -\frac{d^\pm}{\lambda},$$

whence the unique tangent TW. If $\lambda = 0$, then $C_\pm = d^\pm$. □

Properly posed free-boundary conditions (8.51), (8.52) are generated by complete TW-bundles B. Then on smooth solutions the comparison principle is true and new intersections of different solutions cannot occur on interfaces.

Example 8.11: a two-phase Florin-type problem. It is generated by the condition

$$D_x^+ u + D_x^- u = 1 \quad (D_x^\pm u < 1). \tag{8.55}$$

The corresponding right-hand and left-hand dynamic equations

$$s' D_x^\pm u = -D_{xx}^\pm u \quad \implies \quad s' = -(D_{xx}^+ u + D_{xx}^- u) \equiv \mathbf{N}_2(u) \tag{8.56}$$

deal with the second-order interface operators determining two slopes of the solution on the interface given by

$$D^\pm \equiv D_{xx}^\pm u. \tag{8.57}$$

If necessary, they can be written in terms of an algebraic representation using the expansion in the TW-bundle (8.53). One can see that (8.56) assumes that, on the interface, the slopes always satisfy

$$\frac{D^+}{d^+} = \frac{D^-}{d^-}, \qquad \text{where } d^+ = D_x^+ u, \ d^- = D_x^- u - 1 - d^+. \tag{8.58}$$

Proposition 8.4 *The TW-bundle* (8.53) $B = \{f_\lambda\}$ *is complete on the interface for the free-boundary condition* (8.55), *and it is proper, i.e., monotone decreasing with* λ.

Proof. Given the first-order slope d^+ (then $d^- = 1 - d^+$ by (8.55)) and the second-order one D^+ in (8.57) (hence $D^- = d^- D^+/d^+$ by (8.58)), we have that $\lambda \equiv s'$ is uniquely determined by

$$\lambda = -\frac{D^+}{d^+} \equiv -\frac{D^-}{d^-}.$$

It then follows from (8.53) that the gradients $d^\pm$ uniquely determine the parameters of the bundle

$$C_+ = -\frac{d^+}{\lambda} \equiv \frac{(d^+)^2}{D^+}, \quad C_- = -\frac{d^-}{\lambda} = -\frac{1 - d^+}{\lambda} \equiv \frac{(1 - d^+) d^+}{D^+},$$

whence the uniqueness of the tangent solution on the interface. The general TW-diagram

$$s' = \lambda = -(D_{xx}^+ V + D_{xx}^- V) \equiv \mathbf{N}_2(V) \tag{8.59}$$

is monotone decreasing relative to both slopes $D^{\pm}_{xx}V$, and so the TW-bundle is proper. □

On two-phase FBPs for general nonlinear equations

The two-phase FBPs can be posed for a general parabolic PDE

$$u_t = F(u, u_x, u_{xx}) \quad \text{in } (I\!\!R \setminus \{x = s(t)\}) \times I\!\!R_+ \tag{8.60}$$

for the class of solutions satisfying (8.49). First we study the TW-bundles of the corresponding one-phase FBPs and determine the corresponding first and higher-order interface operators and slopes. This poses Stefan, Florin or other free-boundary conditions provided that the TW-bundles are proper, i.e., complete, continuous and monotone decreasing with λ. This guarantees that the standard comparison principle holds and new intersections cannot appear on the interfaces. The statement of proper two-phase free-boundary conditions and the rest of the geometric analysis are similar. In particular, we always have the following conclusion.

Proposition 8.5 *Regularity results obtained by the transversality analysis such as Bernstein estimates, moduli of continuity, instantaneous smoothing, and so on, for a two-phase FBP, coincide with those for the corresponding one-phase problem.*

Proof. Using (8.49), the transversality analysis can be performed in $\{u > 0\}$ by positive TWs of $B^+ = \{f_\lambda > 0\}$ that are steep enough in the positivity domain, i.e., by steep TWs of the one-phase FBP. We next use the obvious fact that new intersections of $u(x,t)$ and the steep TW $V(x,t) = f_\lambda(x - \lambda t - a)$ cannot occur if $s(t)$ does not intersect $s_\lambda(t) = \lambda t + a$. This observation is enough to apply the same transversality techniques as for the maximal solutions or one-phase FBPs. □

FBPs for the sign PME with absorption

As a final example, consider the sign PME with strong absorption

$$\boxed{u_t = (|u|^{m-1}u)_{xx} - |u|^{p-1}u \quad \text{in } (I\!\!R \setminus \{x = s(t)\}) \times I\!\!R_+,} \tag{8.61}$$

where $m > 1$ and $p > -m$. We have shown in Section 8.5 that, in the parameter range

$$\tfrac{1}{2}(1 - 2m) < p < 1 - m,$$

the one-phase FBP has a third-order interface equation. The set of TWs for the two-phase problem has the following asymptotics:

$$f^m(\xi) = C_+\xi + a_1 C_+^{\frac{p}{m}} \xi^{\frac{p+2m}{m}} - \lambda a_2 C_+^{\frac{1}{m}} \xi^{\frac{m+1}{m}} + \ldots, \tag{8.62}$$

$$(-f)^m(\xi) = C_-(-\xi) + a_1 C_-^{\frac{p}{m}} (-\xi)^{\frac{p+2m}{m}} + \lambda a_2 C_-^{\frac{1}{m}} (-\xi)^{\frac{m+1}{m}} + \ldots \tag{8.63}$$

as $\xi \to 0^+$ and $\xi \to 0^-$ respectively, where $C_\pm > 0$ are arbitrary parameters. The second expansion is obtained from the first by the symmetry $f \mapsto -f$, $\xi \mapsto -\xi$ and $\lambda \mapsto -\lambda$ of the ODE (8.36). We now choose a two-phase free-boundary condition for which (8.62) and (8.63) give a complete and monotone TW-bundle. Such conditions can include interface operators of different orders. For instance, the two-phase Stefan problem arises with the condition

$$s' = -[(|u|^{m-1}u)_x],$$

obtained in terms of the first-order interface operators $\mathbf{M}_1(u)$. Then the slopes $d^\pm = D_x^\pm(|u|^{m-1}u)$ on the interface uniquely determine the speed $\lambda = -[d^+ - d^-]$ and both parameters $C_\pm$ of the unique profile f_λ. On the other hand, in a Florin-like problem with the free-boundary condition

$$D_x^+ u^m + D_x^-(|u|^{m-1}u) = 1,$$

we have that a unique tangent profile $f_\lambda \in B$ is obtained by the first-order slope $d^+ = D_x^+ u^m$ (then $d^- = 1 - d^+ > 0$) and the right-hand third-order slope $\lambda = \mathbf{N}_2^+(u)$ (then, by continuity, $D^- \equiv \mathbf{N}_2^-(u) = \lambda$). Bernstein estimates and continuity moduli in x and t stay the same as for the one-phase FBP.

Two-phase FBPs as the least well posed problems for second-order parabolic PDEs. It follows from the geometric analysis that a proper setting of two-phase FBPs uses a whole complete proper TW-bundle that depends on all three parameters $\{\lambda, C_+, C_-\}$, unlike the one-phase problem where a suitable choice of functions $C = C(\lambda)$ was allowed. This implies that two-phase FBPs are the least problems that can be well posed for second-order parabolic PDEs admitting proper (maximal or non-maximal) solutions and extended limit semigroups defined in sufficiently dense functional classes.

Remarks and comments on the literature

§ **8.1–8.6.** Classical 1D Stefan problems have been studied since 1831; see [332]. References to some of J. Stefan's original papers are available in [91]. See also L.I. Rubinstein's classic book [302]. A mathematical theory of Stefan problems and other types of FBPs are presented in the books [104], [122], [265], and in the survey [279].

Singularities in the supercooled Stefan problem were discovered in 1970 [313]. Blow-up can be "non-essential" with a continuation [111] or otherwise complete; see further references in the detailed survey [279, Section 6]. A classification of singular free-boundary patterns in the supercooled case was done in [189]. A cusp formation mechanism and a formal construction of generic patterns in two and three dimensions are given in [339], [340]. Cusp formation also occurs for the 2-dimensional Hele-Shaw model with suction, cf. [197], [198]. Complex variable methods are important in studying these flows and their singularities, [196].

Quasilinear parabolic equations with second-order dynamic boundary conditions were considered in [58], [107], [108]. A general functional approach to parabolic equations with nonlinear boundary conditions can be found in [7] and [8]; see also references therein. The approach is developed in such a generality that the main results extend to fully nonlinear equations with dynamic boundary conditions. Note that these fall into the scope of the Ya.B. Lopatinskii condition

(1953) [252] (or Lopatinskii–Z.Ya. Shapiro condition) after known suitable modification to parabolic equations, which is necessary and sufficient for the normal solvability of the boundary-value problems for linear PDEs. This gives the local existence of a solution. Then *a priori* bounds by comparison with TWs can be useful for deriving estimates for global existence results.

Finite-time extinction may occur for flows in porous media, where the model consists of the heat equation supplied with the free-boundary conditions

$$u = 0, \quad \tfrac{\partial u}{\partial n} = -1,$$

which is the multi-dimensional Florin problem, [118]. The problem is related to the propagation of equi-diffusional flames. Existence and uniqueness of classical solutions for a setting in a half-line was first proved by T.D. Ventsel' (1960) [342] and then was extended to local existence for a problem in a half strip in two space dimensions in [264]. Local classical solutions of a two-phase version of the problem with $N \geq 1$ were constructed in [9]. A global study of TWs for such a free-boundary model and for more general systems of combustion was done in a series of papers; see [48] and [55]. The study of global weak solutions was proposed in [78], where a radial self-similar finite-time vanishing pattern was constructed. Such a pattern is stable in radial geometry as proved in [142], where the corresponding focusing problem was also studied. See further details on this problem in [337] and [142].

CHAPTER 9

Regularity of Solutions of Changing Sign

We now apply intersection techniques of the geometric theory (Chapter 7) to describe the regularity and singular propagation properties of solutions of changing sign. We show that such solutions of various 1D singular parabolic equations belong to the same regularity class as the typical one-phase FBPs studied in the previous Chapter 8.

9.1 Introduction: Solutions of changing sign and the phenomenon of singular propagation

We have shown that the typical regularity for the maximal solutions (Chapter 7) and classes of 1D FBPs (Chapter 8) for a general singular parabolic equation

$$u_t = \mathbf{F}(u) \equiv F(u, u_x, u_{xx}) \tag{9.1}$$

is induced by complete subsets of TW profiles. The two-phase FBPs deal with both positive and negative TWs, and therefore these geometric techniques can be applied to the solutions of changing sign.

We consider the Cauchy problem for the parabolic equation (9.1) with initial function oscillating (changing sign) about the singular level $\{u = 0\}$. Then we lose a FBP but arrive at a singularity propagation problem of zero curves (again called singular interfaces) accompanied by similar regularity questions. The geometric intersection analysis is then a slight modification to that for the FBPs and maximal solutions. In what follows we will mainly concentrate on some distinctive features of the intersection analysis.

We keep the same assumptions on the operator $\mathbf{F}$ and solutions as in the previous two chapters. We assume that a proper solution $u(x, t)$ can be constructed by regular approximation of the equation, with

$$F_n(p, q, r) \to F(p, q, r) \quad \text{as} \quad n \to \infty$$

uniformly on any compact subset bounded away the singularity level $\{u = 0\}$, and, if necessary, with approximation of the initial data, $u_{0n} \to u_0$ uniformly on compact subsets. We suppose that a proper solution of the Cauchy problem is unique (this is well known for wide classes of singular equations of divergent form where classical concepts of weak solutions apply). Dealing with global classical solutions $\{u_n(x, t)\}$ of non-singular uniformly parabolic equations, by passing to the limit we justify that intersection comparison principles and concepts can be applied to such solutions.

Let us recall the main steps of the corresponding geometric theory.

(i) TW solutions. We assume that the second-order ODE for TWs $V(x, t) = f(\xi)$,

$\xi = x - \lambda t$,

$$F(f, f', f'') + \lambda f' = 0 \quad \text{for } \xi \neq 0, \quad f(0) = 0, \tag{9.2}$$

admits a two-dimensional subset of solutions

$$B = \{f = f(\xi, \lambda, C), \lambda \in \Delta_0 \neq \emptyset, C \in I\!R_+\},$$

where the second parameter C (the constant of integration) is the same as in the geometric analysis of FBPs in Chapter 8. We consider proper TWs that are obtained by regular approximations,

$$f = \lim f_n,$$

where f_n satisfies the ODE (9.2) with the regularized operators $\mathbf{F}_n$. Thus we have a two-parameter subset B and assume that TW profiles depend continuously on parameters λ and C.

(ii) Pressure, slopes, gradient function and TW interface operators. As in Section 7.2, in order to apply geometric techniques, we need to introduce a number of notions.

The classification of interface operators can be done similarly to that for maximal solutions (corresponding to values $C = 0$) as in Section 7.2. For instance, consider a simpler case of the *second-order interfaces* that can occur for the sign PME with absorption to be studied in detail later on (though the interfaces can be of higher orders).

Dealing now with this case in less generality than in Section 7.2, we consider a particularly interesting case. Assume that there exists a strictly monotone pressure function $Q : I\!R \to I\!R$ such that, near the singularity point as $\xi \to 0$, the TW profiles exhibit the following expansion:

$$Q(f(\xi)) = C\xi - \lambda H(C)\rho(\xi) + ..., \quad C > 0, \;\; \lambda \in \Delta_0, \tag{9.3}$$

where $\rho(\xi) > 0$ for $\xi \neq 0$, $\rho(\xi) = o(\xi)$ as $\xi \to 0$, is a given sufficiently smooth function and $H(C) > 0$ is a smooth function for $C > 0$. Taking positive C, we restrict our attention to monotone increasing TW profiles $f(\xi)$ (at least near the interface). The expansion (9.3) assumes that, due to the choice of the pressure Q, the first term of the expansion contains the same $C > 0$ for $\xi > 0$ and $\xi < 0$. This is equivalent to the general expansion for second-order interfaces

$$Q_\pm(f(\xi)) = C_\pm \xi - \lambda H_\pm(C_\pm)\rho_\pm(\xi) + ...,$$

where the subscripts $\pm$ correspond to $\xi > 0$ ($f > 0$) and $\xi < 0$ ($f < 0$).

It is of crucial importance that, for solutions of changing sign, the second parameter C_- *is not arbitrary* and the dependence $C_-(C_+)$ is obtained from approximation or necessary regularity of $f(\xi)$ at the singularity point $\{f = 0\}$ or follows from the definition of weak solutions of the ODE (of course, related to the definition of proper solutions of the PDE). In many typical cases of odd operators satisfying $\mathbf{F}(-u) = -\mathbf{F}(u)$, it is natural to demand that $C_- = C_+$ as in some examples studied below.

It is important that then, due to (9.3), the TW bundle is monotone decreasing with λ for $\xi \approx 0$ (we recall that proper solutions are always assumed to satisfy the usual comparison). This expansion determines two interface operators

$$\begin{aligned} \mathbf{M}_1(f) &= \lim_{\xi\to 0} \tfrac{Q(f(\xi))}{\xi} \quad (= C), \\ \lambda = \mathbf{N}_2(f) &= \lim_{\xi\to 0} \tfrac{C\xi - Q(f(\xi))}{H(C)\rho(\xi)}. \end{aligned} \tag{9.4}$$

Given a fixed $C = Q(f)'(0) > 0$, we define the gradient function

$$G(\lambda, C, \varepsilon) = \frac{\mathrm{d}}{\mathrm{d}\xi} Q(f(\xi)) \,|_{Q(f(\xi))=\varepsilon}, \quad \varepsilon \neq 0. \tag{9.5}$$

Assuming that we can differentiate expansion (9.3),

$$[Q(f(\xi))]' = C - \lambda H(C)\rho'(\xi) + ...,$$

where

$$\rho'(\xi) = o(1) \quad \text{as} \quad \xi \to 0,$$

we deduce that, for small $\varepsilon > 0$ (or for $\varepsilon < 0$), the gradient function is estimated as follows:

$$G(\lambda, C, \varepsilon) = C - \lambda H(C)\rho'\left(\tfrac{\varepsilon}{C}\right) + \dots . \tag{9.6}$$

Let us now apply geometric techniques to the study of proper solutions of the Cauchy problem.

(iii) Existence, nonexistence and further geometric notions. We apply geometric techniques to the class of proper solutions for which new intersections with necessary TWs do not occur for $|x| \gg 1$. Existence and nonexistence results stay the same as in Sections 7.3–7.5. Roughly speaking,

$$\Delta \neq \emptyset$$

implies the local existence of a nontrivial solution (Section 7.3), while $\Delta = \emptyset$ plus a steep limit of approximation in the ODE for any $\lambda \in I\!R$ means that, for any suitable initial data u_0, the solution is trivial, $u(x,t) \equiv 0$, for arbitrarily small $t > 0$ (Section 7.5). Several results from Section 7.6 can be extended to solutions of changing sign. In particular, since the TW subset B is two-parametric, choosing a proper (complete) subset $B(C) \subset B$, for instance, by taking a suitable monotone decreasing function $C = C(\lambda)$ as in Chapter 8, we obtain a proper TW-bundle. Then we define the B-concavity property preserved in time (or happening eventually for a class of initial data), the corresponding sign-invariant and B-number satisfying typical properties.

(iv) Regularity in B-classes. Local functional classes $B^{\pm}_{\rm loc}(c)$ are defined as in Section 7.7 relative to any TW profile $f_c(\xi, C)$ with a fixed value of $C > 0$. To this end we impose the condition of local single intersection,

$$\mathrm{Int}_\nu(t_0, V) = 1$$

with TWs $V(x,t) = f_c(x - a, C)$ for all $a \approx s(t_0) = 0$, and we count intersections (with their character different for B^+ and B^-) in a two-sided neighbourhood $(-\nu, \nu)$ of the unique singular point $x = 0$.

We first assume that the initial function $u_0(x)$ is monotone increasing and intersects the zero level $\{u = 0\}$ at the origin $x = 0$ only. By the MP, a nontrivial proper solution $u(x,t)$, which is assumed to exist, is monotone increasing with x. We assume that the intersection with $\{u = 0\}$ is transversal,

$$C_0 = [Q(u_0)]'(0) > 0, \tag{9.7}$$

and $Q(u_0(x))$ is sufficiently smooth near $x = 0$. It follows from (9.3) that fixing neighbouring TWs with sufficiently large $C > C_0$ and sufficiently small $C < C_0$, we have that Proposition 7.23 holds establishing the main Bernstein gradient bound that is a basis of further regularity theory. Moreover, covering the profile $Q(u_0(x))$ by TWs $V(x,0)$ with the slopes $C_0 \pm \varepsilon$ (then locally $\mathrm{Int}_\nu(0,V) = 1$), we conclude that the slope $[Q(u(x,t))]_x$ changes continuously with time for all $x \approx s(t)$.

The phenomenon of instantaneous smoothing has the same geometric interpretation as in Theorem 7.25. In particular, the proof of the most important Bernstein bound for $t > \delta > 0$ (part (i) in Theorem 7.25) remains the same. It uses a subset $B_{c,\delta}$ of steep enough TWs $V(x,t)$ intersecting $u(x, \frac{1}{2}\delta)$ exactly once and in the right-hand interval $x \in [\frac{1}{2}\delta, \frac{1}{2}]$. Since $\lambda \ll -1$ for steep profiles, at $t = \delta$ these TWs cover a neighbourhood of the singular interface at $x = s(\delta)$, thus creating the upper gradient estimate via the intersection argument

$$\mathrm{Int}_\nu(\delta, V) = 1.$$

Gradient estimates guarantee the optimal moduli of continuity of solutions near singular interfaces that coincide with typical ones for TW profiles f satisfying the ODE (9.2) with $\lambda \ll -1$.

(v) Interface operators and equations. Here we follow the lines of Section 7.10. In order to prove that the first-order operator in (9.4) is well defined on the solution profile $u(x,t_0)$, with $t_0 > 0$ sufficiently small, we fix a $\lambda_0 \in \Delta_0$ and consider the TW subset

$$B(\lambda_0) = \{f(\xi, \lambda_0, C), C > 0\},$$

which is assumed to be complete in the class of transversal profiles. Then, as in the proof of Proposition 7.38, the assumption that $\mathbf{M}_1(u(\cdot, t_0))$ is not defined from the right, i.e., there exist different particular limits $C_1 < C_2$ along two sequences

$$\{x_k^{1,2}\} \to s(t_0) = 0$$

implies that the profile $u(x,t_0)$ has infinitely many intersections in $\{x > 0\}$ with any function $f(x, \lambda_0, C)$ with $C \in (C_1, C_2)$ and many others corresponding to slight perturbations of the TW speed λ_0. Under natural assumptions on the initial data, this contradicts Sturm's Theorem (Proposition 7.38). Thus there exists a finite limit as $x \to 0^+$

$$C_0 = C_0(t_0) = \mathbf{M}_1(u),$$

and (9.6) implies the gradient bound for $t \in [t_0 - \delta, t_0 + \delta]$

$$|(Q(u))_x - C| \le c\,\rho'\left(\tfrac{1}{C}\,Q(u)\right), \quad C = (Q(u))_x|_{x=s(t)}.$$

The same argument applies, to prove that a finite limit C_0^- as $x \to 0^-$ exists. Geometric techniques do not guarantee that both limits coincide in general, i.e., $C_0 = C_0^-$, even if it is true for wide subsets of TWs. Recall that this is not the case for proper solutions of two-phase FBPs in Chapter 8, where basic intersection principles of transversality and convexity of geometric evolution can be applied. Actually, the equality of slopes is connected with definitions of proper solutions or their construction that may exclude discontinuous slopes for classes of singular equations.

For convenience, from now on we assume that the left-hand and right-hand slopes coincide. It is easy to exclude a class of "non-evolution" ("non-entropy") discontinuities, contradicting comparison with regular proper TWs. Other ones corresponding to occurrence of new intersections with TWs are excluded by Sturm's Theorem.

In order to justify the second-order operator in (9.4), we perform the same Sturmian intersection analysis in the proper complete bundle

$$B(C_0) = \{f(x, \lambda, C_0(t_0)), \lambda \in \Delta_0\}.$$

Then we obtain that there exists a finite limit as $x \to 0^+$

$$\lambda_0 = \lambda_0(t_0) = \mathbf{N}_2(u).$$

Similarly, the limit as $x \to 0^-$ also exists and is equal to the same value λ_0, since, as in the two-phase FBPs, by regularity analysis in B-classes, for transversal intersections with the singular level $\{u = 0\}$, the slope $[Q(u(x,t)]_x$ changes continuously with t in a neighbourhood of $x = s(t)$. Hence,

$$\lambda_0^- < \lambda_0$$

would mean that new intersections with arbitrarily flat TWs would occur for $t > t_0$. This can happen and new interfaces can occur from non-transversal intersections, i.e., from tangency or inflection points at $\{u = 0\}$.

The proof that

$$\lambda_0 = D^+ s(t_0)$$

is given by comparison with perturbed TWs with fixed C_0 and arbitrarily small perturbed $\lambda_0 \pm \varepsilon$, $\varepsilon > 0$. Since such TWs are proper and by (9.3) are strictly monotone decreasing with λ, we perform local comparison to obtain that

$$(\lambda - \varepsilon)(t - t_0) \le s(t) - s(t_0) \le (\lambda + \varepsilon)(t - t_0) \quad \text{for small } t - t_0 > 0,$$

whence the result by passing to the limit $t \to t_0^+$ and $\varepsilon \to 0^+$. Then $s(t)$ is Lipschitz continuous. The rest of the interface regularity analysis is performed as in Section 7.10.

(vi) Eventual transversality. Consider the case when the transversality condition (9.7) does not hold, i.e., $C_0 = 0$. We may also assume that $u_0(x) \equiv 0$ on a bounded interval $x \in [-1, 1]$ and $u_0(x) > 0$ for $x > 1$ and $u_0(x) < 0$ for $x < -1$. Assume that there exists a global in time proper solution. In the more interesting case of finite propagation on the singularity level $\{u = 0\}$, these assumptions lead to the extremal solutions that are the maximal in $\{u \ge 0\}$ and minimal in $\{u \le 0\}$, which have to be studied first as in Chapter 7. If the corresponding interfaces are not localized, we always find a finite $T > 0$ such that the unique intersection becomes transversal at $t = T$, the profile $u(x, T)$ satisfies the transversality condition and, as a consequence, for $t > T$, the above regularity results apply.

On the other hand, it is easy to formulate such an *eventual transversality* result in terms of suitable subsets of TWs. In fact, this analysis is easier than the eventual concavity one. Namely, assume that, under given hypotheses on u_0, the initial function satisfies

$$u_0(x) \geq f(x-a, \bar{\lambda}, \bar{C}) \quad \text{in } I\!R,$$

where $a > 1$, $\bar{\lambda}$ and $\bar{C}$ are some constants. Then, by comparison,

$$u(x,t) \geq f(x-\bar{\lambda}t-a, \bar{\lambda}, \bar{C}) \quad \text{in } I\!R \times I\!R_+.$$

Assume in addition that, for some $\lambda_1 > \bar{\lambda}$, the number of intersections between $u_0(x)$ and $V(x,0) = f(x-b, \lambda_1, C)$ with some $C > 0$ satisfies

$$\text{Int}(0,V) = 1 \quad \text{for all } \; b < -1,$$

and every intersection lies in the domain $\{x > 1\}$ (no intersections are available in the domain $\{x < 1\}$ of negativity of u_0). This means that $V(x,0)$ in $\{x > 1\}$ is sufficiently flat to intersects u_0 from above. By Sturm's Theorem

$$\text{Int}(t,V) = 1 \quad \text{for } t > 0 \text{ and for all } b < -1,$$

and the character of each intersection (always from below) cannot change with time. Let us show that there exists a finite $T > 0$ such that the transversality condition holds,

$$\mathbf{M}_1(u(\cdot,T)) > 0.$$

Indeed, by assumption $\lambda_1 > \bar{\lambda}$, the TW profiles

$$V(x,t) = f(x-\lambda_1 T - b, \lambda_1, C)$$

for sufficiently large $T > 0$ will cover the solution profile $u(x,T)$ in a neighbourhood of the singularity point $x = s(T)$ forming intersections from above. Therefore, $u(x,t)$ has a larger slope than the TW $V(x,T)$ with interface at $x = s(T)$ and this implies the transversality. This eventual transversality analysis repeats the steps of the proof of Theorem 7.25 on eventual smoothing in the local class B^-_{loc}.

(vii) Regularity for arbitrary non-monotone initial data. The results remain true for any transversal intersection. If initial data are non-monotone, multiple intersections with $\{u = 0\}$ (multiple singular zeros) can occur when two or more singular interfaces collide with each other. Such processes of formation of multiple zeros as a focusing-like behaviour cannot be described by TW intersection analysis and the actual regularity of a singular curve at such focusing moments are not known in general. It is expected that, for particular scaling invariant equations, intersection comparison with other families of self-similar solutions of special multi-zero structure can play an important part. This difficulty is similar to that in Chapter 7 for maximal solutions, where we exclude the possibility of focusing of different singularity levels (a singularity formation phenomenon, which cannot be covered by the geometry of TW subsets).

On the other hand, for some classes of essentially non-monotone initial functions the eventual transversality can be proved following the lines of intersection comparison presented above, i.e., the regularity will be achieved eventually when a unique transversal intersection is created and the geometric regularity theory applies.

Once necessary proper TW subsets are well understood for a general singular parabolic equation (9.1), we translate regularity TW properties to solutions of changing sign under the monotonicity assumption. Rephrasing Proposition 8.5, we state that *under the monotonicity assumption, the regularity results of the transversality analysis such as Bernstein estimates, moduli of continuity, instantaneous smoothing, etc., for solutions of changing sign, coincide with those for typical one-phase and two-phase FBPs.*

Indeed, this is true because for both classes of problems, the same TW subsets B are involved in the geometric analysis.

9.2 Application: the sign porous medium equation with singular absorption

Consider the sign PME with absorption

$$\boxed{u_t = (|u|^{m-1}u)_{xx} - |u|^{p-1}u \quad \text{in } I\!R \times I\!R_+,} \tag{9.8}$$

where $m > 1$, $p > -m$ and initial function u_0 satisfies necessary monotonicity and transversality conditions. The FBPs for equation (9.8) were studied in Section 8.5 where all necessary computations are already available. Consider the equation in the parameter range

$$\tfrac{1}{2}(1-2m) < p < 1-m,$$

where the one-phase FBP has a third-order interface equation. TWs of changing sign have the following expansion near the singularity at $\xi = 0$ (cf. (8.38)):

$$f^m(\xi) = C\xi + a_1 C^{\frac{p}{m}} \xi^{\frac{p+2m}{m}} - \lambda a_2 C^{\frac{1}{m}} \xi^{\frac{m+1}{m}} + \dots, \tag{9.9}$$

and

$$(-f)^m(\xi) = C(-\xi) + a_1 C^{\frac{p}{m}} (-\xi)^{\frac{p+2m}{m}} + \lambda a_2 C^{\frac{1}{m}} (-\xi)^{\frac{m+1}{m}} + \dots \tag{9.10}$$

as $\xi \to 0^+$ and $\xi \to 0^-$ respectively, with the parameters $\lambda \in \Delta_0 = I\!R$ and $C > 0$. In comparison with expansion (9.3) we have three terms on the right-hand side here, leading to three interface operators. These TWs are proper since each function f_λ is monotone decreasing with λ and such TWs do not violate the comparison principle on the singularity level $\{u = 0\}$.

The asymptotic behaviour as $\xi \to \pm\infty$ is different for the non-stationary case $\lambda \neq 0$. For instance, for $\lambda > 0$,

$$f(\xi) \sim \xi^{\frac{1}{1-p}} \quad \text{as } \xi \to \infty$$

and

$$f(\xi) \sim -(-\xi)^{\frac{1}{m-1}} \quad \text{as } \xi \to -\infty.$$

We assume that a unique (weak or proper) solution of the problem can be obtained by approximation, $u = \lim u_n$, where $\{u_n\}$ are solutions of regularized uniformly parabolic equations. A typical regularization with uniformly Lipschitz continuous absorption term is as follows

$$u_t = ((u^2 + \tfrac{1}{n^2})^{\frac{m-1}{2}} u)_{xx} - (u^2 + \tfrac{1}{n^2})^{\frac{p-1}{2}} u$$

with the same sufficiently smooth initial data u_0. Assuming that such a regular approximation can be performed for both $u(x,t)$ and any TW $V(x,t) \in B$, it follows that intersection comparison techniques apply providing us with a number of regularity results from Chapter 7. For simplicity we suppose that $u_0(x)$ exhibits a slow growth as $x \to \pm\infty$, so that the extinction time due to singular absorption is infinite and

$$u(x,t) \not\equiv 0 \quad \text{for all} \quad t > 0.$$

Assume that the unique intersection at $x = 0$, where $u_0(0) = 0$, is transversal, i.e.,

$$C_0 = (|u_0|^{m-1} u_0)'(0) > 0,$$

and $|u_0|^{m-1} u_0$ is sufficiently smooth in a neighbourhood of the origin. By constructing steep and flat TWs, we deduce that, for sufficiently small $t \in [0, \delta]$, the solution $u(x,t)$ belongs to the functional classes $B^{\pm}_{\rm loc}(c)$ with a $c \gg 1$. The gradient function (9.5) for $\varepsilon > 0$ is

$$G(\lambda, C, \varepsilon) = (|f|^{m-1} f)'_\xi)|_{\{|f^m| = \varepsilon\}} = C + a_3 \tfrac{1}{C}\, \varepsilon^{\frac{p+m}{m}} - \lambda a_4 \varepsilon^{\frac{1}{m}} + \dots .$$

Using the parameter C in the intersection comparison with the TW-bundle, we obtain the first gradient bound near the interface,

$$|(|u|^{m-1} u)_x| \le c.$$

Taking into account all three expansion terms in the gradient function and varying the velocity λ, we arrive at the optimal gradient estimate (cf. a similar one (8.44) in Section 8.5)

$$\left|(|u|^{m-1} u)_x - C - a_3 \tfrac{1}{C}\, |u|^{p+m}\right| \le c\,|u|, \tag{9.11}$$

where

$$C(t) = (|u|^{m-1} u)_x|_{x = s(t)}.$$

This implies optimal moduli of continuity of solutions.

In the case $C_0 = 0$, the maximal TWs corresponding to $C = 0$ should be used first in the intersection comparison.

Since $\Delta_0 = I\!R$, by other necessary hypothesis of the behaviour of TW profiles as $\lambda \to -\infty$ (f_λ becomes arbitrarily steep) and $\lambda \to +\infty$ (f_λ gets flatter) and a result similar to Theorem 7.25, we conclude that the instantaneous smoothing effect occurs for classes of initial data. For arbitrary monotone *non-transversal* initial data forming a multiple singular zero, we can guarantee the instantaneous smoothing in B-classes corresponding to nonnegative maximal TWs ($C = 0$ according to our classification). Note that maximal TWs $V \ge 0$ can be compared with proper solutions $u(x,t)$ of changing sign and the proof is the same as in Section 7.7.

Finally, under the monotonicity and transversality assumptions, we obtain that the singular propagation of the unique zero curve $x = s(t)$ is governed by a system of three equations

$$\begin{cases} \mathbf{M}_1(u) = C \equiv (|u|^{m-1}u)_x|_{x=s(t)}, \\ \mathbf{N}_2(u) = C^{\frac{p}{m}}, \\ D^+ s = \mathbf{N}_3(u), \end{cases} \tag{9.12}$$

where the operators are given by the same formulae as in Section 8.5. In general, the geometric arguments establish the existence of the limits as $x \to s(t)^+$ and $x \to s(t)^-$ in three operators, but do not guarantee that the right and left-hand limits coincide in the first two expressions. In the third operator both limits coincide since, by regularity of the solutions, each slope must correspond to the interface velocity. The interface is Lipschitz continuous. We do not mention other results proved by intersection comparison by transversality and concavity approaches.

For more general initial data u_0, the same geometric TW local regularity estimates hold provided that the solution is strictly monotone in a neighbourhood of the singularity point under consideration (then the intersection with $\{u = 0\}$ instantaneously becomes transversal and the regularity results apply). If the monotonicity is violated at a given moment of time, a focusing-like behaviour of two or more neighbouring singularity curves can occur, leading to the formation of multiple zeros. In these non-generic cases the geometric regularity theory based on intersection comparison with TWs does not apply, and other similarity solutions can play a similar role.

On interior gradient blow-up of bounded solutions

In Chapter 7 we have already presented some examples of maximal solutions with $u_x = \infty$ on singular interfaces. This phenomenon often occurs for solutions of changing sign. For instance, $u_x = \infty$ on the singular interfaces for the TWs for the sign PME with absorption (see expansions (9.9) and (9.10)), and hence, it follows from the geometric analysis that the gradient blow-up is a generic phenomenon occurring for any transversal intersection with the singular level. This conclusion applies to general singular parabolic equations (9.1). If the phenomenon of instantaneous "smoothing" is not available (this means a waiting time effect), the finite-time formation of singularities can be proved by intersection comparison with specially chosen TWs as explained in Section 9.1 (the phenomenon of eventual transversality). The eventual transversality argument establishes an estimate of the time of transversality, when the required phenomenon of interior gradient blow-up occurs.

Singularity phenomena of interior gradient blow-up occur for many quasilinear and nonlinear parabolic PDEs with u_0 changing sign. The eventual occurrence of such singularities is checked by constructing suitable TWs and intersection comparison. Interior gradient blow-up (or other similar types of singularity formation) can occur for equations ($m > 1, p > 1, q \geq 1$)

$$u_t = (|u|^{m-1}u)_{xx} \pm |u|^{q-1}u|u_x|^{p-1}u_x,$$

$$u_t = (|u_x|^{m-1}u_x) \pm |u|^{q-1}u|u_x|^{p-1}u_x,$$

$$u_t = |u_{xx}|^{m-1}u_{xx} \pm |u|^{q-1}u|u_x|^{p-1}u_x,$$

and so on. For instance, replacing u_t by $|u_t|^{\gamma-1}u_t$ with exponent $\gamma > 1$ creates fully nonlinear equations similar to that studied in Section 7.11. Then u_x can blow up at the interface $x = s(t)$, on which $u(s(t), t) = 0$ in finite time $t = T$ and for $t > T$ we observe a singular interface propagation according to the regularity of the corresponding TW-bundles. It gives optimal Bernstein estimates, moduli continuity in x and t and the interface operators and equations. The analysis does not differ from that for the sign PME and inherits basic results from the regularity theory for the 1D FBPs in Chapter 8.

9.3 On propagation of singularity curves

We briefly comment on some problems and difficulties that can occur if the singularity subsets essentially depend on the solutions.

It is easy to derive parabolic models creating singularities on various curves in the hodograph plane $I\!R^2 = \{u, u_x\}$. For instance, consider the HE with a singular absorption term

$$u_t = u_{xx} - |u_x - u|^{p-1}(u_x - u), \quad p < 0. \tag{9.13}$$

The singularity occurs on the straight line $L = \{u_x = u\}$ in $I\!R^2$. Then the singular TWs are constructed starting at any point of L, and we check how to define complete and proper TW-bundles. In the present case this is easy and the geometric analysis in TW-bundles reduces to the standard one. Setting

$$w = u_x - u,$$

one obtains a parabolic equation for w. We have

$$w_t = u_{tx} - u_t \equiv (u_{xx} - |w|^{p-1}w)_x - (u_{xx} - |w|^{p-1}w),$$

and substituting

$$u_{xx} = w_x + u_x$$

yields the equation

$$w_t = w_{xx} - (|w|^{p-1}w)_x + |w|^{p-1}w. \tag{9.14}$$

This is a standard problem of singularity formation and propagation on the fixed singular level $\{w = 0\}$ for the HE with singular source and convection terms, where solutions change sign.

In the general case, this can yield more complicated parabolic equations and ODEs that need special detailed analysis. We do not review such technical applications of the geometric theory, and instead finish this discussion with a class of nonlinear parabolic equations creating special singularity subsets.

Very singular equations with dense singularity subsets.* Consider the following singular semilinear heat equation:

$$u_t = u_{xx} - \sum_{k=1}^{\infty} c_k |u - \alpha_k|^{p_k - 1}(u - \alpha_k), \tag{9.15}$$

* The author would like to thank P.I. Plotnikov and J.F. Toland for a discussion concerning such models.

where $\{\alpha_k\}$ is a bounded sequence consisting of distinct terms satisfying $0 < \alpha_k < 1$, and $\{c_k > 0\}$ are small positive constants such that

$$c_k \to 0 \quad \text{as} \quad k \to \infty$$

sufficiently (say, exponentially) fast. We take bounded initial data satisfying

$$-1 \le u_0(x) \le 2 \quad \text{on } I\!\!R, \quad \text{and } \operatorname{Im} u_0 = [-1, 2].$$

Exponents $p_k \in (-1, 0)$ are chosen so that on each singularity level $\{u = \alpha_k\}$, the equation is locally solvable and the corresponding extinction singularity in incomplete (the complete one occurs if $p \le -1$ as proved in Theorem 6.17). Using regular approximations, $\{u_n\}$ can be constructed, for instance, by the following regularized equation:

$$u_t = u_{xx} - \sum_{k=1}^{n} c_k |(u-\alpha_k)^2 + \tfrac{1}{n^2}|^{\frac{p_k-1}{2}} (u - \alpha_k) \tag{9.16}$$

with the same initial function u_0. We then expect to obtain a proper solution $u(x,t)$ in the limit (possibly, along a subsequence), which turns out to be singular on the subset $\mathcal{B}[u] = \{u = \alpha_k,\ k = 1, 2, ...\}$ at least for sufficiently small $t > 0$ with $\operatorname{Im} u(\cdot, t) \supset [0, 1]$.

Let $\{\alpha_k\}$ be the set of all rational numbers on $(0, 1)$. We then obtain a *very* singular PDE (9.15) that possesses dense singularity subsets satisfying $\overline{\mathcal{B}}[u] = [0, 1]$. If this is true and the proper solution has finite regularity, it may happen that, in the differential form, equation (9.15) makes no sense, and the derivatives u_{xx} and u_t are not defined at any point within the range $[0, 1]$ of the solution. Observe that, in this case, a definition of the solution to the original equation (9.15) in the standard weak sense by using integration by parts is also difficult since $p_k < 0$, so the singular absorption terms are not known *a priori* to be locally integrable.

In this case the only "real" characterization of the proper solution and the original equation itself is the sequence $\{u_n\}$ of the regular approximation satisfying the uniformly parabolic equation (9.16). Actually, the original equation (9.15) disappears from the analysis and can be dispensed with.

Other types of dense singularity subsets are exhibited by equations

$$u_t = u_{xx} - \sum_{k=1}^{\infty} c_k |u_x - \alpha_k u|^{p_k - 1} (u_x - \alpha_k u), \tag{9.17}$$

where the singularity subset is

$$\mathcal{B}[u] = \{u_x = \alpha_k u,\ k = 1, 2, ...\}.$$

One can replace the linear diffusion operator u_{xx} by the nonlinear ones $(|u|^{m-1}u)_{xx}$ (the sign PME), $(|u_x|^{m-1}u_x)$ (the p-Laplacian), $|u_{xx}|^{m-1}u_{xx}$ (the dual PME) or others, creating more complicated models with dense singularity subsets.

For instance, the following model is obtained by distributing the singularity levels of the operators of the PME with absorption

$$u_t = \sum_{k=1}^{\infty} [(|u-\alpha_k|^{m_k-1}(u-\alpha_k))_{xx} - c_k|u-\alpha_k|^{p_k-1}(u-\alpha_k),$$

where $m_k \geq 1$. By Theorem 6.18, the local solvability (incomplete singularity) at each level $\{u = \alpha_k\}$ is guaranteed by the assumption

$$p_k > -m_k, \quad k = 1, 2, \dots .$$

The analysis of existence and regularity properties of proper solutions becomes more involved.

Therefore, in some special nonlinear singular models, where the behaviour close to dense singularity subsets is obtained via nonlinear interaction of many singular operators, it appears that the original PDE cannot be understood in a "weak" sense for proper solutions within the singularity range. The only "trace" of the equation is regular approximations determining the proper solutions by $u = \lim u_n$.

Remarks and comments on the literature

§ **9.1, 9.2.** More detailed results on regularity of the singular interfaces for 1D filtration equations with solutions of changing sign can be found in [54] and [305]. The eventual monotonicity result in [305] (a key point in the regularity analysis [54]) is proved by intersection comparison via the Sturm Theorem. Various types of eventual monotonicity results for quasilinear parabolic equations were established by using intersection comparison with subsets of exact solutions in [273], [162], [166].

§ **9.3.** Interior gradient blow-up of bounded solutions was studied in a number of papers; see [59], [18] (singular TWs were exploited), [173] and references therein.

CHAPTER 10

Discontinuous Limit Semigroups for the Singular Zhang Equation

In the next two chapters we continue the geometric study of some special parabolic PDEs with a more sophisticated structure of singularity subsets. The main feature of such equations is that the corresponding semigroups are discontinuous. We consider a number of such nonlinear models associated with various applications.

10.1 Introduction: New nonlinear models with discontinuous semigroups

We will study some special properties of extended limit semigroups of proper maximal solutions in the cases that were not covered by the geometric theory in Chapter 7. Let us introduce our main new parabolic model exhibiting unusual discontinuity properties. It is known that the singular limit as $q \to 0^+$ of the generalized equation [216], [230]

$$u_t = u_{xx} + \tfrac{1}{q}\,|u_x|^q \tag{10.1}$$

is the singular *Zhang equation* [349]

$$\boxed{u_t = u_{xx} + \log|u_x|.} \tag{10.2}$$

It occurs as a by-product in the study of complex directed polymer structures and is considered as a "prototype of much more complicated systems" [184]. Its simpler "hydrodynamic version" takes the form

$$v_t = v_{xx} + \frac{v_x}{v}. \tag{10.3}$$

These are two examples of semilinear parabolic equations with strong singularities. Equation (10.3) follows from (10.2) by differentiating and setting $u_x = v$. Such parabolic equations with standard singularities in the convection-like terms on the level $\{v = 0\}$ have been considered in Chapter 7.

On the other hand, equation (10.2) possesses the singularity subset $\{u_x = 0\}$ depending on the solution under consideration.

Though equations (10.2) and (10.3) are directly related to each other, their solutions exhibit different properties. It is important that only the solutions of (10.2) with a non-empty set of extremal singular points, where $u_x = 0$, are of physical interest. These correspond to white noise initial conditions that are supposed to create special asymptotic patterns. We pay more attention to equation (10.2) describing a new type of singularity formation on the set $\{u_x = 0\}$ of unknown *a priori* structure. We first describe main discontinuity properties of its more standard hydrodynamic version to be used in the analysis of (10.2).

Finally, we study the limit semigroup of the corresponding N-dimensional equation

$$\boxed{u_t = \Delta u + \log|\nabla u|.} \tag{10.4}$$

All equations in this chapter belong to a class of singular, but, formally, uniformly parabolic equations. Though singularities are given by lower order operators, they are so strong that the corresponding semigroups are not continuous at the initial instant $t = 0$.

Our analysis of (10.2) and (10.4) uses geometric ideas. It is important that, for (10.2), the singularity formation occurs at the set $\{u_x = 0\}$, which is not a fixed level like $\{u = 0\}$. Therefore, instead of the TW solutions, we need to use special self-similar solutions describing effects of singularity formation for this PDE, and, as the next step, we apply such particular solutions in the intersection comparison to obtain the existence and regularity properties of general proper solutions.

10.2 Existence and nonexistence results for the hydrodynamic version

We consider the Cauchy problem for the hydrodynamic version of the Zhang equation (10.3) with nonnegative bounded continuous initial data

$$v(x,0) = v_0(x) \geq 0 \quad \text{in } I\!R. \tag{10.5}$$

Equation (10.3) has the divergent structure

$$v_t = (v_x + \log|v|)_x \quad \text{in } S = I\!R \times I\!R_+ \tag{10.6}$$

and contains a single strong singularity at $v = 0$. Therefore, as usual, we consider proper maximal solutions that are constructed by regular monotone approximations of the problem as in Section 7.3. Similar to the PME, this can only be done for (10.3) by a suitable uniformly positive approximation of the initial data. Regular approximations of the equation for a subset of uniformly positive solutions are not necessary.

Let $\{v_\varepsilon\}$ for $\varepsilon > 0$ be a family of global classical solutions of (10.3) with strictly positive bounded initial data

$$v_{0\varepsilon}(x) = v_0(x) + \varepsilon \quad \text{in } I\!R.$$

By the MP, $v_\varepsilon \geq \varepsilon > 0$ everywhere, so that the family consists of non-singular solutions uniformly bounded away from 0. Therefore, the classical parabolic theory applies, to guarantee the global existence of smooth bounded v_ε. The family $\{v_\varepsilon\}$ is monotone decreasing with ε by the standard comparison, and is uniformly bounded above: for small $\varepsilon > 0$, $v_\varepsilon \leq v_1 \leq \sup v_0 + 1$. There exists a finite limit

$$v = \lim_{\varepsilon \to 0^+} v_\varepsilon \geq 0,$$

which is a unique proper (maximal) solution. As in the general case considered in Section 7.3, it does not depend on the type of monotone decreasing approximation. This defines the extended limit semigroup of maximal solutions

$$v(x,t) = \mathcal{T}(t)v_0(x) \geq 0 \quad \text{for } t > 0.$$

As usual, this definition does not use any regularity properties of the solutions v. We even do not know if, for smooth initial data u_0, $v(x,t)$ is a continuous function (in fact, it is not). As usual in the study of singular nonlinear evolution equations with blow-up or extinction, the first existence question consists of the following preliminary classification of singularities. As in previous chapters, we distinguish the two cases:

(i) *complete singularity*: for arbitrarily small $t > 0$, the maximal solution is entirely singular, i.e., $v(x,t) \equiv 0$, and

(ii) *incomplete singularity*: $v(x,t) \not\equiv 0$ for all small $t > 0$.

Such a classification concerns continuity properties of maximal solutions at $t = 0$. In the classical parabolic theory, the case (i) is treated as *nonexistence* (and (ii) as local *existence* if $\mathcal{T}(0^+)v_0 = v_0$) of a solution of the Cauchy problem, since the maximal solution $v(x,t) \equiv 0$ for any $t > 0$, obviously, cannot satisfy any nontrivial initial condition. On the other hand, as we have seen in the previous three chapters, the nonexistence treatment is not suitable for those singular equations admitting essentially discontinuous semigroups. Nevertheless, for convenience, we sometimes keep using the classical term "nonexistence" where the context dictates.

Subset of travelling waves

Substituting the TW solution

$$V(x,t) = f(\xi) \geq 0, \quad \xi = x - \lambda t, \quad \lambda \in I\!R, \tag{10.7}$$

into (10.3), we obtain the ODE for the continuous profile $f = f_\lambda$

$$f'' + \lambda f' + \tfrac{f'}{f} = 0. \tag{10.8}$$

Setting $f' = P(f)$ and integrating once yields the integral curves

$$P(f) = C - \log f - \lambda f \quad \text{for } f > 0, \tag{10.9}$$

where $C \in I\!R$ is a free constant of integration. We study the singular TW profiles $f \geq 0$ of (10.8) satisfying

$$0 \in \overline{\mathrm{Im}}\, f, \tag{10.10}$$

i.e., $f(\xi)$ attains the singular 0-level at a finite or infinite point. We next define the set

$$\Delta = \{\lambda \in I\!R \,:\, (10.8),\ (10.10) \text{ has a maximal solution}\}. \tag{10.11}$$

By maximal solutions of the ODE we mean those that can be constructed by a regular strictly positive approximation (similar to the definition for the PDE in Section 7.2). Fixing an $n \gg 1$, we consider the orbit $P = P_n(f)$ satisfying

$$P_n(\tfrac{2}{n}) = 0. \tag{10.12}$$

It follows from (10.9) that such an orbit exists with the constant

$$C = C_n = \tfrac{2}{n}\lambda + \log \tfrac{2}{n} \to -\infty \quad \text{as } n \to \infty. \tag{10.13}$$

Hence, for any fixed $\lambda \in I\!R$ and $n \gg 1$, the approximating orbit

$$P_n(f) = \tfrac{2}{n}\,\lambda + \log \tfrac{2}{n} - \log f - \lambda f > 0 \tag{10.14}$$

is monotone decreasing for $f \in (0, \frac{1}{n})$ and

$$P_n(f) \to +\infty \quad \text{as} \quad f \to 0^+.$$

This corresponds to the nonexistence case $\Delta = \emptyset$ with the flat limit of approximation; see Section 7.2.

Let us describe some properties of the corresponding TW profiles $f_n(\xi)$ for large n. For any $\lambda \in I\!R$, we obtain from (10.9), (10.10) a small monotone increasing (almost "flat") profile f_n^+ satisfying

$$\begin{aligned} f_n^+(0) = 0, \;\; f_n^+(\xi) > 0, \\ (f_n^+)'(\xi) > 0 \;\text{ for }\; \xi > 0, \\ f_n^+(\xi) \to \left(\tfrac{2}{n}\right)^- \;\text{ as }\; \xi \to \infty. \end{aligned} \tag{10.15}$$

Therefore, uniformly in ξ,

$$f_n^+(\xi) \to 0 \quad \text{as } n \to \infty. \tag{10.16}$$

On the other hand, for any $\lambda \in I\!R$, there exists a "steep" profile f_n^- in $\{f > 0, P < 0\}$ satisfying

$$f_n^-(\xi) > \tfrac{1}{n}, \;\; (f_n^-)'(\xi) < 0 \;\text{ for } \xi \in I\!R, \quad f_n^-(\xi) \to \left(\tfrac{2}{n}\right)^+ \text{ as } \xi \to \infty. \tag{10.17}$$

Due to (10.14), we have that on any fixed level $\delta > 0$, there holds

$$f'(\xi)|_{f(\xi)=\delta} \to -\infty \quad \text{as} \quad n \to \infty. \tag{10.18}$$

This is the steep limit of nonexistence in the class of decreasing orbits $P = f' < 0$. One can choose a sequence $\{f_n^-\}$ such that

$$\lim_{n\to\infty} f_n^-(\xi) = H_\infty(\xi) = \begin{cases} \infty & \text{for } \xi \le 0, \\ 0 & \text{for } \xi > 0, \end{cases} \tag{10.19}$$

where H_∞ is the infinite-step Heaviside function.

Discontinuity: first example of complete singularity

The steep limit as $n \to \infty$ of negative orbits $P_n(f) < 0$ implies nonexistence for the initial data exhibiting the corresponding negative slope behaviour; see Section 7.5. For convenience, we present a simple geometric proof of this singular property of nonexistence. We introduce the positivity subset as follows:

$$\operatorname{supp} v_0 = \{x \in I\!R : \;\; v_0(x) > 0\}.$$

Theorem 10.1 *Let* $\sup\{\operatorname{supp} v_0\} < \infty$. *Then*

$$v(x,t) \equiv 0 \quad \textit{for any} \quad t > 0. \tag{10.20}$$

Proof. Fix an $\varepsilon = \frac{1}{n} \ll 1$ and choose the corresponding steep TW profile f_n^- satisfying the properties (10.17)–(10.19). Since $\operatorname{supp} v_0$ is bounded from the right-hand side, there exists a constant a independent of n such that the regularized

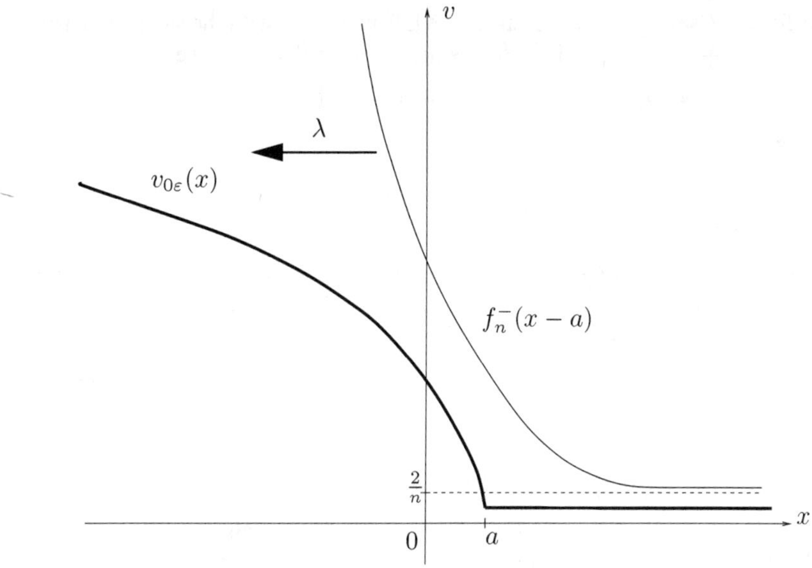

Figure 10.1 *Illustration of the proof of Theorem* 10.1*: the steep TW* $f_n^-(x-\lambda t-a)$ *moving to the left with a large speed* $\lambda \ll -1$*, by comparison destroys any solution below.*

initial data satisfy (see Figure 10.1)

$$v_{0\varepsilon}(x) \equiv v_0(x) + \tfrac{1}{n} \le f_n^-(x-a).$$

Then, by comparison of bounded classical solutions of a uniformly parabolic equation, we obtain that

$$v_\varepsilon(x,t) \le f_n^-(x-\lambda t-a) \quad \text{in } S.$$

Passing to the limit $\varepsilon = \frac{1}{n} \to 0$, we arrive at the inequality $v(x,t) \le H^\infty(x - \lambda t - a)$. Now passing to the limit $\lambda \to -\infty$ (then $x-\lambda t-a \to +\infty$ for any fixed x), we have that $v(x,t) \le 0$ for $t>0$ and hence (10.20) holds. □

Nonexistence for solutions changing sign

The comparison technique also applies to any bounded solutions of changing sign.

Proposition 10.2 *If* v_0 *is compactly supported, then* $v(x,t) \equiv 0$ *for any* $t>0$.

Proof. By the same comparison from the right-hand side with the steep decreasing TWs $f_n^-(x-\lambda t)$, where $\lambda \ll -1$, we first establish that the positive part of the solutions vanish identically:

$$[v(x,t)]_+ \equiv 0 \quad \text{for } t>0.$$

Since (10.3) is invariant under the symmetry transformation

$$v \mapsto -v, \;\; x \mapsto -x,$$

we perform the same comparison from the left-hand side with the steep increasing TWs $-f_n^-(x+\lambda t)$ for speeds $\lambda \gg 1$, which in a similar way gives that

$$[v(x,t)]_- \equiv 0 \quad \text{for } t > 0.$$

□

Singular finite-time limit of self-similar solutions

We now show that the singular behaviour of the solutions depend on the behaviour of the initial data $v_0(x)$ for $x \gg 1$. Consider self-similar solutions depending on the standard blow-up similarity variables

$$v_*(x,t) = \sqrt{T-t}\, g(\eta), \quad \eta = x/\sqrt{T-t}, \quad t \in (0,T), \tag{10.21}$$

where $T > 0$ is the blow-up time. Then $g \geq 0$ solves the ODE

$$g'' - \tfrac{1}{2}\, g'\eta + \tfrac{1}{2}\, g + \tfrac{g'}{g} = 0, \quad \eta \in I\!R. \tag{10.22}$$

By a standard local analysis of (10.22) for $\eta \gg 1$, one can see that it admits a family of solutions $g(\eta)$ satisfying

$$g(\eta) = \tfrac{1}{\eta} - \tfrac{2\log\eta}{\eta^3} + \ldots \quad \text{as } \eta \to \infty. \tag{10.23}$$

Here we omit a bundle of exponential perturbations of order $O(e^{-\eta^2/4})$. Therefore, near the blow-up time $t \to T^-$, for any fixed $x > 0$, there holds $\eta = x/\sqrt{T-t} \to \infty$ so that uniformly on any subset $[\delta, \infty)$ with $\delta > 0$, the self-similar solution has the asymptotic behaviour

$$v_*(x,t) = (T-t)\frac{1}{x}\,(1+o(1)) \to 0 \quad \text{as } t \to T^-. \tag{10.24}$$

We thus observe the effect of disappearance of the solution in a finite time. This depends on the special spatial structure of the corresponding initial data:

$$v_*(x,0) = T\frac{1}{x}\,(1+o(1)) \quad \text{as } x \to \infty. \tag{10.25}$$

If (10.25) is violated and

$$v_0(x) = o\left(\tfrac{1}{x}\right) \quad \text{as } x \to \infty,$$

the maximal solution disappears instantaneously, i.e., $T = 0$. As in the above nonexistence theorem, the proof is based on the careful comparison of the maximal solution with the steep TW solutions $f_n^-(x - \lambda_n t - a_n)$ and passage to the limit $n \to \infty$, where both parameters $\lambda_n \to -\infty$ and $a_n \to \infty$ eventually become unbounded and can be chosen such that $|\lambda_n| \gg a_n$. On the other hand, the shifted self-similar maximal solutions $v_*(x - a_n, t)$ with $T_n \to 0$ can also be used for comparison.

Positivity

The flat limit $n \to \infty$ of nonexistence in the ODE implies existence and positivity of the solutions for another class of initial data.

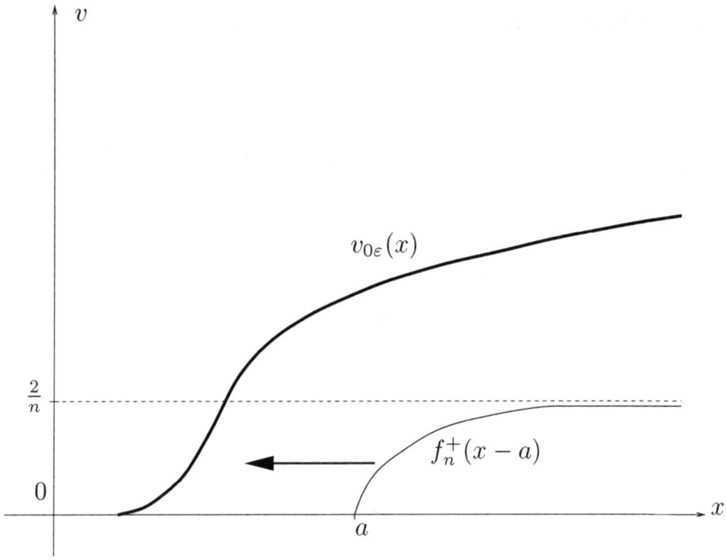

Figure 10.2 *Illustration of the proof of Theorem* 10.3*: the small flat TW* $f_n^+(x-\lambda t-a)$ *moving to the left with a large speed* $\lambda \ll -1$ *makes the solution strictly positive everywhere.*

Theorem 10.3 *Let there exist a constant* $\delta > 0$ *such that*

$$v_0(x) \geq \delta \quad \text{for } x \gg 1. \tag{10.26}$$

Then

$$v(x,t) \geq \delta \quad \text{in } S. \tag{10.27}$$

Proof. We again set $\varepsilon = \frac{1}{n} \ll 1$, choose $a \gg 1$ and $\lambda = \lambda_n \ll -1$ so that, by (10.26) and (10.16), we guarantee that

$$v_{0\varepsilon}(x) \geq f_n^+(x-a) \quad \text{for } x \geq a,$$

as shown in Figure 10.2. Then, by comparison, since $v_\varepsilon(x,t) > 0$,

$$v_\varepsilon(x,t) \geq f_n^+(x-\lambda t-a) \quad \text{in } S. \tag{10.28}$$

We first fix $n = M \gg 1$ on the right-hand side and pass to the limit $\varepsilon = \frac{1}{n} \to 0$ to obtain that

$$v(x,t) \geq f_M^+(x-\lambda t-a) \not\equiv 0.$$

The positivity follows from (10.28) by setting

$$\lambda = \lambda(M) \to -\infty \quad \text{as} \quad M \to \infty.$$

In the limit the perturbations propagate to the left with infinite speed. $\square$

Continuous self-similar collapse of singularity

We now construct a global in time self-similar solution of the form

$$\tilde{v}(x,t) = \sqrt{t}\, h(\zeta), \quad \zeta = x/\sqrt{t}, \tag{10.29}$$

which describes a continuous collapse of singularity for monotone increasing initial data. The function $h > 0$ solves the ODE (cf. (10.22) with the opposite sign in the linear first-order operator)

$$h'' + \tfrac{1}{2}\, h'\zeta - \tfrac{1}{2}\, h + \frac{h'}{h} = 0,$$

which admits solutions satisfying (cf. (10.23))

$$h(\zeta) = -\tfrac{1}{\zeta}\,(1 + o(1)) > 0 \quad \text{as } \zeta \to -\infty.$$

Such monotone increasing profiles have the following asymptotic behaviour:

$$h(\zeta) = C\zeta + \tfrac{1}{\zeta} + ... \quad \text{as } \zeta \to +\infty,$$

where $C > 0$ is a constant. Passing to the limit $t \to 0^+$ in (10.29) and using the above expansions, we obtain the corresponding initial data

$$\tilde{v}(x,0^+) = \begin{cases} 0 & \text{for } x \le 0, \\ Cx & \text{for } x > 0. \end{cases} \tag{10.30}$$

These self-similar solutions exhibit the continuous behaviour at $t = 0^+$:

$$\tilde{v}(x,t) = \begin{cases} -\frac{t}{x} + o(1) \to 0 = \tilde{v}(x,0^+) & \text{for } x < 0, \\ h(0)\sqrt{t} \to 0 = \tilde{v}(0,0^+) & \text{for } x = 0, \\ Cx + o(1) \to Cx = \tilde{v}(x,0^+) & \text{for } x > 0. \end{cases}$$

In particular, the uniform estimate for $x < 0$ is

$$\tilde{v}(x,t) \le \tilde{v}(0,t) = h(0)\sqrt{t} \to 0 \quad \text{as } t \to 0^+.$$

This limit analysis as $t \to 0^+$ proves that the self-similar solutions are the maximal ones. Each one $\tilde{v}$ can be obtained as the limit of the regularized sequence $v_\varepsilon(x,t) = \tilde{v}(x, t+\epsilon) > 0$ uniformly on compact subsets from S.

Continuity: local comparison with similarity solutions

We now show that the usual local comparison applies to ensure a continuous collapse of singularity for more general maximal solutions. This approach will be used later on in other problems and here we discuss it in greater detail.

First, given arbitrary increasing initial data

$$v_0(x) \equiv 0 \quad \text{for } x \le 0, \;\; v_0(x) > 0 \quad \text{for } x > 0, \tag{10.31}$$

and assuming first that

$$v_0(x) = Cx(1 + o(1)) \quad \text{as } x \to 0^+,$$

by the local comparison with self-similar solutions $\tilde{v}(x \pm \delta, t)$ shifted in x with slightly perturbed expansion coefficients $C \mp \delta$, it is not difficult to prove that, for any data like that, the singularity at $x = 0$ collapses continuously as $t \to 0^+$.

Second one can see that the same is true for any monotone continuous initial data $u_0(x)$. In fact, such local comparison shows that once the problem admits a particular continuous (self-similar) solution, any solution from the corresponding class is continuous at the singularity. Indeed, if a solution was be discontinuous at $x = 0,\, t = 0^+$, this would contradict the local comparison from above with the self-similar solution $\tilde{v}(x+\delta, t)$ with $C = C(\delta) \gg 1$. Indeed, using the asymptotic properties of $\tilde{v}$, we conclude that

$$v(0,t) \le \tilde{v}(\delta, t) = C(\delta)\delta + o(1) \le 2C(\delta)\delta \quad \text{for small } t > 0,$$

and since $C(\delta)\delta \to 0$ as $\delta \to 0$ for continuous initial data v_0, the continuity at $x = 0,\, t = 0^+$ follows.

In this analysis we use the fact that the parabolic flow is regular and smooth on any compact subset where the solution is uniformly bounded away from zero, so that the only discontinuity would occur from the singularity at the origin $x = 0,\, t = 0^+$ (and this discontinuity can give an essential change of the solution beyond the singularity in the smooth part of the flow). This singularity point can be controlled by the local comparison with continuous similarity solutions.

Theorem 10.4 *For continuous monotone increasing initial data* (10.31), *the maximal solution $v(x,t)$ is continuous at $t = 0^+$, and so the solution satisfies the initial condition in the usual pointwise sense*

$$v(x,0) = v_0(x) \quad \textit{in } I\!\!R.$$

It is easy to see that, in this case, even for discontinuous initial data such as the step-function

$$v_0(x) = \begin{cases} 1 & \text{for } x > 0, \\ 0 & \text{for } x \le 0, \end{cases}$$

the singularity collapses in a continuous manner though we cannot prove this by comparison with similarity solutions (10.29) with continuous piecewise linear initial data (10.30). Indeed, here $C(\delta) = \frac{1}{\delta}$ so that

$$C(\delta)\delta = 1 \not\to 0 \quad \text{as} \quad \delta \to 0.$$

Nevertheless, this collapse is described by the standard similarity solution

$$\bar{v}(x,t) = \theta(\zeta), \quad \zeta = x/\sqrt{t}, \tag{10.32}$$

of the heat equation $v_t = v_{xx}$. Then $\theta > 0$ solves the ODE

$$\theta'' + \tfrac{1}{2}\theta'\zeta = 0 \quad \text{in } I\!\!R, \quad \theta(-\infty) = 0, \;\; \theta(\infty) = 1.$$

We have that on such scaling structures, as $t \to 0^+$,

$$|\bar{v}_{xx}| = \frac{1}{t}\,|\theta''(\zeta)| \gg \left|\frac{\bar{v}_x}{\bar{v}}\right| = \frac{1}{\sqrt{t}}\left|\frac{\theta'(\zeta)}{\theta(\zeta)}\right|,$$

so that the lower-order singular term is negligible in a small neighbourhood of initial singularity in comparison with the diffusion operator. This makes it possible to construct a supersolution for $t \approx 0^+$ with a slightly perturbed asymptotic scaling structure given in (10.32). See examples of such a construction in [306, Chapters 2, 4].

Discontinuous limit semigroup

Theorem 10.1 implies that the maximal solution is not a solution of the Cauchy problem and cannot satisfy the initial condition in any "weak" sense. Therefore, the limit semigroup $\mathcal{T}(t)$ of maximal solutions is not continuous at $t = 0$.

To emphasize this once more, consider the initial data with a zero gap on an interval of non-zero measure, say, $[-a, a]$ with a constant $a > 0$:

$$v_0(x) = 0 \quad \text{for } |x| \le a. \tag{10.33}$$

Consider two continuous bounded functions, $\phi(x) > 0$ that is monotone increasing for $x > a$, $\phi(a) = 0$ and $\psi(x) > 0$ that is monotone decreasing for $x < -a$ and $\psi(-a) = 0$.

Thus we take initial data

$$v_0(x) = \begin{cases} \psi(x) & \text{for } x < -a, \\ 0 & \text{for } |x| \le a, \\ \phi(x) & \text{for } x > a, \end{cases} \tag{10.34}$$

satisfying the gap condition (10.33). With a slight modification to the proofs of Theorems 10.1 and 10.3, we obtain the following result.

Theorem 10.5 *The maximal solution $v(x,t) > 0$ of the Cauchy problem* (10.3), (10.34) *satisfies*

$$v(x, 0^+) \equiv \lim_{t \to 0^+} v(x,t) = \phi(x) \quad \text{in } \mathbb{R}, \tag{10.35}$$

i.e., $v(x,t)$ does not depend on $\psi(x)$.

Proof. In order to see that the left-hand structure of the initial data given by $\psi(x)$ for $x \le -a$ disappears instantaneously, we apply the proof of Theorem 10.1.

Fix an $\varepsilon \ll 1$ and consider the classical solution $v_\varepsilon \ge \varepsilon$. By continuity (see below), there exists $t_\varepsilon > 0$ such that

$$v_\varepsilon(0,t) \le 2\varepsilon \quad \text{on} \quad (0, t_\varepsilon].$$

Choosing $n \gg 1$ such that $\frac{2}{n} \ge 2\varepsilon$, we can now compare, on $P_\varepsilon = \mathbb{R}_- \times (0, t_\varepsilon]$, the solution v_ε and the steep TW $f_n^-(x - \lambda t)$ with $\lambda = \lambda_\varepsilon \ll -1$ which gives

$$v_\varepsilon(x,t) \le f_n^-(x - \lambda t) \quad \text{in} \quad P_\varepsilon.$$

Passing to the limit $\varepsilon \to 0$ and $n \to \infty$, since $\lambda_\varepsilon \to -\infty$, we obtain that

$$v(x, 0^+) \equiv 0 \quad \text{for} \quad x < 0.$$

The rest of (10.35), the continuity for $x > 0$, is proved by the local comparison with the continuous maximal solutions $\tilde{v}(x+\delta, t)$ as in the previous subsection. □

10.3 A generalized model with complete and incomplete singularities

The analysis applies to more general equations. For instance, consider the generalized equation

$$v_t = v_{xx} + \frac{v_x}{\varphi(v)}, \tag{10.36}$$

where $\varphi(v) > 0$ for $v > 0$ is a smooth function satisfying the singularity condition

$$\varphi(v) \to 0 \quad \text{as} \quad v \to 0^+.$$

We then define a unique maximal solution by the positive approximation of initial data. Further theory of equation (10.36) depends on the properties of the TW solutions

$$V(x,t) = f(\xi), \quad \xi = x - \lambda t,$$

where $f \geq 0$ solves the ODE the ODE

$$f'' + \lambda f' + \frac{f'}{\varphi(f)} = 0. \tag{10.37}$$

Integrating it, we obtain (cf. (10.9))

$$f' \equiv P = C + \Psi(f) - \lambda f, \tag{10.38}$$

where

$$\Psi(f) = \int_f^1 \frac{\mathrm{d}s}{\varphi(s)} > 0 \quad \text{for } f < 1.$$

Complete singularity (nonexistence)

If

$$\Psi(0) = \int_0^1 \frac{\mathrm{d}s}{\varphi(s)} = \infty, \tag{10.39}$$

then as in Section 10.2, we construct steep approximating TW profiles f_n^- exhibiting similar fast propagation properties. This gives Theorem 10.1 on complete singularity for any initial data with a finite right-hand interface.

Positivity and finite propagation

Since $\varphi(0) = 0$, we can always prove Theorem 10.3 using the appropriate "flat" profiles f_n^+ with $\lambda = \lambda_n \ll -1$ satisfying (10.15). Let the integral converge:

$$\int_0^1 \frac{\mathrm{d}s}{\varphi(s)} < \infty. \tag{10.40}$$

Given arbitrary $\lambda \in I\!R$, we define the orbit $P_{\min}(f) < 0$ for small $f > 0$ with $C = -\Psi(0)$ such that

$$P_{\min}(0) = 0.$$

This minimal orbit can be constructed by a regular limit $P_{\min}(f) = \lim P_n(f)$, where the approximations $P_n(f) < 0$ satisfy $P_n(\frac{2}{n}) = 0$. Then solving the corresponding first-order equation

$$\begin{aligned} f' = P_{\min}(f) &\equiv -\int_0^f \frac{\mathrm{d}s}{\varphi(s)} - \lambda f \\ &= -\int_0^f (1 + o(1)) \frac{\mathrm{d}s}{\varphi(s)} \quad \text{as } f \to 0, \end{aligned}$$

one can derive the criterion of finite propagation of the right-hand interface of the TWs

$$\int_0 \left(\int_0^z \frac{\mathrm{d}s}{\varphi(s)} \right)^{-1} \mathrm{d}z < \infty. \tag{10.41}$$

By comparison of maximal solutions $v(x,t)$ and TWs $f(\xi)$ we have

Theorem 10.6 *Let* (10.40) *and* (10.41) *hold. Then if* $\sup\{\operatorname{supp} v_0\} < \infty$, *the maximal solution has the right-hand interface*

$$s(t) = \sup\{\operatorname{supp} v(\cdot,t)\} < \infty \quad \textit{for } t > 0.$$

Regularity and the interface propagation properties are studied by the complete set $B = \{V(x,t),\ \lambda \in I\!R\}$ of such proper TWs as in Chapter 7.

10.4 Complete singularity in the Cauchy problem for the Zhang equation

Consider the Cauchy problem for the semilinear equation

$$u_t = u_{xx} + \log|u_x| \quad \text{in } S = I\!R \times I\!R_+ \tag{10.42}$$

with bounded continuous initial data u_0. The singular term $\log|u_x|$ in the equation has a standard single-well shape but it is not bounded and attains $-\infty$ at $u_x = 0$.

In order to determine the maximal solution, we introduce a monotone regular approximation of the equation

$$u_t = u_{xx} + F_\varepsilon(u_x), \tag{10.43}$$

where, for any $u_x \in I\!R$, a monotone decreasing convergence takes place

$$F_\varepsilon(u_x) = \log(\varepsilon^2 + (u_x)^2)^{1/2} \to \log|u_x| \quad \text{as } \varepsilon \to 0^+ \tag{10.44}$$

including the singularity value $u_x = 0$.

Let $\{u_\varepsilon\}$ be the global classical solutions of the regularized equation with the same initial data u_0. Then $\{u_\varepsilon\}$ is monotone decreasing with ε and there exists the limit

$$u = \mathcal{T}(t)u_0 \equiv \lim_{\varepsilon \to 0} u_\varepsilon \quad \text{in } S,$$

which is the maximal solution independent of the type of monotone decreasing approximation and $\mathcal{T}(t)$ is the corresponding limit semigroup. The maximal solutions is unique and satisfies the MP (Section 7.3).

Concerning actual properties of the limit semigroup, we easily obtain the complete singularity (nonexistence) result.

Theorem 10.7 *For any initial data* u_0 *that is bounded from above, the maximal solution of equation* (10.42) *is*

$$u(x,t) \equiv -\infty \quad \textit{in } S. \tag{10.45}$$

Proof. Let $u_0 \le C_0$. We compare $u_\varepsilon(x,t)$ with the flat solution $U_\varepsilon(t)$ satisfying the ODE

$$U_\varepsilon' = F_\varepsilon(0) \quad \text{for } t > 0,\ U_\varepsilon(0) = C_0.$$

Then

$$U_\varepsilon(t) = F_\varepsilon(0)t + C_0.$$

By comparison of classical solutions, there holds

$$u_\varepsilon(x,t) \le F_\varepsilon(0)t + C_0 \quad \text{for } t > 0.$$

Passing to the limit $\varepsilon \to 0$ and using approximation (10.44), where $F_\varepsilon(0) \to -\infty$, yields (10.45). □

Thus, for initial data that is bounded above, the Cauchy problem makes no sense and admits the trivial solutions (10.45) only. On the other hand, for unbounded data, nontrivial solutions do exist. For instance,

$$u(x,t) \equiv x \quad \text{in } I\!\!R$$

is a global stationary solution of (10.45). Actually, this means that nontrivial solutions in the Cauchy problem can occur if a certain special behaviour of solutions is prescribed at infinity for $|x| \gg 1$ prohibiting the behaviour

$$u_\varepsilon \to -\infty \quad \text{as } \varepsilon \to 0^+.$$

This shows that nontrivial solutions are also expected to exist in the initial-boundary problem posed in a bounded domain to be studied below.

10.5 Instantaneous shape simplification in the Dirichlet problem for the Zhang equation in one dimension

Existence for bounded initial data

We consider equation (10.42) in $S^1 = (-1,1) \times I\!\!R_+$ with the Dirichlet boundary conditions

$$u(\pm 1, t) = 0 \quad \text{for } t \ge 0 \tag{10.46}$$

and bounded continuous initial data u_0. We will establish the global existence of nontrivial maximal solutions constructed by the same monotone approximation (10.44) and show that, at $t = 0^+$, instantaneously, the solution takes an inverse bell-shaped form.

Let us describe this phenomenon of the *instantaneous shape simplification* of the solution. Given a function $w \in C([-1,1])$ satisfying $w(\pm 1) = 0$, we first define the left- and right-hand monotone extension

$$w_-(x) = \inf_{s \in (-1,x)} w(s)$$

and

$$w_+(x) = \inf_{s \in (x,1)} w(s).$$

We next introduce the *operator of the shape simplification*:

$$\hat{w}(x) = \mathcal{M}w(x) \equiv \max\{w_-(x),\, w_+(x)\} \quad \text{in } (-1,1), \tag{10.47}$$

see Figure 10.3. The existence result is as follows.

Theorem 10.8 *The Dirichlet problem* (10.42), (10.46) *with bounded continuous initial data* u_0 *admits a bounded maximal solution* $u(x,t)$, *which for* $t > 0$ *has*

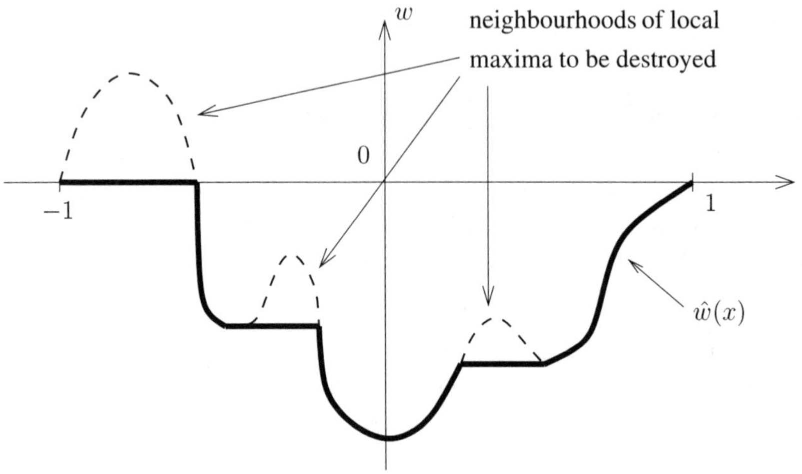

Figure 10.3 *Operator of shape simplification cuts of local maxima of $w(x)$ forming function $\hat{w}(x)$ with a single absolute minimum.*

the inverse bell-shaped form, i.e., has a unique minimum and no local maxima in x, and satisfies the following initial condition:

$$\mathcal{T}(0^+)u_0(x) = \hat{u}_0(x) \quad in \ [-1,1]. \tag{10.48}$$

The proof is based on comparison with special stationary and similarity solutions of the equation.

A priori bound

In order to construct a uniform lower bound on $\{u_\varepsilon\}$, we consider a family of symmetric stationary solutions $U_\mu = U_\mu(|x|)$ of (10.42) satisfying the ODE

$$U'' + \log|U'| = 0 \quad \text{for } x > 0, \ \ U(0) = \mu, \ \ U'(0) = 0, \tag{10.49}$$

where $\mu \in I\!R$ is a parameter. By translation, we have that $U_\mu = U_0 + \mu$, where U_0 is given by the quadrature

$$\int_0^{U_0'(x)} \frac{\mathrm{d}z}{(-\log z)} = x.$$

It follows that $U_0'(x) \to 1^-$ as $x \to \infty$ and $U_0(x)$ is monotone increasing with the behaviour

$$U_0(x) = x + O(1) \quad \text{ as } x \to +\infty. \tag{10.50}$$

Observe that $U_0 \in C^1$ for $x \geq 0$ and $U_0 \in C^\infty$ at any non-singular point $x > 0$. At the singularity point it exhibits the following behaviour:

$$U_0(x) = \tfrac{1}{2}\,x^2|\log|x|| - \tfrac{1}{2}\,x^2 \log|\log|x||(1+o(1)) \quad \text{ as } x \to 0. \tag{10.51}$$

This means that $U_0 \in C^{2-\varepsilon}$ for any $\varepsilon > 0$, and more precisely, at $x = 0$,

$$\left(\frac{U_0}{-\log U_0}\right)^{1/2} \quad \text{and} \quad \frac{1}{U_0''}\,e^{-U_0''} \quad \text{are Lipschitz continuous.} \tag{10.52}$$

We now prove a lower bound on the approximating solutions.

Proposition 10.9 *There exists a constant* μ *such that*

$$u_\varepsilon \ge \mu \quad \text{in} \quad S^1 \quad \text{for any} \ \ \varepsilon > 0 \ \ \text{small}.$$

Proof. Since $U_\mu(x)$ is a convex function, it is a subsolution of the regularized equation (10.43), (10.44). Choosing $\mu < 0$ such that $u_{0\varepsilon} \equiv u_0 \ge U_\mu$ in $[-1, 1]$, we deduce by comparison that $u_\varepsilon \ge U_\mu$ in S^1. □

Passing to the limit $\varepsilon \to 0^+$, we have that the maximal solution that is bounded above by approximation, $u \le u_1$, is also uniformly bounded from below, $u \ge \mu$ in S^1.

Self-similar solutions: local singularity formation

As a natural extension of the study of differential properties of stationary profiles, we now consider the question of formation of evolution singularities from a smooth profile for equation (10.42). Consider the following self-similar solution of equation (10.42):

$$u_*(x,t) = \tfrac{1}{2}\,t\log t + t\theta(\eta), \quad \eta = x/\sqrt{t} \tag{10.53}$$

where a symmetric profile $\theta = \theta(|\eta|)$ satisfies the ODE

$$\mathbf{D}(\theta) \equiv \theta'' + \tfrac{1}{2}\,\theta'\eta - \theta + \log|\theta'| = \tfrac{1}{2} \quad \text{for } \eta > 0, \ \ \theta'(0) = 0. \tag{10.54}$$

By the Banach Contraction Principle applied to the equivalent integral equation, one can show that the solution with $\theta(0) = \mu$ exists locally for small $\eta > 0$. The solutions obey the same behaviour (10.51) and (10.52) as the stationary ones. The global properties of the self-similar profiles are listed below.

Proposition 10.10 *For any* μ*, the solution* $\theta(\eta)$ *of* (10.54) *with* $\theta(0) = \mu$ *is strictly monotone increasing and has the asymptotic behaviour*

$$\theta(\eta) = C\eta^2 + \log\eta + 2C + \log(2C) + o(1) \quad \text{as } \eta \to \infty, \tag{10.55}$$

where $C = C(\mu) > 0$ *is strictly monotone increasing with* μ.

Proof. Obviously, due to the weak nonlinearity in (10.54), θ is a global solution defined for all $\eta > 0$. Monotonicity of θ follows from the MP. The expansion follows from the local analysis of the singular point $\eta = \infty$. Monotonicity of $C(\mu)$ is guaranteed by comparison of the solutions. □

We will need self-similar subsolutions of the same structure

$$\underline{u}_*(x,t) = \tfrac{1}{2}\,t\log t + t\underline{\theta}(\eta) \tag{10.56}$$

for $\mu \ll -1$ and $C(\mu) \approx 0^+$. Then $\underline{\theta}$ must satisfy the differential inequality

$$\mathbf{D}(\underline{\theta}) \ge \tfrac{1}{2}$$

and we choose (cf. (10.55))

$$\underline{\theta}(\eta) = C\eta^2 + \log \eta + B \quad \text{for} \quad \eta \gg 1,$$

where the constant B satisfies

$$B < 2C + \log 2C. \tag{10.57}$$

For sufficiently small $\eta > 0$, the transitional structure of the subsolution $\underline{\theta}$ is chosen in accordance with the stationary profiles. It is important that, by (10.57), $C \to 0^+$ implies that $B \to -\infty$, so, for any fixed $\eta \gg 1$, $\underline{\theta}(\eta) \to -\infty$. This means that, after a suitable extension up to the origin, we have that $\underline{\theta}(0) \to -\infty$. This shows that the similarity profiles with behaviour (10.55) satisfy

$$C(\mu) \to 0^+ \quad \text{as} \ \mu \to -\infty. \tag{10.58}$$

Using a similar supersolution $\overline{u}_*$ with $C \gg 1$, one can show that

$$C(\mu) \to +\infty \quad \text{as} \ \mu \to +\infty. \tag{10.59}$$

Fixing μ and hence $C = C(\mu)$, we have that the self-similar solution (10.53) is continuous at $t = 0$ and satisfies the following initial condition: as $t \to 0^+$ uniformly on any compact subset in x,

$$u_*(x,t) = Cx^2 + O(t),$$

so that the initial condition is

$$u_*(x, 0^+) = Cx^2 \quad \text{for} \ x \in I\!R. \tag{10.60}$$

The same is true for the corresponding subsolutions $\underline{u}_*$ and supersolutions $\overline{u}_*$.

It follows by comparison with the stationary solutions from below that the self-similar solution is a maximal solution, i.e., it can be approached by regular monotone approximations. It describes the generic local singularity formation for equation (10.42) from the C^∞-smooth initial profile (10.60). The local singularity formation on compact subsets in $\eta = x/\sqrt{t}$ is then given by

$$u_*(x,t) \approx \tfrac{1}{2}\, t \log t + \mu t + \tfrac{1}{2}\, x^2 |\log(|x|/\sqrt{t})| + ...,$$

where $\mu = \theta(0)$ determines the parameter $C(\mu) > 0$ in the initial condition (10.60). By comparison, it also shows the instability of a uniform pattern $u \equiv \text{const.}$, which is destroyed faster than the above singular patterns. This means that the piecewise constant in x parts of the maximal solution profile $u(x,t)$ do not exist for $t > 0$, cf. the statement of Theorem 10.8.

This singularity formation means that the singular operator $\log |u_x|$ produces for $t > 0$ bounded C^1-perturbations of initial profiles in a neighbourhood of smooth minima like for the parabola (10.60). This conclusion is important for the further construction of the discontinuous limit semigroup.

Instant shape simplification of initial data

We begin with the following simple initial configuration that explains the first principle of the initial shape simplification.

Proposition 10.11 *Let $u_0(x) \geq 0$ in $I\!R$ (hence $\hat{u}_0(x) \equiv 0$). Then the maximal solution satisfies*

$$u(x, 0^+) \equiv 0. \tag{10.61}$$

Proof. As in the proof of Theorem 10.7, by comparison from above with the flat solution, we deduce that

$$u_\varepsilon(x,t) \leq F_\varepsilon(0)t + C_0 \quad \text{for } t \in (0, C_0|F_\varepsilon(0)|^{-1}], \quad C_0 = \sup u_0 > 0.$$

Passing to the limit $\varepsilon \to 0$ yields

$$u(x, 0^+) \leq 0.$$

On the other hand, using the self-similar subsolution (10.56) in the form $\underline{u}_*(x,t) - C$ with a positive $C(\mu) \ll 1$ such that

$$u_0(x) \geq \underline{u}_*(x, 0) - C = -C(1 - x^2) < 0 \quad \text{for } x \in (-1, 1),$$

we conclude that

$$u(x,t) \geq \underline{u}_*(x,t) - C$$

in S^1 by comparison and letting $\varepsilon \to 0^+$. Thus, taking $t = 0^+$ yields

$$u(x, 0^+) \geq -C(1 - x^2) \quad \text{on } (-1, 1).$$

Passing to the limit $C \to 0^+$, we obtain the converse inequality $u(x, 0^+) \geq 0$ completing the proof. □

We now consider general continuous initial data u_0 and prove the main result.

Proof of Theorem 10.8. Given arbitrary bounded continuous initial data, exactly as in the proof of Proposition 10.11, we deduce that $u(x, 0^+) \leq 0$. We now prove that

$$u(x, 0^+) \leq \hat{u}_0(x). \tag{10.62}$$

We need a slight modification to the standard construction of the maximal solution. Consider the regularized solution $u_{\varepsilon\rho}(x,t)$ with smooth regularized initial data $u_{0\rho}$ having a finite number of isolated minima. Here $\rho > 0$ is the new parameter, $\{u_{0\rho}\}$ is monotone decreasing with ρ and $u_{0\rho} \to u_0$ uniformly as $\rho \to 0^+$. Since the maximal solutions does not depend on the type of monotone decreasing regularization (Section 7.3), we have that

$$u = \lim_{\rho \to 0^+} \lim_{\varepsilon \to 0^+} u_{\varepsilon\rho}.$$

Denote

$$u_\rho(x,t) = \lim_{\varepsilon \to 0^+} u_{\varepsilon\rho}(x,t).$$

Fix a minimum point $x_1 \in (-1, 1)$ of $u_{0\rho}(x)$ so that $u_{0\rho}(x) \geq u_{0\rho}(x_1)$ in a neighbourhood $(x_1 - \nu, x_1 + \nu)$ with $\nu = \nu(\rho) > 0$ small. Let us compare from above $u_\rho(x,t)$ with the self-similar supersolution

$$\bar{V}(x,t) \equiv \overline{u}_*(x - x_1, t) + u_{0\rho}(x_1) + \delta$$

with positive arbitrarily small δ and $C = C(\delta) \gg 1$. Then, since C can be arbitrarily large for small $\delta > 0$, by construction, this supersolution is steep enough for $x \approx x_1$, $x \neq x_1$, i.e., there exist $x_{1\pm} \approx x_1$ such that $x_{1-} < x_1 < x_{1+}$ and $\bar{V}(x_{1\pm}, t) \geq 0$ for small $t > 0$. This implies necessary comparison on the parabolic boundary of this domain, and hence we obtain that $u_\rho(x,t)$ is bounded above for $x \approx x_1$ and $t \approx 0^+$,

$$u_\rho(x,t) \leq u_{0\rho}(x_1) + \mu, \quad x \in (x_1 - \nu, x_1 + \nu), \;\; t \in (0, \delta), \tag{10.63}$$

where $0 < \mu = \mu(\delta, \rho) \ll 1$ for small δ and ρ. We derive similar estimates at all other minimal points $x_2, ..., x_k$ of $u_{0\rho}(x)$, and obtain upper bounds such as (10.63) at each minimum point.

Consider now a pair of neighbouring minimum points $-1 < x_1 < x_2 < 1$ and assume that $u_{0\rho}(x) \not\equiv \text{const.}$ on (x_1, x_2). We temporarily consider the regularized solutions $u_{\varepsilon\rho}(x,t)$ with an $\varepsilon \ll 1$. Then estimates from above of the type (10.63) are true for $u_{\varepsilon\rho}(x_1,t)$ and $u_{\varepsilon\rho}(x_2,t)$, and hence we can compare $u_{\varepsilon\rho}(x,t)$ and the flat solution $U_\varepsilon(t) = F_\varepsilon(0)t$ in the domain $(x_1, x_2) \times (0, \delta_1)$ since $u_{\varepsilon\rho} \leq U_\varepsilon$ on its parabolic boundary provided that

$$U_\varepsilon(\delta_1) \geq \max\{u_{0\rho}(x_1), u_{0\rho}(x_2)\} + \mu.$$

Passing to the limit $\varepsilon \to 0^+$, we conclude that

$$u_\rho(x, 0^+) \leq \max\{u_{0\rho}(x_1), u_{0\rho}(x_2)\} \quad \text{on } (x_1, x_2).$$

Performing this comparison successively with all pairs of minima points, we obtain that $u_\rho(x,0^+) \leq \hat{u}_{0\rho}(x)$. Now passing to the limit $\rho \to 0^+$ and using the fact that $\hat{u}_{0\rho} \to \hat{u}_0$, we arrive at (10.62).

As the last step, we establish the opposite inequality

$$u(x, 0^+) \geq \hat{u}_0(x). \tag{10.64}$$

Consider the left- and right-hand monotone extensions $u_{0-}(x)$ and $u_{0+}(x)$ that are decreasing and increasing functions respectively. Given an arbitrarily small constant $\delta > 0$, we construct a strictly monotone smooth regularization of these functions, $u_{0-\delta}(x)$ and $u_{0+\delta}(x)$ respectively, such that on $[-1, 1]$,

$$u_{0-}(x) - \delta \leq u_{0-\delta}(x) \leq u_{0-}(x), \quad u'_{0-\delta}(x) \leq -\tfrac{1}{2}\,\delta,$$

$$u_{0+}(x) - \delta \leq u_{0+\delta}(x) \leq u_{0+}(x), \quad u'_{0+\delta}(x) \geq \tfrac{1}{2}\,\delta,$$

and

$$u_{0\pm\delta}(x) \to u_{0\pm}(x) \quad \text{as } \delta \to 0^+$$

uniformly on $[-1, 1]$. We now compare the maximal solution $u(x,t)$ with the solutions $u_{\pm\delta}(x,t)$ corresponding to the initial data $u_{0\pm\delta}(x)$ and fixed boundary conditions at $x = 1$ and $x = -1$:

$$u_{\pm\delta}(\pm 1, t) \equiv u_{0\pm\delta}(\pm 1).$$

Observe that due to the strict monotone properties of such regularized initial data, the local in time maximal solutions are not singular, so that they are continuous and can be compared directly with the corresponding approximating sequences $\{u_\varepsilon\}$. Indeed, differentiating the equation, we have that $v = u_x$ solves the hydrodynamic version (10.3) with data v_0 of constant sign, and so singularities cannot occur in finite time, see Section 10.2. This gives that $u_\varepsilon(x,t) \ge u_{\pm\delta}(x,t)$ for all small $t > 0$, so that passing to the limit $\varepsilon \to 0^+$, we obtain

$$u(x,t) \ge \max\{u_{-\delta}(x,t),\, u_{+\delta}(x,t)\}.$$

Observe that the right-hand side is a sufficiently regular subsolution of the singular equation. Letting here $t \to 0^+$ and passing to the limit $\delta \to 0^+$, we arrive at inequality (10.64). For such $u_0(x)$ changing sign, the same comparison guarantees that $u(x,t)$ has a single minimum and no maxima for any $t > 0$. This completes the proof. □

On correspondence with the hydrodynamic version

The solutions of the singular Zhang equation (10.2) and its hydrodynamic version (10.3) are connected via the first-order differential (contact) transformation $u_x = v$. For the Dirichlet problem with conditions (10.46), this means that

$$u(x,t) = \int_{-1}^{x} v(y,t)\,\mathrm{d}y. \tag{10.65}$$

Then $u(1,t) = 0$ provided that $v(x,t)$ satisfies the nonlocal condition (a constraint)

$$\int_{-1}^{1} v(y,t)\,\mathrm{d}y = 0 \quad \text{for } t > 0. \tag{10.66}$$

The unique maximal solution u of the singular Zhang equation generates the unique solution $v = u_x$ of its hydrodynamic version with the nonlocal constraint (10.66) and initial data $v_0 = u_0'$. By Theorem 10.8, $v(x,t)$ exists, is bounded and for any $t > 0$ (and for $t = 0^+$) changes sign exactly once on $[-1,1]$.

The problem (10.3), (10.66) falls in the scope of nonlinear parabolic equation on a subspace. The nonlocal condition means that the solution belongs to the subspace of codimension 1

$$F_{-1} = F \setminus \mathrm{Span}\{1\}$$

of the space of Fourier series

$$F = \mathrm{Span}\{1, \sin(\pi n x), \cos(\pi n x),\; n \ge 1\}.$$

Note that though the equation is parabolic, the subspace F_{-1} is not ordered in the usual sense (no functions $v_1 \le v_2$ are available on F_{-1} unless $v_1 \equiv v_2$) so that the comparison principle makes no sense.

Generalized models

The properties of the maximal solutions described above for the singular Zhang equation are available for other equations with a single-well singularity such as

$$u_t = u_{xx} + f(|u_x|), \tag{10.67}$$

where $f(q)$ is a continuous monotone increasing function satisfying

$$f(q) \to -\infty \quad \text{as } q \to 0^+ \quad (f(0) = -\infty).$$

Then the maximal solutions are constructed by monotone approximations via smooth solutions of the equation

$$u_t = u_{xx} + f(\sqrt{\varepsilon^2 + (u_x)^2}).$$

For instance, consider the equation

$$u_t = u_{xx} - \frac{1}{|u_x|^\alpha}, \quad \text{where } \alpha > 0, \tag{10.68}$$

with a single-well singularity $f(|q|) = -|q|^{-\alpha}$. Then given initial data in the Dirichlet problem, the limit semigroup $\mathcal{T}(t)$ is not continuous at $t = 0^+$ and produces the instant shape simplification

$$\mathcal{T}(0^+)u_0 = \hat{u}_0.$$

The proof is essentially the same with a slightly different particular mechanism of the local singularity formation. The symmetric stationary solutions have the representation $U_\mu = U_0 + \mu$, where U_0 solves the equation

$$U_0'' - \frac{1}{|U_0'|^\alpha} = 0.$$

It is integrated explicitly and the solution

$$U_0(x) = C_0|x|^\gamma$$

with constants

$$\gamma = \tfrac{\alpha+2}{\alpha+1} \quad \text{and} \quad C_0 = \tfrac{1}{\alpha+2}\,(\alpha+1)^\gamma$$

explains the actual regularity of the solutions, $U_0 \in C^{1+\nu}$ with $\nu = \frac{1}{\alpha+1}$. By comparison, these stationary solutions determine a uniform lower bound on $\{u_\varepsilon\}$ similar to Proposition 10.9 establishing that the maximal solution is nontrivial. A generic singularity formation is described by symmetric self-similar solutions

$$u_*(x,t) = t^{\gamma/2}\theta(\eta), \quad \eta = \frac{x}{\sqrt{t}},$$

where θ solves the ODE

$$\theta'' + \tfrac{1}{2}\,\theta'\eta - \tfrac{1}{2}\,\gamma\theta - \frac{1}{|\theta'|^\alpha} = 0 \quad \text{for } \eta > 0, \quad \theta'(0) = 0,$$

with similar local and global properties. In particular,

$$\theta(\eta) = C\eta^\gamma(1 + o(1)) \quad \text{as } \eta \to \infty,$$

and $C = C_0$ corresponds to the stationary solution: $\theta(\eta) \equiv U_0(\eta)$. If $C > C_0$, then $\theta(0) > 0$, and $C < C_0$ implies $\theta(0) < 0$. Therefore, fixing $-\mu = \theta(0) < 0$ so that

$$u_*(x, 0^+) \equiv Cx^\gamma < C_0 x^\gamma \quad \text{for } x > 0,$$

we obtain for $t > 0$,

$$\inf_x u_*(x,t) \equiv u_*(0,t) = -\mu t^\gamma.$$

Hence, $O(t^\gamma)$ is the actual rate of such a slow ordered evolution of the maximal solutions from the stationary ones. The analysis of the instant shape simplification is done similarly.

10.6 Discontinuous limit semigroups and operator of shape simplification for singular equations in $I\!R^N$

Consider similar discontinuous properties of maximal solutions of the multi-dimensional version of the Zhang equation

$$u_t = \Delta u + \log |\nabla u| \quad \text{in } S = \Omega \times I\!R_+, \tag{10.69}$$

with the Dirichlet boundary condition

$$u = 0 \quad \text{on } \partial\Omega \times I\!R_+,$$

where $\Omega \subset I\!R^N$ is a bounded domain with a smooth boundary $\partial\Omega$. Given bounded smooth data $u_0(x)$, $u_0 = 0$ on $\partial\Omega$, we construct a unique maximal solution by a monotone decreasing approximation via regular solutions of the smooth uniformly parabolic PDE

$$u_t = \Delta u + \log(\varepsilon^2 + |\nabla u|^2)^{1/2}$$

with same initial data. A uniform bound on $\{u_\varepsilon\}$ from below ensuring that the maximal solution is nontrivial is obtained by comparison with radial stationary solutions as in Proposition 10.9.

Let us concentrate on the construction of the corresponding operator $\hat{u}_0 = \mathcal{M}u_0$ of the shape simplification. It is defined as follows. First of all, Proposition 10.11, which is true in this geometry, implies that

$$\hat{u}_0 < (u_0(x))_- \quad \text{in } \Omega,$$

so that we may assume that $u_0 \le 0$. Next, for any $\mu \in (a_0, 0)$, where $a_0 = \inf u_0 < 0$, we define the open subset

$$L_\mu = \{x \in \Omega : \; u_0(x) > \mu\}.$$

For smooth u_0, L_μ consists of a countable number of connected components

$$L_\mu = \textstyle\sum_{(k)} \Gamma_\mu^k,$$

where Γ_μ^1 denotes the first outer component satisfying $\partial\Omega \subset \overline{\Gamma}_\mu^1$ (if necessary, we perform an extra approximation of u_0, to ensure this property). We next choose all components $\{\Gamma_\mu^{k_j}, \; j = 1, 2, ...\}$ such that

$$\text{dist}(\Gamma_\mu^1, \Gamma_\mu^{k_j}) > 0,$$

and set

$$u_0^\mu = \begin{cases} \mu & \text{in } \Gamma_\mu^{k_j}, \; j = 1, 2, \dots, \\ u_0 & \text{otherwise.} \end{cases}$$

Finally, we present the operator of shape simplification $\mathcal{M}$, yielding nonpositive (regularized) functions

$$\hat{u}_0 = \mathcal{M}u_0 \equiv \inf_\mu u_0^\mu \quad \text{in } \Omega. \tag{10.70}$$

Then, in a similar manner, we establish that the limit semigroup of maximal solutions satisfies

$$u(x, 0^+) = \mathcal{T}(0^+)u_0 = \hat{u}_0. \tag{10.71}$$

The first ingredient of the proof is an upper bound on $u(x,t)$ for $t \approx 0^+$ in small neighbourhoods of minimum points, which by approximation are assumed to be isolated. Here we use supersolutions of the form (10.53), where θ solves the radial ordinary differential inequality

$$\mathbf{D}(\theta) \equiv \Delta\theta + \tfrac{1}{2}\, \nabla\theta \cdot \eta - \theta + \log|\nabla\theta| \le \tfrac{1}{2}\,,$$

which can be studied as in 1D. We prove the existence of a family of steep profiles $\{\theta(|\eta|, C), C \gg 1\}$. Such supersolutions are used as upper bounds on $u(x,t)$ in a neighbourhood of strict local minima of u_0.

We next perform the evolution analysis by comparison from above of $u_\varepsilon(x,t)$ with the flat solution $U_\varepsilon(t) = F_\varepsilon(0)t$ establishing, roughly speaking, that all isolated local maxima of u_0 are destroyed for small $t > 0$. Actually, such evolution comparison mimics the steps of construction of operator (10.70), where we use the time-parameterization $\mu \mapsto F_\varepsilon(0)t$. This proves that $u(x, 0^+) \le \hat{u}_0(x)$.

The reverse inequality is proved by an extra δ-approximation since any function $\hat{u}_0$ of such simple shape can be approximated from above and below by functions $u^{\pm}_{0\delta}$ having the gradient uniformly bounded away from zero, i.e.,

$$|\nabla u^{\pm}_{0\delta}| \ge \delta > 0 \quad \text{in } \Omega,$$

excluding a small neighbourhood of a unique minimum point. The equation is not singular for such initial data a.e. excluding the unique strict absolute minimum point, where, as we have seen, the continuous behaviour from above and below in the geometry of simple shapes is controlled by similarity super and subsolutions. Hence, $u^{\pm}_{\delta}(x,t)$ are continuous solutions satisfying $u^-_\delta \le u \le u^+_\delta$. Passing to the limit $t \to 0^+$ and $\delta \to 0^+$ yields $u(x, 0^+) = \hat{u}_0(x)$.

Remarks and comments on the literature

The main results on the singular Zhang equation in 1D are given in [144]. The hydrodynamic version was also studied in [143]. Equation (10.3) is the limit case, as $\alpha \to 0^+$, of the diffusion equation with singular convection

$$v_t = v_{xx} + v^{\alpha-1} v_x. \tag{10.72}$$

For $\alpha > 0$, the mathematical theory was well-developed in the 1970s and 1980s (we refer to the survey [213]) and exhibits several interesting properties such as infinite one-sided propagation, existence of back front, instantaneous shrinking of the support, etc.

CHAPTER 11

Further Examples of Discontinuous and Continuous Limit Semigroups

We continue the study of discontinuous limit semigroups and consider two new classes of problems with unusual features of proper solutions.

First we study discontinuous semigroups for a class of non-autonomous singular reaction-diffusion equations describing an interaction between spatial singularities and standard combustion terms creating the usual finite-time blow-up singularities. Second we present an optimal characterization of singular parabolic PDEs admitting proper maximal solutions with supports that do not move with time.

11.1 Equations in $I\!R^N$ with blow-up and spatial singularities: discontinuous semigroups and singular initial layers

We study a class of non-autonomous parabolic equations

$$u_t = F(x, u, |\nabla u|, \Delta u) \quad \text{in } S = I\!R^N \times I\!R_+, \tag{11.1}$$

where $F(x, p, q, r)$ satisfies typical assumptions on singular blow-up behaviour as $p \to \infty$ and regularity assumptions on compact subsets in the variables p, q, r. In addition, we add an extra *blow-up singularity in* x posed at a single point, say, at the origin $x = 0$, so

$$F(x, p, q, r) \to +\infty \quad \text{as } x \to 0. \tag{11.2}$$

We thus include into the parabolic PDE a nonlinear interaction between two different types of singularities, in which the last one is localized at $x = 0$ but can affect some key properties of the solutions due to the blow-up phenomenon for $u \gg 1$. In the examples below, the minimal solution $u = \lim u_n$ is constructed by monotone approximating sequences $\{u_n \geq 0\}$ of global bounded solutions of the regular uniformly parabolic equations

$$u_t = F_n(x, u, |\nabla u|, \Delta u) \quad \text{in } S,$$

where the monotone operator approximation $F_n \uparrow F$ also includes a suitable truncation of the spatial singularity at $x = 0$. We present some special properties of the extended limit (minimal) semigroup

$$u(x, t) = \mathcal{T}(t)u_0(x) : S \to \bar{I\!R}_+ = [0, \infty]. \tag{11.3}$$

Critical non-autonomous singularity for the PME with source

We begin with the most familiar model, namely, the porous medium equation (PME) with an extra singular term

$$u_t = \Delta u^m + \frac{u^p}{|x|^\alpha}, \quad m > 1, \;\; p > 1, \tag{11.4}$$

where

$$\alpha = 2 \quad \text{is the critical exponent.} \tag{11.5}$$

We take nonnegative continuous compactly supported initial data u_0. This is precisely the case, where the reaction term in the equation

$$q(x,u) = \frac{u^p}{|x|^\alpha}$$

can create blow-up by the following two mechanisms:

(i) because of the superlinear term u^p, where blow-up can occur far away from the origin $x = 0$ (a standard blow-up phenomenon studied before), and

(ii) explosion close to the origin $x = 0$, where the spatial singularity $\frac{1}{|x|^\alpha}$ becomes crucial (a new type of blow-up).

Let us first discuss some easy properties of the possible interaction between the diffusion and the reaction terms in equation (11.4). Assume first that the nonnegative, continuous and bounded initial data u_0 are compactly supported and

$$\mathrm{dist}(0, \mathrm{supp}\, u_0) > 0. \tag{11.6}$$

Then, due to finite propagation driven by the degenerate porous medium operator, a unique bounded continuous weak solution exists locally in time on some finite interval $(0, T)$ and it is compactly supported. Since the support of any nontrivial solution of the PME is known to increase without bound in any direction, we first assume that T corresponds to the situation when $\mathrm{supp}\, u(\cdot, t)$ reaches the origin as $t \to T^-$ and no blow-up happens on $(0, T)$ due to the superlinear source term u^p. Then the proper minimal solution is bounded on $(0, T)$.

What happens to the proper continuation of the weak solution for $t > T$, when the non-autonomous singularity starts to play an important role?

In order to answer this question, we next consider a simpler initial configuration assuming that the initial data are already positive at $x = 0$, i.e., there exists a small radius $\rho > 0$ such that

$$B_\rho = \{|x| < \rho\} \subset \mathrm{supp}\, u_0. \tag{11.7}$$

Then $T = 0$.

Let us prove that, in the critical case $\alpha = 2$, the proper minimal extension is entirely singular, i.e.,

$$u(x,t) \equiv \infty \quad \text{for any } \; t > 0. \tag{11.8}$$

Note that this result is trivial in dimensions $N = 1$ and 2 since $\frac{1}{|x|^2} \notin L^1_{\mathrm{loc}}(I\!R^N)$. This implies the nonexistence in the parabolic problem (see e.g. the proof of Proposition 6.10).

The proper minimal solution is constructed as the limit of a monotone increasing sequence $\{u_\varepsilon\}$ of solutions satisfying the regularized equation that is not singular at $x = 0$. Since, by monotonicity, the proper solution does not depend on the type of monotone increasing approximations, we can use the following regularization:

$$u_t = \Delta u^m + \frac{u^p}{|x|^2 + \varepsilon^2}, \quad \varepsilon > 0, \tag{11.9}$$

with the same initial data. Such an approximation gives a rare opportunity to derive an actual asymptotic rate of divergence of u_ε as $\varepsilon \to 0^+$, i.e., to describe the structure of the *singular initial layer* occurring as $\varepsilon \to 0^+$. In the most of the models studied in Chapters 7–9, we often managed to avoid analysis of approximating sequences and proved results by the actual comparison of different proper solutions.

In order to prove (11.8), we argue by contradiction and first assume that (11.9) admits a bounded solution $u_\varepsilon(x,t)$ in $I\!R^N \times [0,\delta]$ for all small $\varepsilon > 0$. Let us derive step-by-step lower bounds on u_ε for small $\varepsilon > 0$ ensuring (11.8) in the limit $\varepsilon \to 0^+$. We introduce the rescaled variables

$$y = \frac{x}{\varepsilon}, \quad \tau = \frac{t}{\varepsilon^2} \quad \text{and} \quad v_\varepsilon(y,\tau) = u_\varepsilon(\varepsilon y, \varepsilon^2 \tau), \tag{11.10}$$

where v_ε solves the PME with the ε-independent source term

$$v_\tau = \Delta_y v^m + \frac{v^p}{1+|y|^2} \quad \text{with data } v(y,0) = u_0(\varepsilon y). \tag{11.11}$$

Recall that $v_\varepsilon(y,\tau)$ is strictly monotone increasing with ε, which is important for the future asymptotic analysis. We have that

$$v_\varepsilon(y,0) = u_0(\varepsilon y) \to \mu_0 = u_0(0) > 0 \quad \text{as } \varepsilon \to 0^+ \tag{11.12}$$

uniformly on compact subsets, i.e., in the limit $\varepsilon = 0$, the initial function becomes a positive constant. We now argue by contradiction assuming that $\{v_\varepsilon\}$ is a bounded sequence composed of classical solutions of the uniformly parabolic equation (11.11). Then, by the standard parabolic theory, the sequence is a compact subset in C so that we may assume that, as $\varepsilon \to 0$,

$$v_\varepsilon(y,\tau) \to v_0(y,\tau) \quad \text{uniformly on compact subsets,} \tag{11.13}$$

where $v_0(y,\tau)$ denotes the unique solution of (11.11) with constant initial data

$$v_0(y,0) = \mu_0 \quad \text{in } I\!R^N. \tag{11.14}$$

Obviously, $v_0 = v_0(|y|,\tau)$ is radially symmetric and, by the MP, is decreasing in $|y|$. It is also easy to see that, in the case $p \ge m$, $v_0(y,\tau)$ blows up in finite time $T_0 = T_0(\mu_0) > 0$. First of all, by the MP applied to the linear parabolic equation for the derivative $w = v_\tau$ (observe then that $w(y,0) > 0$), it follows that

$$(v_0)_\tau > 0 \quad \text{everywhere} \quad (\text{where } v_0 \text{ exists}). \tag{11.15}$$

Therefore, it v_0 is a global in time solution, then as $\tau \to \infty$,

$$v_0(y,\tau) \to +\infty \quad \text{uniformly on any subset } \{|y| \le c\}. \tag{11.16}$$

If (11.16) were not true, the time-monotonicity (11.15) would mean that $v_0(y, \tau)$ converges as $\tau \to \infty$ to a stationary solution V of (11.11) from below (the standard stability Lyapunov argument applies since (11.15) determines a strong Lyapunov function),

$$v_0(y, \tau) \to V(|y|) \quad \text{as } \tau \to \infty,$$

and moreover, the radial solution $V = V(|y|)$ is large enough and $V \geq \mu_0$ by monotonicity (11.15). An elementary asymptotic analysis of the radial ODE

$$\frac{1}{|y|^{N-1}}(|y|^{N-1}U')' + \frac{U^q}{1+|y|^2} = 0, \quad U = V^m, \ q = \tfrac{p}{m},$$

shows that the assumption on the existence of a finite limit, say, $U(|y|) \to \mu_0^m$ as $|y| \to \infty$, leads to the logarithmic divergence of U, which follows by integrating the ODE

$$(|y|^{N-1}U')' = -\mu_0^p |y|^{N-3}(1 + o(1)) \quad \text{as } |y| \to \infty.$$

Hence, for $N \geq 3$, such a positive stationary solution does not exist since

$$U(|y|) = -\frac{\mu_0^p}{N-2} \log|y|(1 + o(1)) \to -\infty \quad \text{as } |y| \to \infty.$$

If $N = 2$, we have the divergence $U(|y|) \sim -\ln^2|y|$ instead. For $N = 1$, the nonexistence is obvious.

Second, in the case $p \geq m$, for unbounded solutions satisfying (11.16), finite time blow-up is obvious (use any technique based on ordinary differential inequalities or blow-up subsolutions; see references in Remarks and a blow-up subsolution below). Hence, $v_0(y, \tau)$ blows up at some $t = T_0 < \infty$. By Theorem 6.14 in Section 6.9, the monotone with time blow-up is always complete for such equations in the radial geometry, so that we have

$$v_0(y, \tau) \equiv \infty \quad \text{for } \tau > T_0. \tag{11.17}$$

It is clear now that (11.13) actually remains true for $\tau \in (0, T_0)$ and v_ε becomes arbitrarily large for $\tau > T_0$ and hence also blows up in finite time T_ε that is close to T_0 for small $\varepsilon > 0$. Actually, this is enough to conclude that the sequence $\{u_\varepsilon(x, t)\}$ is unbounded and divergent for arbitrarily small $t > 0$. Moreover, by continuous dependence, we can guarantee that the divergence occurs on intervals $t \in [0, T_0\varepsilon^2]$ vanishing as $\varepsilon \to 0$. To prove (11.8) and to know more about the initial layer, we continue our asymptotic analysis.

At the first stage of the analysis, we need to know that (11.16), (11.13) imply the divergence at the origin only: $v_\varepsilon(0, \tau) \to +\infty$ as $\tau \to \infty$. Hence, for any fixed arbitrarily small $t_1 > 0$,

$$u_\varepsilon(0, t_1) \equiv v_\varepsilon(0, \tfrac{t_1}{\varepsilon^2}) \to \infty \quad \text{as } \varepsilon \to 0.$$

We see that the minimal solution $u(x, t)$ satisfies

$$u(0, t_1) = \infty \quad \text{and} \quad u(x, t_1) \to \infty \quad \text{as } x \to 0. \tag{11.18}$$

The second infinite limit in (11.18) follows from (11.16) and the monotonicity of u_ε relative to ε.

Next, we consider equation (11.4) with initial data u_0 already satisfying (11.18). Then performing the same ε-scaling, we arrive at equation (11.11) with initial data satisfying (11.12), where $\mu_0 = \mu_0(\varepsilon)$ can be chosen arbitrarily large as $\varepsilon \to 0$. Obviously, then v_ε must blow-up as $\tau \to T_\varepsilon \ll 1$. For instance, this is easily seen by constructing a natural similarity blow-up subsolution of the form

$$\underline{v}(y,\tau) = (T-\tau)^{-\frac{1}{p-1}}\theta(\xi), \quad \xi = y/(T-\tau)^{\frac{p-m}{2(p-1)}}, \tag{11.19}$$

where

$$\theta(\xi) = A\left(1 - |\xi|^2/a^2\right)_+^{\frac{1}{m-1}}.$$

Given any small blow-up time $T > 0$, one can fix $\mu_0 \gg 1$ and find positive values of A and a such that (11.19) is a subsolution of the equation with initial data $\underline{v}(y,0) \le \mu_0$. Such constructions are effective in quasilinear blow-up reaction-diffusion problems; see [306, p. 215] for the present problem without the singular potential $\frac{1}{|x|^2}$ that obviously improves the blow-up capacity of the equation close to the origin $x = 0$. Thus, according to our construction (notice that here v_ε is currently not the rescaled function in (11.10)), given arbitrarily small $\tau_2 > 0$, there exists $\varepsilon_0 > 0$ such that

$$v_\varepsilon(y,\tau) \equiv \infty \quad \text{for } \tau \ge \tau_2 \text{ and } \varepsilon \in (0,\varepsilon_0). \tag{11.20}$$

Hence, $u_\varepsilon(x,t) \equiv v_\varepsilon(\frac{x}{\varepsilon}, \frac{t}{\varepsilon^2}) \equiv \infty$ for any $t \ge \varepsilon^2\tau_2$ and $\varepsilon \in (0,\varepsilon_0)$. Since $u_\varepsilon \uparrow u$, passing to the limit $\varepsilon \to 0$ yields (11.8).

Returning to the original rescaled variables (11.10), we now describe the initial layer occurring in this approximation as $\varepsilon \to 0^+$. According to (11.13), the layer is given by the radially symmetric blow-up solution of the problem (11.11), (11.14). Formation of stable blow-up patterns in such quasilinear heat equations is well understood, see [306, Chapter 4] and references therein. Let $p > m$ where single-point blow-up occurs. Then the asymptotic behaviour of this blow-up solution is self similar and in the rescaled sense, we have for $t \approx T_0^-$ (cf. (11.19)),

$$v_0(y,\tau) \approx v_*(y,\tau) = (T_0-\tau)^{-\frac{1}{p-1}}\theta(\xi), \quad \xi = y/(T_0-\tau)^{\frac{p-m}{2(p-1)}}, \tag{11.21}$$

where now $\theta > 0$ is a nontrivial, non-constant solution of the radial ODE (such a solution is known to exist for any $m > 1$ and $p > 1$)

$$\Delta\theta^m - \tfrac{p-m}{2(p-1)}\theta'\xi - \tfrac{1}{p-1}\theta + \theta^p = 0.$$

Observe that the reaction term here θ^p is now autonomous since, according to (11.21), blow-up as $t \to T_0^-$ occurs in vanishing neighbourhoods in the variable y (single point blow-up at $y = 0$ only). Using (11.17), we claim that the function

$$V_*(y,\tau) = \begin{cases} v_*(y,\tau), & \tau \le T_0, \\ \infty, & \tau > T_0, \end{cases}$$

describes the formation of the singular initial layer in the Cauchy problem for equation (11.9), i.e., as $\varepsilon \to 0^+$,

$$u_\varepsilon(x,t) \approx v_\varepsilon(\tfrac{x}{\varepsilon}, \tfrac{t}{\varepsilon^2}) \approx V_*(\tfrac{x}{\varepsilon}, \tfrac{t}{\varepsilon^2}).$$

In the case of the regional blow-up $p = m$, the solution $v_0(y,\tau)$ is described as

$t \to T_0^-$ by a simpler separable solution

$$v_*(y,\tau) = (T_0 - \tau)^{-\frac{1}{m}}\theta(y) \quad (\xi = y),$$

where θ solves the ODE

$$\Delta\theta^m - \tfrac{1}{m}\,\theta + \frac{\theta^m}{1+|y|^2} = 0.$$

For $p \in (1, m)$, where (11.16) and hence (11.8) hold, the structure of the initial layer (possibly generated by blow-up or global in time asymptotic patterns of equation (11.11) similar to those discussed below) remains an open problem.

We summarize the results on complete singularity in part (i) of the following statement, in which, as the second half (ii), we include a related local non-solvability of the corresponding stationary equation to be proved next.

Theorem 11.1 *(i) The proper minimal solution of equation* (11.4) *in the critical case* $\alpha = 2$*, with any continuous initial data satisfying* $u_0(0) > 0$ *is*

$$u(x,t) = \mathcal{T}(t)u_0(x) \equiv \infty \quad \mathit{in}\;\; S. \tag{11.22}$$

(ii) The stationary equation

$$\Delta U^m + \frac{U^p}{|x|^2} = 0 \tag{11.23}$$

does not admit a nonnegative solution $U \not\equiv 0$ *in an arbitrarily small deleted neighbourhood of the origin* $B_\rho(0) \setminus \{0\}$.

Proof of (ii). Indeed, if $U(x)$ were such a solution ($U \not\equiv \infty$), then, by a monotone approximation in $B_\rho \times I\!R_+$ of solutions of the parabolic equation, this nontrivial $U(x)$ would give a uniform upper bound on the regularizing time-increasing sequence $\{u_\varepsilon\}$ with sufficiently small initial data. This contradicts (11.22). □

In other words, in the critical case, the *minimal* (stable from below) solution $U(x) \geq 0$ of (11.23) in any neighbourhood of the singular point $x = 0$ is

$$U(x) \equiv \infty.$$

This means that the stationary equation (11.23) is not locally solvable near $x = 0$ in the class of positive functions, and, *vice versa*, this implies the nonexistence result (11.22) in the nonstationary parabolic problem.

Such a local non-solvability of the critical stationary equation is easily seen in the radial geometry where $V = U^m$ solves the equation

$$V_{rr} + \tfrac{N-1}{r}\,V_r + \frac{V^q}{r^2} = 0, \quad q = \tfrac{p}{m} > 1. \tag{11.24}$$

Setting $-\log r = z$, as for linear Euler's equations, leads to the autonomous ODE

$$V'' - (N-2)V' + V^q = 0. \tag{11.25}$$

On the phase-plane $P = V'$ we have

$$\frac{\mathrm{d}P}{\mathrm{d}V} = N - 2 - \frac{V^q}{P}. \tag{11.26}$$

In the half-plane $\{V \geq 0\}$, the origin $(0,0)$ is an unstable node with the 1D orbit bundle

$$P = \frac{1}{N-2}V^q + \frac{q}{(N-2)^3}V^{2q-1} + ... + C_0 e^{-\beta V^{1-q}} + ...,$$

where $\beta = \frac{(N-2)^2}{q-1} > 0$ and $C_0 \in I\!R$ is an arbitrary parameter in the exponential bundle. One can see from the phase-plane that any orbit of (11.26) starting at the point $V = 0$, $P = P_0 > 0$ ($P > 0$ means $V_r < 0$) intersects the P-axis again at some $P_1 < 0$ and hence the solution $V(r)$ must change sign (see more comments below).

On oscillatory solutions of changing sign

The above analysis of stationary solutions suggests posing a question on the existence of solutions of changing sign that are oscillatory at $x = 0$.

The semilinear heat equation. Without loss of generality we discuss this extension for the semilinear equation

$$u_t = \Delta u + \frac{|u|^{p-1}u}{|x|^2} \quad \text{in} \quad S = I\!R^N \times I\!R_+ \tag{11.27}$$

with bounded continuous initial data $u_0 \not\equiv 0$.

We say that a continuous function $w : I\!R^N \setminus \{0\} \to I\!R$ is *oscillatory* at $x = 0$ if $w(x)$ changes sign in any arbitrarily small open deleted neighbourhood of the origin. The following conclusion explains the existence features for such singular nonlinear models.

Proposition 11.2 *Let $u \in C((I\!R^N \setminus \{0\}) \times (0,T))$ be a nontrivial solution of equation (11.27). Then, for any $t_0 \in (0,T)$, $u(x,t_0)$ is oscillatory at $x = 0$.*

Proof. Obviously, if $u(x,t_0)$ is not oscillatory at $x = 0$, and hence we may assume that $u(x,t_0) \geq 0$ ($\not\equiv 0$) in $B_\rho \setminus \{0\}$ for a sufficiently small $\rho > 0$, then, by the strong MP, $u(x, t_0 + \delta) > 0$ in $B_{\rho/2} \setminus \{0\}$ for arbitrarily small $\delta > 0$. Hence, by Theorem 11.1 (i), $u(x, t_0 + 2\delta) = +\infty$ in $B_{\rho/4} \setminus \{0\}$ contradicting the imposed continuity assumption. □

It is easy to show that global solutions that are oscillatory at $x = 0$ do exist. For instance, these are stationary solutions given by the ODE (11.25), which now has the form

$$V'' - (N-2)V' + |V|^{q-1}V = 0 \quad \text{for} \quad z = -\log r \gg 1. \tag{11.28}$$

In particular, any periodic solution V of (11.28) (a limit cycle on the $\{V, V'\}$-plane) shows the following oscillatory behaviour of stationary solutions:

$$|U|^{m-1}U(x) = V(-\log|x|) \quad \text{as} \quad x \to 0.$$

These solutions are uniformly bounded near the origin; see Figure 11.1.

As for non-stationary oscillating solutions, we introduce the self-similar ones

$$u_*(x,t) = g(\eta), \quad \eta = x/\sqrt{t},$$

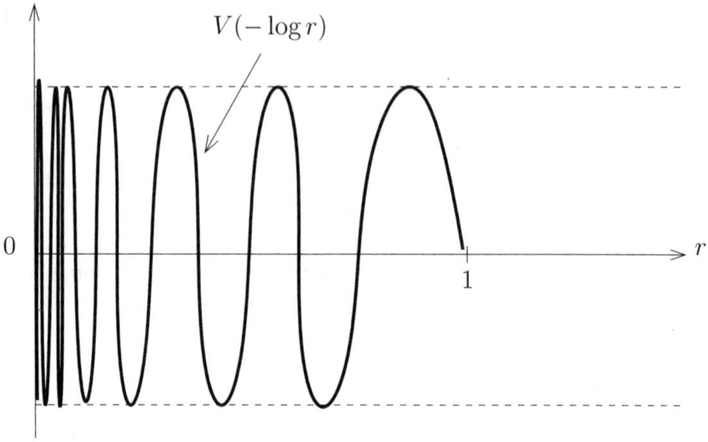

Figure 11.1 *Such oscillatory solutions of* (11.27) *do exist.*

where g is a solutions of the elliptic equation

$$\Delta g + \tfrac{1}{2}\eta \cdot \nabla g + \frac{|g|^{p-1}g}{|\eta|^2} = 0.$$

In the radial geometry, this ODE can be studied and admits oscillatory solutions at $\eta = 0$. The zeros of the oscillatory solution $u_*(x,t)$ then expand with time as $\sqrt{t}$. On the other hand, one can take blow-up solutions with the similarity variable

$$\eta = x/\sqrt{T-t}$$

(then the second term in the ODE changes sign) describing blow-up focusing of zeros to the origin $x = 0$ as $t \to T^-$ like $\sqrt{T-t}$.

Nevertheless, one can expect that the existence of nontrivial oscillatory solutions is not a generic phenomenon for equation (11.27). The stable generic behaviour is *instantaneous blow-up* of solutions, where the limit semigroup is discontinuous and $\mathcal{T}(0^+)u_0 = +\infty$ or $-\infty$ depending on the initial data. In this case oscillatory solutions form an unstable border behaviour between those two generic classes.

The porous medium equation with source. A similar result can be formulated for the equation with the sign porous medium operator

$$u_t = \Delta |u|^{m-1}u + \frac{|u|^{p-1}u}{|x|^2} \quad \text{in} \quad S = I\!R^N \times I\!R_+. \tag{11.29}$$

The only difference is due to the finite propagation property. Indeed, the condition

$$u(x, t_0) > 0 \quad \text{in} \quad B_\rho \setminus \{0\} \tag{11.30}$$

in the proof does not guarantee that $u \equiv +\infty$ instantaneously, since a waiting time phenomenon can occur. Then it takes some time to get $u(x,t_1) > 0$ in $B_{\rho/2} \setminus \{0\}$. This is equivalent to the assumption (11.7). Existence (or nonexistence) of this singular waiting time effect at the origin and its sharp characterization are not easy questions, where delicate similarity and approximate similarity solutions can play an important role.

On the other hand, assuming in addition to (11.30) that

$$\liminf_{x \to 0} u(x, t_0) > 0$$

guarantees the complete blow-up, i.e., $\mathcal{T}(t_0^+)u_0 \equiv +\infty$. Note that under assumption (11.6), the solution can be bounded and continuous on intervals $t \in (0, t_0)$, where $t = t_0$ becomes the instant of the blow-up discontinuity.

Supercritical and subcritical ranges

Let us return to the nonnegative solutions of the original model (11.4). For completeness of our nonexistence analysis, consider other cases with non-critical exponents $\alpha \neq 2$. Performing a similar construction of the monotone approximation $\{u_\varepsilon\}$ of the regularized equation

$$u_t = \Delta u^m + \frac{u^p}{|x|^\alpha + \varepsilon^\alpha},$$

by rescaling (11.10), we obtain

$$v_\tau = \Delta_y v^m + \varepsilon^{2-\alpha} \frac{v^p}{1 + |y|^\alpha}.$$

We observe an extra factor $\varepsilon^{2-\alpha}$ in the reaction term that was equal to 1 in the critical case $\alpha = 2$.

Then, obviously, in the *supercritical* range $\alpha > 2$, where $\varepsilon^{2-\alpha} \to \infty$ as $\varepsilon \to 0$, any nontrivial solution v_ε blows up in finite time and completely. This ensures the existence of the trivial minimal solutions (11.22).

On the other hand, in the *subcritical* range $\alpha < 2$, the approximation v_ε is globally bounded for sufficiently small initial data, and this gives bounded minimal solutions for small $t > 0$. Furthermore, it is easy to see that, for $\alpha < 2$, the radial stationary equation

$$\Delta_r U + \frac{U^q}{|x|^\alpha} = 0$$

is locally solvable near the origin and

$$U(r) = 1 - \tfrac{1}{(2-\alpha)(N-\alpha)}\, r^{2-\alpha} + \dots \quad \text{as} \quad r \to 0.$$

These local in r solutions can be used as uniform bounds on the sequence $\{u_\varepsilon\}$ approximating the maximal solutions $u(x,t)$ of equation (11.4) in a sufficiently small domain Ω, $0 \in \Omega$, with the Dirichlet boundary conditions $u = 0$ on $\partial\Omega$. In this case we obtain that $u = \lim u_\varepsilon$ is uniformly bounded in S.

Examples of incomplete critical singularity

The assumption $m > 1$ in the porous medium operator (or $m = 1$ in the linear heat operator) is essential for completeness of the critical singularity at $x = 0$ as the following example shows. Consider a similar equation with the fast diffusion operator

$$u_t = \Delta u^m + \frac{u^{2-m}}{|x|^2} \quad \text{in } S, \quad m \in (0,1). \tag{11.31}$$

The reaction term u^{2-m} is known to correspond to the critical case of incomplete blow-up (Chapter 5). Therefore, given any compactly supported data $u_0 \geq 0$ or with sufficiently fast decay as $x \to \infty$, one can see that any kind of blow-up generated by the superlinear source u^{2-m} or by the inverse square potential $\frac{1}{|x|^2}$ is not complete. Obviously, this spatial singularity at $x = 0$ does not affect finite propagation of blow-up interfaces when they stay away from the origin.

The incompleteness of blow-up is easily proved by the local comparison with singular TWs of the equation without the non-autonomous singularity

$$u_t = \Delta u^m + u^{2-m}.$$

See Section 6.4. Indeed, such TWs are blow-up supersolutions of (11.31) in $\{|x| \geq 1\}$ and this guarantees the following result.

Proposition 11.3 *Let* $m \in (0,1)$. *For any compactly supported data* u_0, (11.31) *admits a nontrivial minimal solution* $u(x,t) \not\equiv \infty$ *for any* $t > 0$.

This shows that the criterion of incomplete blow-up for the general equation with the critical spatial singularity

$$u_t = \Delta\varphi(u) + \frac{\psi(u)}{|x|^2} \tag{11.32}$$

is the same as for the one-dimensional equation without the non-autonomous singular multiplier $\frac{1}{|x|^2}$. Therefore, Theorem 4.8 guarantees the global existence of nontrivial solutions of equation (11.32).

Other examples of critical complete and incomplete blow-up

Similar results are true for other nonlinear models. For instance, consider the equation with the p-Laplacian operator

$$u_t = \nabla \cdot (|\nabla u|^\sigma \nabla u) + \frac{u^p}{|x|^\alpha}, \tag{11.33}$$

where $\sigma > 0$, $N > 2 + \sigma$ and $p \geq 1 + \sigma$. Then

$$\alpha = \sigma + 2 \quad \text{is the critical exponent}$$

and Theorem 11.1 is valid (the range $p \in (1, m)$ included) and we can specify a similar initial ε-layer for $p \geq 1 + \sigma$. For $\alpha < 2 + \sigma$ there exist nontrivial local solutions.

On the other hand, if $\sigma \in (-1, 0)$ and $p = \frac{1}{1+\sigma} > 1$, even for $\alpha = 2+\sigma$, (11.33) admits nontrivial minimal solutions $u \not\equiv \infty$ for all $t > 0$. In the autonomous case these values of parameters create the limit case of incomplete blow-up, and the proof for (11.33) is based on the direct comparison in $\{|x| \geq 1\}$ with singular blow-up TW supersolutions from Section 6.10.

For the singularly perturbed dual PME with source

$$u_t = |\Delta u|^{m-1}\Delta u + \frac{u^p}{|x|^\alpha}, \tag{11.34}$$

where $m > 1$, $N > 2$ and $p \ge m$, the critical exponent is $\alpha = 2m$ and (11.22) holds, as well as other existence ($\alpha < 2m$) and nonexistence ($\alpha > 2m$) properties. In the critical case

$$p = p_c = \tfrac{m}{2m-1} > 1, \quad \text{where} \;\; m \in (\tfrac{1}{2}, 1),$$

the nonlinear coefficients in (11.34) are such that the autonomous equation

$$u_t = |\Delta u|^{m-1}\Delta u + u^{\frac{m}{2m-1}}$$

admits explicit singular plane TWs $U = f(\xi)$, $\xi = x_1 - \lambda t$, where

$$f(\xi) = A_0(\lambda)(-\xi)_+^{-\frac{2m-1}{1-m}}.$$

Therefore, by comparison in $\{|x| \ge 1\}$, blow-up for equation (11.34) in this parameter range is incomplete, the blow-up set is bounded (or empty) for all $t > 0$ and expands not faster than linearly with time as $t \to \infty$.

Without essential changes, the oscillatory results on solutions of changing sign near $x = 0$ can be extended to these new models.

The scaling approximation analysis can be applied to more general equations such as

$$u_t = F(|x|, u, |\nabla u|, \Delta u),$$

where the right-hand side is singular at $x = 0$ and admits blow-up due to the superlinear growth of $F(\cdot, u, 0, 0)$ as $u \to \infty$. A naturally truncated equation is then

$$u_t = F_\varepsilon((|x|^2 + \varepsilon^2)^{1/2}, ...).$$

This suggests the same rescaled spatial variable $x = \varepsilon y$ initiating estimates of the approximating sequence of regular solutions $\{u_\varepsilon\}$ and establishing completeness or incompleteness of the initial singularity.

On local non-solvability of critical stationary equations

All above problems with critical exponents α exhibit local non-solvability in an arbitrarily small neighbourhood of the origin in the class of non-oscillating functions. Let us show that for the stationary elliptic equations, this can be done by an evolution argument, i.e., without using an ODE reduction. We consider equation (11.33) as a typical example.

Proposition 11.4 *For an arbitrarily small $\rho > 0$, the equation*

$$\nabla \cdot (|\nabla U|^\sigma \nabla U) + \frac{U^p}{|x|^{(2+\sigma)}} = 0 \quad in \;\; B_\rho \setminus \{0\}, \tag{11.35}$$

with $\sigma > 0$, $N > 2 + \sigma$, $p > 1 + \sigma$ does not admit a solution $U > 0$.

Proof. Assume that there exists a solution $\bar U > 0$, which is a weak one on any subset where it is bounded. Then, since $u \equiv 0$ is unstable in the parabolic problem, we consider equation (11.33) in $B_\rho \times I\!R_+$, $u = 0$ on $\partial B_\rho \times I\!R_+$ with sufficiently small radial initial data $u_0(r) > 0$ in B_ρ such that $u_t(r,t) \ge 0$ for $t > 0$. Then $u(r,t) \le \bar{U}(x)$ for $t > 0$ by approximation and comparison. By the time monotonicity, $u(r,t)$ must converge to a stationary radial profile U satisfying the stationary radial equation in domain of boundedness

$$\frac{1}{r^{N-1}}(r^{N-1}|U'|^\sigma U')' + \frac{1}{r^{2+\sigma}}\, U^p = 0, \quad r \in (0,\rho). \tag{11.36}$$

The case when $U(r)$ is uniformly bounded as $r \to 0$ is easier. Let it be unbounded near $r = 0$. Then there exists a sequence $\{r_k\} \to 0$ such that $U'(r_k) \le 0$. Integrating (11.36) over (r_k, r), we obtain

$$r^{N-1}|U'|^\sigma U' = r_k^{N-1}|U'|^\sigma U' - \int_{r_k}^{r} s^{N-(\sigma+3)} U^p \mathrm{d}s \le -U^p \int_{r_k}^{r} s^{N-(\sigma+3)} \mathrm{d}s,$$

since $U(s) \ge U(r)$ for $s \in (r_k, r)$. Passing to the limit $r_k \to 0$ yields

$$(-U')^\sigma U' \le -U^p \tfrac{1}{N-(2+\sigma)} \tfrac{1}{r^{1+\sigma}} \quad \text{for } r \in (0,\rho),$$

and hence $U' \le -cU^q \frac{1}{r}$ for $r \in (0,\rho)$, where $q = \frac{p}{1+\sigma} > 1$. The second integration over (r,ρ) of the inequality $U^{-q}U' \le -\frac{c}{r}$ gives the logarithmic divergence as $\rho \to 0$ in the right-hand side,

$$U(r) \ge [U^{1-q}(\rho) - c_1 \log(\tfrac{\rho}{r})]^{-\frac{1}{q-1}}.$$

Therefore, $U(r)$ is not defined on $(0,\rho)$ and blows up at a finite $r_\rho > 0$. □

Similar to Proposition 11.2, the ODE (11.35) describes oscillatory stationary solutions as nontrivial global solutions of the corresponding singular parabolic equation

$$u_t = \nabla \cdot (|\nabla u|^\sigma \nabla u) + \frac{|u|^{p-1}u}{|x|^{2+\sigma}}.$$

11.2 When do singular interfaces not move?

For singular parabolic PDEs, the problem of the optimal regularity of solutions and moving interfaces in $I\!R^N$ is known to be very difficult. Indeed, optimal estimates are often supposed to include a detailed study of possible types of different singularity formation phenomena (types of self-focusing formation of "multiple zeros" in $I\!R^N$), which are still unknown for most multi-dimensional nonlinear parabolic PDEs with rather arbitrary nonlinearities.

We consider the opposite aspect of the problem. Obviously, these difficult regularity problems disappear if singular interfaces do not move at all. We will show that the question of non-moving interfaces for nonnegative proper maximal solutions is covered by typical techniques of the geometric theory and will perform such a geometric analysis, in both $I\!R$ and $I\!R^N$.

One-dimensional problems

We begin with a simple 1D example, which can be classified using the previous results.

Example 11.1 Consider nonnegative weak solutions of the filtration equation written down in the equivalent form

$$u_t = \varphi(u)u_{xx} \quad \text{in } S = I\!R \times I\!R_+ \tag{11.37}$$

with bounded initial data

$$u_0(x) > 0 \ \text{ for } x > 0 \quad \text{and} \quad u_0(x) \equiv 0 \ \text{ for } x < 0.$$

Hence, the initial position of the interface is $s(0) = 0$. We assume that the coefficient $\varphi(u)$ is continuous for $u \geq 0$ and is sufficiently smooth and positive for $u > 0$. The maximal solution $u \geq 0$ is to be constructed by the regularization of initial data by taking $u_{0n} = u_0 + \frac{1}{n}$.

Consider the TWs $V(x,t) = f(\xi)$, where $f = f_\lambda$ satisfies

$$\varphi(f)f'' + \lambda f' = 0.$$

Hence, setting $P = f'$ and integrating yields

$$P(f) = P_0 - \lambda \int_1^f \frac{\mathrm{d}z}{\varphi(z)}, \quad \text{where } P_0 = P(1).$$

It follows that, if

$$\int_0^1 \frac{\mathrm{d}z}{\varphi(z)} = \infty, \tag{11.38}$$

then $\Delta = \emptyset$. Moreover, any $\lambda > 0$ corresponds to the flat limit of approximation, and so there exists a sequence $\{f_\lambda^n(\xi) > 0, \xi > 0\}$ of local "flat" TW profiles satisfying

$$f_\lambda^n(0) = 0, \quad f_\lambda^n(\xi) \to (\tfrac{2}{n})^- \quad \text{as } \xi \to \infty, \tag{11.39}$$

so that $f_\lambda^n(\xi) \to 0$ as $n \to \infty$. For $\lambda = 0$ we obtain a family of stationary solutions $f_0^n(\xi) = \frac{1}{n}\xi$ for $\xi > 0$. For any $\lambda < 0$, the approximation exhibits the steep limit and there exists $\{\bar{f}_\lambda^n > \frac{2}{n}\}$ such that, on any level $\delta > 0$, there holds

$$\frac{\mathrm{d}}{\mathrm{d}\xi}\bar{f}_\lambda^n \Big|_{\bar{f}_\lambda^n = \delta} \to \infty \quad \text{as } n \to \infty. \tag{11.40}$$

Using small flat solutions (11.39) with any $\lambda \approx 0^+$ as bounds of positive regular solutions u_n from below, and the steep ones (11.40) with any $\lambda \approx 0^-$ for similar comparison from above, as on Figure 11.2, by passing to the limit $n \to \infty$, we conclude that the interface $s(t)$ of maximal solutions does not move and

$$s(t) \equiv 0 \quad \text{for all } t > 0.$$

If on the contrary, the integral in (11.38) converges,

$$\int_0^1 \frac{\mathrm{d}z}{\varphi(z)} < \infty,$$

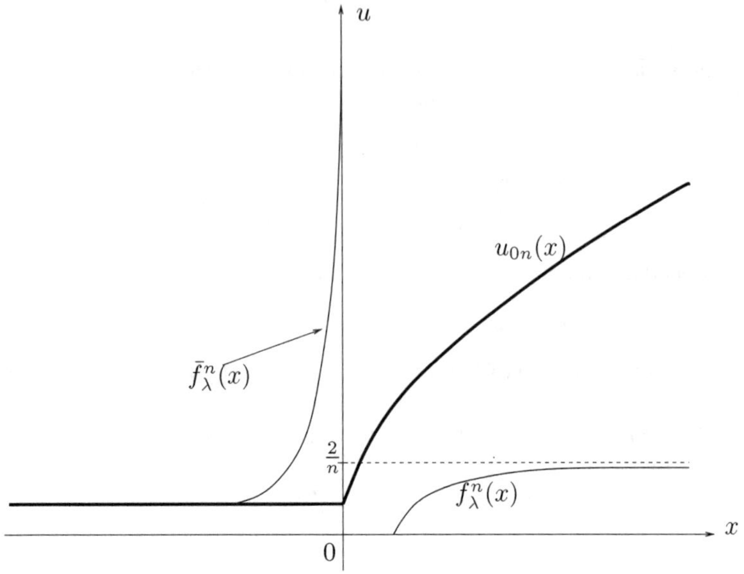

Figure 11.2 *Flat TW* f_λ^n *with* $\lambda \approx 0^+$ *proves by* $n \to \infty$ *that* $s(t) \le \lambda t$*, while the steep one* $\bar f_\lambda^n$ *with* $\lambda \approx 0^-$ *establishes that* $s(t) \ge \lambda t$*. Setting* $\lambda \to 0^\pm$ *yields* $s(t) \equiv 0$.

then there exist maximal TWs and

$$\int_0^1 \left(\int_0^s \frac{\mathrm{d}z}{\varphi(z)} \right)^{-1} \mathrm{d}s < \infty \implies \Delta_0 = (-\infty, 0),$$

$$\int_0^1 \left(\int_0^s \frac{\mathrm{d}z}{\varphi(z)} \right)^{-1} \mathrm{d}s = \infty \implies \Delta = (-\infty, 0), \;\; \Delta_0 = \emptyset.$$

In the first case we have moving interfaces, while in the second one the propagation is infinite. Both cases are studied as in Chapter 7.

The criterion (11.38) on stationary interfaces of nonnegative maximal solutions can be re-formulated for the general parabolic equation

$$u_t = F(u, u_x, u_{xx}) \quad \text{in } S, \tag{11.41}$$

under usual assumptions of F and for the same initial data u_0 as above.

Proposition 11.5 *Let the ODE associated to the PDE* (11.41),

$$F\left(f, P, P\frac{\mathrm{d}P}{\mathrm{d}f}\right) + \lambda P = 0, \;\; f > 0, \;\; P \ge 0,$$

is such that any $\lambda \approx 0^+$ *is in the flat limit of positive approximation and any* $\lambda \approx 0^-$ *is in the steep one. Then the left-hand interface of the maximal solution does not move.*

Non-moving singular interfaces in $I\!R^N$

The criterion on non-moving singular interfaces in 1D is based on the usual comparison with approximating regular TWs and no advanced intersection techniques are involved. We then need some slight modification to extend the result to interfaces in $I\!R^N$. In order to explain the geometric analysis in $I\!R^N$, we begin with the N-dimensional version of the filtration equation written in the same form

$$u_t = \varphi(u)\Delta u \quad \text{in } S = I\!R^N \times I\!R_+, \tag{11.42}$$

where φ satisfies (11.38). We need to construct two local bounds on the interface velocity, from below and from above. We will use the radial form of the filtration equation

$$u_t = \varphi(u)\left(u_{rr} + \tfrac{N-1}{r}\, u_r\right). \tag{11.43}$$

Note that using different geometric shapes of TW sub and supersolutions, in both cases, we need to perform the constructions for negative speeds $\lambda \approx 0^-$, which will be explained in the proof of stationary interfaces.

Estimate from below: a TW subsolution. Fix a small $\rho > 0$ and let $A_\rho = \{\rho < r = |x| < 2\rho\}$ be the annulus. Consider the following equation:

$$V_t = \varphi(V)\left(V_{rr} + \tfrac{N-1}{\rho}\, V_r\right), \tag{11.44}$$

and let

$$V(r,t) = f(\xi), \quad \xi = r - \lambda t$$

be its local solution with monotone decreasing profile $f(\xi) > 0$ on $(\rho, 2\rho)$, $f(2\rho) = 0$ and arbitrary $\lambda \approx 0^-$. Since $V_r < 0$, one can see comparing (11.44) and (11.43) that V is a subsolution. It is clear that any $\lambda < 0$ is in the flat limit of approximation in the class of decreasing solutions $P = f' < 0$ for the ODE corresponding to (11.44),

$$\varphi(f)\left(f'' + \tfrac{N-1}{\rho}\, f'\right) + \lambda f' = 0.$$

Setting $P = f'$ yields

$$\frac{\mathrm{d}P}{\mathrm{d}f} = -\frac{\lambda}{\varphi(f)} - \tfrac{N-1}{\rho}, \quad f > 0, \;\; P < 0,$$

where we observe the extra constant $-\frac{N-1}{\rho}$ on the right-hand side, which is not essential for $0 < f \ll 1$.

Estimate from above: a TW supersolution. We now construct a steep TW solution of (11.44) in A_ρ,

$$\bar V(r,t) = f(\xi), \quad \xi = r - \lambda t, \tag{11.45}$$

where $f(\xi)$ is monotone increasing for $\xi \geq \rho$, $f(\rho) = \frac{2}{n}$, $f'(\rho) = 0$ (i.e., $f = f_\lambda^n$), so that $\bar V_r > 0$ and (11.45) is a supersolution of (11.43) in this domain. In a similar way we show that any $\lambda \approx 0^-$ is in the steep limit in the class of increasing orbits $P = f' > 0$.

This construction is easily extended to general parabolic equations

$$u_t = F(u, |\nabla u|, \Delta u) \quad \text{in } I\!R^N \times I\!R_+. \tag{11.46}$$

We impose the same assumptions as for the filtration equation:

(i) For an arbitrarily small $\rho > 0$, the ODE

$$F\left(f, P, P\frac{\mathrm{d}P}{\mathrm{d}f} + \frac{N-1}{\rho}P\right) + \lambda P = 0 \tag{11.47}$$

exhibits the flat limit of approximation for any $\lambda \approx 0^-$ in the class of negative orbits $P = f' < 0$.

(ii) For a small $\rho > 0$, (11.47) exhibits a steep limit of approximation for any $\lambda \approx 0^-$ in the class of positive orbits $P = f' > 0$.

Proposition 11.6 *Let (i) and (ii) above hold and $u(x,t)$ be a nontrivial maximal solution of* (11.46) *with smooth compactly supported initial data. Assume that* $\operatorname{supp} u_0$ *has a smooth boundary. Then, for any $t > 0$,*

$$\operatorname{supp} u(\cdot, t) = \operatorname{supp} u_0.$$

Proof. Take an interior point $x_0 \in \operatorname{supp} u_0$ that is arbitrarily close to its boundary and fix $\rho > 0$ so small that $B_{3\rho}(x_0) \subset \operatorname{supp} u_0$. We next locally compare $u(x,t)$ (or $u_n(x,t)$ if necessary) from below with arbitrarily small TW subsolution $V(r,t)$ in A_ρ, $\lambda \approx 0^-$, to obtain in the limit that $A_\rho(x_0) \supset \operatorname{supp} u(\cdot,t)$ for any $t > 0$.

Similarly, we take x_1 as any interior point of $I\!R^N \setminus \operatorname{supp} u_0$, which can be arbitrarily close to $\partial \operatorname{supp} u_0$. We choose $A_\rho(x_1)$, which does not intersect the free boundary of u_0. We next compare from above the solution $u(x,t)$ and the steep supersolution $\bar V(r,t)$ in A_ρ, $\lambda \approx 0^-$. Unlike the comparison from below that is essentially local in space, we need to take care over the comparison assumption on $u_n(x,t)$ at the inner boundary at $|x - x_1| = \rho$. We then assume that $\{u_n\}$ is a monotone sequence of smooth positive solutions of the initial-boundary value problems in a smooth domain $\Omega_\rho \times I\!R_+$, where $\operatorname{supp} u_0 \subset \Omega_\rho$ with

$$\operatorname{dist}(\partial\Omega_\rho, \operatorname{supp} u_0\} \le \rho$$

and $u_n = \frac{1}{n}$ on $\partial\Omega_\rho$ for $t > 0$. Then, since by continuity $\frac{1}{n} < f(\rho) = \frac{2}{n}$, we guarantee the necessary comparison $u_n \le V$ on $\partial\Omega_\rho$, $t > 0$, and the result follows. □

In the most of the cases, the extra lower-order term $\frac{N-1}{\rho}P$ in the ODE (11.47) does not change the flat and steep limit of approximation for $\lambda \approx 0^-$ of the purely one-dimensional ODE

$$F\left(f, P, P\frac{\mathrm{d}P}{\mathrm{d}f}\right) + \lambda P = 0.$$

If this is true and interfaces do not move in the 1D geometry, we claim that bounded smooth supports of proper solutions in $I\!R^N$ remain stationary as well.

Remarks and comments on the literature

§ **11.1.** The main results are inspired by the detailed nonexistence analysis for the singular semilinear parabolic equation

$$u_t = \Delta u + \frac{u^p}{|x|^2}, \quad p > 1, \tag{11.48}$$

performed by H. Brezis and X. Cabré in [65]. As a distinguishing feature, we consider only the models with quasilinear or even fully nonlinear diffusion operators, where we cannot use the advanced techniques associated with the linear semigroup of the heat operator in (11.48), with which more detailed results can be achieved.

The nonexistence and existence results for such nonlinear parabolic equations have a counterpart in the theory of linear parabolic equations with singular potentials. The remarkable result by P. Baras and J.A. Goldstein (1984) [38] establishes that the linear parabolic equation

$$u_t = \Delta u + \frac{a}{|x|^2} u$$

is well posed in $L^2(I\!R^N)$ for any $a \leq \frac{(N-2)^2}{4}$ and it is not well posed for $a > \frac{(N-2)^2}{4}$. This is associated with the Hardy inequality for the operator $H_a = -\Delta - \frac{a}{|x|^2}$ that is nonnegative in $L^2(I\!R^N)$ iff $a \leq \frac{(N-2)^2}{4}$. Therefore, by using a regular approximation of the equation, it is proved that the only nonnegative solutions in the ill posed case $a > \frac{(N-2)^2}{4}$ are the trivial ones $u \equiv 0$. On the other hand, it is easy to construct nontrivial solutions oscillatory at $x = 0$. For details, about more recent extensions see [180].

The nonexistence results similar to that in Theorem 11.1 (i) have been extended to various quasilinear equations and inequalities with divergent differential operators of different types by a general "nonlinear capacity" approach explained in the book by E. Mitidieri and S.I. Pohozaev [267], where extended related blow-up literature is discussed. The classes of equations include the quasilinear parabolic equations with a degenerate diffusion operator such as [267, Part II]

$$u_t = |x|^\gamma \Delta u^m + \frac{u^p}{|x|^\alpha}.$$

The extensions to higher-order equations where the Laplacian Δ is replaced by $-(-\Delta)^l$ with an integer $l \geq 1$ are also available. The nonexistence approach applies to the semilinear and quasilinear *hyperbolic* equations [267, Part III], [291] such as

$$u_{tt} = \Delta u + \frac{|u|^p}{|x|^\alpha}$$

and to their higher-order counterparts. The existence problem (continuous limit semigroups in other parameter ranges) and the structure of the initial singular layer as $\varepsilon \to 0$ are not treated by those approaches and often represent interesting and difficult problems.

§ **11.2.** Various delicate properties including nonuniqueness and discontinuity of viscosity maximal solutions (and some others not mentioned here) of the non-divergence equation

$$u_t = u\Delta u - \gamma|\nabla u|^2$$

with a parameter $\gamma \in I\!R$, were studied in detail in [334], [95], [52] and [53] .

On limit minimal semigroups for singular initial data

Below we present brief final comments on discontinuous semigroups for some well-known parabolic models. As we have seen, the discontinuity of a proper limit semigroup at the initial moment $t = 0$ means that the Cauchy or the initial-boundary value problem does not admit a solution in the usual sense. In the theory of evolution equations this case is treated as the nonexistence of solutions.

It is curious that this is precisely the case of the classical area of the theory of PDEs such as existence and uniqueness of solutions in classes of unbounded (increasing) functions. This direction of the theory has a remarkable history in the twentieth century.

The heat equation. For the heat equation

$$u_t = \Delta u \quad \text{in } \mathbb{R}^N \times \mathbb{R}_+, \quad u(x,0) = u_0(x) \quad \text{in } \mathbb{R}^N, \tag{11.49}$$

the first uniqueness theorems in classes of solutions unbounded as $|x| \to \infty$ were proved by E.E. Levi [241], E. Holmgren [194], A.N. Tikhonov [333] and S. Täklind [331]. Tikhonov's uniqueness class has the form

$$\{|u(x,t)| \le Ae^{a|x|^2}\} \quad \text{with constants } A,\, a > 0. \tag{11.50}$$

The optimal Täklind class of uniqueness

$$\{|u(x,t)| \le e^{|x|h(|x|)}\},$$

is characterized by a positive increasing function $h(s)$, satisfying Osgood's criterion

$$\int^{\infty} \frac{ds}{h(s)} = \infty.$$

We refer to results and the literature in [280], where a detailed analysis of uniqueness classes is performed for second and higher-order parabolic equations and system by energy estimates based on Saint-Venant's principle.

In the uniqueness class, the solution of the Cauchy problem (11.49) is given by Poisson's formula for the semigroup $\{e^{\Delta t}\}$

$$u(x,t) = \mathcal{T}(t)u_0(x) \equiv (4\pi t)^{-N/2} \int_{\mathbb{R}^N} e^{-|x-y|^2/4t} u_0(y)\, dy. \tag{11.51}$$

Therefore, one can check the continuity of the limit (minimal) semigroup $\mathcal{T}(t)$ at $t = 0$. Convolution (11.51) gives a sharp estimate on the initial data, for which the Cauchy problem has a global in time solution and the semigroup is continuous. A key role is then played by the explicit solution (easily obtained from the fundamental one)

$$u_*(x,t) = (T-t)^{-N/2} e^{|x|^2/4(T-t)} \quad \text{with a constant } T > 0.$$

The corresponding initial data $u_{0*}(x) = T^{-N/2} e^{|x|^2/4T}$ has critical exponential growth as $|x| \to \infty$, the solution is local in time and blows up, $u_*(x,t) \to \infty$ as $t \to T^-$, uniformly in $\mathbb{R}^N$. It follows from (11.51) that, for initial functions with faster growth such as

$$u_0(x) \ge e^{|x|^{2+\varepsilon}} \quad \text{as } |x| \to \infty$$

with an arbitrarily small exponent small $\varepsilon > 0$, the semigroup becomes discontinuous at $t = 0$ in the sense that

$$\mathcal{T}(0^+)u_0(x) \equiv +\infty \quad \text{in } \mathbb{R}^N. \tag{11.52}$$

The porous medium equation. For the PME

$$u_t = \Delta u^m, \quad m > 1 \quad (u \ge 0),$$

a similar class of global solvability and continuity of the corresponding semigroup is described by the separate variable solution

$$u_*(x,t) = c_0(T-t)^{-1/(m-1)}|x|^{2/(m-1)},$$

where $c_0^{1-m} = 2m\,\frac{N(m-1)+2}{m-1}$, blowing up as $t \to T^-$ globally in $\mathbb{R}^N$. It corresponds to initial data with the critical growth

$$u_0(x) \sim |x|^{-2/(m-1)} \quad \text{as } |x| \to \infty.$$

If the growth is faster, we observe discontinuity (11.52) of the semigroup. Concerning existence and uniqueness classes for the PME from the 1950s, we refer to detailed Kalashnikov's survey [213] and general results in [46].

Similar to the Tikhonov-Täcklind classes, for the one-dimensional sign PME

$$u_t = (|u|^{m-1}u)_{xx},$$

the uniqueness class [314]

$$\{|u|^{m-1} \le C(1+|x|^2)h(|x|)\}$$

includes a positive function $h(s)$ satisfying

$$\int^{\infty} \frac{\mathrm{d}s}{h(e^s)} = \infty.$$

When a nonlinear parabolic equation contains two or more different nonlinear operators having different scaling properties, the discontinuity classes of limit (minimal) semigroups can be affected by several operators and the continuity problem becomes more delicate. We briefly describe such an example.

The PME with strong absorption. Consider the Cauchy problem for the 1D PME with absorption

$$u_t = (u^m)_{xx} - u^p \quad \text{in } S = \mathbb{R} \times \mathbb{R}_+ \tag{11.53}$$

in the parameter range $1 < p < m$ with unbounded nonnegative continuous initial data $u_0(x)$. It is convenient to state the results in the current variable u without reducing the singularity level $\{u = \infty\}$ to the standard zero one $\{v = 0\}$ by setting $u = \frac{1}{1+v}$. Equation (11.53) admits the stationary solution

$$U_1(x) = c_0|x|^{\gamma}, \quad \gamma = \tfrac{2}{m-p},$$

where $c_0 = \left[\frac{(m-p)^2}{2m(m+p)}\right]^{\frac{\gamma}{2}}$, showing the optimal growth of the initial data as $x \to \infty$ for which the Cauchy problem makes sense. The definition of a unique proper minimal solution is standard and is performed by monotone increasing truncation of the data.

Discontinuous limit semigroup. The limit semigroup $\mathcal{T}(t)$ can be discontinuous in the class of initial data [263], [214]

$$u_0(x) = \begin{cases} cx^\gamma & \text{for } x > 0, \\ 0 & \text{for } x \le 0, \end{cases}$$

where $c > 0$ is a parameter. Namely, there exists a constant $c_1 > c_0$ such that, for any $c \le c_1$, the proper solution takes a standard self-similar form and satisfies the initial condition in the usual pointwise sense:

$$\mathcal{T}(0^+)u_0 = u_0.$$

On the other hand, for any $c > c_1$, we obtain a complete initial singularity so that

$$\mathcal{T}(0^+)u_0(x) \equiv +\infty.$$

The asymptotic analysis in [263] uses a general structure of the bundle of similarity solutions of equation (11.53)

$$V(x,t) = t^\alpha f(\eta), \quad \eta = xt^\beta,$$

with exponents $\alpha = -\frac{1}{p-1}$ and $\beta = \frac{m-p}{2(p-1)}$, where $f \ge 0$ solves a non-autonomous second-order ODE

$$(f^m)'' - \beta\eta f' - \alpha f - f^p = 0.$$

Continuous limit semigroup. It is easy to see by the geometric theory that the limit semigroup of minimal solutions is continuous in the other parameter range $p > m > 1$ even in the class of extremely singular initial data u_0 bounded and continuous on $(-\infty, 0)$ and

$$u_0(x) = +\infty \quad \text{for } x > 0. \tag{11.54}$$

Proposition 11.7 *Let $p > m > 1$. Then, for data $u_0(x)$ that are bounded for $x < 0$ and satisfy* (11.54), $\mathcal{T}(0^+)u_0 \not\equiv \infty$.

Proof. Similar to examples in Section 7.11 (see also Chapter 4) one can show that equation (11.53) with $p > m$ admits singular TWs $V = f(\xi)$, $\xi = x - \lambda t$, where f solves the autonomous ODE

$$(f^m)'' + \lambda f' - f^p = 0 \quad \text{for } \xi < 0, \quad f(0^-) = \infty,$$

i.e., $\Delta_0 \ne \emptyset$. Then a standard comparison and the classical parabolic regularity theory (applied in domains where solutions are uniformly bounded) imply the result. $\square$

The limit semigroup is continuous at $t = 0$ for such singular initial data.

References

[1] A.D. Aleksandrov, Certain estimates for the Dirichlet problem, *Soviet Math. Dokl.*, **1960** (1960), 1151–1154.

[2] W. Allegretto, A comparison theorem for nonlinear operators, *Ann. Scuola Norm. Sup. Pisa, Cl. Sci (4),* **25** (1971), 41–46.

[3] W. Allegretto, Sturm type theorems for solutions of elliptic nonlinear problems, *Nonl. Differ. Equat. Appl.,* **7** (2000), 309–321.

[4] W. Allegretto, Sturm's theorem for degenerate elliptic equations, *Proc. Amer. Math. Soc.,* **129** (2001), 3031–3035.

[5] S. Altschuler, S. Angenent, and Y. Giga, Mean curvature flow through singularities for surfaces of rotation, *J. Geom. Anal.,* **5(3)** (1995), 293–358.

[6] O. Alvarez, J.-M. Lasry, and P.-L. Lions, Convex viscosity solutions and state constraints, *J. Math. Pures Appl.,* **76** (1997), 265–288.

[7] H. Amann, *Linear and Quasilinear Parabolic Problems*, Volume I: *Abstract Linear Theory*, Birkhäuser, Basel, 1995.

[8] H. Amann and J. Escher, Strong continuous dual semigroups, *Ann. Mat. Pura Appl.,* **CLXXI** (1996), 41–62.

[9] D. Andreucci and R. Gianni, Classical solutions to a multidimensional free boundary problem arising in combustion theory, *Comm. Partial Differ. Equat.,* **19** (1994), 803–826.

[10] S.B. Angenent, The Morse-Smale property for a semi-linear parabolic equation, *J. Differ. Equat.,* **62** (1986), 427–442.

[11] S. Angenent, The zero set of a solution of a parabolic equation, *J. reine angew. Math.,* **390** (1988), 79–96.

[12] S. Angenent, Solutions of the one-dimensional porous medium equation are determined by their free boundary, *J. London Math. Soc. (2),* **42** (1990), 339–353.

[13] S. Angenent, Parabolic equations for curves on surfaces. Part I. Curves with p-integrable curvature, *Ann. Math.,* **132** (1990), 451–483.

[14] S. Angenent, Parabolic equations for curves on surfaces. Part II. Intersections, blow-up and generalized solutions, *Ann. Math.,* **133** (1991), 171–215.

[15] S. Angenent, Inflection points, extatic points and curve shortening, In: *Proc. Conf. on Hamilt. Sys. (S'Agarro, 1995)*, NATO Adv. Sci. Int. Ser. C Math. Phys. Sci, **533**, Kluwer Acad. Publ., Dordrecht, 1999, pp. 3–10.

[16] S.B. Angenent and D.G. Aronson, The focusing problem for the radially symmetrical porous-medium equation, *Comm. Partial Differ. Equat.,* **20** (1995), 1217–1240.

[17] S.B. Angenent and B. Fiedler, The dynamics of rotating waves in scalar reaction diffusion equations, *Trans. Amer. Math. Soc.,* **307** (1988), 545–568.

[18] S.B. Angenent and M. Fila, Interior gradient blow-up in a semilinear parabolic equation, *Differ. Integr. Equat.,* **9** (1996), 865–877.

[19] S.B. Angenent and J.J.L. Velazquez, Degenerate neckpinches in mean curvature flow, *J. reine angew. Math.,* **482** (1997), 15–66.

[20] V.I. Arnold, On the characteristic class entering in quantization condition, *Funct. Anal. Appl.,* **1** (1967), 1–14.

[21] V.I. Arnold, The Sturm theorems and symplectic geometry, *Funct. Anal. Appl.,* **19** (1985), 251–259.

[22] V.I. Arnold, *Topological Invariants of Plane Curves and Caustics*, A.M.S. University Lecture Series **5**, 1994.

[23] V.I. Arnol'd, The geometry of spherical curves and the algebra of quaternions, *Russian Math. Surveys,* **50** (1995), 3–68.

[24] V.I. Arnold, Topological problems of the theory of wave propagation, *Russian Math. Surveys,* **51** (1996), 1–47.

[25] V.I. Arnold, Topological problems in wave propagation theory and topological economy principle in algebraic geometry, *Third Lecture by V. Arnold at the Meeting in the Fields Institute Dedicated to His 60th Birthday*, Fields Inst. Commun., 1997.

[26] V.I. Arnold, Symplectic geometry and topology, *J. Math. Phys.,* **41** (2000), 3307–3343.

[27] V.I. Arnold, Asteroidal geometry of hypocycloids and the Hessian topology of hyperbolic polynomials, *Russian Math. Surveys,* **56** (2001), 1019–1083.

[28] V.I. Arnold, The longest curves of given degree and the quasicrystallic Harnack theorem in pseudoperiodic topology, *Funct. Anal. Appl.,* **36** (2002), 165–171.

[29] D. Aronson, Regularity properties of flows through porous media, *SIAM J. Appl. Math.,* **17** (1969), 461–467.

[30] D.G. Aronson, Density-dependent interaction systems, In: *Dynamics and Modelling of Reactive Systems* (W.E. Stuart, W.H. Ray, and C.C. Conley, Eds.), Academic Press, New York, 1980, 161–176.

[31] D.G. Aronson and P. Bénilan, Régularité des solutions de l'equation des milieux poreux dan $\mathbf{R}^n$, *C. R. Acad. Sci. Paris, Série I Math.,* **288** (1979), 103–105.

[32] D.G. Aronson and J.L. Vazquez, Eventual C^∞-regularity and concavity for flows in one-dimensional porous media, *Arch. Rat. Mech. Anal.,* **99** (1987), 329–348.

[33] F.V. Atkinson, *Discrete and Continuous Boundary Problems*, Acad. Press, New York/London, 1964.

[34] F.V. Atkinson and L.A. Peletier, Similarity profiles of flows through porous media, *Arch. Rat. Mech. Anal.,* **42** (1971), 369–379.

[35] F.V. Atkinson and L.A. Peletier, Similarity solutions of the nonlinear diffusion equation, *Arch. Rat. Mech. Anal.,* **54** (1974), 373–392.

[36] P. Baras, L. Cohen, and J.L. Lions, Complete blow-up after $T_{\max}$ for the

solution of a semilinear heat equation, *C. R. Acad. Sci. Paris, Série I Math.*, **300** (1985), 295–298.

[37] P. Baras and L. Cohen, Complete blow-up after $T_{\max}$ for the solution of a semilinear heat equation, *J. Funct. Anal.*, **71** (1987), 142–174.

[38] P. Baras and J.A. Goldstein, The heat equation with a singular potential, *Trans. Amer. Math. Soc.*, **284** (1984), 121–139.

[39] G.I. Barenblatt, On some unsteady motions of a liquid and a gas in a porous medium, *Prikl. Mat. Mekh.*, **16** (1952), 67–78.

[40] R. Bari and B. Rynn, Solution curves and exact multiplicity results for $2m$th order boundary value problems, *J. Math. Anal. Appl.*, to appear.

[41] G. Barles, H.M. Soner, and P.E. Souganidis, Front propagation and phase field theory, *SIAM J. Cont. Opt.*, **4** (1991), 271–283.

[42] G. Barles and P.E. Souganidis, A new approach to front propagation problem: theory and applications, *Arch. Rat. Mech. Anal.*, **141** (1998), 237–296.

[43] J. Bebernes and D. Eberly, *Mathematical Problems in Combustion Theory*, Appl. Math. Sci., Vol. **83**, Springer-Verlag, Berlin, 1989.

[44] Ph. Bénilan, *Equations d'évolution dans un espace de Banach quelconque et applications*, Thése Doctorat d'Etat, Orsey, 1972.

[45] Ph. Bénilan and M. Crandall, The continuous dependence on φ of solutions of $u_t - \Delta\varphi(u) = 0$, *Indiana Univ. Math. J.*, **30** (1981), 161–177.

[46] Ph. Bénilan, M.G. Crandall, and M. Pierre, Solutions of the porous media equation in $\mathbb{R}^N$ under optimal conditions on initial values, *Indiana Univ. Math. J.*, **33** (1984), 51–87.

[47] Ph. Bénilan and J.L. Vazquez, Concavity of solutions of the porous medium equation, *Trans. Amer. Math. Soc.*, **299** (1987), 81–93.

[48] H. Berestycki, B. Nikolaenko, and B. Scheurer, Travelling waves solutions to combustion models and their singular limits, *SIAM J. Math. Anal.*, **16** (1985), 1207–1242.

[49] F. Bernis, J. Hulshof, and J.L. Vazquez, A very singular solution for the dual porous medium equation and the asymptotic behaviour of general solutions, *J. reine angew. Math.*, **435** (1993), 1–31.

[50] S.N. Bernstein, The boundedness of the moduli of successive derivatives of the solutions of parabolic equations, *Doklady Acad. Nauk SSSR*, **18** (1938), 385–388.

[51] S.N. Bernstein, The basis of a Chebyshev system, *Izv. Akad. Nauk SSSR, Ser. Mat.*, **2** (1938), 499–504.

[52] M. Bertsch, R. dal Passo, and M. Ughi, Nonuniqueness and irregularity results for a nonlinear degenerate parabolic equation, In: *Nonl. Diff. Equat. Equil. States*, Vol. **I**, J. Serrin, W.-M. Ni and L.A. Peletier, Eds., Springer-Verlag, New York, 1988, pp. 147–159.

[53] M. Bertsch, R. dal Passo, and M. Ughi, Nonuniqueness of solutions of a degenerate parabolic equation, *Ann. Mat. Pura Appl.*, **161** (1992), 57–81.

[54] M. Bertsch and D. Hilhorst, The interface between regions where $u < 0$ and $u > 0$ in the porous medium equation, *Appl. Anal.*, **41** (1991), 111–130.

[55] M. Bertsch, D. Hilhorst, and C. Schmidt-Lainé, The well-posedness of a

free-boundary problem arising in combustion theory, *Nonl. Anal., TMA,* **23** (1994), 1211–1224.

[56] G. Birkhoff and G.-C. Rota, *Ordinary Differential Equations*, John Wiley & Sons, New York/Singapore, 1989.

[57] M.S. Birman and M.Z. Solomjak, *Spectral Theory of Self-Adjoint Operators in Hilbert Space*, D. Reidel Publ. Comp., Dordrecht/Tokyo, 1987.

[58] G.I. Bizhanova and V.A. Solonnikov, On solvability of initial-boun-dary value problem for a second order parabolic equation with time derivative in the boundary condition in a weighted Hölder functional space, *Algebra and Analysis,* **5** (1993), 109–142.

[59] Ph. Blanc, Existence de solutions discontinues pour des équations paraboliques, *C. R. Acad. Sci. Paris, Série I Math.,* **310** (1990), 53–56.

[60] J. Bochnak, M. Coste, and M.-F. Roy, *Real Algebraic Geometry*, Springer-Verlag, Berlin/New York, 1998.

[61] R. Bott, On the iteration of closed geodesics and the Sturm intersection theory, *Comm. Pure Appl. Math.,* **70** (1959), 313–337.

[62] H.J. Brascamp and E.H. Lieb, On extension of the Brunn-Minkowski and Prekoja-Leindler theorems, including inequalities for log concave functions, and with application to diffusion equation, *J. Funct. Anal.,* **22** (1976), 366–389.

[63] A. Bressan, Stable blow-up patterns, *J. Differ. Equat.,* **98** (1992), 57–75.

[64] H. Brezis, *Opérateurs Maximaux Monotones et Semi-Groupes de Contractions dans les Espaces de Hilbert*, North-Holland Math. Studies No. **5**, North-Holland, Amsterdam/London; American Elsevier, New York, 1973.

[65] H. Brezis and X. Cabreé, Some simple non-linear PDEs without solutions, *Boll. U.M.I., serie VIII,* **I-B** (1998), 223–262.

[66] P. Brunovský and B. Fiedler, Number of zeros on invariant manifolds in reaction-diffusion equations, *Nonl. Anal., TMA,* **10** (1986), 179–193.

[67] P. Brunovský and B. Fiedler, Simplicity of zeros in scalar parabolic equations, *Nonl. Anal., TMA,* **10** (1986), 179–193.

[68] P. Brunovský and B. Fiedler, Connecting orbits in scalar reaction diffusion equations. II: The complete solution, *J. Differ. Equat.,* **81** (1989), 106–135.

[69] P. Brunovský, P. Polácik, and B. Sandstede, Convergence in general periodic parabolic equations in one space dimension, *Nonl. Anal., TMA,* **18** (1992), 209–215.

[70] J.D. Buckmaster, A theory for triple point spacing in overdriven detonation waves, *Combustion and Flame,* **77** (1989), 219–228.

[71] J.D. Buckmaster and G.S.S. Ludford, *Theory of Laminar Flames*, Cambridge University Press, Cambridge, 1982.

[72] C. Budd and V. Galaktionov, Stability and spectra of blow-up in problems with quasi-linear gradient diffusivity, *Proc. Roy. Soc. London A,* **454** (1998), 2371–2407.

[73] L.A. Caffarelli and X. Cabré, *Fully Nonlinear Elliptic Equations*, Coll. Publ. No. **43**, Amer. Math. Soc., Providence, RI, 1995.

[74] L.A. Caffarelli and A. Friedman, Regularity of the free boundary of the one-dimensional flow of gas in a porous medium, *Amer. J. Math.,* **101**

(1979), 1193–1218.

[75] L.A. Caffarelli and A. Friedman, Convexity of solutions of semilinear elliptic equations, *Duke Math. J.,* **52** (1985), 431–456.

[76] L.A. Caffarelli and A. Friedman, Blow-up of solutions of nonlinear heat equations, *J. Math. Anal. Appl.,* **129** (1988), 409–419.

[77] L.A. Caffarelli and J. Spruck, Convexity properties of solutions of some classical variational problems, *Comm. Partial Differ. Equat.,* **7** (1982), 1337–1379.

[78] L.A. Caffarelli and J.L. Vazquez, A free-boundary problem for the heat equation arising in flame propagation, *Trans. Amer. Math. Soc.,* **347** (1995), 411–441.

[79] L.A. Caffarelli and J.L. Vazquez, Viscosity solutions for the porous medium equation, In: *Differential Equations,* La Pietra, 1996 (Florence), pp. 13-26; *Proc. Sympos. Pure Math.,* **65**, 1999.

[80] T. Carleman, Sur un problème d'unicité pour le système d'équations aux dérivées partielles à deux variables indépendantes, *Ark. Mat. Astr. Fys.,* **26B** (1939), 1–9.

[81] L. Cesari, *Asymptotic Behaviour and Stability Problems in Ordinary Differential Equations*, Springer-Verlag, Berlin/Heidelberg, 1963.

[82] S.A. Chaplygin, *A New Method of Approximate Integration of Differential Equations*, GTTI, Moscow/Leningrad, 1932.

[83] X.-Y. Chen, A strong unique continuation theorem for parabolic equations, *Math. Ann.,* **311** (1998), 603–630.

[84] M. Chen, X.-Y. Chen, and J.K. Hale, Structural stability for time-periodic one-dimensional parabolic equations, *J. Differ. Equat.,* **96** (1992), 355–418.

[85] X.-Y. Chen and H. Matano, Convergence, asymptotic periodicity, and finite-point blow-up in one dimensional semilinear parabolic equations, *J. Differ. Equat.,* **78** (1989), 160–190.

[86] X.-Y. Chen and P. Polácik, Asymptotic periodicity of positive solutions of reaction diffusion equations on a ball, *J. reine angew. Math.,* **472** (1996), 17–51.

[87] Y.-G. Chen, Y. Giga, and S. Goto, Uniqueness and existence of viscosity solutions of generalized mean curvature flow equations, *J. Differ. Geom.,* **33** (1991), 749–786.

[88] S.-N. Chow, K. Lu, and J. Mallet-Pare, Floquet theory for parabolic differential equations, *J. Differ. Equat.,* **109** (1993), 147–200.

[89] S.-N. Chow, K. Lu, and J. Mallet-Pare, Floquet bundles for scalar parabolic equations, *Arch. Rat. Mech. Anal.,* **129** (1995), 245–304.

[90] R. Courant and D. Hilbert, *Methods of Mathematical Physics*, Vol. **1**, Intersci. Publ., Inc., New York, 1953.

[91] J. Crank, *Free and Moving Boundary Problems*, Clarendon Press, Oxford, 1984.

[92] M.G. Crandall, An introduction to evolution governed by accretive operators, In: *Dynamical Systems* (Proc. Internat. Sympos., Brown Univ., Providence, RI, 1974), I.L. Cesari et al., Eds., Academic Press, New York, 1976, pp. 131–165.

[93] M.G. Crandall, H. Ishii, and P.-L. Lions, User's guide to viscosity solutions of second order partial differential equations, *Bull. Amer. Math. Soc.,* **27** (1992), 1–67.

[94] M.G. Crandall, M. Kocan, and A. Śiech, L^p-theory for fully nonlinear uniformly parabolic equations, *Comm. Partial Differ. Equat.,***25** (2000), 1997–2053.

[95] R. dal Passo and S. Luckhaus, A degenerate diffusion problem not in divergence form, *J. Differ. Equat.,* **69** (1987), 1–14.

[96] C. de la Vallée-Poissin, Sur l'équation différentielle linéaire du second ordre. Détermination d'une intégrale par deux valeurs assignées. Extension aux équations d'ordre n, *J. Math. Pures Appl.,* **8** (1929), 125–144.

[97] K. Deng and H.A. Levine, The role of critical exponents in blow-up theorems: the sequel, *J. Math. Anal. Appl.,* **243** (2000), 85–126.

[98] E. DiBenedetto, Regularity results for the porous media equation, *Ann. Mat. Pura Appl.,* **121** (1979), 249–262.

[99] E. DiBenedetto, *Degenerate Parabolic Equations,* Series Universitext, Springer-Verlag, New York, 1993.

[100] G. Dong, *Nonlinear Partial Differential Equations of Second Order,* Transl. Math. Monogr., Vol. **95**, Amer. Math. Soc., Providence, RI, 1991.

[101] H.M. Edwards, A generalized Sturm theorem, *Ann. Math.,* **80** (1964), 22–57.

[102] Yu.V. Egorov, V.A. Galaktionov, V.A. Kondratiev, and S.I. Pohozaev, Global solutions of higher-order semilinear parabolic equations in the supercritical range, *C. R. Acad. Sci. Paris, Série I Math.,* **335** (2002), 805–810.

[103] U. Ellias, Eigenvalue problems for the equation $Ly + \lambda p(x)y = 0$, *J. Differ. Equat.,* **29** (1978), 28–57.

[104] C.M. Elliott and J.R. Ockendon, *Weak and Variational Methods for Moving Boundary Problems*, Pitman, London, 1982.

[105] R. Emden, *Gaskugeln, Anwendungen der mechanischen Warmentheorie auf Kosmologie und meteorologische Probleme*, Chap. XII, Teubner, Leipzig, 1907.

[106] C.L. Epstein and M.I. Weinstein, A stable manifold theorem for the curve shortening equations, *Comm. Pure Appl. Math.,* **40** (1987), 119–139.

[107] J. Escher, Quasilinear parabolic systems with dynamic boundary conditions, *Comm. Partial Differ. Equat.,* **18** (1993), 1309–1364.

[108] J. Escher, On quasilinear fully parabolic boundary value problems, *Differ. Integr. Equat.,* **7** (1994), 1325–1343.

[109] L.C. Evans, H.M. Soner, and P.E. Souganidis, Phase transition and generalized motion by mean curvature, *Comm. Pure Appl. Math.,* **45** (1992), 1097–1123.

[110] L.C. Evans and J. Spruck, Motion of level sets by mean curvature. I, *J. Differ. Geom.,* **33** (1991), 635–681.

[111] A. Fasano, M. Primicerio, S.D. Howison, and J.R. Ockendon, Some remarks on the regularization of supercooled one-phase Stefan problems in one dimension, *Quart. Appl. Math.,* **48** (1990), 153–168.

[112] E. Fereisl and P. Polácik, Structure of periodic solutions and asymptotic behavior for time-periodic reaction-diffusion equations on $I\!R$, *Adv. Differ. Equat.,* **5** (2000), 583–622.

[113] B. Fiedler, Discrete Ljapunov functionals and ω-limit sets, *Math. Model. Numer. Anal.,* **23** (1989), 415–431.

[114] B. Fiedler and J. Mallet-Paret, A Poincaré-Bendixson theorem for scalar reaction diffusion equations, *Arch. Rat. Mech. Anal.,* **107** (1989), 325–345.

[115] B. Fiedler and J. Mallet-Paret, Connections between Morse sets for delay-differential equations, *J. reine angew. Math.,* **397** (1989), 23–41.

[116] S. Filippas and R.V. Kohn, Refined asymptotics for the blow-up of $u_t - \Delta u = u^p$, *Comm. Pure Appl. Math.,* **45** (1992), 821–869.

[117] G. Floquet, Sur les équations differetielles linéaires à coefficients periodiques, *Ann. Sci. École Norm. Sup.,* **12** (1883), 47–89.

[118] V.A. Florin, Earth compaction and seepage with variable porosity taking into account the inflience of bound water, *Izvestiya Akad. Nauk SSSR, Otdel. Tekhn. Nauk,* No. **11** (1951), 1625–1649.

[119] R.H. Fowler, The form near infinity of real continuous solutions of a certain differential equation of the second order, *Quart. J. Math.,* **45** (1914), 289–305.

[120] D.A. Frank-Kamenetskii, Towards temperature distributions in a reaction vessel and the stationary theory of thermal explosion, *Doklady Acad. Nauk SSSR,* **18** (1938), 411–412.

[121] A. Friedman, *Partial Differential Equations of Parabolic Type*, Prentice Hall, Englewood Cliffs, NJ, 1964.

[122] A. Friedman, *Variational Principles and Free-Boundary Problems,* Wiley-Interscience Publication, New York, 1982.

[123] A. Friedman and B. McLeod, Blow-up of positive solutions of semilinear heat equations, *Indiana Univ. Math. J.,* **34** (1985), 425–447.

[124] A. Gabriaelov and N. Vorobjov, Complexity of stratification of semi-Pfaffian sets, *Discr. Comput. Geom.,* **14** (1995), 71–91.

[125] M. Gage and R.S. Hamilton, The heat equation shrinking convex plane curves, *J. Differ. Geom.,* **23** (1986), 69–96.

[126] V.A. Galaktionov, Some properties of travelling waves in a medium with non-linear thermal conductivity and a source of heat, *USSR Comput. Math. Math. Phys.,* **21** (1981), 167–176.

[127] V.A. Galaktionov, On localization conditions for unbounded solutions of quasilinear parabolic equations, *Soviet Math. Dokl.,* **25** (1982), 775–780.

[128] V.A. Galaktionov, Conditions for global non-existence and localization of solutions of the Cauchy problem for a class of nonlinear parabolic equations, *USSR Comput. Math. Math. Phys.,* **23** (1983), 36–44.

[129] V.A. Galaktionov, Proof of the localization of unbounded solutions of the nonlinear parabolic equation $u_t = (u^\sigma u_x)_x + u^\beta$, *Differ. Equat.,* **21** (1985), 11–18.

[130] V.A. Galaktionov, Asymptotics of unbounded solutions of the nonlinear equation $u_t = (u^\sigma u_x)_x + u^\beta$ in a neighbourhood of a "singular" point, *Soviet Math. Dokl.,* **33** (1986), 840–844.

[131] V.A. Galaktionov, On new exact blow-up solutions for nonlinear heat conduction equations with source and applications, *Differ. Integr. Equat.,* **3** (1990), 863–874.

[132] V.A. Galaktionov, Best possible upper bound for blow-up solutions of the quasilinear heat conduction equation with source, *SIAM J. Math. Anal.,* **22** (1991), 1293–1302.

[133] V.A. Galaktionov, On a blow-up set for the quasilinear heat equation $u_t = (u^\sigma u_x)_x + u^{\sigma+1}$, *J. Differ. Equat.,* **101** (1993), 66–79.

[134] V.A. Galaktionov, Quasilinear heat equations with first-order sign-invariants and new explicit solutions, *Nonl. Anal., TMA,* **23** (1994), 1595–1621.

[135] V.A. Galaktionov, Blow-up for quasilinear heat equations with critical Fujita's exponents, *Proc. Royal Soc. Edinburgh,* **124A** (1994), 517–525.

[136] V.A. Galaktionov, Invariant subspaces and new explicit solutions to evolution equations with quadratic nonlinearities, *Proc. Roy. Soc. Edinburgh,* **125A** (1995), 225–246.

[137] V.A. Galaktionov, Concavity and B-concavity of solutions of quasilinear filtration equations, *J. London Math. Soc. (2),* **59** (1999), 955–977.

[138] V.A. Galaktionov, On a spectrum of blow-up patterns for a higher-order semilinear parabolic equations, *Proc. Royal Soc. London A,* **457** (2001), 1–21.

[139] V.A. Galaktionov, Geometric theory of one-dimensional nonlinear parabolic equations I. Singular interfaces, *Adv. Differ. Equat.,* **7** (2002), 513–580.

[140] V.A. Galaktionov, Sturmian nodal set analysis for higher-order parabolic equations and applications, *Trans. Amer. Math. Soc.,* submitted.

[141] V.A. Galaktionov, S. Gerbi, and J.L. Vazquez, Quenching for a one-dimensional fully nonlinear parabolic equation in detonation theory, *SIAM J. Appl. Math.,* **61** (2000), 1253–1285.

[142] V.A. Galaktionov, J. Hulshof, and J.L. Vazquez, Extinction and focusing behaviour of spherical and annular flames described by a free boundary problem, *J. Math. Pures Appl.,* **76** (1997), 563–608.

[143] V.A. Galaktionov and R. Kersner, On a discontinuous parabolic semigroup, In: *Free Boundary Problems: Theory and Applications,* GAKUTO Int. Series, Mathematical Sciences and Applications, Vol. **14**, 2000, pp. 135–145.

[144] V.A. Galaktionov and R. Kersner, Discontinuous limit semigroups for the singular Zhang equation and its hydrodynamic version, *Adv. Differ. Equat.,* **6** (2001), 1493–1516.

[145] V.A. Galaktionov, S.P. Kurdyumov, S.A. Posashkov, and A.A. Samarskii, Unbounded solutions of the Cauchy problem for the para-bolic equation $u_t = \nabla(u^\sigma \nabla u) + u^\beta$, *Soviet Phys. Dokl.,* **25** (1980), 458–459.

[146] V.A. Galaktionov, S.P. Kurdyumov, S.A. Posashkov, and A.A. Samarskii, A nonlinear elliptic problem with a complex spectrum of solutions, *USSR Comput. Math. Math. Phys.,* **26**, (1986), 48–54.

[147] V.A. Galaktionov, S.P. Kurdyumov, S.A. Posashkov, and A.A. Samarskii, Quasilinear parabolic equation with a complex spectrum of unbounded

self-similar solutions, In: *Math. Model., Processes in Nonlinear Media,* A.A. Samarskii, S.P. Kurdyumov and V.A. Galaktionov, Eds., Nauka, Moscow, 1986, pp. 142–182; English transl.: Boca Raton, CRC Press, 1992.

[148] V.A. Galaktionov, S.P. Kurdyumov, and A.A. Samarskii, On the method of stationary states for quasilinear parabolic equations, *Math. USSR Sbornik,* **67** (1990), 449–471.

[149] V.A. Galaktionov and A.A. Lacey, Monotonicity in time of large solutions to a nonlinear heat equation, *Rocky Mountain J. Math.,* **28** (1998), 1279–1301.

[150] V.A. Galaktionov and S.A. Posashkov, Application of new comparison theorem in the investigation of unbounded solutions of nonlinear parabolic equations, *Differ. Equat.,* **22** (1986), 809–815.

[151] V.A. Galaktionov and S.A. Posashkov, A method of investigating unbounded solutions of quasilinear parabolic equations, *USSR Comput. Math. Math. Phys.,* **28** (1988), 148–156.

[152] V.A. Galaktionov and S.A. Posashkov, Lower bounds for unbounded solutions of a parabolic system of quasilinear equations, *Math. Notes,* **47** (1990), 111–116.

[153] V.A. Galaktionov and S.A. Posashkov, Single point blow-up for N-dimensional quasilinear equation with gradient diffusion and source, *Indiana Univ. Math. J.,* **40** (1991), 1041–1060.

[154] V.A. Galaktionov and S.A. Posashkov, Any large solution of nonlinear heat conduction equation becomes monotone in time, *Proc. Roy. Soc. Edinburgh,* **118A** (1991), 13–20.

[155] V.A. Galaktionov and S.A. Posashkov, Explicit solutions and invariant subspaces for nonlinear equations with gradient-dependent diffusivity, *Comput. Maths Math. Phys.,* **34** (1994), 313–321.

[156] V.A. Galaktionov and S.A. Posashkov, Examples of nonsymmetric extinction and blow-up for quasilinear heat equations, *Differ. Integr. Equat.,* **8** (1995), 87–103.

[157] V.A. Galaktionov and S.A. Posashkov, New explicit solutions of quasilinear heat equations with general first-order sign-invariants, *Physica D*, **99** (1996), 217–236.

[158] V.A. Galaktionov and S.A. Posashkov, Maximal sign-invariants of quasilinear parabolic equations with gradient diffusivity, *J. Math. Phys.,* **39** (1998), 4948–4964.

[159] V.A. Galaktionov, S.I. Shmarev, and J.L. Vazquez, Regularity of interfaces in diffusion processes under the influence of strong absorption, *Arch. Rat. Mech. Anal.,* **149** (1999), 183–212.

[160] V.A. Galaktionov, S.I. Shmarev, and J.L. Vazquez, Second-order interface equations for nonlinear diffusion with very strong absorption, *Comm. Contemp. Math.,* **1** (1999), 51–64.

[161] V.A. Galaktionov and J.L. Vazquez, Regional blow-up in a semilinear heat equation with convergence to a Hamilton-Jacobi equation, *SIAM J. Math. Anal.,* **24** (1993), 1254–1276.

[162] V.A. Galaktionov and J.L. Vazquez, Extinction for a quasilinear heat equation with absorption I. Technique of intersection comparison, *Comm. Partial Differ. Equat.*, **19** (1994), 1075–1106.

[163] V.A. Galaktionov and J.L. Vazquez, Extinction for a quasilinear heat equation with absorption II. A dynamical systems approach, *Comm. Partial Differ. Equat.*, **19** (1994), 1107–1137.

[164] V.A. Galaktionov and J.L. Vazquez, Geometrical properties of the solutions of one-dimensional nonlinear parabolic equations, *Math. Ann.*, **303** (1995), 741–769.

[165] V.A. Galaktionov and J.L. Vazquez, Necessary and sufficient conditions for complete blow-up and extinction for one-dimensional quasilinear heat equations, *Arch. Rat. Mech. Anal.*, **129** (1995), 225–244.

[166] V.A. Galaktionov and J.L. Vazquez, Blow-up for quasilinear heat equations described by means of nonlinear Hamilton-Jacobi equations, *J. Differ. Equat.*, **127** (1996), 1–40.

[167] V.A. Galaktionov and J.L. Vazquez, Continuation of blow-up solutions of nonlinear heat equations in several space dimensions, *Comm. Pure Appl. Math.*, **50** (1997), 1–68.

[168] V.A. Galaktionov and J.L. Vazquez, Incomplete blow-up and singular interfaces for quasilinear heat equation, *Comm. Partial Differ. Equat.*, **22** (1997), 1405–1452.

[169] V.A. Galaktionov and J.L. Vazquez, The problem of blow-up in nonlinear parabolic equations, *Discr. Cont. Dyn. Syst.*, **8** (2002), 399-433.

[170] V.A. Galaktionov and J.L. Vazquez, *A Stability Technique for Evolution Partial Differential Equations. A Dynamical Systems Approach*, Birkhäuser, Boston/Berlin, 2004.

[171] F.R. Gantmakher, Non-symmetric Kellogg kernels, *Doklady Akad. Nauk SSSR,* **1** (1936), 3–5.

[172] F.R. Gantmakher and M.G. Krein, *Oscillation Matrices and Kernels and Small Vibrations of Mechanical Systems*, Gostekhizdat (Izdat. Tekhn. Teor. Lit.), Moscow/Leningrad, 1950; English transl. from the 1950 2nd Russian ed.: AEC-tr-4481, U.S. Atomic Energy Commission, Oak Ridge, Tenn., 1961.

[173] Y. Giga, Interior derivative blow up for quasilinear parabolic equations, *Discr. Cont. Dyn. Syst.*, **7** (1994), 811–821.

[174] B.H. Gilding, Hölder continuity of solutions of parabolic equations, *J. London Math. Soc. (2),* **13** (1976), 103–106.

[175] B.H. Gilding, The correspondence between travelling-wave solutions of a nonlinear reaction-convection-diffusion equation and an integral equation, *Differ. Integr. Equat.*, **9** (1996), 917–947.

[176] B.H. Gilding and R. Kersner, The characterization of reaction-convection-diffusion processes by travelling waves, *J. Differ. Equat.*, **124** (1996), 27–79.

[177] B.H. Gilding and L.A. Peletier, On a class of similarity solutions of the porous media equation. II, *J. Math. Anal. Appl.*, **57** (1977), 522–538.

[178] A. Givental, Sturm theorem for hyperelliptic integrals, *Leningrad J. Math.*,

1 (1990), 1157–1163.

[179] E.K. Godunova and V.I. Levin, Certain qualitative questions of heat conduction, *USSR Comp. Math. Math. Phys.*, **6** (1966), 212–220.

[180] J.A. Goldstein and Qi S. Zhang, Linear parabolic equations with strong singular potentials, *Trans. Amer. Math. Soc.*, **355** (2002), 197–211.

[181] M.A. Grayson, The heat equation shrinks embedded plane curves to round points, *J. Differ. Geom.*, **26** (1987), 285–314.

[182] M.A. Grayson, Shortening embedded curves, *Ann. Math.*, **129** (1989), 71–111.

[183] J.K. Hale, Dynamics in parabolic equations–an example, In: *Syst. Nonl. Part. Differ. Equat.*, J.M. Ball, Ed., D. Reidel Publ. Comp., Dordrecht/Lancaster, 1983, pp. 461–472.

[184] T. Halpin-Healy and Y.-C. Zhang, Kinetic roughening phenomena, stochastic growth, directed polymers and all that, *Physics Reports*, **254** (4-6) (1995), 215–414.

[185] P. Hartman, *Ordinary Differential Equations*, Birkhäuser, Boston/Stuttgart, 1982.

[186] D.B. Henry, Some infinite-dimensional Morse-Smale systems defined by parabolic partial differential equations, *J. Differ. Equat.*, **59** (1985), 165–205.

[187] M.A. Herrero and J.L. Vazquez, The one-dimensional nonlinear heat equation with absorption: regularity of solutions and interfaces, *SIAM J. Math. Anal.*, **18** (1987), 149–167.

[188] M.A. Herrero and J.J.L. Velázquez, Blow-up behaviour of one-dimensional semilinear parabolic equations, *Ann. Inst. Henri Poincaré, Analyse non linéaire*, **10** (1993), 131–189.

[189] M.A. Herrero and J.J.L. Velázquez, Singularity formation in the one-dimensional supercooled Stefan problem, *Euro J. Appl. Math.*, **7** (1996), 119–150.

[190] P. Hess and P. Polácik, Symmetry and convergence properties for non-negative solutions of nonautonomous reaction-diffusion problems, *Proc. Royal Soc. Edinburgh*, **124A** (1994), 573–587.

[191] E. Hille, *Ordinary Differential Equations in the Complex Domain*, Dover Publications, Mineola, New York, 1997.

[192] J. Hulshof, Similarity solutions of the porous medium equation with sign change, *J. Math. Anal. Appl.*, **157** (1991), 75–111.

[193] J. Hulshof, J.R. King, and M. Bowen, Intermediate asymptotics of the porous medium equation with sign changes, *Adv. Differ. Equat.*, **6** (2001), 1115–1152.

[194] E. Holmgren, Sur les solutions quasianalytiques de l'équation de la chaleur, *Ark. Mat.*, **18** (1924), 64–95.

[195] E. Hopf, A remark on linear elliptic equations of the second order, *Proc. Amer. Math. Soc.*, **3** (1952), 791–793.

[196] S.D. Howison, Complex variable methods in Hele-Shaw moving boundary problems, *Euro J. Appl. Math.*, **3** (1992), 209–224.

[197] S.D. Howison, A.A. Lacey, and J.R. Ockendon, Singularity development

in moving-boundary problems, *Quart. J. Mech. Appl. Math.,* **38** (1985), 343–360.

[198] S.D. Howison, A.A. Lacey, and J.R. Ockendon, Hele-Shaw free-boundary problems with suction, *Quart. J. Mech. Appl. Math.,* **41** (1988), 183–193.

[199] G. Huisken, Flow by mean curvature of convex surface into spheres, *J. Differ. Geom.,* **20** (1984), 237–266.

[200] A. Hurwitz, Über die Fourierschen Konstanten integrierbarer Funktionen, *Math. Ann.,* **57** (1903), 425–446.

[201] N.H. Ibragimov, On the group classification of differential equations of second order, *Soviet Math. Dokl.,* **9** (1968), 1365–1369.

[202] Y. Ilýashenko and S. Yakovenko, Counting real zeros of analytic functions satisfying linear ordinary differential equations, *J. Differ. Equat.,* **126** (1996), 87–105.

[203] E.L. Ince, *Ordinary Differential Equations*, Dover Publ., New York, 1956.

[204] I.K. Ivanov, A relation between the number of changes of sign of the solution of the equations $\partial[a(t,x)\partial u/\partial x]/\partial x - \partial u/\partial t = 0$ and the nature of its diminution, *Godisnik Viss. Tehn. Ucebn. Zaved., Mat.,* **1** (1964), kn. 1, 107–116 (1965).

[205] C.G.J. Jacobi, *Vorlesungen über Dynamik*, G. Reiner, Berlin, 1884.

[206] J. Jones, Jr. and T. Mazumdar, On zeros of solutions of certain Emden-Fowler like differential equations in Hilbert space, *Nonl. Anal., TMA,* **12** (1988), 365–373.

[207] D.D. Joseph and T.S. Lundgren, Quasilinear Dirichlet problems driven by positive sources, *Arch. Rat. Mech. Anal.,* **49** (1973), 241–269.

[208] A.S. Kalashnikov, The Cauchy problem in a class of growing functions for equations of nonstationary filtration type, *Vestnik Moscow Univ., Ser. 1, Math., Mech.,* **6** (1963), 17–27.

[209] A.S. Kalashnikov, Equations of the unsteady filtration type with infinite rate of propagation of perturbations, *Moscow Univ. Math. Bull.,* **27** (1972), 104–108.

[210] A.S. Kalashnikov, On the Cauchy problem in classes of growing initial functions for some quasilinear degenerate parabolic equations, *Differentsial'nye Uravneniya,* **9** (1973), 682–691.

[211] A.S. Kalashnikov, The nature of the propagation of perturbations in problems of non-linear heat conduction with absorption, *USSR Comput. Math. Math. Phys.,* **14** (1974), 70–85.

[212] A.S. Kalashnikov, On the differential properties of the generalized solutions of equations of unsteady filtration type, *Moscow Univ. Math. Bull.,* **29** (1974), 48–53.

[213] A.S. Kalashnikov, Some problems of the qualitative theory of non-linear degenerate second-order parabolic equations, *Russian Math. Surveys,* **42** (1987), 169–222.

[214] S. Kamin, L.A. Peletier, and J.L. Vazquez, A nonlinear diffusion-absorption equation with unbounded initial data, In: *Nonl. Diff. Equat. Equil. States*, Vol. **3**, N.G. Lloyd, W.-M. Ni, L.A. Peletier and J. Serrin, Eds., Birkhäuser, Boston/Berlin, 1992, pp. 243–263.

[215] S. Kamin and Ph. Rosenau, Thermal waves in an absorbing and convecting medium, *Physica D,* **8** (1983), 273–283.

[216] M. Kardar, G. Parisi, and Y.-C. Zhang, Dynamic scaling of growing interfaces, *Phys. Rev. Letters,* **56**(9) (1986), 889–892.

[217] S. Karlin, Total positivity, absorption probabilities and applications, *Trans. Amer. Math. Soc.,* **111** (1964), 33–107.

[218] B. Kawohl, *Rearrangements and Convexity of Level Sets in PDE,* Lecture Notes in Math., Vol. **1150**, Springer-Verlag, New York, 1985.

[219] B. Kawohl and R. Kersner, On degenerate diffusion with very strong absorption, *Math. Meth. Appl. Sci.,* **15** (1992), 469–477.

[220] O. Kellog, The oscillation of functions on an orthogonal set, *Amer. J. Math.,* **38** (1916), 1–5.

[221] O.D. Kellogg, On the existence and closure of sets of characteristic functions, *Math. Ann.,* **86** (1922), 14–17.

[222] R. Kersner, On the behaviour of temperature fronts in media with nonlinear heat conductivity under absorption, *Moscow Univ. Math. Bull.,* **33** (1978), 35–41.

[223] A.G. Khovanskii (A.G. Hovanskii), On a class of systems of transcendental equations, *Soviet Math. Dokl.,* **22** (1980), 762–765.

[224] A.G. Khovanskii, *Fewnomials*, AMS Transl. Math. Monogr., Vol **88**, Amer. Math. Soc., Providence, Rhode Island, 1991.

[225] A. Kneser, Festschrift zum 70. Geburtstag von H. Weber, Leipzig, 1912, pp. 170–192.

[226] A.N. Kolmogorov, I.G. Petrovskii, and N.S. Piskunov, Study of the diffusion equation with growth of the quantity of matter and its application to a biological problem, *Byull. Moskov. Gos. Univ., Sect. A,* **1** (1937), 1–26. See [286], pp. 105–130 for an English transl.

[227] N.J. Korevaar, Convex solutions to nonlinear elliptic and parabolic boundary value problems, *Indiana Univ. Math. J.,* **32** (1983), 603–614.

[228] M.A. Krasnosel'skii and P.P. Zabreiko, *Geometrical Methods of Nonlinear Analysis*, Springer-Verlag, Berlin/Tokyo, 1984.

[229] M.G. Krein, Sue les fonctions de Green non-symétriques oscillatories des opérateurs différentiels ordinaires, *Doklady Akad. Nauk SSSR,* **25** (1939), 643–646.

[230] J. Krug and H. Spohn, Kinetic roughening of growing surfaces, In: *Solids Far from Equilibrium*, C. Godreche, Ed., Cambridge Univ. Press, 1992, pp. 479–582.

[231] S.N. Kruzhkov, Results concerning the nature of the continuity of solutions of parabolic equations and some of their applications, *Math. Notes,* **6** (1969), 517–523.

[232] S.N. Kruzhkov, First order quasilinear equations in several independent variables, *Math. USSR Sbornik,* **10** (1970), 217–243.

[233] N.V. Krylov, *Nonlinear Elliptic and Parabolic Equations of the Second Order,* D. Reidel, Dordrecht, 1987.

[234] K. Kunisch and G. Peichl, On the shape of the solutions of second-order parabolic differential equations, *J. Differ. Equat.,* **75** (1988), 329–353.

[235] A.A. Lacey, Global blow-up of a nonlinear heat equation, *Proc. Roy. Soc. Edinburgh,* **104A** (1986), 161–167.

[236] A.A. Lacey and D. Tzanetis, Complete blow-up for a semilinear diffusion equation with a sufficiently large initial condition, *IMA J. Appl. Math.,* **41** (1988), 207–215.

[237] A.A. Lacey and D.E. Tzanetis, Global, unbounded solutions to a parabolic equation, *J. Differ. Equat.,* **101** (1993), 80–102.

[238] L.D. Landau and E.M. Lifschitz, *Fluid Mechanics*, Pergamon Press, Oxford, 1959.

[239] E.M. Landis, A property of solutions of a parabolic equation, *Soviet Math. Dokl.,* **7** (1966), 900–903.

[240] S. Lang, *Algebra*, Addison-Wesley Publ. Comp., Reading/Tokyo, 1984.

[241] E.E. Levi, Sull'equazione del calore, *Ann. Mat. Pura Appl., Ser.3,* **14** (1908), 187–264.

[242] A.Yu. Levin, Non-oscillation of solutions of the equation $x^{(n)} + p_1(t)x^{(n-1)} + ... + p_n(t)x = 0$, *Russian Math. Surveys,* **24** (1969), 43–99.

[243] H.A. Levine, Quenching, nonquenching and beyond quenching for solutions of some parabolic equations, *Ann. Mat. Pura Appl.,* **155** (1989), 243–290.

[244] H.A. Levine, The role of critical exponents in blow-up problems, *SIAM Rev.,* **32** (1990), 262–288.

[245] H.A. Levine, S.R. Park, and J. Serrin, Global existence and nonexistence theorems for quasilinear evolution equations of formally parabolic type, *J. Differ. Equat.,* **142** (1998), 212–229.

[246] B.M. Levitan and I.S. Sargsian, *Sturm-Liouville and Dirac Operators*, Kluwer Acad. Publ., Dordrecht/London, 1991.

[247] V.B. Lidskii, Oscillatory theorems for a canonical system of differential equations, *Doklady Acad. Nauk SSSR,* **102** (1955), 877–880.

[248] G.M. Lieberman, *Second Order Parabolic Differential Equations*, World Scientific, Singapore/Hong Kong, 1996

[249] P.L. Lions, Two geometrical properties of solutions of semilinear problems, *Appl. Anal.,* **12** (1981), 267–272.

[250] J. Liouville, Démonstration d'un théorème dû à M. Sturm, et relatif à une classe de funcions transcendantes, *J. Math. Pures Appl.,* **1** (1836), 269–277.

[251] C. Loewner and L. Nirenberg, Partial differential equations invariant under conformal or projective transformations, In: *Contributions to Analysis*, Acad. Press, New York, 1974, pp. 145–272.

[252] Ya.B. Lopatinskii, On a method for reducing boundary-value problems for a system of differential equations of elliptic type to regular integral equations, *Ukrain. Math. Zh.,* **5** (1953), 123–151.

[253] A. Lunardi, *Analytic Semigroups and Optimal Regularity in Parabolic Problems,* Progr. in Nonl. Differ. Equat. Appl., Vol. **16**, Birkhäuser, Basel/Berlin, 1995.

[254] L. Lusternik and L. Schnirelman, Sur le problème de trois géodésiques fermées sur les surfaces de genre O, *C. R. Acad. Sci. Paris, Série I Math.,*

189 (1929), 269–271.

[255] A.M. Lyapunov, *The General Problem of the Stability of Motion*, Kharkov, 1892 (in Russian); Taylor & Francis, London, 1992. A. Liapunoff, Problème général de la stabilité du mouvement, Ann. Fac. Sciences de Touluse, Second series, **9** (1907), 203–469; Reprinted as Ann. of Math. Studies, No. **17**, Princeton Univ. Press, 1947.

[256] J. Mallet-Paret, Morse decomposition for delay-differential equations, *J. Differ. Equat.*, **72** (1988), 270–315.

[257] J. Mallet-Paret and H.L. Smith, The Poincaré-Bendixson theorem for monotone cyclic feedback systems, *J. Dyn. Differ. Equat.*, **2** (1990), 367–421.

[258] W.A. Markov, Über Polynome, die in einem gegebenen Intervalle möglichst wenig von Null abweichen, *Math. Ann.*, **77** (1916), 213–258.

[259] V.P. Maslov, *Théorie des Perturbations et Méthodes Asymptotiques*, Thesis, Moscow State Univ., 1965; Dunod, Paris, 1972.

[260] H. Matano, Convergence of solutions of one-dimensional semilinear parabolic equations, *J. Math. Kyoto Univ. (JMKYAZ)*, **18** (1978), 221–227.

[261] H. Matano, Nonincrease of the lap-number of a solution for a one-dimensional semilinear parabolic equation, *J. Fac. Sci. Univ. Tokyo, Sect. IA, Math.*, **29** (1982), 401–441.

[262] H. Matano, Asymptotic behaviour of solutions of semilinear heat eqautions on S^1, In: *Nonl. Diff. Equat. Equil. States*, Vol. **II**, J. Serrin, W.-M. Ni and L.A. Peletier, Eds., Springer-Verlag, New York, 1988, pp. 139–162.

[263] J.B. McLeod, L.A. Peletier, and J.L. Vazquez, Solutions of a nonlinear ODE appearing in the theory of diffusion with absorption, *Differ. Integr. Equat.*, **4** (1991), 1–14.

[264] A.M. Meirmanov, On a free boundary problem for parabolic equations, *Math. USSR Sbornik*, **43** (1982), 473–484.

[265] A.M. Meirmanov, *The Stefan Problem*, Walter de Gruyter, Berlin, 1992.

[266] F. Merle and H. Zaag, Optimal estimates for blowup rate and behavior for nonlinear heat equations, *Comm. Pure Appl. Math.*, **51** (1998), 139–196.

[267] E. Mitidieri and S.I. Pohozaev, *Apriori Estimates and Blow-up of Solutions to Nonlinear Partial Differential Equations and Inequalities*, Proc. Steklov Inst. Math., Vol. **234**, Intern. Acad. Publ. Comp. Nauka/Interperiodica, Moscow, 2001.

[268] N. Mizoguchi and E. Yanagida, Critical exponents for the decay rate of solutions in a semilinear parabolic equation, *Arch. Rat. Mech. Anal.*, **145** (1998), 331–342.

[269] M. Morse, A generalization of the Sturm theorems in n space, *Math. Ann.*, **103** (1930), 52–69.

[270] M. Morse, *The Calculus of Variation in the Large*, AMS Colloquium Publications, Vol. **18**, New York, 1934.

[271] S. Mukhopadyaya, New methods in the geometry of a plane arc I, *Bull. Calcutta Math. Soc.*, **1** (1909), 31–37.

[272] A.D. Myschkis, *Lineare Differentialgleichungen mit Nacheilendem Argument*, Deutscher Verlag Wiss., Berlin, 1955.

[273] W.-M. Ni and P.E. Sacks, The number of peaks of positive solutions of semilinear parabolic equations, *SIAM J. Math. Anal.*, **16** (1985), 460–471.

[274] K. Nickel, Gestaltaussagen über Lösungen parabolischer Differentialgleichungen, *J. reine angew. Math.*, **211** (1962), 78–94.

[275] K. Nickel, Einige Eigenschaften von Lösungen der Prandtlschen Grenzschicht-Differentialgleichungen, *Arch. Rat. Mech. Anal.*, **2** (1958), 1–31.

[276] O.A. Oleinik, On properties of some boundary problems for equations of elliptic type, *Mat. Sbornik, N.S.*, **30** (72) (1952), 695–702.

[277] O.A. Oleinik, A.S. Kalashnikov, and Chzhou Yui-Lin', The Cauchy problem and boundary-value problems for equations of unsteady filtration type, *Izv. Akad. Nauk SSSR, Ser. Mat.*, **22**, No. 5 (1958), 667–704.

[278] O.A. Oleinik and S.N. Kruzhkov, Quasilinear second-order parabolic equations with many independent variables, *Russian Math. Surveys,* **16** (1961), 105–146.

[279] O.A. Oleinik, M. Primicerio, and E. Radkevich, Stefan-like problems, *Meccanica,* **28**, 129–143.

[280] O.A. Oleinik and E.V. Radkevich, Method of introducing of a parameter for evolution equations, *Russian Math. Surveys,* **33** (1978), 7–84.

[281] W.F. Osgood, Beweis der Existenz einer Lösung der Differentialgleichung $dy/dx = f(x,y)$ ohne Hinzunahme der Cauchy-Lipschitzschen Bedingung, *Monatshefte für Mathematik und Physik (Vienna)*, **9** (1898), 331–345.

[282] S. Osher and J. Sethian, Fronts moving with curvature dependent speed: algorithms based on Hamilton-Jacobi equations, *J. Comp. Phys.*, **79** (1988), 12–49.

[283] A. de Pablo and J.L. Vazquez, The balance between strong reaction and slow diffusion, *Comm. Partial Diff. Equat.*, **15** (1990), 159–183.

[284] A. de Pablo and J.L. Vazquez, An overdetermined initial and boundary-value problem for a reaction-diffusion equation, *Nonl. Anal., TMA,* **19** (1992), 259–269.

[285] A. de Pablo and J.L. Vazquez, Travelling waves and finite propagation in a reaction-diffusion equation, *J. Differ. Equat.*, **93** (1991), 19–61.

[286] *Dynamics of Curved Fronts,* P. Pelcé, Ed., Acad. Press, New York, 1988.

[287] L.A. Peletier, A necessary and sufficient condition for the existence of an interface in flows through porous media, *Arch. Rat. Mech. Anal.*, **56** (1974), 183–190.

[288] D. Phillips, Existence of solutions of quenching problems, *Appl. Anal.,* **24** (1987), 253–264.

[289] M. Picone, Un teorema sulle soluzioni delle equazioni ellittiche autoaggiunte alle derivate parziali del secondo ordine, *Atti Accad. Naz. Lincei Rend.,* **20** (1911), 213–219.

[290] S.I. Pohozaev, On the eigenfunctions of the equation $\Delta u + \lambda f(u) = 0$, *Soviet Math. Dokl.*, **6** (1965), 1408–1411.

[291] S.I. Pohozaev and A. Tesei, Instantaneous blow-up of solutions to a class of hyperbolic inequalities, *Electr. J. Differ. Equat.*, **08** (2002), 155–165.

[292] P. Polácik, Transversal and nontransversal intersections of stable and un-

stable manifolds in reaction diffusion equations on symmetric domains, *Differ. Integr. Equat.*, **7** (1994), 1527–1545.

[293] P. Polácik, Parabolic equations: asymptotic behaviour and dynamics on invariant manifolds, In: *Handbook of Dynamical Systems*, Vol. **2**, North-Holland, Amsterdam, 2002, pp. 835–883.

[294] P.Ya. Polubarinova-Kochina, On a nonlinear differential equation encountered in the theory of filtration, *Dokl. Akad. Nauk SSSR,* **63**, No. 6 (1948), 623–627.

[295] G. Pólya, On the mean-value theorem corresponding to a given linear homogeneous differential equation, *Trans. Amer. Math. Soc.,* **24** (1924), 312–324.

[296] G. Pólya, Qualitatives über Wärmeausgleich, *Z. Angew. Math. Mech.,* **13** (1933), 125–128.

[297] C.-C. Poon, Unique continuation for parabolic equations, *Comm. Partial Differ. Equat.*, **21** (1996), 521–539.

[298] G. Da Prato and P. Grisvard, Equation d'évolutions abstraites de type parabolique, *Ann. Mat. Pura Appl.,* **IV(120)** (1979), 329–396.

[299] M.H. Protter and H.F. Weinberger, *Maximum Principles in Differential Equations*, Springer-Verlag, New York/Tokyo, 1984.

[300] R.M. Redheffer and W. Walter, The total variation of solutions of parabolic differential equations and a maximum principle in unbounded domains, *Math. Ann.,* **209** (1974), 57–67.

[301] W.T. Reid, *Sturmian Theory for Ordinary Differential Equations*, Springer-Verlag, Berlin/New York, 1980.

[302] L.I. Rubinstein, *The Stefan Problem*, Transl. Math. Monograph, Vol. **27**, Amer. Math. Soc., Providence, RI, 1971.

[303] B. Rynn, Global bifurcation for $2m$th-order boundary value problems and infinitely many solutions of superlinear problems, *J. Differ. Equat.,* **188** (2003), 461–472.

[304] S. Sakaguchi, The number of peaks of nonnegative solutions to some nonlinear degenerate parabolic equations, *J. Math. Anal. Appl.,* **203** (1996), 78–103.

[305] S. Sakaguchi, Regularity of the interfaces with sign changes of solutions of the one-dimensional porous medium equation, *J. Differ. Equat.,* **178** (2002), 1–59.

[306] A.A. Samarskii, V.A. Galaktionov, S.P. Kurdyumov, and A.P. Mikhailov, *Blow-up in Quasilinear Parabolic Equations,* Nauka, Moscow, 1987; English transl.: Walter de Gruyter, Berlin/New York, 1995.

[307] A.A. Samarskii, N.V. Zmitrenko, S.P. Kurdyumov, and A.P. Mikhailov, Thermal structures and fundamental length in a medium with non-linear heat conduction and volumetric heat sources, *Soviet Phys. Dokl.,* **21** (1976), 141–143.

[308] G. Sansone and R. Conti, *Ordinary Differential Equations*, New York, Pergamon, 1971.

[309] G. Sapiro and A. Tannenbaum, An affine curve evolution, *J. Funct. Anal.,* **119** (1994), 79–120.

[310] D.H. Sattinger, On the total variation of solutions of parabolic equations, Math. Ann., **183** (1969), 78–92.

[311] J.A. Sethian, Curvature and the evolution of fronts, *Comm. Math. Phys.*, **101** (1985), 487–499.

[312] W. Shen and Y. Yi, Asymptotic almost periodicity of scalar parabolic equations with almost periodic time dependence, *J. Differ. Equat.*, **122** (1995), 373–397.

[313] B. Sherman, A general one-phase Stefan problem, *Quart. Appl. Math.*, **28** (1970), 377–382.

[314] A.E. Shishkov, Uniqueness classes for the solutions of the Cauchy problem for nonlinear degenerate parabolic equations, *Math. Notes*, **48** (1990), 1253–1258.

[315] S.I. Shmarev, On a degenerate parabolic equation in filtration theory: monotonicity and C^∞-regularity of interface, *Adv. Math. Sci. Appl.*, **5** (1995), 1–29.

[316] S.I. Shmarev and J.L. Vazquez, The regularity of solutions of reaction-diffusion equations via Lagrangian coordinates, *Nonlin. Differ. Equat. Appl.*, **3** (1996), 465–497.

[317] J. Smoller, *Shock Waves and Reaction-Diffusion Equations*, Springer-Verlag, New York, 1983.

[318] H.M. Soner and P.E. Souganidis, Singularities and uniqueness of cylindrically symmetric surfaces moving by mean curvature, *Comm. Partial Differ. Equat.*, **18** (1993), 859–894.

[319] P.E. Souganidis, *Front Propagation: Theory and Applications*, CIME Course on Viscosity Solutions, Lecture Notes in Math., Vol. **1660**, Springer, 1997.

[320] A.N. Stokes, Intersections of solutions of nonlinear parabolic equations, *J. Math. Anal. Appl.*, **60** (1977), 721–727.

[321] A.N. Stokes, Nonlinear diffusion waveshapes generated by possibly finite initial disturbances, *J. Math. Anal. Appl.*, **61** (1977), 370–381.

[322] C. Sturm, Mémoire sur la résolution des équations numériques, *Inst. France Sc. Math. Phys.*, **6** (1835).

[323] C. Sturm, Mémoire sur les équations différentielles du second ordre, *J. Math. Pures Appl.*, **1** (1836), 106–186.

[324] C. Sturm, Mémoire sur une classe d'équations à différences partielles, *J. Math. Pures Appl.*, **1** (1836), 373–444.

[325] *Sturm-Liouville Theory, Past and Present*, W.O. Amrein, A. Hinz and D.B. Pearson, Eds., Birkhäuser, Basel, 2005.

[326] R. Suzuki, Complete blow-up for quasilinear degenerate parabolic equations, *Proc. Roy. Soc. Edinburgh*, **130A** (2000), 877–908.

[327] C.A. Swanson, *Comparison and Oscillation Theory of Linear Differential Equations*, Acad. Press, New York/London, 1968.

[328] G. Szegö, *Orthogonal Polynomials*, Amer. Math. Soc., Providence, Rhode Island, 1975.

[329] S.L. Tabachnikov, Around four vertices, *Russian Math. Surveys*, **45** (1990), 229–230.

[330] M. Tabata, A finite difference approach to the number of peaks of solutions for semilinear parabolic problems, *J. Math. Soc. Japan,* **32** (1980), 171–191.

[331] S. Täklind, Sur les classes quasianalytiques des solutions des équa-tions aux dérivees partielles du type parabolique, *Nova Acta Regalis Societatis Scientiarum Uppsaliensis (4),* **10**, No. 3 (1936), 3–55.

[332] D.A. Tarzia, *A Bibliography on Moving-free Boundary Problems for the Heat Equation*, Instituto Matematico "U. Dini", Firenze, 1988.

[333] A.N. Tikhonov, Uniqueness theorem for the equation of heat conduction, *Mat. Sborhik,* **42** (1935), 199–215.

[334] M. Ughi, A degenerate parabolic equation modeling the spread of an epidemic, *Ann. Mat. Pura Appl.,* **143** (1986), 385–400.

[335] J.L. Vazquez, New selfsimilar solutions of the porous medium equation and the theory of solutions of changing sign, *Nonl. Anal., TMA,* **15** (1990), 932–942.

[336] J.L. Vazquez, Convexity properties of the solutions of nonlinear heat equations, In: *Contr. Nonl. PDEs*, Vol. II, J.I. Díaz and P.L. Lions, Eds., Pitman Research Notes in Math., Longman, 1987, pp. 267–275.

[337] J.L. Vazquez, The free boundary problem for the heat equation with fixed gradient condition, In: *FBPs, Theory and Appl.*, Pitman Research Notes in Math. Series **363**, M. Niezgodka and P. Strzelecki, Eds., Longman, 1996.

[338] J.J.L. Velazquez, Estimates on $(N-1)$-dimensional Hausdorff measure of the blow-up set for a semilinear heat equation, *Indiana Univ. Math. J.,* **42** (1993), 445–476.

[339] J.J.L. Velazquez, Cusp formation for the undercooled Stefan problem in two and three dimensions, *Euro J. Appl. Math.,* **8** (1997), 1–21.

[340] J.J.L. Velazquez, Corner formation for the undercooled Stefan problem, *SIAM J. Appl. Math.,* **61** (2001), 1156–1201.

[341] J.J.L. Velazquez, V.A. Galaktionov, and M.A. Herrero, The space structure near a blow-up point for semilinear heat equations: a formal approach, *Comput. Maths Math. Phys.,* **31** (1991), 46–55.

[342] T.D. Ventsel', A free boundary-value problem for the heat equation, *Dokl. Akad. Nauk SSSR,* **131** (1960), 1000–1003; English transl.: *Soviet Math. Dokl.,* **1** (1960).

[343] W. Walter, *Differential- und Integral-Ungleichungen*, Springer Tracts in Natural Philosophy, Vol. **2**, Springer, Berlin/New York, 1964.

[344] W. Walter, *Differential and Integral Inequalities*, Springer-Verlag, Berlin/New York, 1970.

[345] L. Wang, On the regularity theory of fully nonlinear equations I, II, *Comm. Pure Appl. Math.,* **45** (1992), 27–76, 141–178.

[346] Ya.B. Zel'dovich, G.I. Barenblatt, V.B. Librovich, and G.M. Makhviladze, *The Mathematical Theory of Combustion and Explosions*, Consultants Bureau (Plenum), New York/London, 1985.

[347] Ya.B. Zel'dovich and A.S. Kompaneetz, Towards a theory of heat conduction with thermal conductivity depending on the temperature, In: *Collection of Papers Dedicated to 70th Birthday of Academician A.F. Ioffe*, Izd.

Akad. Nauk SSSR, Moscow, 1950, pp. 61–71.

[348] T.I. Zelenyak, Stabilization of solutions of boundary value problems for a second order parabolic equation with one space variable, *Differ. Equat.*, **4** (1968), 17–22.

[349] Y.-C. Zhang, Singular interface equation from complex directed polymers, *J. Phys. 1. France,* **2** (1992), 2175–2180.

[350] N.F. Zhukovskii, Conditions of finiteness of integrals of the equation $\frac{d^2y}{dx^2} + py = 0$, *Mat. Sbornik,* **3** (1892), 582–591.

List of Frequently Used Abbreviations

1D - one-dimensional

FBP - free-boundary problem

G-theory - geometric theory

HE - heat equation

KPP - Kolmogorov-Petrovskii-Piskunov

MP - Maximum Principle

ODE - ordinary differential equation

PDE - partial differential equation

PME - porous medium equation

TW - travelling wave

ZKB - Zel'dovich-Kompaneetz-Barenblatt

ZND - Zel'dovich-von Neumann-Doering

Index